Medicinal plants in tropical West Africa

BEP OLIVER-BEVER

Medicinal plants in tropical West Africa

WITH ILLUSTRATIONS BY THE AUTHOR

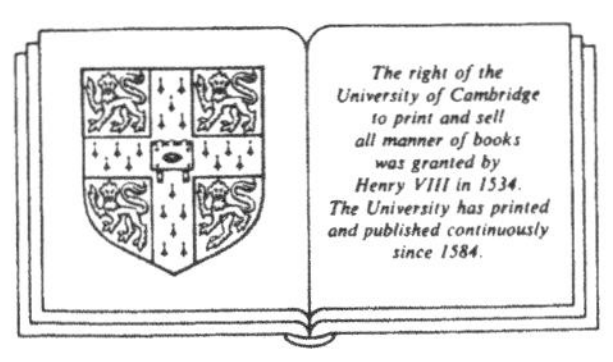

CAMBRIDGE UNIVERSITY PRESS
Cambridge
London New York New Rochelle
Melbourne Sydney

CAMBRIDGE UNIVERSITY PRESS
Cambridge, New York, Melbourne, Madrid, Cape Town, Singapore, São Paulo, Delhi

Cambridge University Press
The Edinburgh Building, Cambridge CB2 8RU, UK

Published in the United States of America by Cambridge University Press, New York

www.cambridge.org
Information on this title: www.cambridge.org/9780521105446

First published 1986
This digitally printed version 2009

A catalogue record for this publication is available from the British Library

Library of Congress Catalogue Card Number: 85–5952

ISBN 978-0-521-26815-8 hardback
ISBN 978-0-521-10544-6 paperback

CONTENTS

FOREWORD

In 1975 I began to be interested in the study of the active principles of African medicinal plants and searched for relevant literature. At the time I was particularly interested in receiving more information on West African plants as we were developing experimental work in collaboration with the Chemistry Department of the Nigerian University of Nsukka and I became aware of the great difficulty in finding fairly reliable documentation even on some of the best-known traditional herbal remedies.

Factors which may account for this are that the local uses were very numerous and often differed from one tribe, village or healer to another. Also, not only did superstition play an important part (often both magical purposes and empirical beliefs were attributed to the plants) but purgatives, diuretics and emetics were often used to chase the evil influences the people did not understand.

The patient work of some distinguished scholars working in the field in Africa provided an important contribution to our acquisition of knowledge on the traditional uses but only a few publications gave a more selective view on the subject. Among these one of the most relevant to me was the book of Dr Bep Oliver (Oliver-Bever), *Medicinal Plants in Nigeria*. She selected uses which were confirmed by the use of the same plants as cures by primitive populations in other parts of the world with similar climate, and also those which were likely to have real therapeutic value from a consideration of the then known chemical and pharmacological information.

Dr Oliver's book, which was published in Nigeria in 1960, was the best I could expect for what I needed, but it was out of print and completely unavailable. By courtesy of the author, who showed me her copy, I managed to obtain photocopies of some parts and I considered myself very lucky. At the same time I thought that this important material had to be republished so that this information on the Nigerian plants could be available to scientists and therefore suggested to Dr Oliver that she prepare a second edition of her publication.

It was a real pleasure when I learned that Dr Oliver was preparing not just a second edition of the first book but a new book on medicinal plants in West Africa including much of the information present in the earlier book but covering a larger area and with up-to-date bibliographical information on botanical, pharmacological and chemical aspects and properties of the plants.

The book covers a large number of plants. Confronted by the difficulty of finding a proper classification of the very abundant material, the Author has chosen a simplified pharmacological approach, presenting in different chapters the plants with constituents which act on the cardiovascular system, on the nervous system, on infectious diseases and on hormone secretions in man.

The great interest of this publication is that it contributes to the knowledge of the basic principles of the plants used in traditional therapy in tropical West Africa. It provides the exact botanical identity and synonyms of the plants mentioned: many people involved in the study of African plants found most of their difficulties in obtaining the exact identification. The book also gives the known chemical aspects of the active constituents of the plants, based on recent published data.

Parallel to the traditional uses of the mentioned plants a modern pharmacological appreciation or interpretation is given, and where traditional medicinal uses may lead to quoting a number of claimed and non-documented results of the treatments with the plants, the data reported in the above-mentioned sections are presented scientifically and based on abundant literature.

The interest of the book exists not only in its multidisciplinary aspect but also because it suggests areas for further research.

In my position as a chemist devoted for many years to research into the biologically active principles of plants, I found in Dr Oliver's book a great deal of very important information for research and a basis for an important aspect of traditional medicine in Africa. This work, to my opinion, fits perfectly in the WHO program on Traditional Medicine, for a better knowledge of plants used in a vast area of Africa, and will surely contribute to better health care of these populations.

Professor G. B. Marini Bettolo
Professor of Chemistry, University of Rome
'La Sapienza', Roma

Chairman, Scientific Committee, WHO Coordinating Centre
for Traditional Medicine, Instituto Italo Africano, Roma

PREFACE

Plants and herbs have been used by man to cure disease and heal injuries since time immemorial. In recent years, renewed interest has been shown in the use of medicinal plants, and scientific studies are beginning to explain some of the curative phenomena associated with traditional herbal remedies. There has also been growing awareness by governments, and the scientific and medical communities, of the importance of medicinal plants in health care systems in many developing countries. This has led the World Health Organization to develop an international programme which will, *inter alia*, review available scientific data relating to the efficacy of medicinal plants in the treatment of specific conditions and diseases. A major task therefore will be to identify those plants suitable for use in primary health care, and to identify simple and/or intermediate technology that will produce enough drugs and therapeutic agents at low cost.

This work presents clear and concise scientific data on the pharmacology of West African plants and extends our knowledge of medicinal plants in West Africa. It will be of particular value to those interested in specific drug applications and will further encourage research into local herbs which in its turn will generate technology locally; this is more reliable and more relevant than introduced technology.

The flora of tropical West Africa has for centuries provided a wealth of material for healing purposes, and its further investigation presents a challenge to scientists who seek to contribute to the search to find new means of alleviating suffering and disease.

The author has put many years of labour and meticulous research into this work, the findings of which are presented clearly and succinctly in this book.

Dr T. A. Lambo
Deputy Director-General
World Health Organization
Geneva

9 January 1984

ACKNOWLEDGEMENTS

I want to thank Dr Lambo, Deputy Director of the World Health Organization in Geneva for the interest and encouragement he showed all these years for my work and also want to thank him and his coworkers for helping me with some practical problems and arranging for some photographs to be made of a few of my watercolours.

I also am grateful to Dr Laurent Rivier and Mr Ian Holmes, editors of the *Journal of Ethnopharmacology* in Lausanne, in which the first chapters of this book were published in order to raise an interest, for encouragement, useful criticism and minor corrections.

In addition I want to acknowledge the kind assistance of Professor Norman Farnsworth, Professor of the Department of Pharmacognosy and Pharmacology at Illinois University, USA, and Dr David Griffin, Manager of Research in Human Reproduction at the WHO, in providing recent documentation allowing me to select the more important information in the very extensive field of antifertility plants.

My thanks also go to Dr Norman Langford, Freelance interpreter at the United Nations for reading through part of the text to check my English, which is not my mother tongue.

I would also like to acknowledge the great kindness of Dr Dinah James, OBE, former Professor of Pharmacology in Nigeria, in reading through a great part of my text and for her extremely helpful suggestions, queries and corrections.

My thanks also go to Mrs Louise Sanders, subeditor of Cambridge University Press, for her help in preparing the text for publication.

Throughout the text, the descriptions of the plants include some of the local medicinal uses (**L**), the chemical constituents (where known) (**C**) and the pharmacological and clinical actions of the plant concerned (**P**), when this knowledge is available. As a short and therefore incomplete botanical description could be misleading, no botanical details have been given. These can be found in the revised edition of Hutchinson and Dalziel's *Flora of West Tropical Africa* (1954–72), which provided the information for the occurrence of the plants described and their names and synonyms. Non-indigenous plants currently cultivated or grown in the area have also been included in the book, this being mentioned in the text or indicated in the tables by c. before or after the botanical name.

Thus:

L indicates local uses
C indicates chemistry
P indicates pharmacology
c. indicates non-indigenous plants

Within each therapeutic group, plants are, as far as possible, assembled by chemical constituents. A chemical relationship often exists in the same botanical family, hence the plants are assembled by families within these groups. The descriptions, when dealing with well-known plants appearing in most Pharmacopoeias, have been restricted to a few essentials; details are available in most standard textbooks. On the other hand, West African plants which, while less well known, seem of potential medicinal interest have been treated in greater detail, in the belief that such details might prove useful in further scientific investigations. Some plants with weaker pharmacological action or with higher toxicity have also been included as further research and chemical separation might enable their use.

Many plants contain a number of different constituents and if these are employed for different purposes the plant may appear under several pharmacodynamic groups. To avoid repetition for plants already described, their action(s) will be mentioned only briefly and for further details reference will be made to earlier information.

1

Introduction

This book is a sequel to the monograph *Medicinal Plants in Nigeria*, written in 1960 (Oliver, 1960), which was a critical survey of the scattered information available about drug plants found in Nigeria; it suggested a first choice of the plant material which seemed potentially most important, and made suggestions concerning points requiring further scientific investigation (constituents, pharmacology, etc.).

As medical science develops and becomes more organized in the West African countries, the time would seem to have come to reassemble and update our knowledge of the subject and extend it to the whole of tropical West Africa. Furthermore, greater importance is now being attached to the use of locally available medicines as a means of reducing reliance on expensive imported drugs.

Since the first book appeared, a number of papers dealing with the chemical analysis, pharmacology and clinical action of West African plants have been published. Supplementary information now available about individual plants will be included here, and the range of plants considered can thus be more selective. This time an attempt is made to classify the drugs according to their established or possible medical uses, this being the best way of rapidly assessing the medical interest of any particular drug.

The value of a drug will depend on several factors:

(1) whether it is the only drug, or one of the few drugs, used in the treatment of a disease;
(2) whether the disease in the treatment of which it is used is a common one;
(3) whether it is less toxic than other existing drugs for a particular purpose;
(4) whether it can be produced more cheaply than alternative drugs used in treating the same disease.

This last criterion may hold good for plants grown for other commercial uses (fibres, timber, fixed oils (liquid fats or fatty oils), essential oils, gums, resins, etc.), for in such cases the drug may be available as a by-product, with a substantial cut in production costs.

The following points, which were mentioned in my earlier publications, still hold good:

(a) Recognized drug plants which also grow in West Africa and are officially listed in various Pharmacopoeias or are already in use elsewhere have to be checked by well-established methods to ascertain whether the yield in active constituents of the plant when grown in West Africa reaches a suitable standard. If it does, the plant can, if not indigenous, be grown on a greater scale and the relevant drugs produced by standard procedures.

(b) The therapeutic action of a plant depends on its chemical constituents and can often be forecast and easily investigated pharmacologically if these constituents are known.

(c) The botanical relationship of a particular plant to well-known drug plants, or to plants containing therapeutically active constituents, may be an indication of a potential therapeutic interest. Indeed, chemical relationship, based on secondary substances specifically found in certain genera and families, has been observed and is made use of in botanical taxonomy. Several genera of one botanical family may thus have a similar action. Many Solanaceae contain alkaloids with a parasympatholytic action, and many Labiatae contain essential oils, while cardiac heterosides are often found among the Asclepiadaceae and Apocynaceae.

(d) As many plants found in tropical West Africa also grow in other areas of similar climate, such as parts of India, Sri Lanka and Indonesia, their use in such countries will require investigation too. However, an attempt ought to be made to ascertain whether the African plants have the same constituents in equivalent quantities, and the same properties, for in these other countries the content of active principles may not be the same. Such differences may be attributed to differences in climate, soil or other ecological conditions, but are more likely to be due to varying degrees of enzymatic destruction of the chemical principles (Debray, 1966, p. 51, quoted in Oliver-Bever, 1968).

(e) Local medicinal usage may provide useful information about lesser-known plants. Unfortunately, local uses can be very numerous and often differ completely from one tribe to another for one and the same plant. It should not be forgotten that superstition plays a considerable part in folk medicine. Vesicants, purgatives, diuretics and emetics are often used because they 'oppose strong action' or 'expel evil influences'. However, the herbalist is sometimes right and then his medicine has to be investigated further.

In some cases certain local plants are used in the same way by many different tribes, or for similar ailments in other parts of the world where such plants are also found. This would seem to make this use more likely to be accurate. It could, however, be empirical and might be based, for instance on the 'Law of Signatures', which has adepts among several under-developed peoples existing without contacts. Hence in 1960 the local uses indicated for tropical West Africa (Dalziel, 1937) were compared with the uses to which the same species are put in India (Chopra *et al.*, 1956), Sri Lanka (Jayaweera, 1945, 1952, 1954), Indonesia (van Steenis-Kruseman, 1953), the Ivory Coast (Kerharo and Bouquet, 1950), Ghana (Irvine, 1930), Senegal

(Sébire, 1899; Chevalier, 1905–13), Guinea (Pobéguin, 1912), the Congo (Staner and Boutique, 1937), Nigeria (Holland, 1908–29), Africa in general (Githens, 1949), etc. This resulted in a fairly rigorous selection of local medicinal uses and this information was partly made use of in the preparation of the present text. Some interesting indications may thus well have been overlooked, but a rapid survey of the existing knowledge seemed to be the first requirement. More detailed and more up-to-date information on local medicinal uses can be found in the 'Mémoires' published by the Office de la Recherche Scientifique et Technique Outre Mer (ORSTOM) (see e.g. Bouquet, 1969) in Kerharo and Adam (1974) and in Ayensu (1978). These latter books (and some of the others) also give the vernacular names of the plants they deal with.

Throughout the text, the phrase 'plants acting on . . .' is used as convenient terminology for 'plants whose leaves (or roots, extracts, active principles, etc.) act on . . .' . The chapters themselves are named on the basis of the physiological system affected.

Early traditional medicine

Treatment has been provided in West Africa by the native 'doctors' (*ifas*, *juju* men and herbalists). The *juju* man is believed to be able to get the support of the gods (through magic and tribal rituals) for the numerous problems affecting his applicants. He not only treats diseases, which the people consider to be an adversity imposed upon them by outside forces they do not comprehend, but is also required to be a rainmaker or to perform rites to ensure good crops, prosperity, etc., or to help in calamities caused by offended dead relatives or evil spirits (Oliver-Bever, 1983). Therefore his aims, being so closely related to mystic practices, are often more concerned with the spirit than with the body. It seems that early civilizations felt the importance of a psychosomatic approach to illness long before this received attention in modern medicine!

A great quantity and variety of 'medicines' based on plants, or parts thereof (Oliver, 1960), are given by the different herbalists throughout West Africa. Often they are sold in the markets to people in search of cures and a great number are 'assured' to heal almost every disease under the sun: others have a definite use. Of course, this *materia medica* is by no means limited to plants, and frequently in 'strong medicines' components like the heart of chicken, animal remains, human saliva and even flesh and blood are part of the preparation. Generally, the drugs are made up in the shape of small rissoles or balls of mastic, but liquid potions, ointments or powder can also be found and even enemas or fumigations are used in a local fashion.

The knowledge of the properties of the drug plants shown by some local *juju* men may either have been passed on to them by their elders or be based on experience. Frequently, neither the 'doctor' nor the patient attributes the action to the plant itself, a situation reminiscent of a similar attitude in Europe in the Middle Ages. Indeed, disease in old Anglo-Saxon times was attributed to 'possession by devils' or to 'flying venom' or to 'the loathed things that rove through the land' (Rohde, 1922). To counteract these evils, religious rites, together with herbs and charms of traditional value, were employed not only for man but also for his cattle (Rohde, 1922) (religion was the outward sign of man's appeasement of forces that he did not understand). Also, it was superstitiously believed that when a plant was pulled out of the ground it uttered shrieks and caused death or at least insanity to the gatherer if he heard them (Lloyd, 1921). Shakespeare refers to this belief in *Romeo and Juliet* when he writes (Act 4, Scene 2): 'shrieks like mandrakes torn out of earth, that living mortals, hearing them, run mad . . .'. The problem of gathering the root, therefore, was overcome by tying a dog to the plant while the gatherer stopped his ears lest he should hear anything.

The *ifa* may say that he has discovered a plant possessing a spirit stronger than the disease spirit, and he and his patients believe that the power of this spirit, or the soul of the medicine, is not manifested before the healer has spoken some magic words or has chanted an incantation over the plant. Before doing so the *ifa* himself may appeal for advice to gods or worship idols which in Yoruba country (Western Nigeria) are often small carved figures of a man or woman and sometimes also of animals. The patients in turn should not only take or apply the medicine but also appeal to and make offerings to communal and household gods, which may also be carved statues or other objects blessed by the local priests in ceremonies that generally last for days. In a school in Badagry (Nigeria, near the frontier with Dahomey) there were in a dark corner places of sacrifice consisting of cones of clay with an irregular shiny surface. They were streaked in white, turquoise, yellow and brown by the numerous offerings that had been made, and had the odd piece of eggshell and feather glued to them. Trees can also be idols, for example the iroko tree, *Chlorophora excelsa*, which is regarded as a sacred tree by the Ibos (Eastern Nigeria) and is credited for 'furnishing souls for the newborn'. Sacrifices are made to this tree and offerings are often found at its base. The household gods (*Ikenja*) are always carved from iroko-wood and pieces of its bark are added to many medicines to increase their action.

Another rare but interesting tree used by the *juju* men is *Okoubaka aubrevillei*. The bark is used by the Binis (Eastern Nigeria) to drive away evil from a house or to inflict a curse upon an enemy. The bark, according to Hardie (1963), may be removed but never at sunset or sunrise when it 'spits poison' (the foliage then exudes a dark poisonous liquid). Before removal of the bark, however, the spirit of the tree must be propitiated by the offering of gifts. These usually consist of portions of kola nut, white yam, coco yam and plantain, two cowrie shells, a piece of white drill cloth and a quantity of chalk. With these it is possible to approach an *Okuobisi* (its local name) after having stripped off all clothes at a safe distance. The gifts are laid at the foot of the tree and at the same time the spirit is begged for whatever help may be required. A small piece of bark can now be removed with the aid of a wooden batten (under no circumstances may a machete or metal implement be used) after which it is advisable to run away quickly 'out of sight' and to re-dress. The reason for this is that the spirit of the tree may not have been sufficiently appeased by the gifts and if it pursues the applicant it will fail to recognize him fully clothed.

An Ibibio in trouble may say to the iroko: 'Oh tree! you who are a strong man and to whom heavy things are light, I am only a small weak creature, and my worry is so big that I cannot carry it, will you, who are strong, take it from me? It would be a straw to you.' And he will sacrifice to the tree and leave in peace, convinced that his burden will be taken from him.

Formerly, plants were used not only for healing but also for killing. Arrow poisons were prepared by rubbing certain seeds between two stones until they formed a paste to which was added saliva and the juice of different toxic plants. A vesicant latex, for example from *Euphorbia* spp., was often used as this damaged the skin, thus facilitating penetration and absorption. Often the remains of animals were also added for magic purposes. Another method consisted of extracting the active constituent, generally with water or palm wine, and concentrating the extract until it formed a paste. If the poison was part of the latex of a plant then the latex might be dried out until it had the right consistency.

In trials by ordeal, a man suspected of evil influence or action was forced with much ceremony to swallow a dose of poison. If he survived, this was the wish of the tribal gods and he was considered innocent; if he died this was evidence of guilt. A classical example is the trial of Lander in Badagry in 1827. Fortunately, Lander, who had been given a decoction of a portion of bark from *Erythrophleum guineense* (Sassy bark), which contains erythrophleine, a strong heart poison, was wise enough to take a violent emetic immediately afterwards and so survived.

Apart from the local *juju* men, who treat the more complicated and resistant cases, the more common ailments are treated by villagers, generally elderly women, with knowledge of the local plants. The reasons for the local selection of drug plants is varied. In a number of cases prescriptions are based on the observation of what happened to animals and men who had eaten certain plants accidentally. In other cases it was noticed that the plants produced, for example, a local irritation of the skin but at the same time relieved a pain or cleared up a sore on persons who touched them, and so local inhabitants used such plants in this way.

Other uses are just empirical, for example those based on the 'Law of Signatures'. This is an old belief which says that nature has provided a plant for every disease and has indicated by an obvious sign for which disease or which part of the body each drug plant is to be used. Thus the shape of a plant or of one of its components may suggest a cure. This belief existed in many parts of the world, including Europe in the Middle Ages. The classical example was a walnut, which having the shape of a brain, should thus be used for diseases affecting the brain. Grier (1937) cites other examples: 'Plants with red flowers were to be used in blood disorders and those with yellow flowers, also turmeric, in jaundice. Saxifrages, which grow on rocks and break them up, would be useful for stones in the bladder, a belief in the Middle Ages in England (Grier, 1937). Euphrasia was to be given in eye diseases, because a black spot in the flower resembles the pupil of the eye.' Similar beliefs are prevalent in primitive West African medicine and have been documented by many authors. . . . plants with white latex are used to increase milk production; those with big swollen fruits to favour fertility (Githens, 1949): *Commelina*, with its bright blue flowers like eyes, for ophthalmic treatment (Dalziel, 1937, p. 465): *Eryngium foetidum*, a plant with a powerful odour, is supposed to bring a person to his senses; the stems of *Palisota hirsuta*, with joints which are swollen and bend like a knee, are used for sprained knees; and the bark of *Pentaclethra macrophylla*, a tree which never grows straight but always has a hump in the trunk, is part of a preparation applied to the hump of a hunchback. The leaves of *Ficus exasperata* (sandpaper leaves) are cooked with salt fish and eaten with the idea that these scratchy leaves will scrape out whatever is causing the trouble (Harley, 1941).

On similar lines it is believed that 'the administration of owl's feathers makes the disease fly silently away' (Githens, 1949, p. 2). In the Cameroons, for the treatment of migraine, a spider's web spun on the grass is found, and grass, web and all are mixed with white clay and rubbed on the patient's head. As the spider runs away on its web, so will the headache run away (Talbot, 1926). Also, abuse exists like everywhere else, and Gerarde's comment on the use of henbane seeds to cure toothache (Woodward, 1931) is reminiscent of practices used in West Africa by some unconscientious healers. He writes: 'The seed is used by Mountibank toothdrawers which run about the country, to cause worms come forth of the teeth, by burning in a chafing dish or coles, the party holding his mouth over the fume thereof; but some crafty companions to gain money convey small lute-strings into the water, persuading the patients that those small creepers came out of his mouth and other parts which he intended to ease.'

Even in those parts of the world where different populations communicated, often through Greek or Latin texts, and where drugs were received by overland or sea routes from China, India and the Far East (Gunther, 1934), it took from ancient times until the eighteenth century before the causes and treatment of illness began to be understood. The folklore of young isolated communities, still based on a scheme similar to that of Anglo-Saxon medicine in the Middle Ages, is therefore not surprising but it is rapidly disappearing with the development of communications and education.

Practical therapeutic indications and mechanisms of action of the drugs

A distinction should be made between the practical use of a drug and the way in which it acts. The therapeutic effects of a number of plants are the result of their action on the nervous system.

First brief mention should be made of the mode of action of the drugs on the nervous system. Their activity may be the result of:

(a) stimulant or depressant effects on the central nervous system, activity being exerted at various levels from the higher centres to peripheral nerve terminals;
(b) modulating effects on autonomic nervous system activity.

The therapeutically useful effects are those that are selectively induced at important sites mainly by substances which simulate (mimic) neurotransmitters or interact with them or their receptors.

Plants acting on the cardiovascular system (Chapter 2) mainly produce their effects through the autonomic nervous system (ANS). Autonomically innervated structures are regulated at a subconscious level by nerve fibres from the sympathetic and parasympathetic divisions. The influence of each division varies with each tissue, i.e. sympathetic activity augments the heartbeat but inhibits the tone of intestinal and bronchiolar muscles. Constituents which stimulate the release of the neurotransmitter (noradrenaline) will increase the adrenergic effects but drugs which antagonize its activity give prominence to the cholinergic division and an exaggeration of the cholinergic effects (Fig. 1.1).

The cardiovascular plants have been grouped into:

(a) *cardiotonics*, which are mainly used for their positive inotropic effects, produce reinforcement of the contractibility of the heart;

(b) *cardiac depressants*, which are mainly used for their positive or negative chronotropic effects, regulate the rhythm in tachycardia and fibrillation;

(c) *vascular agents*, the action on the blood pressure being treated here as well as their action on vascular solidity, the permeability of capillaries and blood coagulation and formation.

The ANS intervenes in many different functions of the organism. I have given the descriptions of the plants under their principal effect, which is likely to control the practical demand, rather than by their mode of action. Drugs affecting bronchial, intestinal or uterine motility are described under their stimulating or antispasmodic effect on the smooth muscles.

In Chapter 3 (The nervous system) I discuss mainly those plants used in mental treatment, including sedatives, hypnotics, tranquillizers, anticonvulsants and hallucinogens having a stimulating or depressant action on the central nervous system (CNS) or the ANS. Analgesics, antipyretics, anaesthetics and antispasmodics are also included in this chapter.

The mechanism of action of the drugs acting via the nervous system is believed to be based on their interference with the action of chemical substances such as acetylcholine and the catecholamines, the chemical mediators of nervous transmission. This interference may occur through affinity of the plant constituents for specific receptors, which can be cholinesterase, adenylcyclase or other enzymes or

Fig. 1.1. Action and sites of action on the ANS.

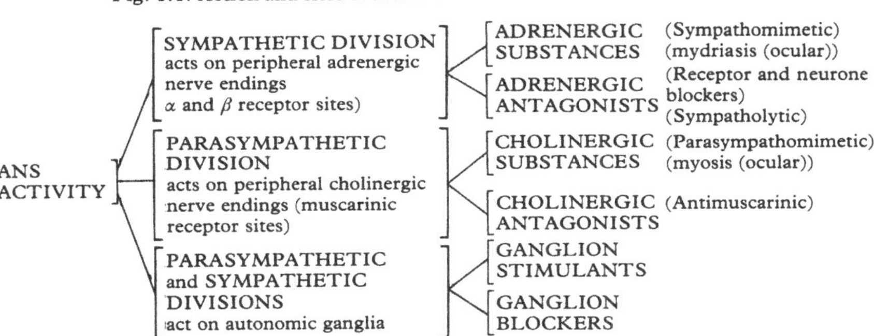

other macromolecules. Blocking at different peripheral levels leads to more or less specific physiological effects. For more information on pharmacodynamic properties, specialized literature should be consulted.

Higher plants used in anti-infection therapy (Chapter 4) need to have different properties as they must be toxic towards organisms that are infectious to or parasites of Man but without notable action on the human beings. As could be expected, the mechanism of action of the plant components varies not only from group to group but also within several groups: there are different mechanisms of action often at a cellular level, through enzymatic cell-receptors. Their supposed mode of action, where known, is indicated under the corresponding groups of plant components.

Where plants acting on hormone secretion in Man are concerned (Chapters 5, 6 and 7), it has been noted that some plant constituents can directly replace certain hormones in their biological functions because they have an almost identical or very similar chemical structure to that of the hormones concerned. Other plant constituents exert their action indirectly by stimulating or inhibiting the secretion of the hypothalamus or the pituitary or of enzymes which intervene with the secretion of certain hormones.

However, as future research will no doubt reveal, there are many more ways in which the secretion of hormones can be stimulated or inhibited. This is illustrated by details found under sections dealing with hormone secretion such as plants with anti-inflammatory, oestrogenic, antifertility controlling, hypoglycaemic and other activities on human hormone secretions.

2

The cardiovascular system

I Plants in tropical West Africa with an action on the cardiovascular system

In the particular field of cardiovascular drugs, plants still provide the basis of treatment, even in orthodox pharmacy. However, some of the plants accepted by most Pharmacopoeias, such as *Digitalis*, *Convallaria*, *Adonis*, *Helleborus* and *Crataegus*, which act mainly on the heart, and *Hydrastis*, *Veratrum*, *Amni visnagi*, *Viscum album* and *Aesculus hippocastanum*, which act more specifically on the blood vessels, do not grow in West Africa. On the other hand, the possibilities of many plants that are locally available have not yet been fully investigated. Also, some of the currently used cardiotonics have a high toxicity; less toxic but yet active constituents might be found amongst the West African plants. As mentioned in the general introduction only a limited number of local uses have been indicated.

Most herbalists will know that many plants in this group (several formerly used as arrow poisons[1] or even in ordeals) are very toxic and will avoid using them. A few healers, however, may, in view of the fact that they are also emetics, purgatives or diuretics, be tempted to make use of them. But these plants should be employed only after complete extraction and with very exact dosages of the active constituents, and then only by physicians in possession of a full clinical diagnosis. In this, these plants differ from many others, which may be given as a decoction, an infusion or in dried or powder form.

Cardiovascular activities are mainly controlled by the ANS. The ANS can be divided into two main divisions. One, which through the influence of noradrenaline on the corresponding nerve endings has a stimulating effect on the heart and produces vasoconstriction, is called the adrenergic or sympathetic division. The other, which through the influence of acetylcholine slows down the heartbeat and produces a fall in blood pressure and vasodilatation, is called the cholinergic or parasympathetic division. Both these divisions can stimulate or antagonize (block) the autonomic ganglia (Fig. 1.1, p.7).

The actions of acetylcholine at the peripheral cholinergic nerve endings are known as its muscarinic actions, because they are mimicked by muscarine (a mushroom alkaloid).

Stimulation of the preganglionic nerve fibres to the ganglia results in liberation of acetylcholine (physiological neurotransmitter in autonomic ganglia). This action is almost immediately counterbalanced by cholinesterase, which destroys acetylcholine through hydrolysis. The cholinergic actions on the ganglia are referred to as the nicotinic actions of acetylcholine because the effects of acetylcholine on the ganglia are similar to those produced by nicotine. There is initial stimulation and then blockade of the ganglion cells (Turner and Richens, 1978).

Changes in the force of contraction of the myocardium are called inotropic effects while changes in the heart rate are called chronotropic effects. The myocardium, which contains the β receptors for noradrenaline, responds to this by increasing the frequency and amplitude of the heartbeat (Lechat *et al.*, 1978).

The plants which act on the cardiovascular system can be divided into three groups:

(a) Cardiotonics. Cardiotonic drugs act on the force, the rate and the rhythm of the heartbeat. They have a stimulating effect on the cardiac muscle and thus increase the contractile force (inotropic effect), decrease the heart rate and regularize the heartbeat (chronotropic effect). By increasing the renal bloodflow, cardiotonics can have a diuretic action. They often produce nausea and vomiting as they irritate the gastric mucosa, and are sometimes used in small doses for their expectorant action, which precedes the vomiting. By increasing the pulse rate, cardiotonics can also increase the blood pressure.

(b) Cardiac depressants. These drugs have a depressant effect on the heart muscle and some are particularly suited to the treatment of arrhythmias (anti-arrhythmic drugs). By slowing the cardiac rhythm they often have an antihypertensive action, either through vasodilatation of the coronary arteries or through direct control by the nervous system.

(c) Vascular agents. These are plant constituents which act primarily on the blood vessels.

(a) Cardiotonics

Today the plant cardiotonics are generally used in orthodox pharmacy as isolated active principles. Many of the plants formerly used in Africa as arrow poisons have been shown to contain cardenolides and to be valuable in minute doses in treating heart conditions. Cardenolides are steroid heterosides. Their aglycones (or genins) are responsible for the specific action but do not act by themselves as they are insoluble and have a low power of fixation on the heart muscle (McIlroy, 1950, p. 79). The fixation on the tissues of the isolated frog heart could be attributed for certain components like flavotannins from *Paullinia pinnata* (see below) to the formation of a complex with calcium on the surface of the heart tissues (Bowden, 1962; Broadbent, 1962).

Table 2.1. *Apocynaceae in tropical West Africa*

In the leaves of the members of the Apocynaceae free ursolic acid is frequently found (*Alstonia boonei*, *Rauvolfia vomitoria*, *Pleioceras barteri*, *Thevetia neriifolia*, etc.), whereas in the coagulum of the latex the triterpenic alcohols β-amyrin are often present. As most of the plants contain a very great number of alkaloids or heterosides, only the main constituents and their most important uses have been indicated.

It appears from the table that only the Plumeroideae contain indole alkaloids and steroid alkaloids, whilst the Echitoideae and Cerberoideae contain cardiac glycosides. However, from studies of the way in which the constituents are built up it appears that the Apocynaceae are able, starting from a steroid nucleus, to produce either cardiac heterosides or steroid alkaloids, thus bringing the members of this family nearer than they might appear at first sight (Goutarel, 1964; Paris and Delaveau, 1966). On the other hand, Paris and Delaveau (1966) mention the fact that the same 'specific' chemical constituents are sometimes found in families which are far apart in their morphological classification. Thus in West Africa cardiac glycosides are found not only in plants of the Apocynaceae and Asclepiadaceae but also in members of the Liliaceae (*Urginea indica*), Moraceae (*Antiaris africana*), Tileaceae (*Corchorus olitorius*), Sterculiaceae (*Mansonia altissima*) and even Compositae (*Vernonia colorata*). Similarly, indole alkaloids are also found in Rubiaceae (*Mitragyna inermis*, *M. macrophylla*, *Corynanthe pachyceras*, *Pausinystalia johimbe*, etc.), and steroid alkaloids also occur in some *Solanum* species (*Solanum nigrum*, *S. lycopersicum* (Oliver-Bever, 1968).

Plant	Part used	Active constituent(s)	Chemical group	Recognized or possible medicinal action[a]
Subfamily Plumeroideae				
Carisseae				
Carissa edulis Vahl	Roots	Carissin	Cardenolide	Oncolytic (sarcoma 180) (Abbot *et al.*, 1966)
Hunteria eburnea Pichon	Seeds	Burnamine	Indole alkaloids	Hypotensive
Picralima nitida Stapf	Seeds	Akuammine, akuammiline, akuammidine, akuammigine, etc.	Indole alkaloids	Hypotensive, local anaesthetic, sympatholytic
Tabernaemontaneae[b]				
Tabernaemontana crassa Benth. syn. (*Conopharyngia durissima* Stapf)	Roots, bark	Isovoacangine, conopharyngine, conodurine, conoduramine	Indole alkaloids	Sympathomimetic
	Seeds	Voacamidine, tabersonine, coronaridine		

(*Table continued*)

Table 2.1. (*Continued*)

Plant	Part used	Active constituent(s)	Chemical group	Recognized or possible medicinal action[a]
Tabernaemontana pachysiphon Stapf syn. (*Conopharyngia pachysiphon* Stapf)	Roots	Pachysiphine	Aminosteroid glycoside	Hypotensive (Hegnauer, 1962–8, vol. 3, p. 129)
		Voacangine	Indole alkaloid	
Hedranthera barteri (Hook.) Pichon. syn. (*Callichilia barteri* Stapf, *C. monopodiales* (Schum.) Stapf)	Roots, stems	Callichine, vobtusine	Indole alkaloids	Cardiotonic (Patel and Rowson, 1964)
	Roots, stems	Callichine	Indole alkaloids	Cardiotoxic (Patel and Rowson, 1964)
	Leaves	? Ursolic acid	Triterpene	Cardiotonic
Voacanga africana Stapf	Stembark and rootbark	Voacamine, voacangine, voacristine, voacorine, voacamidine, vobasine, vobtusine, etc.	Indole alkaloids	Cardiotonic, sympatholytic, hypotensive
Voacanga bracteata Stapf	Stembark and rootbark	Voacamine, voacangine, voacorine, epivoacorine	Indole alkaloids	Same as *V. africana*
Alstonieae				
Alstonia boonei de Wild. syn. (*A. congensis* Chev. & Aubrev.) c.	Bark	Echitamine, echitamidine, alstonine, reserpine	Indole alkaloids	Hypotensive? (Raymond-Hamet, 1934, 1941)
	Leaves	Amyrin, lupeol, ursolic acid	Triterpenes	
Catharanthus roseus (L.) Don. syn. (*Lochnera rosea* Reichb.) c.	Roots, twigs	Catharanthine, lochnerine, vindoline	Indole alkaloids	Hypoglycaemic
		Vincristine, vinblastine	Indole alkaloids	Oncolytic (Hodgkin's disease, leukaemia)
		Reserpine, ajmalicine	Indole alkaloids	Hypotensive, tranquillizer

Holarrhena floribunda (Don.) Dur. & Schinz. syn. (*H. africana*, *H. wulfsbergii*)	Stembark and rootbark	Conessine, conkurchine	Steroid alkaloids	Antibiotic (*Entamoebia histolytica*, *Trichomonas*)
	Bark	Holarrhenine	Steroid alkaloids	Hypotensive, local anaesthetic, spasmolytic
	Leaves	Holarrhimine, holaphyllamine, holaphylline		
Allamanda cathartica L. c.[b]	Seeds, stems, roots	Plumeriede = plumeroside, allamandin	Glycoside of iridoid lactone	Cardiotonic, antitumour agent, cardiotoxic
	Leaves	Ursolic acid		
Plumeria rubra c.	Latex, leaves and bark	Plumieric acid, plumieride	Glycoside of cinnamic acid lactone	Local anaesthetic, cardiotonic
	Bark	Fulvoplumierin		Bacteriostatic
Rauvolfieae				
Rauvolfia vomitoria Afzel.	Rootbark and stembark	Reserpine, rescinnamine, raumitorine	Indole alkaloids	Tranquillizer Sedative Hypotensive
		Reserpiline, rauvanine Ajmaline, rauvanine Ajmalicine	Indole alkaloids	Hypotensive Anti-arrhythmic Raynaud's disease vasodilating
Funtumia africana (Benth.) Stapf	Bark, leaves	Funtumine	Steroid alkaloid	Hemisynthesis of corticosteroids
Pleioceras barteri Baill. syn. (*Wrightia parviflora* Stapf.)	Rootbark, seeds	Alkaloids	? Steroid alkaloid	Local use, emmenagogue, abortifacient
	Leaves	Ursolic acid	Triterpene	Produces sodium retention like desoxycorticosterone (Kerharo and Adam, 1974, p. 157)

(*Table continued*)

Table 2.1. (*Continued*)

Plant	Part used	Active constituent(s)	Chemical group	Recognized or possible medicinal action[a]
Malouetia heudelotii DC.	Bark	Malouetine	Steroid alkaloid	Curare action (Hegnauer, vol. 3, p. 129)
Subfamily Echitoideae				
Adenium obesum (Forsk.) Roem. & Schult. syn. (*Adenium honghel* DC.) c.	Latex	Honghelosides A–G	Cardenolides	Cardiotonic (toxic)
Baissea leonensis Benth.	Leaves	Baisseoside = esculetol-6-rutinoside	Coumarin glycoside	Vitamin P action
Nerium oleander L. c.	Leaves	Oleandrin, digitalin, adynerin, neriantin	Cardenolides	Cardiotonic
Strophanthus gratus Franch.	Seeds	Strophanthins K, g, etc.	Cardenolides	Cardiotonic
Strophanthus hispidus DC.	Seeds			
Strophanthus gracilis Schum.	Seeds			
Strophanthus sarmentosus DC.	Seeds	Heterosides	Steroid heterosides	Hemisynthesis of corticosteroids and oral contraceptives
Strophanthus spp.	Leaves	Quercetol- and kaempferol-heterosides		
Subfamily Cerberoideae				
Thevetia neriifolia Juss. ex Steud. syn. (*T. peruviana*) c.	Roots, bark	Thevetins A and B, peruvoside	Cardenolides	Cardiotonic
	Leaves, seeds	Aucubine	Iridoid heteroside	Insecticide

[a]References are indicated only when they are not mentioned in the text.
[b]A number of *Conopharyngia* and *Tabernaemontana* species also contain voacangine and vobtusine. However, the principal use of those species is based on their content of alkaloids of the ibogaine group, which act on the nervous system, and the plants will therefore be described with those acting on the nervous system. Likewise *Allamanda* will be described with antitumour agents.

The plants containing the cardiotonics can be divided into two groups:

(A) Plants containing cardiotonic steroid heterosides. This group includes plants belonging to the Apocynaceae, those belonging to the Asclepiadaceae and Periplocaceae, and those belonging to other botanical families.

(B) Plants containing cardiotonic alkaloids.

Group A: Plants containing cardiotonic steroid heterosides

APOCYNACEAE. As we see from Table 2.1, which lists the main medicinal Apocynaceae in tropical West Africa, the family includes a number of plants containing cardenolides. We also find amongst the West African Apocynaceae a few of the more important cardiotonics, which appear next to the non-African *Digitalis* in most Pharmacopoeias. I start by mentioning the better-known ones.

Strophanthus gratus (Hook.) Franch. APOCYNACEAE

L The seeds and wood, like those of all *Strophanthus* species, were used in arrow[1] and fish poisons (Dalziel, 1937).

C The seeds yield 3–7% of g-strophanthin or ouabain, first isolated in crystallized form in 1877 (Paris and Moyse, 1971, vol. 3). On hydrolysis, rhamnose and ouabagenin are obtained (Euw and Reichstein, 1950a, b). The seeds also contain several minor alkaloids such as acolongo floroside K and strogoside (0.4%) (Geiger *et al.*, 1967).

P Ouabain is used in preference to digitalis when a more rapid action is required. The effect is more potent but is of shorter duration and non-cumulative (Martindale, 1958, p. 580). As it is badly absorbed when given orally, it is mostly administered intravenously or intramuscularly. It does not cause peripheral vasoconstriction as does digitalis. In toxic doses ouabain produces hypertension, tachycardia, auricular and ventricular dissociation and, finally, cardiac arrest, in the dog.

Strophanthus hispidus DC. APOCYNACEAE

C The seeds of this species yield 4–8% of amorphous strophanthoside H which, although less important than g- or K-strophanthin, is also used for cardiac insufficiencies. The seeds were also found to contain 1.47% of K-strophanthoside-α, found originally in *S. kombe* (which contains 5–10% of active cardenolides). In addition, sarmentocymaroside, saponosides and flavonosides have been reported to be present in the seeds (Euw and Reichstein, 1950a; Keller and Tamm, 1959).

P The use of the seeds of *S. hispidus*, official in the 1949 French Codex, is limited to strophanthus tincture (1/10), whilst *S. gratus* is used for the extraction of ouabain for intravenous injections (doses 0.12–0.25 mg).

Strophanthus gracilis Schum. & Pax APOCYNACEAE

C The seeds of this species, which is also indigenous, contain the largest quantities of total glycosides, including strophanthidin, strophanthidol, emicymarin, odoriside H and G and graciloside. However, *S. gracilis* is not used in pharmacy, having less active constituents than *S. gratus* and *S. kombe*.

Strophanthus sarmentosus DC. (Fig. 2.1) APOCYNACEAE

C The heterosides of this species are of two different types according to the geographical origin of the plants. In those found in southern Nigeria, Congo and Togo the genin is sarvogenin, whilst in those of the savannah areas of northern Nigeria and Mali, it is sarmentogenin. The sugars are in both cases sarmentose and digitalose (Fechtig *et al.*, 1960; Fuhrer *et al.*, 1969).

P The heterosides, which are of no therapeutic interest, were examined with those of other related species as a source of steroids to be used in the hemisynthesis of corticosteroids and sex hormones, but the results were disappointing (Wall *et al.*, 1961; Reichstein, 1963).

Fig. 2.1. *Strophanthus sarmentosus* DC.

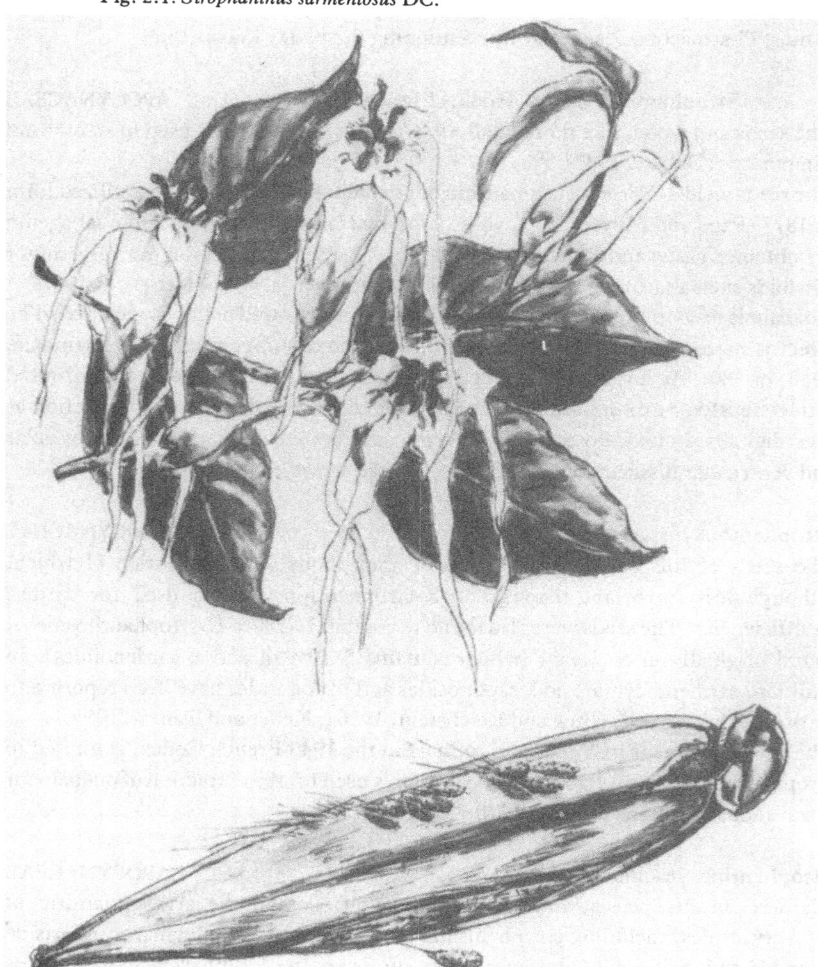

Thevetia neriifolia Juss. syn. (*T. peruviana* Pers., *Cerbera thevetia* L., *C. peruviana* Pers. APOCYNACEAE

Yellow oleander

L Largely cultivated as an ornamental plant. The bark is bitter and said in Ghana and southern Nigeria to be a powerful antipyretic for intermittent fevers, but it is also an emetic and poisonous in excess (Oliver, 1960).

C The roots, stems and kernels (the latter also yield up to 57% of a yellow oil) contain 1–5% of a bitter heteroside, thevetin or thevetoside (a mixture of thevetin A and B), and peruvoside. On hydrolysis, thevetin A yields cannogenin, gentiobiose and thevetose; thevetin B, also called cerberoside, yields gentiobiose and neriifolin, which on further hydrolysis yields one molecule of thevetose and the aglycone digitoxigenin. Acetylneriifolin has also been isolated from the seeds (Frèrejacque, 1947; Bloch *et al.*, 1960; Bisset, 1961; Bisset *et al.*, 1962; Frèrejacque and Durgeat, 1971).

P Thevetin has a short digitalis-like action on the heart and has the advantage of rapid elimination. Peruvoside and neriifolin are more active (action about equal to that of ouabain) and are more rapidly eliminated, but there is little difference between effective and toxic doses. The LD_{50} in the cat is 147 μg/kg for peruvoside as against 1106 μg/kg for thevetin B (Kohli and Vohra, 1960; Chen and Henderson, 1962; Datta and Datta, 1977).

Thevetin is used to a limited extent clinically in cases of intolerance to digitalis and where oedema persists after digitalis therapy. It is recommended by Russian authors for cardiac insufficiency with dyspnoea and for ventricular insufficiency due to hypertension and atherosclerosis (dose 0.5–2 mg daily, orally or intravenously); it is effective 4–6 h after oral administration (Ambrosia and Mangieri, 1955; Aleshkina and Berezhinskajà, 1962; Arora *et al.*, 1967).

An extract of the leaves and fruits of the plant yielded aucubine, an iridoid heteroside. The extract has been found to give excellent results in killing larvae and insects (Heal and Rogers, 1950; Paris and Etchepare, 1966).

Nerium oleander L. APOCYNACEAE

Ornamental shrub often grown all over West Africa.

C The leaves of this species contain several heterosides; the most important of them, representing up to 90% of the total heterosides, is oleandrin or oleandroside. On hydrolysis oleandrin produces a sugar, oleandrose, and oleandrogenin, which is identical with 16-acetyl-gitoxigenin (Abisch and Reichstein 1960, 1962a).

P The plant is used for the extraction of oleandrin, which is an orally active cardiotonic and diuretic and is listed as such in the Russian Pharmacopoeia. It can be given to elderly patients with cardiac deficiencies who cannot tolerate digitalis or ouabain. The dosage is similar to that of digitoxin with a maximum dose of 0.2 mg/day (tablets of 0.1 mg = 3–4 frog doses). It also regularizes cardiac flutter and fibrillation. Oleandrin is rapidly eliminated, producing a stronger diuresis than digitalin, and is only weakly cumulative. The leaves also contain flavonoids (rutoside and 3-rhamnoglucoside of kaempferol), which contribute to the diuretic action. A

Table 2.2. *Asclepiadaceae and Periplocaceae (formerly part of the Asclepiadaceae) in tropical West Africa*

Plant	Part used	Active constituent(s)	Chemical group	Recognized or possible medicinal action
PERIPLOCACEAE				
Cryptolepis sanguinolenta (Lindl.) Schltr.	Root	Cryptolepin	Indole alkaloid	Hypotensive, anti-microbial
		In related *C. apiculata*, cryptoleposide	Cardenolide	
Cryptostegia grandiflora (Roxb.) R.Br. c.	Leaves	Cryptograndosides A, B, etc.	Cardiotoxic heterosides	Cardiotoxic, oncolytic
Parquetina nigrescens (Afzel.) Bullock syn. (*Periploca nigrescens* Afzel., *Omphalogonus nigritans* N.E.Br.)	Bark, roots leaves	Periplocoside, periplocymarin = nigrescigenin, etc.	Cardiac glycosides of digitalis group	Cardiotonic, diuretic
		Cymarin, strophanthidin		Cytotoxic
ASCLEPIADACEAE				
Subfamily Secamonoideae[a]				
Asclepiadeae				
Asclepias curassavica L. c.	Roots	Curassavicin, calotroposide	Cardenolides	Cardiotonic?
	Leaves		Polyphenols[b]	Cytotoxic
Pachycarpus lineolatus (Decne.) Bullock syn. (*Asclepias lineolata* (Decne.) Schltr.)	Roots	Asclepiadin, calotroposide, uzarigenin, corotoxigenin	Cardenolides	Ipeca substitute (India) cytotoxic
Calotropis procera (Ait.) Ait.	Latex, bark	Calotroposide, uscharin, calotropin, calotoxin, etc.	Cardenolides	Cardiotonic, cytotoxic
		Calotropain	Proteolytic enzyme	Anthelmintic

Pergularia daemia (Forsk.) Chiov. syn. (*Asclepias daemia* Forsk., *Pergularia extensa* N. E. Br.)	Plant	Pergularin (related to tomentogenin); saponification to two stigmasterols	Steroid glycosides	Spasmolytic, pituitrin-like action on uterus
	Stems, seeds	Uzarigenin, calactin, calotropin, coroglaucigenin, etc. (in Indian plants)	Cardenolides	Cardiotonic
Xysmalobium heudelotianum Decne.	Roots	Uzarosides, genin = transdigitoxigenin, xysmalogenin, coroglaucigenin, etc.	Steroid glycosides (pregnane derivatives)	Dysmenorrhoea, antispasmodic
Sarcostemma viminale R.Br.	Stems, plant	Friedelin, derivatives of viminolon, metaplexin, sarcostin (Schaub *et al.*, 1968; Stöckel *et al.*, 1969)	Pregnane glycosides	Increase lactation (Watt and Breyer-Brandwijk, 1962)
Marsdenieae				
Gymnema sylvestre R.Br.	Leaves	Gymnemic acid = 9 glycosides of related constitution	Glycosides[b]	Reduces glycosuria diabetes (US Disp. 1926)
Gongronema latifolium Benth. syn. (*Marsdenia latifolia* (Benth.) Schum.)	Stems	In related spp. condurangoside (cyramose + thevetose + glucose + aglycones derived from fluorene)	Glycosides[b]	Vermifuge
Leptadenia hastata (Pers.) Decne. syn. (*Cynanchum hastatum* Pers.)	Leaves	Glycosides related to condurangine In related *Cyn. vincetoxicum*, tylophorine	Glycosides[b] Alkaloid	Diuretic, expectorant, emetic (in French Codex, 1908)

[a]The subfamily Secamonoideae has been divided into four tribes: the Secamoeae, the Asclepiadeae and Marsdenieae (represented here) and the Ceropegieae.

[b]A number of Asclepiadaceae contain condurangin and vincetoxin glycosides (*Asclepias curassavica*, *Sarcostemma viminale*, *Gymnema sylvestre*, *Marsdenia conduranga*, *Cynanchum vincetoxicum* (*Vincetoxicum officinale*) etc.). In *A. curassavica* both cardenolides and the above-mentioned glycoside esters are present: this is probably also the case in *Sarcostemma australe* R.Br.

cytostatic effect of the leafy stems on adenocarcinoma 755 has been reported (Fauconnet and Pouly, 1962; Dykman *et al.*, 1966; Paris and Moyse, 1971, pp. 54 and 55).

Adenium obesum (Forsk.) Roem. & Schult. syn. (*A. honghel* DC., *Nerium obesum* Forsk.) (Fig. 2.2) APOCYNACEAE

L The leaves and stem exude a latex which is used in Adamawa in northern Nigeria as a fish poison and which was formerly used to poison arrows. In local medicine the latex is applied to chronic wounds and ulcers or to carious teeth.

C Seven heterosides, honghelosides A–G, were isolated from the stems and roots by Hunger and Reichstein (1950) and by Hess and Hunger (1953). Hongheloside B is identical with digitalinum verum from *Digitalis purpurea*. Hydrolysis of hongheloside A yields cymarose and oleandrogenin. Hongheloside G is identical with somalin (found in *A. somalense* in East Africa) (Hess and Hunger, 1953).

P The plant acts as a cardiac poison in the same way as digitalin, but it also has an effect on the central nervous system (CNS), on the nerve mechanism of the heart and even

Fig. 2.2. *Adenium obesum* (Forsk.) Roem. & Schult.

on the heart muscle (Perrot and Leprince, 1909). It does not appear to have been used pharmaceutically.

ASCLEPIADACEAE AND PERIPLOCACEAE. Cardenolides and allocardenolides (in the latter the A–B ring fusion is *trans* instead of *cis*, which considerably decreases the cardiotonic efficacy) are also found in a number of Asclepiadaceae and Periplocaceae (Table 2.2). They were formerly used as arrow poisons (Oliver-Bever, 1968).

Parquetina nigrescens (Afzel.) Bullock syn. (*Periploca nigrescens* Afzel., *P. calophylla* (Baill.) Roberty, *Omphalogonus nigritanus* N.E.Br.)

PERIPLOCACEAE

L The whole plant is used to stupefy fish, and the leaves and latex are used in Ghana and Liberia for the treatment of rickets, diarrhoea and skin lesions (Githens, 1949; Oliver, 1960).

C Reichstein and co-workers isolated in 1954 a series of digitalis heterosides from the fresh wood of this species and identified several aglycones including strophanthidin, strophanthidol and nigrescigenin. Later they also isolated, *inter alia*, 6-dehydroxy-strophanthidin, strophanthigenin and convallotoxin (Mauli and Tamm, 1957; Patel and Rowson, 1964; Berthold *et al.*, 1965).

Calocin, a pregnane glycoside has been reported in this species by Sravasta *et al.* (1982).

The barks of the related *Periploca graeca* and of *P. aphylla* (not found in West Africa) yield the glycosides periplocin and periplocymarin of the digitalin group (Paris and Moyse, 1971, p. 97).

P In the USSR periplocin and periplocymarin are used as cardiotonics and diuretics and are said to be better suited for slow intravenous injection than strophanthin (in the Russian Pharmacopoeia the active dose is stated as 0.02–0.05 mg) (see also Martindale, 1969).

Cryptostegia grandiflora (Roxb.) R.Br. ex. Lindley (Fig. 2.3) PERIPLOCACEAE

L The latex of this widely cultivated ornamental shrub has been used as a source of rubber, and has been considered to be oncolytic (Chopra *et al.*, 1956; Paris and Moyse, 1971).

C In earlier investigations leaves and stems were found to contain two cardenolides, cryptograndoside A and B, which are glycosides of oleandrogenin with sarmentose and glucose, respectively, and thus are similar to hongheloside A (Aebi and Reichstein, 1950; McIlroy, 1950; Abisch and Reichstein, 1962b). In 1972 five cardenolides were isolated and identified as oleandrogenin, gitoxigenin, rhodexin B, 16-propionylgitoxigenin and a new natural product, 16-anhydrotoxigenin (Doskotch *et al.*, 1972).

P An alcoholic extract of the above-ground portion of the plant had been found to have an inhibitory action against the cell culture (KB) of human carcinoma of the nasopharynx (Abbot *et al.*, 1966). In a further study at the Cancer Chemotherapy National Service Centre (CCNSC), systematic fractionation showed that mainly oleandrogenin, gitoxigenin and rhodexin B were significantly active (Abbot *et al.*,

Fig. 2.3. *Cryptostegia grandiflora* (Roxb.) R. Br. ex. Lindley.

(*a*) Flower

(*b*) Fruit

1967). A parallel appears to exist between cytotoxicity towards KB cells, a heart action and inhibition of the ATPase-mediated active transport of K^+ and Na^+ (Kupchan *et al.*, 1967; Doskotch *et al.*, 1972).

Asclepias curassavica L. ASCLEPIADACEAE

Swallow wort, wild ipecacuanha

L The roots of this West Indian species, widely cultivated in West Africa as an ornamental shrub, are used in the West Indies and in India, in decoction or pulverised, as an expectorant and emetic. They have similar effects to *Ipecacuanha* roots but are more strongly purgative. The roots of an indigenous related species, **Pachycarpus lineolatus** (Decne.) Bullock syn. (*A. lineolata* (Decne.) Schltr.), are

Fig. 2.4. *Calotropis procera* (Ait.) Ait.

given in northern Nigeria in decoction with native soda for intestinal troubles. In East Africa the roots are used to stimulate digestion (Dalziel, 1937).

C The roots of both *Asclepias* spp. contain cardenolides of which the most important aglycones are uzarigenin, corotoxigenin, and coroglaucigenin, but asclepogenin, curassavogenin and ascurogenin have also been reported as well as the cardenolides asclepin (in the Indian plants) and calotroposide which are both also found in *Calotropis* (Tschesche *et al.*, 1958; Patel and Rowson, 1964; Singh and Rastogi, 1969; Patnaik and Dhawan, 1971; Hocking, 1976). The leaves contain polyphenols (quercetin and kaempferol) (Bate-Smith, 1962).

P The alcoholic extract of the plant and asclepin have a digitoxin-like cardiotonic action and the total extract is used as a diuretic, expectorant and emetic (Paris and Moyse, 1971, p. 98).

Calotroposide shows an inhibiting action on malignant tumors (Kupchan *et al.*, 1964; Bezanger-Beauquesne and Pinkas, 1971).

Calotropis procera (Ait.) Ait. (Fig. 2.4) ASCLEPIADACEAE
Mudar, apple of Sodom, swallow wort

L Local healers use the acid latex of this plant as a rubefacient and to extract guinea worms (*Dracunculus medinensis*). Others use the dried rootbark in soup to treat colic and as a stomachic, and the burnt root is made up as an ointment for skin eruptions, foul ulcers, etc. (Dalziel, 1937).

C The very toxic latex contains a cardiac heteroside, calotroposide, with an aglycone identical to ouabain, and seven other heteroside alkaloids; calotropin, calactin, calotoxin, uscharin, uscharidin, voruscharin and proceroside. Apart from calactin and proceroside these all have the same genin (calotropa H genin) (Hesse and Ludwig, 1960; Crout *et al.*, 1963, 1964). Besides the heterosides the latex is reported to contain amyrin, traces of glutathione and a proteolytic enzyme, 'calotropain'.

P The aqueous and alcoholic extracts of the roots initially produce a slight depression, followed by a stimulation, of the rate and force of myocardial contractions in isolated frog and rabbit hearts (0.2 ml/kg). They also produce marked vasoconstriction in frog and rat and a persistent rise in blood pressure in the dog, which cannot be altered by any sympathetic drug (Derasari and Shah, 1965; Indian Council of Medical Research, 1976). In the cat the cardiotonic actions of calotroposide, calotoxoside and uscharin are 83%, 76% and 58%, respectively, of the action of ouabain (Chen *et al.*, 1942). The cardioactive effect was confirmed by Patel and Rowson (1964).

The enzyme calotropain is said to be more active than papain, bromelin or ficin (Atal and Sethi, 1962); it also has an anthelmintic action (Garg *et al.*, 1963).

Pergularia daemia (Forsk.) Chiov. syn. (*P. extensa* N.E.Br., *Daemia extensa* R.Br., *Asclepias daemia* Forsk.) ASCLEPIADACEAE

L Locally, anthelmintic properties are attributed to the leaves of the plant. The latex or a poultice of the leaves is also applied to boils and abscesses, and the plant is said to have emmenagogic action. In Ghana a soup made with the leaves is given to women immediately after childbirth (Dalziel, 1937).

C The stems of the plant contain uzarigenin, coroglaucigenin and calactin (India), whilst in the seeds calactin, calotropin and eight further cardenolides are found. The plant also contains a bitter resin called pergularin, which is structurally near to tomentogenin from *Marsdenia tomentosa* (Mittal *et al.*, 1962). Patel and Rowson (1964) established that in the Nigerian species only the seeds have cardiotonic action; the leaves, roots and stems do not (Rowson, 1965; Paris and Moyse, 1971).

P The plant has a physiological action on the uterus similar to that of pituitrin but mainly limited to the upper part of the uterus and its use as a pituitrin substitute in delivery has been suggested (Dutta and Gosh, 1947). This action is not inhibited by progesterone. A general stimulating effect on involuntary muscles and an increase of the arterial blood pressure has also been observed (Gupta *et al.*, 1950; Unesco, 1960).

Xysmalobium heudelotianum Decne. ASCLEPIADACEAE

L The tuber of this plant is used as a bitter tonic and is eaten by the Hausas (Northern Nigeria) as a remedy for stomach troubles (Dalziel, 1937).

C The tubers contain, like those of the East African *X. undulatum* R.Br., uzarosides (uzarin is a monoglycoside of uzarigenin, an isomer of digitoxigenin) and glycosides of xysmalogenin and of 17α-uzarigenin (Kuritzkes *et al.*, 1963; Paris and Moyse, 1971, p. 97).

Glucosides of pregnane derivatives were reported in the roots of *X. undulatum* (L.) Ait. by Tschesche and Snatzke in 1960 (Paris and Moyse, 1971, p. 96).

P The uzarosides of *X. heudelotianum* have a weak cardiotonic action. They are mainly used as antispasmodics and antidiarrhoeal agents and in dysmenorrhoea (Paris and Moyse, 1971, p. 97).

PLANTS BELONGING TO OTHER BOTANICAL FAMILIES. Cardenolides related to those found in the above-mentioned Apocynaceae and Asclepiadaceae are also occasionally found in members of other botanic families.

Corchorus olitorius L. TILIACEAE

L The bark provides a fibre (jute) and the mucilagenous leaves are used in food and as a vegetable. In indigenous medicine in India the seeds are used as a purgative and the leaves as a tonic and diuretic (Chopra *et al.*, 1956; Oliver, 1960).

C From the seeds, 11–15% of a fixed oil and several steroid heterosides could be isolated (Chakrabasti and Senn, 1954). The corresponding aglycones, at first named corchorin, corchorogenin, corchsularin and olitorigenin, were later identified as strophanthidin (Senn *et al.*, 1957). The main heterosides are corchoroside, olitoroside and helveticoside. The latter can be hydrolysed to give one molecule of strophanthidin and one molecule of D-digitoxose. Olitoroside is strophanthidin-β-D-boivinopyranoside-3β-D-glucopyranoside (Chakrabasti and Senn, 1954; Senn *et al.*, 1957; Schmersal, 1969).

P Strophanthidin has an action comparable to that of ouabain. The corresponding corchorosides A and B were found to contain lethal doses of 0.0768 and 0.1413 mg/kg, respectively, which thus makes corchoroside A one of the most potent heart poisons (Frèrejacque and Durgeat, 1954). Pharmacological and clinical trials have been carried out by Russian research workers. In pharmacological tests on rabbits with experimental myocarditis and myocardiosclerosis, and in dogs with acute coronary insufficiency a therapeutic effect was obtained with doses of 0.05 units/kg (one cat unit = 14 μg/kg) (Kiteava, 1966).

In clinical trials favourable results have been obtained in cases of chronic cardiac insufficiency both with the corchorosides and with olitoroside by several Soviet authors. The clinical condition and electrocardiogram of the patients were improved (increase in amplitude of T-wave) (Turova, 1962; Umarova *et al.*, 1968).

Urginea indica (Roxb.) Kunth. syn. (*Scilla indica* Roxb.) LILIACEAE

L In South Africa *U. altissima* is known to be fatal to stock (Watt and Breyer-Brandwijk, 1962) and this probably also applies to *U. indica*. In northern Nigeria the scorched and dried bulbs are included in a liniment for rheumatic knees and sometimes the crushed bulbs are applied to bruises and aches (Dalziel, 1937).

C From the bulbs of these species and from the Mediterranean *U. maritima* Bak. a crystalline glycoside, scillaren A, and a mixture of amorphous glycosides, scillaren B, have been extracted (Chopra *et al.*, 1938, p. 251). The genin of scillaren A is scillaridin. The glycosides are chemically related to the digitalis and strophanthus glycosides. They are steroid heterosides of the bufanolid type (with a hexagonal lactonic cycle) (Seshadri and Subramanian, 1950).

P The bulbs are official in the British Pharmaceutical Codex (1949) and US National Formulary. Scillaren, unlike digitalis, does not accumulate, but is bound more to the myocardium than is strophanthin. Scillaren A acts quickly but is rapidly hydrolyzed in the blood. Squill is mainly a diuretic which acts by increasing the renal circulation, but excessive doses may cause irritation and obstruction of the kidneys. Average doses in heart insufficiency are 6 mg/24 h with maintenance doses of 2 mg/24 h (Paris and Moyse, 1967, p. 50). The glycosides have the advantage of being useful in the treatment of conditions refractory to, or no longer responsive to, digitalis and strophanthus therapy (Darwish, 1980).

Urginea extracts have also been found to have antiprotozoal, hypoglycaemic (Bapat *et al.*, 1970) and oncolytic actions (Dhar *et al.*, 1968).

Mansonia altissima (Chev.) Chev. var. *altissima* STERCULIACEAE

L The bark of this tree has been used in the Ivory Coast as an arrow poison and in West African local medicine for the treatment of leprosy and as an aphrodisiac (Oliver, 1960).

C It contains mansonin which is a 2,3-di-(*O*-methyl)-6-deoxy-β-D-glucopyranoside of strophanthidin. Strophothevoside is the corresponding 3-*O*-methyl-6-deoxy-β-glucopyranoside. Besides mansonin, minute quantities of as many as 30 other cardenolides have been traced in the seeds by paper chromatography. They are all derived from three genins: strophanthidin, nigrescigenin and an undetermined genin (Algeier *et al.*, 1967).

P An amorphous fraction of mansonin was found to have a cardiotonic activity comparable to that of strophanthin G. Unfortunately, the yield of active substance is small and variable (Terrioux, 1952).

Antiaris africana Engl. syn. (*A. kerstingii* Engl., *A. toxicaria* (Rumph. ex Pers.) Lesch. var. *africana*) MORACEAE

Bark cloth tree

L The tree is called 'bark cloth tree' as in Ashanti (Ghana) a strong white cloth is made from the bark. In the Ivory Coast the bark has been used as a purgative and in the treatment of leprosy (Dalziel, 1937).

C Seven heterosides of the digitalis type are reported in the latex including α- and β-antiarin. On hydrolysis they produce antiarigenin and antiarose, and antiarigenin and L-rhamnose respectively. The seeds contain evenomoside, antioside and several other glycosides and aglycones (Bisset, 1962; Wehrli *et al.*, 1962; Mühlrad *et al.*, 1965).

P The West African *Antiaris* appears to be less toxic than the Asiatic *A. toxicaria* (Pers.) Lesch. The dried latex of the Asian species (water-soluble extract) is a violent heart poison, causing fibrillation and a drastic fall in the blood pressure. In smaller doses it appears to be a stimulant for the heart and circulation (Chopra *et al.*, 1938; Patel and Rowson, 1964).

Schwenkia americana L. syn. (*S. hirta* Wright, *S. guineensis* Schum. & Thonn.) SOLANACEAE

L The root of this solanaceous shrub is a common remedy for rheumatic pains and swellings in northern Nigeria; in Ghana it is used as a cough medicine and in Angola for chest complaints (Dalziel, 1937).

C According to Rabaté (1940), the leaves, roots and stems of the plant contain a glycoside, schwenkioside, which has a phenolic aglycone, schwenkiol, and also traces of alkaloids. However, Patel and Rowson (1964) could not identify schwenkioside but found a steroid sapogenin to be the main constituent of this herb in Nigeria.

P This sapogenin seems to behave physiologically as a cardiac glycoside. On a toad's heart it has been found to cause, after initial inhibition, a prolonged stimulation. All parts of the plant produce haemolysis of the red blood cells, probably due to the saponin (Patel and Rowson, 1964).

Paullinia pinnata L. SAPINDACEAE

L In West Africa, the juice of the leaves and seedpods is known for its haemostatic action and is used as an infusion in dysentery and fever, as a tonic and in acute infectious disease. The root and seeds are said to be highly toxic (Dalziel, 1937).

C *P. pinnata* collected in West Africa is found to contain no alkaloids but a saponin with a triterpenic aglycone (Kerharo and Adam, 1974). It contains quebrachitol in both the leaves and bark in Madagascar (Plouvier, 1948) and Bowden (1962) extracted a flavotannin from the leaves of the West African plant.

P The flavotannin has a cardiotonic effect on the isolated frog's heart (Bowden, 1962) and on the heart of mammals (Broadbent, 1962). When calcium is absent from the perfusion liquid the tannin has no cardiotonic action. The tannin is shown to be antagonistic to ouabain, probably by preventing the fixation of ouabain on the heart surface and is said to act in forming a calcium–tannin complex on the surface of the heart tissues. The action of the flavotannin on the mammalian heart is to increase the strength of the diastole and the coronary flux (Broadbent, 1962). The saponin was shown to be toxic to *Paramecia*, which were killed in 1 h by a concentration of 1 : 500 (Kerharo *et al.*, 1960–2) (see Chapter 4).

Vernonia colorata (Willd.) Drake syn. (*Eupatorium coloratum* Willd., *V. senegalensis* Less.) COMPOSITAE
Bitter leaf

L A decoction of the leaves is used in local medicine as an antipyretic, expectorant and laxative. The bark of roots and stems is astringent, and is used against fever and diarrhoea. The root without the bark is taken as a tonic (Oliver, 1960).

C A bitter glycoside, vernonin, was first isolated from the roots by Heckel and Schlagdenhaufen (1888) and was also detected in the roots of **V. nigritiana** Oliv. & Hiern. **V. amygdalina** Del. and **V. cinerea** (L.) Less., and a 'bitter principle' was reported in **V. conferta** Benth. and **V. guineensis** Benth. All these species have similar local uses. Later Patel and Rowson (1964) found a cardiac glycoside in the leaves, stems and roots of *V. colorata* and *V. nigritiana* collected in Nigeria. In addition, Toubiana (1969) has isolated two sesquiterpenic lactones (with cytotoxic action in vitro) from *V. colorata*, and from *V. guineensis* vernodalin and vernolepin, which have activity against Wilme's myeloma and KB tumours, respectively (Toubiana, 1975).

P When injected intravenously in dogs vernonin produces hypotension and has an action on the heart comparable to that of digitalin, but is much less toxic. The cardiac glycosides isolated by Patel and Rowson (1964) similarly have a distinct cardiotonic action but no cardiotoxic action. Jawalekar reports (in Caiment-Leblond, 1957) that a leaf extract of *V. amygdalina* reduces the rate and force of contraction of the isolated frog heart. In cats it causes a marked fall in the blood pressure, reduces the heart rate and blocks the transmission of heart contractions. Further, it strongly stimulates contractions of the isolated rabbit intestine. These effects can be blocked by atropine (Kerharo and Bouquet, 1950; Caiment-Leblond, 1957). The LD_{50} in mice for *V. colorata* is 10 g/kg. The leaves of *V. cinerea* have a slight antibiotic action (Kerharo, 1968).

Group B: Plants containing cardiotonic alkaloids

Erythrophleum guineense G. Don CAESALPINIACEAE
Ordeal tree, sasswood, red water tree, sassy bark

L The bark has been used in arrow poisons; its toxic aqueous decoction or cold infusion is called 'red water' and was used in fetish trials and ordeals.[2] Rigal (1941) observed that in guinea pigs death occurred 3 h after a dose of 0.5 g/kg was administered and 55 min after a dose of 1 g/kg. The seeds were found to be more toxic than the bark in spite of a lower alkaloid content. Probably this is due to the simultaneous presence of a strongly haemolytic saponin (Rigal, 1941). In local medicine the bark is sometimes used as a diuretic, emetic and sternutatory.

C Bark and seeds contain 0.1–0.5% of total alkaloids, mainly erythrophleine (Paris and Rigal, 1940, 1941), also cassaine, cassaidine, norcassaidine, coumingine and erythrophleguine (Dalma, 1939; Lindwall *et al.*, 1965). Apart from the alkaloids, a catechuic tannin, a saponin and a flavonoside (luteolin glycoside) have been isolated from the bark, as well as a wax with a high proportion of hexacosanol. **Erythrophleum ivorense** Chev. syn. (*E. micranthum* Harms.) is less toxic than *E. guineense* probably because of a higher tannin and a lower alkaloid content (up to 0.3%).

The cardioactive properties of the alkaloids can be destroyed by saponification and can be changed by chemical modifications. Thus, modification on C-3 of cassenic acid produces a stronger and longer-lasting action (Hauth, 1971).

P Erythrophleine has a digitalis-like action, whilst the action of coumingine, which is the most active of the alkaloids, is similar to that of scillaren A in the cat. Cassaine and cassaidine are less potent than erythrophleine, which raises the blood pressure, slows the pulse whilst increasing the force of the heartbeat and decreases respiration (Cotten *et al.*, 1952). Overdoses of erythrophleine produce symptoms of circulatory depression, breathing difficulties, vomiting and, through direct action on the medulla, convulsions. Erythrophleine is said to be of use in spasmodic asthma (Rigal, 1941). It also has a local anaesthetic action similar to that of cocaine but more powerful and longer lasting. However, no use has been made so far of this action, probably because of the general toxic effect of these alkaloids (Trabucchi, 1937). Cassaine has convulsant action (Santi and Zweifel, 1936). Derivatives of these alkaloids are being prepared in an attempt to decrease their toxicity (Hauth, 1971). In low concentrations cassaine and coumingine increase the translocation of K^+ from the plasma to the cells (Kerharo, 1968).

A bark extract of **Erythrophleum sauveolens** (Guill. & Perr.) Brenam has been shown to have a strong spasmogenic effect on smooth muscles. It also has a chronotropic and isotropic effect on the heart and shows a potent hypotensive action which is probably mediated through release of catecholamines (Bamgbose, 1974).

Voacanga africana Stapf syn. (*V. glabra* Schum., *V. Schweinfurtii* var. *parviflora Schum.*, *V. magnifolia* Wernham, *V. talbotti* Wernham, *V. eketensis* Wernham, *V. glaberrima* Wernham, *V. africana* var. *glabra* (Schum.) Pichon) (Fig. 2.5)

APOCYNACEAE

L Locally the latex is used as a rubber adulterant and is applied to carious teeth (Oliver, 1960).

C Since 1955 this plant has aroused considerable interest and has been the subject of numerous publications. From the bark of the stem and the root many alkaloids have been isolated (4–5% total alkaloids from the stembark and 5–10% from the rootbark). These include voacamine (the main alkaloid), voacangine, voacristine (= voacangarine), voacorine, vobasine, voacamidine (an isomer of voacamine) and many others. Most of these alkaloids have also been found in **Voacanga thouarsi** Roem. & Schult. and some in other species of *Voacanga*, *Tabernaemontana* and even *Alstonia*. In the leaves of *V. africana* voaphylline and vobtusine are found, and tabersonine is found in the seeds (Blanpin *et al.*, 1961; Puisieux *et al.*, 1965; Oliver-Bever, 1967).

P Voacamine and voacangine act on the heart in a similar way to the cardiac glycosides but their toxicity is very low in comparison with that of other alkaloids with a cardiostimulant action such as the *Erythrophleum* alkaloids. A dose of 100 μg of voacamine sulphate has a cardiotonic action equivalent to that of a dose of 0.25 units of digitalis standard (in isolated rabbit auricles) (Oliver-Bever, 1967). Voacamine does not bind to the cardiac proteins and has no cumulative action, but it has a direct myotonic effect on the cardiac fibre (Quevauviller and Blanpin, 1957a). Lethal doses for guinea pigs (by instillation in the jugular vein) are 313 mg/kg for voacamine sulphate and 348 mg/kg for voacangine sulphate, compared to 2.5 mg/kg for digitalin and 0.9 mg/kg for strophanthin. In mice the LD_{50} of voacangine, given intravenously, is 41–42 mg/kg (La Barre and Gillo, 1955; Vogel and Uebel, 1961). Therapeutic doses (1–3 mg daily) are well tolerated, act rapidly and are quickly

Fig. 2.5. *Voacanga africana* Stapf.

eliminated without any cumulative effect. Voacamine, voacangine and voacorine also have a hypotensive action and are simultaneously mildly parasympatholytic and sympatholytic. Voacangine, which is also said to have analgesic and local anaesthetic action, and vobtusine increase the hypnotic effect of barbiturates. Voacorine is also cardiotonic through direct action on the heart muscle and on the coronary perfusion and seems to contribute considerably to the action of the total alkaloids of *Voacanga* (Quevauviller and Blanpin, 1957b). Its minimum LD_{50} in guinea pigs (by slow intravenous injection) is 228 mg/kg (Blanpin *et al.*, 1961).

For the leaf alkaloid vobtusine, given intravenously, the LD_{50} is 33.75 mg/kg. Vobtusine has a depressive action on the heart and has hypotensive and sedative properties (Quevauviller *et al.*, 1965). Tabersonine from the seeds has a hypotensive action which is equivalent to 25% of that of reserpine (Zetler, 1964).

(b) Cardiac depressants: anti-arrhythmic agents

Argemone mexicana L. (naturalized) (Fig. 2.6) PAPAVERACEAE
Mexican poppy, prickly pepper, prickly poppy

L The plant is used in Nigeria and Senegal mainly for its diuretic, sedative, cholagogic and cicatrizing properties (Oliver, 1960; Kerharo and Adam, 1974). The seeds have a cannabis-like effect and in many countries the herb, juice and flowers are reputed to be narcotic (Watt, 1967).

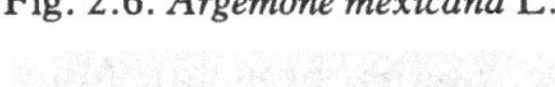

Fig. 2.6. *Argemone mexicana* L.

C Numerous alkaloids have been reported to be present in all parts of the plant. Thus, leaves, stems and seeds contain berberine and protopine and the roots contain coptisine, α-allocryptopine, chelerythrine and dihydrochelerythrine. In the oil of the seeds sanguinarine and dihydrosanguinarine are found (Chakravarti *et al.*, 1954; Bose *et al.*, 1963). Argemonine was isolated from the leaves and capsules and identified as (−)*N*-methylpavine (Martell *et al.*, 1963).

P α-Allocryptopine, (which represents 0.099%) of the roots of the plant in Czechoslovakia, is identical to α-fagarine. It slows down the heart rate and prolongs systole in rats and frogs. In doses of 10–20 mg/kg it also slows down the heartbeat of cats and rabbits. The action is a direct one on the myocardium and is also antifibrillatory, and α-allocryptopine has found clinical applications in this field. Thus it is considered more active than quinidine in cases of arrhythmias with fibrillation and auricular flutter (Alles, 1952; Dhar *et al.*, 1968; Manske and Holmes, 1950–71, vol. 5, pp. 90–91).

Protopine was isolated from the total alkaloid fraction by Bose. He reported that it stimulated the heart, blood pressure and respiration, as well as the striated and smooth muscles on which it appears to act specifically (Manske and Holmes, 1950–71, vol. 5, pp. 92 and 138).

The seed oil is highly toxic due to sanguinarine (when given subcutaneously, the LD_{50} for mice is 1.8 mg/100 g). Sanguinarine can produce experimental glaucomas (Hakim, 1954).

Berberine is relatively non-toxic (when given intravenously, the LD_{50} in cats and dogs is 0.025 mg/100 g). In doses of 2 mg/kg, berberine has a depressant and vasodilating action on the heart. It also depresses breathing but stimulates the smooth muscles of different organs (intestine, uterus, bronchi). Moreover, berberine has marked antibiotic properties on *Mycobacterium tuberculosis*, *Staphylococcus aureus*, *Escherichia coli*, *Eberth typhosa* and *Shigella dysenteriae*, and it also acts at concentrations of 1:80000, on *Leishmania tropica* (Lambin and Bernard, 1953). An alcohol–water extract of the fruits deprived of the seeds proved to be an excellent hypnotic and sedative for convulsions and spasmodic conditions (Martínez, 1959, p. 110). Antifibrillatory action has also been reported for oleandrin from *Nerium oleander*, described earlier under cardiotonics (Fauconnet and Pouly, 1962).

Zanthoxylum zanthoxyloides (Lam.) Watson syn. (*Fagara zanthoxyloides* Lam., *F. senegalense* (DC) Chev., *Z. polyganum* Schum.) RUTACEAE
Prickly ash, candlewood, toothache bark

L The aromatic rootbark is used in Ghana and Nigeria as a decoction, or in application, for its alleged antiseptic and analgesic properties, for example in the treatment of painful conditions, in childbirth, for toothache, etc., sometimes mixed with other ingredients. The decoction is also used as a vermifuge and in Guinea the bark is used to stupefy fish. The fixed oil contained in the rootbark causes salivation, a numbing action on the tongue and paralysis (Pobeguin, 1912; Irvine, 1930; Dalziel, 1937; Oliver, 1960).

C An amorphous alkaloid was isolated from the roots as early as 1887 and called artarine. (Later this was found to be identical with ethoxychelerythrine (Torto *et al.*, 1969.) In 1911 a pungent principle which produced salivation was extracted from the rootbark (Thoms and Thumen, 1911). It was called fagaramide, and proved to be an *N*-isobutylamide of piperonylacrylic acid.

Since 1947 a number of tertiary and quaternary alkaloids have been identified in the bark and rootbark of *Z. zanthoxyloides*. These are, in addition to artarine, β-fagarine (= skimmianine, a dimethoxydictamnine), fagaridine (= erythrofagarine), angoline, angolinine, chelerythrine, dihydrochelerythrine, tembetarine, magnoflorine, *N*-methylcorydine. No α-fagarine (also called aegeline (from *Aegle marmelos*) = allocryptopine) was reported in West African species (Paris and Moyse-Mignon, 1947; Calderwood and Fish, 1966; Torto *et al.*, 1969). The rootbark also contains fagarol (a lignan) and pseudofagarol, and in the fruits two coumarins, xanthotoxin and bergapten, were reported by Paris and Moyse-Mignon (1947). The leaves contain traces of alkaloids and a flavone heteroside.

P α-Fagarine is mainly extracted for pharmaceutical use from the leaves of the Argentinian *Z. coco* as it has proved to be a useful substitute for quinidine in auricular fibrillation. In some cases it has been found to normalize the sinus rhythm within 30 min and is so far the only *Fagara* base to be exploited in medical science. (Aegeline was considered to be a weak vasoconstrictor and in large doses a cardiac depressant (Paris and Moyse, 1967, p. 304).)

Xanthotoxin is ichthyotoxic in concentrations of only 0.1 p.p.m. and the pungent fagaramide has weak local anaesthetic action (Paris and Moyse-Mignon, 1947; Bowden and Ross, 1963).

An aqueous extract of the rootbark has further been reported to bring about a reversal of sickling and crenation in erythrocytes (Sofowora and Isaac-Sodeye, 1971; Murayama and Makyo, 1972). Later the antisickling compound was isolated and identified as 2-hydroxymethylbenzoic acid. On further investigation it was found that the greatest amounts of this acid were contained in the leaves, then in the stembark and lastly in the roots. Also the antisickling fractions varied among six different Nigerian *Zanthoxylum* spp. and also varied amongst parts of the same species (Isaac-Sodeye *et al.*, 1975; Rumen, 1975; Sofowora *et al.*, 1975).

From the roots of the Ghanaian species a new crystalline alkaloid was isolated in 1972 by Messmer *et al.* and called fagaronine. This benzophenanthridine alkaloid has a cytotoxic action, and is believed to be an inhibitor of RNA-directed activity in avian nucleoblastosis and in cases of infection by Rauscher leukaemia virus and Simian sarcoma virus, probably by preventing the elongation reaction (Messmer *et al.*, 1972).

Fagara leprieuri (Guill. & Perr.) Engl., **Zanthoxylum gillettii** (de Wild.) Watson syn. (*F. macrophylla* Engl.) and **Zanthoxylum rubescens** (Planch. ex Hook. f.) Watson syn. (*F. rubescens* (Planch. ex Hook. f.) Engl.) are used in similar ways to *Z. zanthoxyloides* in local medicine. They also contain β-fagarine (skimmianine), fagaridine, xanthofagarine, angoline, angolidine and a few other bases but so far no α-fagarine has been reported (Fish and Waterman, 1971, 1972).

Cinchona spp. (cultivated in the Cameroons and Guinea) RUBIACEAE

P One of the alkaloids of the *Cinchona* bark is used in cardiology. It is a dextrorotary stereoisomer of quinine, quinidine, which is used preferably as the sulphate, in the treatment of auricular arrhythmias as it has a specific depressant effect on the auricular muscle. It should, however, be reserved for the treatment of early persistent fibrillation as it is cumulative in action. Overdoses may cause extrasystole, paroxysmal ventricular tachycardia, ventricular fibrillation, intraventricular block and cardiac arrest (Martindale, 1969).

Rauvolfia vomitoria Afzel. syn. (*R. senegambiae* DC., *Hylacium owariense* P. Beauv.) (Fig. 2.7) APOCYNACEAE

P Ajmaline, one of the sympatholytic alkaloids of this plant (the others are discussed below under hypotensives), is chemically and pharmacologically closely related to quinidine. It is used in the treatment of arrhythmias as it slows down the rhythm and decreases myocardial excitability in doses of mg/kg without influencing the blood pressure. In clinical trials it produced a return to normal sinus rhythm in a high percentage of patients with multiple extrasystole and sinus tachycardia, but results

Fig. 2.7. *Rauvolfia vomitoria* Afzel.

have been more uncertain for atrial fibrillation (Knipel *et al.*, 1971; Lampertico, 1971). The maximum single dose should not exceed 50 mg and constant cardiographic control is required. The action of 10 mg of ajmaline (given intravenously) is approximately equivalent to that of 100 mg of prominamide (Puech *et al.*, 1964).

Rauvanine (also in *R. vomitoria*) has an effect on the cardiovascular system similar to that of ajmaline (antifibrillatory, coronary dilating and slowing down the heartbeat), but it is non-sympatholytic and also has hypotensive action. It is only half as toxic as reserpine and is not ulcer producing (Quevauviller *et al.*, 1963, 1971, 1972).

(c) Vascular agents

These can be divided into three groups:

(A) Hypotensive and some hypertensive plants. Some of the plants discussed in the sections on cardiotonics and cardiac depressants (above) were shown to act on the blood pressure. Haemodynamically the blood pressure depends on (i) the cardiac output and (ii) the peripheral resistance in the capillaries. The sympathetic system controls hypertension through the action of noradrenaline: stimulation of the α-receptors of the small arteries produces vasoconstriction while stimulation of the β-receptors causes vasodilatation. In normal conditions the action on the α-receptors predominates over the action on the β-receptors and a state of semi-contraction is maintained.

Hypotensive treatment can include a depletion of catecholamines in the post-ganglionic fibres of the sympathetic system as well as in the CNS. This is the case e.g. of reserpine. As an increase in blood pressure entails adaptation of glomerular filtration and maximum reabsorption of sodium, requiring an increase in sodium excretion, hypotensive treatment often includes administration of diuretics to produce sodium depletion.

(B) Plants containing compounds that are capable of increasing resistance and decreasing the permeability of capillaries and veins. These compounds are widely distributed in fruit and green leaves and are used on a large scale in capillary and venous insufficiencies. Their action, discovered by Szent Gyorgyi in 1936, was at first attributed to a compound called vitamin P, or sometimes vitamin C_2. This consists of several constituents, also called bioflavonoids, which mainly belong to three groups:

(i) the coumarin or α-benzopyrone group, which includes aesculetol and its glycoside aesculoside;

(ii) the chromane group, including polyhydroxylated derivatives of phenylchromane or flavane (catechins, anthocyanins, leucoanthocyanins;

(iii) the chromone or γ-benzopyrone group, comprising flavanone, flavanol, flavonol and their (polyhydroxylated) derivatives like the flavone derivatives quercitin (3, 3′, 4′, 5, 7-pentahydroxyflavone), kaempferol (3, 4′, 5, 7-tetrahydroxyflavone), or the flavanones eryodictyol, naringetol and hesperetol.

These flavonoids increase the capillary resistance, have an antihistamine and antihyaluronidase action and can protect against radiation, the leucopenic effect of cytostatics and disorders due to an atherogenic diet. They have also been called citroflavonoids as they are found in the pericarp of citrus fruits and are extracted in large quantities from different plants (*Sophora*, *Vaccinium* leaves, *Eucalyptus macrorrhyncha*) for use in pharmacy. They appear to be well tolerated, no serious side-effects having been reported (Paris, 1971; 1977). A few of the plants reported to heal oedemas and piles in indigenous medicine may be found to contain bioflavonoids.

(C) Plants containing constituents which act more specifically on blood coagulation and formation. Some plants promote coagulation and are reputed to have a haemostatic action (group C_1). Amongst the constituents responsible for this action there are some naphthoquinones closely related to vitamin K. In West Africa these are found in, for example, *Lawsonia inermis* and *Diospyros mespiliformis*. Vitamins K and K_1 (phytylnaphthoquinone) improve prothrombin formation and as a result hasten blood coagulation. Vitamin K is very easily prepared synthetically. Other plants are anticoagulants (group C_2), inhibiting prothrombin formation, an action which could be attributed to dicoumarin in the case of dried *Melilotus sativa* in Europe, which produces a haemorrhagic syndrome in cattle. Yet other plants have an anti-anaemic action (group C_3).

Most of the West African plants having a haemolytic action contain saponosides. Coumarin derivatives (calophyllide and inophyllide) are found in a plant introduced into West Africa, *Calophyllum inophyllum* Guttiferae, which increases capillary permeability and has anticoagulant properties (see Chapter 5).

Group A_1: hypotensives

Plants containing hypotensive alkaloids. Amongst these many indole and indoline alkaloids are found. Many of them act through the ANS or through the CNS and, as we will see, most of them are not only hypotensive but are also sympatholytic (yohimbine, akuammidine and corynanthine), sympathomimetic (eserine and *Mitragyna* and *Hunteria* alkaloids), local anaesthetic (*Mitragyna* alkaloids) or sedative (reserpine).

Rauvolfia vomitoria Afzel. (for synonyms see above) APOCYNACEAE

L Ghanaian and Nigerian healers use the rootbark, which in high doses is a powerful purgative and emetic, in cases of infantile convulsions, jaundice and gastrointestinal troubles. The latex or a decoction of the leaves is used in the treatment of parasitic skin diseases, head lice, *etc*. (Dalziel, 1937).

C Although a decoction of the root was used in 1936 by Shapara as a sedative in cases of maniacal symptoms, inducing several hours of sleep (see Dalziel, 1937), it was only in 1952 that the *Rauvolfia* species raised any considerable interest. This was after the isolation of reserpine, with its sedative and hypotensive action, from the Indian *R. serpentina* Benth. (Müller *et al.*, 1952). Since then numerous alkaloids have been isolated from different *Rauvolfia* spp. and their pharmacological properties tested. From *R. vomitoria* rootbark, 4–8% of total alkaloids have been isolated, containing up to 1.7% of reserpine (which has to be separated from accompanying resins). The plant thus contains more reserpine than the Indian *R. serpentina* and is indicated in the 1968 British Pharmaceutical Codex and the 1968 British Pharmacopoeia as a source of reserpine. Harvesting is done by periodically cutting small pieces of root without uprooting the tree. Many other *Rauvolfia* alkaloids have now been isolated; they appear to belong to four main groups. In the first, the yohimbane group, in addition to reserpine, rescinnamine, seredine and yohimbine have been reported. In the second, the heteroyohimbane group, have been found reserpiline, raumitorine, alstonine (also in *Alstonia*), rauvanine and ajmalicine. In the third, the ajmaline group, ajmaline is the main alkaloid, and in the fourth, the oxindole group, rauvoxine. All these alkaloids are accompanied by a number of secondary alkaloids (Woodson *et al.*, 1957; Patel *et al.*, 1964; Delaveau, 1966). A complete and clear table of the numerous *Rauvolfia* alkaloids can be found in Kerharo and Adam (1974, p. 182). Mainly reserpine, rescinnamine, ajmaline, ajmalicine and reserpiline are extracted from the rootbark for therapeutic use by pharmaceutical firms.

From the seeds 2,6-dimethoxybenzoquinone, and from the leaves two flavone heterosides, 3-rhamnoglucoside and the 3-glucoside of kaempferol, have been isolated (Patel *et al.*, 1964; Paris and Etchepare, 1967; Paris and Moyse, 1971).

P Reserpine has a hypotensive action in cases of hypertension and slows down the heartbeat. The alkaloid also has sedative and tranquillizing effects but is not hypnotic. It acts through the CNS and is active only in the presence of the hypothalamus and diencephalon and seems to act as an antimetabolite of serotonin and catecholamines, decreasing considerably the serotonin content of the nerve centres. This explains why, next to its use as a hypotensive agent in arterial hypertension, reserpine (given orally in 0.1–0.25 mg tablets) is currently used as a tranquillizer in anxiety states and in psychoses with hallucinations and delirium. Although it is not very toxic, the action of reserpine is cumulative and after prolonged administration side-effects like nasal congestion, bradycardia, oedema, stimulation of intestinal peristalsis and even ulceration are noticed (Woodson *et al.*, 1957; Smith, 1963; Delaveau, 1966; Fattorusso and Ritter, 1967).

Rescinnamine and, in particular, reserpiline also have a hypotensive action. Reserpiline, representing up to 75% of the remaining alkaloids, produces, in contrast to reserpine, no digestive troubles or ulcers in rats, even in doses of 2 mg/kg. It has, however, no tranquillizing or hypnotic effects (La Barre and Gillo, 1958).

Ajmalicine, which also acts through the central nervous system, is a coronary and peripheral vasodilator and is used in angina pectoris and Raynaud's disease (Fattorusso and Ritter, 1967).

Raumitorine has a hypotensive action similar to that of reserpiline (La Barre and Hans, 1958) and does not act on the digestive tract. It has, however, to a certain extent, the tranquillizing effect of reserpine (La Barre and Demarez, 1958; La Barre and Gillo, 1958; La Barre *et al.*, 1958). **Rauvolfia macrophylla** Stapf, **R. caffra** Sond. and **R. mannii** Stapf also contain alkaloids but are less common than *R. vomitoria* in tropical West Africa. Reserpine and ajmalicine (= vincaine = δ-yohimbine) are also found in *Catharanthus roseus*, now mainly used for the extraction of antileukaemia principles (Oliver-Bever, 1971).

Picralima nitida (Stapf) Th. & H. Dur. syn. (*P. macrocarpa* Chev., *Tabernae montana nitida* Stapf, *P. klaineana* Pierre) APOCYNACEAE

L The seeds are eaten locally as a tonic and excitant and are used in the treatment of malaria and jaundice (Irvine, 1930; Dalziel, 1937).

C They contain 3–5% of total alkaloids many of which have been identified. The main alkaloids are akuammine, akuammidine, and akuammigine, an isomer of ajmalicine. Some further alkaloids reported are pseudoakuammidine and akuammiline (indoline derivatives), akuammicine (indole derivative), etc. (Olivier *et al.*, 1965; Pousset *et al.*, 1965). From the leaves picraphylline and from the roots picracine and melinosime have been isolated (Le Double *et al.*, 1964).

P Akuammine, the main alkaloid, has been found to be inactive in malaria both in pharmacological and clinical trials, but it is a powerful sympathomimetic and has a local anaesthetic action almost equal to that of cocaine (Raymond-Hamet, 1951; Paris and Moyse, 1971, p. 94).

Akuammidine has a sympatholytic action and a hypotensive effect which is weaker but longer lasting than that of yohimbine. Akuammidine also has a strong local anaesthetic action (three times that of cocaine hydrochloride) (Raymond-Hamet, 1944).

Holarrhena floribunda (Don) Dur. & Schinz var. *floribunda* syn. (*H. africana* DC., *H. Wulfsbergii* Stapf, *Rondeletia floribunda* Don) APOCYNACEAE

L Used in Ghana and Nigeria as an antipyretic and antidysenteric (Dalziel, 1937).

C The stembark and rootbark of this small tree have been found to contain 1.2–2.44% and 2.5–3.8% of total alkaloids, respectively. Of these alkaloids at least 50% is

Fig. 2.8(*a*). Holarrhenine.

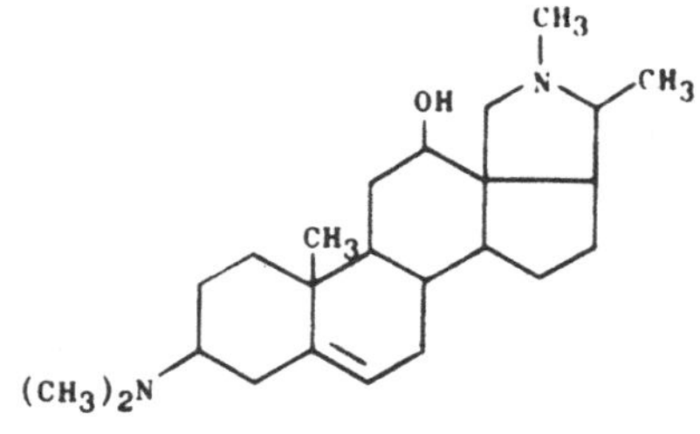

Fig. 2.8(*b*). 5β-pregnane.

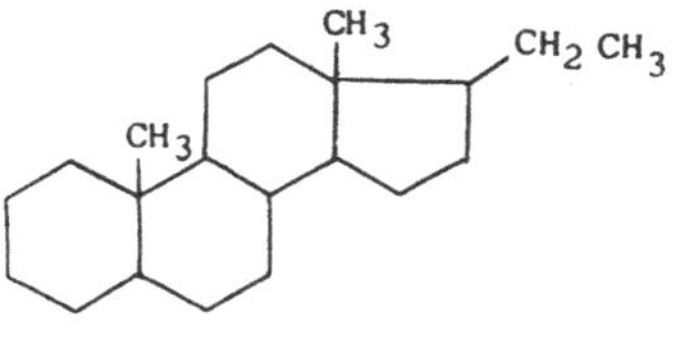

conessine, used mainly for its antibiotic action, which is not relevant here. Most of the main alkaloids of the bark are steroid alkaloids derived from conamine, whilst in the leaves alkaloids derived from pregnane (Janot *et al.*, 1959) plus 0.6% of a non-steroid alkaloid, triacanthine (an adenine derivative), are found (Janot *et al.*, 1959, 1960). Further acid phenols (*p*-hydroxybenzoic, protocatechuic and *p*-coumaric acids) are found in the leaves of *Holarrhena* and quercetol and kaempferol flavonols are found in the leaves and seeds (Paris and Duret, 1973).

P Many of the steroid alkaloids derived from conamine, holarrhenine (Fig. 2.8*a*) or pregnane (Fig. 2.8*b*), like holarrimine, holaphyllamine and holaphylline, have a hypotensive action and are simultaneously local anaesthetic and spasmolytic. Triacanthine also has a hypotensive action, but while conessine, holarrhenine, etc., are cardiotoxic, triacanthine has a cardiotonic action on the heart of the rabbit in doses of 1/30 of the LD_{50}. It produces an important and long-lasting vasodilatation of the coronary arteries and is, in addition, antispasmodic and respiratory analeptic (Quevauviller and Blanpin, 1961). For this reason Foussard-Blanpin *et al.* (1969) considered the possibility of its clinical use for cardiovascular disorders.

In addition, triacanthine appears to stimulate erythropoiesis and has been observed to act on experimental anaemia in rabbits (Foussard-Blanpin *et al.*, 1969) (see also under plants with antibiotic and antiparasitic action).

Hunteria eburnea Pichon syn. (*Picralima gracilis* Chev.) APOCYNACEAE
Hunteria elliotii (Stapf) Pichon syn. (*Picralima elliotii* (Stapf) Stapf, *Polyadoa elliotii* (Stapf) Pichon)
Hunteria umbellata (Schum.) Hallier syn. (*Carpodinus umbellatus* Schum., *Picralima umbellata* (Schum.) Stapf, *Polyadoa umbellata* (Schum.) Stapf)

L The bark of *H. umbellata* is used in Sierra Leone and the Ivory Coast as a bitter tonic and febrifuge (Dalziel, 1937).

C The stembark and rootbark of all three species have a closely related chemical composition. They mostly contain indole alkaloids with cardiovascular effects. In 1978 Le Men and Olivier identified 34 alkaloids in *H. eburnea* of which 18 were in the bark, 9 in the leaves and 7 in the seeds. In *H. elliotii* the authors reported 26 alkaloids of which 7 were in the bark, 12 in the leaves and 7 in the rootbark. Amongst the bark alkaloids eburnamonine, eburnamine, hunterine, vincadifformine, isoburnamine and eburnamenine were found, whilst in the leaves of both species corymine and acetylcorymine, were reported. The leaves of *H. elliotii* also contain tetrahydroalstonine (Morfaux *et al.*, 1978). Eburnamonine and eburnine are also found in the seeds (Bartlett and Taylor, 1963; Bartlett *et al.*, 1959, 1963; Renner, 1963).

P Eburnamonine, eburnamine and hunterine are sympathomimetic and have a strong and lasting hypotensive action (Raymond-Hamet, 1955). Morfaux *et al.* (1969) suggest that the hypotensive properties are mainly due to hunteramine (a quaternary ammonium compound) and that eburnamonine seems to have a favourable effect on the circulation in general. Vincamonine, which is less toxic than its antipode eburnamonine, has recently been introduced in pharmacy (Le Men and Olivier, 1978).

Pausinystalia johimbe (Schum.) Pierre ex Beille syn. (*P. macroceras* Kenn., *Corynanthe johimbe* Schum.) RUBIACEAE

L The longitudinally fissured bark of the trunk of this tall forest tree is considered in the Cameroons to be an aphrodisiac and stimulant (Dalziel, 1937).

C The bark of the tree contains the alkaloids yohimbine, mesoyohimbine and yohimbinine as well as corynanthine (closely related to yohimbine but less toxic and more active as a sympatholytic agent), alloyohimbine and ajmalicine (which has a vasodilating effect on the coronary arteries, as mentioned earlier). The bark is used to extract yohimbine; the main stem gives the best material but is not rich in alkaloids until the tree is 15–20 years of age when it can contain 2–15% (Holland, 1929; Paris and Letouzey, 1960; Poisson, 1964).

P Yohimbine is sympatholytic and hypotensive and has a local anaesthetic action similar to that of cocaine but it is not mydriatic (see Oliver, 1960). It is given in cases of atherosclerosis as it dilates the walls of the small peripheral arteries, thereby increasing the flow of blood and decreasing the blood pressure. It is interesting to note that its action differs from that of reserpine, which is also hypotensive but not sympatholytic.

The vasodilating action of yohimbine is particularly strong on the sex organs, hence its aphrodisiac action. It is mainly used in the form of the hydrochloride, which was in the British Pharmaceutical Codex in 1949, in the French Codex 1949 Table A, and in the Pharmacopoeia Helvetica, 1949 (5–20 mg daily) (Raymond-Hamet and Goutarel, 1965).

Corynanthe pachyceras Schum. syn. (*Pausinystalia pachyceras* (Schum.) de Wild., *Pseudocinchona africana* Chev. ex Perrot) RUBIACEAE

C The bark of the tree is used on the Ivory Coast as an aphrodisiac and antipyretic. It contains corynanthine, the closely related corynanthidine and corynantheidine and several other alkaloids (Poisson, 1964).

P Corynanthine is 4–5 times less toxic than yohimbine whilst its sympatholytic effects are twice as strong (Steinmetz, 1976). It has mild local anaesthetic action, inferior to that of cocaine.

Mitragyna inermis (Willd.) Ktze. syn. (*M. africana* (Willd.) Korth, *Uncaria inermis* Willd., *Nauclea africana* Willd.) RUBIACEAE

Mitragyna stipulosa (DC.) Ktze. syn. (*Nauclea stipulosa* DC., *M. macrophylla* Hiern)

L The bark and leaves of both species are used in Nigeria, Guinea and the Ivory Coast for fever and diarrhoea and also as a diuretic and analgesic (Pobeguin, 1912; Raymond-Hamet and Millat, 1934; Kerharo and Bouquet, 1950). The wood is used in Nigeria for carving small objects as it is easy to work (Dalziel, 1937).

C The oxindole alkaloids rhynchophylline and rotundifoline have been reported to be present in the leaves and in the bark of the stem and roots of both *M. inermis* and *M. stipulosa* (Ongley, 1953; Shellard and Alam, 1968). At first Raymond-Hamet and Millat (1934) had named an alkaloid they isolated from *M. inermis* 'mitrinermine', but it could be shown by Badger *et al.* (1950) that this alkaloid, when purified, was identical with rhynchophylline, isolated earlier by these authors from *M. stipulosa*. From *M. stipulosa* another oxindole alkaloid, mitraphylline, has also been obtained whilst in *M. inermis* speciophylline was also found (Beckett et al., 1963; Shellard and Sarpong, 1969; Shellard *et al.*, 1976). A bitter heteroside, quinovin, which can be split into quinovic acid and quinovose has been isolated from *M. inermis* as well (Badger *et al.*, 1950; Beckett *et al.*, 1963).

In order to try to understand the biogenesis and translocation of these alkaloids, the alkaloid distribution of three West African *Mitragyna* spp. was studied by Shellard and Sarpong (1969; 1970). In the above-mentioned species as well as in *M. ciliata* Aubrev. and Pellegr., the main stembark and rootbark alkaloid is rhynchophylline and in *M. stipulosa* and **M. ciliata** the principal leaf alkaloid is rotundifoline while in *M. inermis* it is isorhynchophylline. The authors conclude that it would appear that the alkaloids are synthetized in the leaves and that conversion of the leaf oxindoles (by dehydroxylation of rotundifoline to rhynchophylline) takes place in the leaves before translocation to the root (Shellard and Sarpong, 1970). The leaves of an Asian species of *Mitragyna*, *M. speciosa*, contain in addition to other alkaloids mitragynine (methoxycorynantheidine), a hallucinogenic agent (Tyler, 1966).

P Already in 1932 Blaise had reported the occurrence of a hypotensive alkaloid in *Mitragyna* spp. It was confirmed later that rhynchophylline, mitraversine and mitraphylline lower the blood pressure by decreasing the rhythm of the heart (Xiao, 1983), and that they also have a local anaesthetic action. Further, these alkaloids strongly stimulate intestinal and uterine contractions and are toxic to protozoa (Massion, 1934; Caiment-Leblond, 1957; Ansa-Asamoa, 1967).

Cryptolepis sanguinolenta (Lindl.) Schltr. syn. (*Pergularia sanguinolenta* Lindl., *C. triangularis* N.E.Br.) PERIPLOCACEAE

L In Nigeria the macerated roots are used for gripe (colic) as a tonic and sometimes in rheumatism and urogenital infections (Boakiji Yiadom, 1979).

C The roots contain a quinoline-derived indole alkaloid, cryptolepine, which is violet in colour, producing yellow salts (Gellert *et al.*, 1951).

P Cryptolepine has a marked hypothermic effect; it also induces prolonged and important vasodilatation, causing marked and durable hypotension (Raymond-Hamet, 1937, 1938). It has a low toxicity (120 mg/kg produce death in guinea pigs about 12 h after administration). An aqueous extract of the root has antimicrobial activity against three urogenital pathogens (Boakiji Yiadom, 1979).

Physostigma venenosum Balfour FABACEAE
Ordeal tree of Calabar

L The poisonous effect of the Calabar bean in trials is caused by its strong sedative action on the spinal cord which results in paralysis of the lower limbs and death by asphyxia, and, in large doses, in paralysis of the heart. It is used by the Bakwiris with other drugs in the local treatment of articular rheumatism (Dalziel, 1937).

C From the seeds an alkaloid, physostigmine or eserine, is obtained (0.15%); in addition, the beans contain geneserine or eseridine and several other alkaloids like eseramine and physovenine (which is also myotic) (Robinson and Spitteler, 1964).

P Eserine is mainly used in ophthalmic medicine for its myotic action (1–2 drops of a 1 : 1000 physiological solution: British Pharmacopoeia, 1934, British Pharmaceutical Codex, Italian Pharmacopoeia, French Pharmacopoeia, 1949, Pharmacopoeia Helvetica, etc.), but it also dilates peripheral blood vessels and slows the pulse. The alkaloid acts by inhibition of cholinesterase, thus allowing acetylcholine to exert its full effect on the smooth muscles, glands and heart. Being antidotal to strychnine, nicotine, curare and atropine, it was used in myasthenia gravis to improve peristalsis in post-operative intestinal atony, but has been replaced by synthetic neostigmine (Martindale, 1969).

Eseridine is used in dyspepsia and as eye drops in glaucoma (Oliver, 1960; Paris and Moyse, 1967, Vol. 2). The solutions must be protected from air, light and moisture; their oxidation can be delayed for some time by adding ascorbic acid (Swallow, 1951).

Thalictrum rhynchocarpum Dill. & Rich. RANUNCULACEAE

C The roots of most *Thalictrum* spp. contain alkaloids of the berberine group with an aporphine nucleus, such as thalictrine (= magnoflorine), as well as flavonoids.

P Some *Thalictrum* alkaloids like thaliadine, adiantifoline and thaliadanine from *T. minus* and an alkaloid fraction from *T. revolutum* DC. have a powerful and prolonged hypotensive effect (at 2 mg/kg) in dogs, cats and rabbits. Further, thaliadanine is antimicrobial to *Mycobacterium smegmatis* (Patel *et al.*, 1963; Wan Tra Liao *et al.*, 1978).

Thalicmine (ocoteine) hydrochloride is also hypotensive in dogs and cats (1–2 mg/kg given intravenously) and inhibits the blood pressure response to adrenaline at 1–3 mg/kg. Detailed information about the constitution of the West African species was not available in 1976 (Farnsworth and Cordell, 1976); none seems to have become available since then.

Carica papaya L. CARICACEAE

Pawpaw, papaya

L Originating from Tropical America, this tree is extensively grown for its fruit. In West Africa the plant is mainly used as a diuretic (roots and leaves), anthelmintic (leaves and seeds) and to treat bilious conditions (fruit) (Dalziel, 1937).

C The milky sap of the unripe fruit yields a complex proteolytic enzyme, papain, which is not destroyed by heating. The crude papain consists of two crystallized enzymes, papain and chymopapain, as well as tryptophan, tyrosine and cysteine, which all seem to be part of the crude enzyme preparation. The enzyme has peptidase, coagulase (acting on milk casein), amylase, pectase and lipase action (Kerharo and Adam, 1974). Vitamins and traces of an alkaloid have also been found in the latex. This alkaloid from the pyridine group, called carpaine, has also been reported in other parts of the plant and particularly in young leaves (0.28%) (Bevan and Ogan, 1964). The seeds contain, apart from fixed oils, carbohydrates, etc., carpasemine (a benzylthiourea), benzyl senevol and a glucoside (Manske and Holmes, 1950–71, vol. 11, p. 491; Watt and Breyer-Brandwijk, 1962).

P Papain has an anticoagulant effect. Intravenous injection of a purified extract in the dog (2 mg/kg) increases prothrombin and coagulation time threefold; an anticoagulant action has also been noticed in rabbits, rats and mice, the maximum effect being achieved half an hour after the injection (Chandrasekhar *et al.*, 1961). Standardization of the enzyme has been suggested (International Commission for the Standardization of Pharmaceutical Enzymes, 1965). The maximum dose tolerated by rats and mice was found to be 50 mg/kg, while a therapeutic action was observed with doses of 1–2 mg/kg. Acute toxic effects at higher doses were similar to those observed with heparin (Chandrasekhar *et al.*, 1961).

The enzyme is also used as a digestive enzyme in dyspepsia and digestive troubles (British Pharmaceutical Codex, 1950, French Pharmacopoeia, 1949, Indian Pharmacopoeia) and has also been used successfully in peritoneal instillation to avoid adherences. In addition, it is claimed to eliminate necrotic tissues in chronic wounds, burns and ulcers (Ravina and Wenger, 1957; Rigaud *et al.*, 1956).

(Crude papain is of considerable commercial importance. In addition to its pharmaceutical applications, great quantities are used in the brewing industry (chill-proofing beer), in the food industry (in pre-cooked foods and in meat-tenderizing preparations), and in the manufacture of chewing gum. It also finds application in the textile industry (shrinkage resistance and other treatment of wool and wool-containing materials), and in the rubber industry to season latex (Oliver, 1960).)

In small doses carpaine slows down the heart and thus reduces the blood pressure. Higher doses produce vasoconstriction. In addition the alkaloid has a spasmolytic action on the smooth muscles (Henry, 1949).

In humans, carpaine hydrochloride, given orally in doses of 0.01–0.02 mg/day or given subcutaneously in doses of 0.006–0.01 mg/day has a digitalis-like action and Noble (1947) recommends its use in hypertension.

Anthelmintic and amoebicide actions of the alkaloid have been reported (Kerharo and Bouquet, 1950; Kerharo and Adam, 1974), and the seeds are also considered to be anthelmintic and carminative (Dar *et al.*, 1965).

Plants containing hypotensive non-alkaloidal constituents.

Anacardium occidentale L. ANACARDIACEAE
Cashew nut tree

L The cashew tree, native of tropical America, is widely cultivated, its kernels and fruit being much appreciated. In Nigerian local medicine the astringent infusion of the bark and leaves is used as a lotion and mouthwash to relieve toothache and sore gums and is given internally in dysentery (Dalziel, 1937).

C The gum exuding from the bark is a mixture of bassorin and true gum (Dispensary of USA, 1955). Cashew 'balsam' is composed of anacardic acid and its decarboxylated derivates, anacardol, cardol and gingkol, which are aromatic phenols.

In the leaves, polyphenols (chiefly hydroxybenzoic acid) and flavonoids which are heteromonosides (glucoside, rhamnoside, arabinoside or xyloside) of kaempferol and in particular quercetin were found by Laurens and Paris (1976) and by Attanasi and Caglioti (1970).

P Ingestion of extracts of the leaves and bark has been found to reduce hypertension and hyperglycaemia to normal values. The effects are believed to be due to peripheral vasodilatation. The hypotensive effect was observed first in rats with three different forms of experimental hypertension (Giono *et al.*, 1971) (see also under plants with antibiotic and antiparasitic action).

Morinda lucida Benth. syn. (*Morinda citrifolia* Chev.) RUBIACEAE
Brimstone tree

L Stem, bark, roots and leaves are bitter and astringent and are used in Nigeria in the treatment of fever, malaria, yellow fever, jaundice and dysentery (Dalziel, 1937; Oliver, 1960).

C Tannins, methylanthraquinones and a heteroside, morindin, have been reported in *M. lucida* and allied species (**M. geminata** DC.) in varying amounts according to the species and the geographical origin of the plant (Caiment-Leblond, 1957).

P The use of a total extract of leaves and stembark of *M. lucida* is recommended by Dang Van Ho (1955) for the treatment and prevention of hypertension and its cerebral complications. Purified extracts produced a strong hypotensive action which, however, when compared to the action of *Rauvolfia vomitoria*, proved to be of shorter duration. Further, the extract showed a distinct diuretic and a slight tranquillizing effect. In view of the complete absence of toxic side-effects, permitting the use of strong and frequent doses, La Barre and Wirtheimer (1962) consider that *M. lucida* may be very useful when strong doses are required to initiate the treatment of hypertension. *M. lucida* bark and leaves have been proved to be effective in the treatment of jaundice, thus justifying one of its local uses (Guedel, 1955).

Allium sativum L. LILIACEAE

L Garlic is widely cultivated and finds broad applications in northern Nigeria (respiratory and infectious diseases, worms, skin diseases, etc.) (Oliver, 1960).

C The strong-smelling juice of the bulbs contains a mixture of mono- and polysulphides. The main compound of these is allicin (diallyl disulphide oxide) which is the result of spontaneous enzymatic degradation of alliin (*S*-allylcysteine sulphoxide). Allicin is unstable and decomposes into polysulphides (Schulz and Mohrman, 1965; Augusti, 1974, 1975, 1976a, b).

P The hypotensive effect in arterial hypertension of a tincture of garlic (20–25 drops daily) is attributed to allicin. This constituent is also antidiabetic and bacteriostatic (Bhandari and Mukerjee, 1959; Jain and Vyas, 1974). For more details on *Allium* see Oliver-Bever and Zahnd (1979).

Sapindus trifoliatus L. SAPINDACEAE

L A native of tropical Asia, the plant is naturalized in many areas of West Africa. It is used as a fish poison in India (Singh *et al.*, 1978).

C The fruit of *S. trifoliatus* contains saponins. Hederagenin-heterosides have been identified by Takagi *et al.* (1980).

P Alcoholic extracts showed, when injected intravenously in cats in doses of 10–20 mg/kg, a dose-dependent decrease in blood pressure and heart rate. Tests demonstrate the direct action of the drug on the vascular smooth muscles to produce a hypotensive effect (Singh *et al.*, 1978).

Mostuea hirsuta (Anders. ex Benth.) Baill. ex Bak. LOGANIACEAE

C The roots of related *M. stimulans* and *M. buchholzii* Engl. contain indole alkaloids identical to, or closely related to sempervirine and probably gelsemine. The alkaloid content of *M. hirsuta* seems poor (0.2%) in the roots (Bouquet, 1975). The LD_{50} of these alkaloids in guinea pigs is 15.4 mg and 250 mg, respectively (Paris and Moyse-Mignon, 1949b; Gellert and Schwartz, 1951).

P In mice 600 mg of alkaloids per kg of body weight, given subcutaneously, produced agitation and convulsions followed by death. The alkaloids were found to be analgesic and cardiac depressant, stimulating respiration in small doses. They can cause death through respiratory paralysis in higher doses, sempervirine being the most toxic of the two. In the chloralosed dog a dose of 0.1–0.2 g of an extract of the root of *M. stimulans* produces a prolonged fall in blood pressure and after a short spell of tachycardia, cardiac and respiratory depressant effects (Chevalier, 1947).

It appears that the *Mostuea* spp. that were investigated, and no doubt also *M. hirsuta*, have similar properties to *Gelsemium* and are used for the same indications as this (Paris and Moyse, 1971, Vol. 3; Kerharo and Adam, 1974).

Adenia cissampeloides (Planch. ex Benth.) Harms syn. (*Modecca cissampeloides* Planch. ex Benth., *Ophiocaulon cissampeloides* (Planch. ex Benth.) Mast) PASSIFLORACEAE

L Used in southern Nigeria and Ghana as a fish poison and as a remedy for lumbago. The Mano of Liberia use the plant to produce amnesia (Dalziel, 1937; Harley, 1941; Watt, 1967).

C In the related South African *A. digitata* Engl. a toxalbumin and a cyanogenetic glycoside were reported by Watt and Breyer-Brandwijk (1962).

P Pharmacological investigations showed that aqueous extracts of the Nigerian climber had a graded depressor effect on the blood pressure of the anaesthetized cat. The effect was neutralized by small doses of atropine. A second active principle might be sympathomimetic with vasoconstrictive action (Adesogan and Olatunde, 1974).

Achyranthes aspera L. AMARANTHACEAE

L A common weed locally used in Nigeria as a diuretic and expectorant (Oliver, 1960).

C Chemical and pharmacological analyses have been undertaken on the Indian plants.

A betaine derived from *N*-methylpyrrolidine 3-carboxylic acid was located by Basu and called achyranthine (Basu *et al.*, 1957; Kapoor and Singh, 1966). The seeds were found to contain a saponin fraction composed of glycosides (glucose, galactose, xylose and rhamnose) of the genin oleanic acid (Gopalachari and Dhar, 1958). Later work led to the isolation of two pure saponins, *Achyranthes* saponins A and B (Hariharan and Ranjaswani, 1970).

P Achyranthine has hypotensive, cardiac depressant, vasodilatating and respiratory analeptic actions. It also has a spasmogenic effect on smooth muscles (guinea pig, rabbit, rat) and is diuretic, purgative and slightly antipyretic (Neogi *et al.*, 1970). The diuretic action of the plant has been attributed to its high potassium content. The total saponosides of the Indian plant significantly increase the tone and force of contraction of isolated frog, guinea pig and rabbit hearts. The effect was quicker in onset and shorter in duration than that exerted by digitoxin (Kapoor and Singh, 1966; Neogi *et al.*, 1970; Gupta *et al.*, 1972a, b). It is suggested that the increased contractility caused by the saponin could be related to its phosphorylase activity (Ram *et al.*, 1971; Indian Council of Medical Research, 1976). It has also been noted that an extract of the plant, when given orally (5 mg/kg), exerts a diuretic, purgative and hypoglycaemic action in rats (Dhar *et al.*, 1968; Neogi *et al.*, 1970; Oliver-Bever and Zahnd, 1979) and that it is also useful in treating subacute and mild reactions in leprosy (Ojha *et al.*, 1966).

Group A_2: hypertensives

Musa sapientum L. syn. (*M. paradisiaca* var. *sapientum* Ktze.) MUSACEAE
Banana

C Analysis of the bracts of ten wild species of bananas has shown the presence of six anthocyanidins (pelargonidin, cyanidin, delphinidin, malvidin, paeonidin and petunidin). The ripe and unripe fruit also contains 5-hydroxytryptamine (= serotonin) (Hood and Lowburry, 1954; Hegnauer, 1962–8; Sinha *et al.*, 1962). Further, dopamine and noradrenaline (adrenaline precursors) have also been reported in banana plants (Harborne *et al.*, 1974, p. 1013). The three amino phenols are sympathomimetic and in other plants (*Sarothamnus scoparius* Koch.) have proved to have marked vasoconstrictive properties and to be hypertensive. Banana flowers also have an oral hypoglycaemic action (Jain, 1968; Oliver-Bever and Zahnd, 1979).

Moringa oleifera Lam. syn. (*M. pterygosperma* Gaertn.) MORINGACEAE
Horseradish tree (native of India, cultivated throughout the tropics)

L In Nigeria root and bark are considered to be antiscorbutic and are used externally as counter-irritants (Dalziel, 1937). In India the root is used as a stimulant in paralytic affections, epilepsy, nervous debility, hysteria, spasmodic affections of the bowel and as a cardiac and circulatory tonic *et al.*, (Chopra *et al.*, 1956; Watt and Breyer-Brandwijk, 1962; Ramachandran *et al.*, 1980).

C In the rootbark the sulphurated aminobases moringinine and spirochine have been reported, as well as benzylamine (first called moringine) and glucotropaeoline.

Further, the root contains two antibiotic constituents, athomine and pterygospermine; the latter is probably a condensation product of two benzylisothiocyanate molecules with one benzoquinone molecule (Kurup and Narasimha Rao, 1954; Hegnauer, 1962–8, vol. 5, p. 130; Kondagbo and Delaveau, 1974; Saluja *et al.*, 1978).

P Moringinine has a sympathomimetic action similar to that of adrenaline; it produces peripheral vasoconstriction, raises the blood pressure and acts as a cardiac stimulant. However, Chopra *et al.* (1938) consider that the quantities present in the plant are too small to make it of interest for cardiology. Moringinine also depresses the smooth muscle fibres; it relaxes the bronchioles and inhibits the tone and movement of the intestine in rabbits and guinea pigs (Das *et al.*, 1957a, b).

Spirochine accelerates and amplifies the heartbeat in Man in doses of 0.035 g/kg and has an opposite effect at a dose of 0.35 g/kg (Watt and Breyer-Brandwijk, 1962). It can produce general paralysis of the CNS. Spirochine also has an antibiotic action, mainly in gram-positive infections, and is used as an external antiseptic and prophylactic in infected wounds (Chatterjee and Mitra, 1951).

Pterygospermine has a powerful antibiotic and antimycotic effect and athomine is particularly active against the cholera vibrion (Kurup and Narasimha Rao, 1954; Sen Gupta *et al.*, 1956; Das *et al.*, 1957a, b; Kurup *et al.*, 1957).

Group B: plants acting on vascular resistance (vitamin P action)

Citrus limonum. C. aurantium, C. decumana RUTACEAE

Largely cultivated

C Citroflavanoids are extracted from the peel of citrus fruit (as a by-product of fruit-juice preparation). They consist of a mixture of which the main constituents are hesperidoside (rhamnoglucoside of hesperetol), naringoside and eryodictyoside (flavanones). The peel also contains essential oils and vitamin C. The inconvenience of the citroflavanones is their low solubility in water, which has led to research with a view to finding more soluble hemisynthetic derivatives like Mg^{2+} chelates (Hörhammer and Wagner, 1962; Ravina, 1964; Paris, 1971 *et al.*, 1972).

P Citroflavonoids control the permeability of the blood vessels by decreasing the porosity of the walls and thus improving the exchange of liquids and the diffusion of proteins. They are therefore used in complaints in which the permeability is increased, such as venous insufficiency (varicose veins, haemorrhoids, capillaritis), and in oedema and ascites in cirrhosis (Paris and Delaveau, 1977; Pourrat, 1977).

The increase in the resistance of the capillaries through the citroflavonoids is based on a complex mechanism including the protective action of *o*-diphenols on catecholamines participating in vascular solidity. When capillary resistance is diminished, citroflavonoids can prevent bleeding in hypertensive or diabetic patients (diabetic retinopathy) or in purpurea and where there is a tendency to haematoma (Paris and Moury, 1964; Vogel and Ströcker, 1966; Paris, 1977).

The citroflavonoids are also said to have anti-inflammatory, antihistamine and diuretic actions and can cause dilatation of the coronaries (Paris and Delaveau, 1977).

Piliostigma reticulatum (DC.) Hochst. syn (*Bauhinia reticulata* DC., *B. benzoin* Kotschy) CAESALPINIACEAE
Bauhinia purpurea L., **B. tomentosa** L., **B. variegata** L. (introduced spp.)
St Thomas tree

L A poultice of the leaves and bark of *Piliostigma* is used in Senegal and northern Nigeria as a haemostatic, and an infusion of the bark and buds of *B. variegata* is given to control bleeding in haematuria and menorrhagia and as an astringent in diarrhoea. The powdered bark of *B. thonningii* is applied to wounds and ulcers (Dalziel, 1937; Kerharo and Adam, 1974).

C In all species flavonoids (quercetol and kaempferol glycosides) have been found in the leaves, bark and flowers. In *B. variegata* and *B.* tomentosa rutosides and isoquercitroside have been reported; from *B. tomentosa* flower petals 4.6% of rutin has been extracted (Visnawadham *et al.*, 1970; Duret and Paris, 1977). *Piliostigma* contains tartaric acid and tartrates in fruit and leaves. In the leaves 5.9% free (−)-tartaric acid and 0.5% quercitroside have been reported. The flavonoids seem to justify their local use as coagulants.

Fig. 2.9(a). *Adansonia digitata* L., flower.

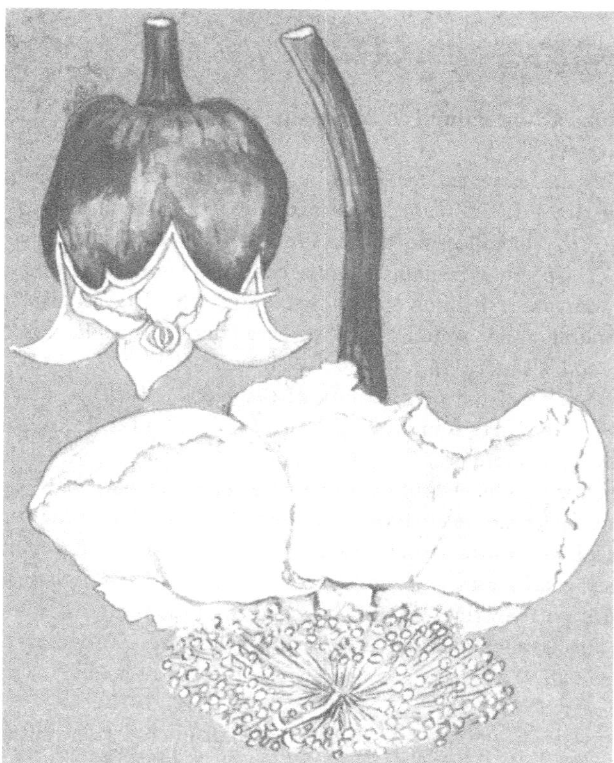

Baissea leonensis Benth. syn. (*B. brachyantha* Stapf) APOCYNACEAE

C The leaves contain no alkaloids but they do contain flavonoids, mainly coumarins. A new crystallized coumarin heteroside, baisseoside, which is a 6-rutinoside of esculetol, was isolated in 1970 by Pousset *et al.* Other heterosides reported are isoquercitroside, kaempferol 3-glucoside and kaempferol heteroside (Duret and Paris, 1972).

Adansonia digitata L. (Fig. 2.9) BOMBACACEAE

Baobab, monkey bread

L The fruit, seeds and leaves of the baobab are used in food and the bark provides fibre. In West African local medicine all parts are used; they are said to be diaphoretic, antipyretic, antidysenteric, emmenagogic, antifilarial, vulnerary, etc. (Sebire, 1899; Dalziel, 1937; Oliver, 1960).

C The leaves contain a mucilage consisting mainly of galacturonic acid and rhamnose, free sugars, tannins, catechins and a dehydroxyflavanol, adansonia flavonoside. They also contain calcium oxalate, potassium tartrate and sodium chloride. Bark and roots also contain a mucilage as well as pectins and an antipyretic agent, adansonin, said to have a strophanthin-like action (Merck Index, 1960; Oliver, 1960; Watt and Breyer-Brandwijk, 1962).

Fig. 2.9(*b*). *Adansonia digitata* L., fruit.

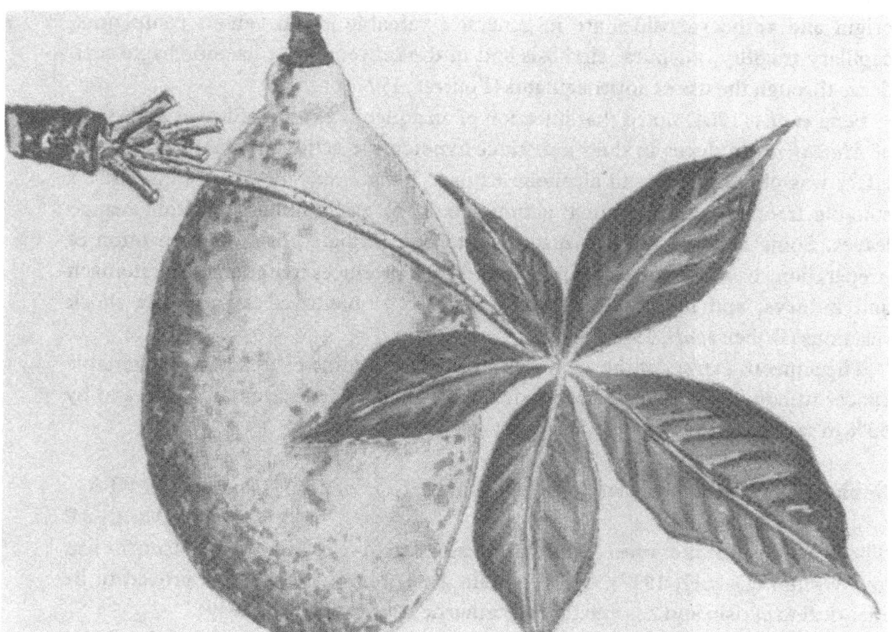

P Adansonia flavonoside showed low toxicity; mice tolerated 1 g/kg given subcutaneously. Given intravenously, 0.01 g/kg produced only slight hypertension in dogs and decreased the permeability of the capillaries in rabbits, but to a lesser extent than does rutoside (Paris and Moyse-Mignon, 1951).

Mangifera indica L. ANACARDIACEAE
Mango tree

L Naturalized in West Africa. The bark and leaves have astringent properties and are used in Nigeria as a lotion to relieve toothache, sore gums, sore throat, etc., or as an infusion in diarrhoea and dysentery (Dalziel, 1937).

C All organs are rich in tannins. In the leaves of the West African species four anthocyanidins (3-monosides of delphinidin, petunidin, paeonidin and cyanidin), leucoanthocyanins, catechic and gallic tannins, mangiferin (flavonic heteroside), kaempferol and quercitin (both free and as glycosides) were reported (Jacquemain, 1970, 1971).

In pharmacological tests anthocyanidins similar to those found in mango leaves but extracted from *Vaccinium myrtillus* leaves increased the resistance and decreased the permeability of capillary vessels, and have been successfully used for over 20 years in treating vascular troubles, eye complaints and diabetes (Pourrat, 1977). In diabetic angiopathy they inhibit, or slow down, the modifications of the capillary wall, and it is believed that the improvement obtained in diabetes with anthocyanosides can well be due to recovery of the vascularization of the pancreas.

Excellent results were also obtained in retinopathy of hypertensive or diabetic origin and anthocyanosides are in general a valuable aid in venous complaints, capillary fragility, purpura, cirrhosis and in the prevention of haemorrhagic accidents through the use of anticoagulants (Pourrat, 1977).

Feng *et al.*, (1964) noted that injection of an aqueous extract of leaves and stems of *Mangifera* produces in dogs a distinct hypotensive action. In rabbits a similar effect was obtained with an alcoholic extract (Feng *et al.*, 1964). It seems that a suitable fraction with vitamin P action should be easily obtainable from mango leaves. Some toxic constituent may have to be eliminated first, as absorption of preparations based on the leaves, stems and bark produces irritation of the stomach and kidneys, and ingestion of the fruit in large quantities can produce shock reactions (Ruben *et al.*, 1965).

The aqueous extract of the stembark showed favourable results in transplantable cancer tumours. These have been reduced by 47% in adenocarcinoma 765, and by 53% in sarcoma 180 (Abbot *et al.*, 1966).

Sophora occidentalis L. syn. (*S. nitens* Schum. & Thonn., *S. tomentosa* (of FTA)
FABACEAE

L The poisonous seeds of most Sophora spp. are used in East Africa as a poison for fish and vermin (Dalziel, 1937). They contain 2% sophorin, which has proved to be identical to cytisin and has emetic and cathartic actions (Henry, 1949).

C The flowerbuds of the Asian 'Pagoda tree' (*S. japonica*) are rich in rutosides (15%) and are used in industry for the extraction of the rutosides. These are also obtained in quantity from the leaves of several *Eucalyptus* spp. and from buckwheat (Paris and Moyse, 1967, p. 375).

No indications as to whether the flowerbuds of the African *Sophora* also contain rutosides have been found.

Camellia thea Link syn. (*Thea sinensis*) TERNSTROEMACEAE
Tea plant

Tea is also cultivated in West Africa.

The leaves contain vitamin P constituents as well as vitamins C and B. The flavonoids are rhamnoglucosides and rhamnodiglucosides of kaempferol, quercitin and myrecetol. Further, catechols and tannins have been reported, as well as 2–4% of alkaloids (caffeine and theophylline) (Roberts and Myers, 1959; Roberts and Williams, 1959).

Tephrosia purpurea (L.) Pers syn. *Cracca purpurea* L., *T. leptostachya* DC.) FABACEAE

L The plant is used in India as a deobstruent, diuretic and cough remedy, and also for bleeding piles (Pandey, 1975).

C The roots, leaves and seeds contain tephrosin, deguelin and quercetin; the roots also contain isotephrosin and rotenone, while 2.5% rutin is found in both the roots and leaves (Krewson and Naghski, 1953).

P Rutin is increasingly used in capillary fragility and may be useful in the treatment of the after-effects of atomic radiation. The roots, like those of *T. vogelii* and *T. densiflora*, are ichthyotoxic but have only a weak insecticide action (Watt and Breyer-Brandwijk, 1962, p. 656).

By improving the peripheral arterial circulation, rutin is also antihypertensive (Paris and Moyse, 1967; Paris, 1977).

Teclea sudanica Chev. RUTACEAE
Paris and Etchepare (1968) found C-flavonosides in the leaves of *T. sudanica*. The leaves are reported to have hypotensive action. They contain 0.5% alkaloids; maculine, skimmianine and kokusaginine (Vaquette *et al.*, 1974).

Arachis hypogaea L. FABACEAE
The skins of the nuts contain bioflavonoids with vitamin P action. As other constituents of the plant promote blood coagulation, its properties have been described below.

Group C: plants promoting blood coagulation and formation

Arachis hypogaea L. FABACEAE
Peanut, groundnut, monkey nut

L The nuts are a valuable local food crop and are also exported on a large scale.

C Peanuts have a high lipid content and contain vitamins B_1, B_2 and B_3 as well as proteins and phytosterols. Aflatoxins can be found mainly in groundnut cakes, formed under the influence of certain fungi under faulty conditions of storage and harvesting. They are toxic, substituted coumarins. The thin skins of the nuts contain bioflavonoids, tannin, phlobaphen and flavanone (Tayeau and Masquelier, 1949; Ravina, 1964; Adrian and Jacquot, 1968).

P The presence of a haemostatic factor in groundnuts, first reported in 1960 (Boudreaux and Frampton, 1960), was later confirmed by several research workers. The factor considerably improves the condition of patients with haemophilia. A daily intake of 180 g of peanut flour, or of 14 g of an alcoholic extract, begins to produce effects after as little as 48 h (Boudreaux and Frampton, 1960; Boudreaux *et al.*, 1960). This action has been attributed to a constituent similar to 5-hydroxytryptamine (= serotonin in animal tissue and in particular in blood platelets and produces vasoconstriction of the blood vessels), which also stimulates smooth muscles (Boudreaux *et al.*, 1960). However, the authors concerned differ as to where the active constituent is located and its mode of action. Adrian and Jacquot (1968) assume that this may be because in the groundnut there are different factors influencing blood coagulation. Some authors consider that only certain forms of haemophilia such as thrombopenic haemorrhage, chronic myelosis, vascular purpura and metrorrhagia can benefit from this treatment, others also include haemophilia A.

The groundnut skins contain, in addition, a lipoxidase and protease inhibitor (Cepelak and Horacova, 1963; Narayanan *et al.*, 1963), 7% pyrocatecholic tannins and a chromogen of anthocyanins, closely related chemically to vitamin P. Injection into guinea pigs of a solution equivalent to 1 mg of this chromogen doubles their capillary resistance. It is very well tolerated; even doses of 20 mg produced no toxic effect (Tayeau and Masquelier, 1949).

Further, an oestrogenic factor, not destroyed by refining, was found in the oil of the nuts and a thermostable goitrogenic factor has been reported in the non-lipid fraction (Buxton *et al.*, 1954; Booth *et al.*, 1960).

Terminalia laxiflora Engl. syn. (*T. elliotii* Engl. & Diels, *T. sokodensis* Engl.)

COMBRETACEAE

Terminalia avicennoides Guill. & Perr. syn. (*T. lecardii* Engl. & Diels, *T. dictyoneura* Diels)

Terminalia macroptera Guill. & Perr. syn. (*T. chevalieri* Diels)

L The powdered bark of *T. macroptera* and *T. laxiflora* is used in wound dressing and in the treatment of piles for its haemostatic and healing effects, whilst a decoction of bark and leaves is said to be diuretic. Antidiarrhoeic and a cholagogic action are also ascribed to different *Terminalia* spp., which are often confused.

C The bark of the three Nigerian species mentioned above contains laxiflorin (a polyhydroxylactone), whilst in the stembark sitosterylpalmitate and trimethylellagic acid are found. In the wood, terminolic, ellagic and tri- and tetramethylellagic acid are reported (Ekong and Idemudia, 1967; Idemudia and Ekong, 1968).

P Ellagic acid (a tannin depside of gallic acid) both reduces the bleeding time in rats and shortens the normal coagulation time without producing any toxic effects (Cliffton *et al.*, 1965; Girolami *et al.*, 1966). In vitro the acid does not seem to act on sections of duodenum or uterus of rats (Bhargava *et al.*, 1968; Bhargava and Westfall, 1969). Chopra *et al.*, (1938) reported a cardiotonic action in a number of Indian *Terminalia* spp. including *T. avicennoides*. In Nigeria the aqueous extract of the stembark and roots of this species proved to have the strongest antibiotic action of the different *Terminalia* spp., mainly on Gram-positive organisms (*Sarcina lutea*, *Staphylococcus aureus* and *Mycobacterium phlei*) (Malcolm and Sofowora, 1969).

Canavalia ensiformis (L.) DC. FABACEAE

Olsen and Liener (1967) report hemagglutinin effects of the Jack bean through concanavalin A. The seed of the bean is a source of urease, an enzyme which finds use as a specific reagent for urea in biological chemistry.

Group C_1: Plants with a vitamin K action (haemostatic action).

Lawsonia inermis L. syn. (*L. alba* Lam.) LYTTHRACEAE

Henna, Egyptian privet, alkanna

L The leaves are used in local northern Nigerian medicine to control perspiration, and both leaves and roots are used as an emmenagogue and anthelmintic. The bark is recommended as an astringent and is administered for jaundice and skin diseases. In Arab medicine the root is considered useful in the treatment of hysteria and of nervous diseases in general (Dalziel, 1937; Oliver, 1960).

C The leaves contain lawsone, a 2-hydroxy-1,4-naphthoquinone (a colouring matter), resin and hennatannin. They can be harvested from the second year onward and the plant may live for 15 years (Latif, 1959).

P The colouring matter is used for colouring oils and ointments. Henna also appeared in the British Pharmaceutical Codex 1934 as a hair dye. Apart from the slight vitamin K action and the antihaemorrhagic properties attributed to lawsone (Almquist and Klose, 1939), it has powerful bactericidal effects comparable to those of sulphonamides and penicillin (Caiment-Leblond, 1957). The naphthoquinone is mainly of interest, however, because of its emmenagogic and oxytocic actions. The effect of a 10% extract on the uterus is comparable to that of pituitary gland preparations (Latour, 1957).

Diospyros mespiliformis Hochst. ex DC. syn. (*D. senegalensis* Perr. ex DC.) EBENACEAE

Swamp ebony, monkey guava

L In northern Nigeria an infusion of the leaves is given for fever and dysentery and is applied in wound dressings as a haemostatic. A decoction of the rootbark is given for skin eruptions and also as an anthelmintic in human and veterinary medicine (Dalziel, 1937; Oliver, 1960).

C Many *Diospyros* spp. were found to contain hydroxynaphthoquinones such as plumbagin (2-methyl-5-hydroxy-1,4-naphthoquinone), diospyron, diospyrol and diosquinone. Plumbagin (originally found in *Plumbago*), occurs in the rootbark of *D. mespiliformis* (0.9%) and of **D. canaliculata** de Wild. (*D. xanthochlamys*) (2.25%) and also in the leaves of these species (traces and 0.12%, respectively). Further, tannins, a saponin and a substance probably identical to scopolamine have been reported in both species (Kerharo and Bouquet, 1950; Paris and Moyse-Mignon, 1949a).

P In doses of 0.2 g/kg a rootbark extract of *D. mespiliformis* produces hypertension and exaggerated respiration in dogs. The rootbark of all the species here noted and of **D. tricolor** (Schum. & Thonn.) Hiern has an antibacterial action on staphylococci, streptococci and diphtheria bacilli probably due to plumbagin. Besides the antibiotic action, insecticide and anthelmintic properties have also been reported in the various species of *Diospyros* (Paris and Moyse-Mignon, 1949a). As the naphthoquinones also have a vitamin K action, their local application in wound dressing seems fully justified (Fieser *et al.*, 1941).

Jatropha curcas L. EUPHORBIACEAE

Physic nut, Barbados nut

L The oil of the seeds has a purgative action and is used all over West Africa in local medicine as a remedy for dropsy, sciatica, paralysis and skin diseases (Dalziel, 1937).

C The seeds contain 50% of a fixed oil, pinhoen oil, as well as a mucilage composed of xylose, galactose, rhamnose, galacturonic acid and a toxalbumin, curcin (Bezanger-Beauquesne, 1956; Mourgue *et al.*, 1961a, b). Glycosides have been detected in an extract of the shell (Bose *et al.*, 1961).

P The mucilage fraction of the seed pulp reduces prothrombin time and coagulation time. Its effect is comparable to that of Russel's viper venom used as a source of thromboplastin. It is four times less active, however. On the other hand, the toxalbumin fraction has been found to increase the prothrombin time (Bose *et al.*, 1961). Curcin has many features in common with ricin but is less toxic. The purgative action of the seed oil has been confirmed. In addition the seeds have a certain value as an insecticide. Fruits and seeds contain a contraceptive principle (Mameesh, 1963). An ethanolic extract of *J. curcas* has shown confirmed in vitro and in vivo action against P388 lymphocytic leukemia (Hufford and Oguntimein, 1978).

Group C_2: plants with an antivitamin K action (haemolytic action).

Balanites aegyptiaca (L.) Del. syn. (*Ximenia aegyptiaca* L., *Agialida senegalensis* v. Tiegh., *A. barteri* v. Tiegh., *B. zizyphoides* Mldbr. & Schltr., *A. tombouctensis* v. Tiegh.) ZYGOPHYLLACEAE

Desert date, soap berry tree, thorn tree

L In Hausa medicine (northern Nigeria), roots and bark are used as a purgative and in colic, and are considered ichthyotoxic and anthelmintic. The oil of the fruit kernels is employed for the dressing of wounds and as an embrocation in rheumatism (Dalziel, 1937; Oliver, 1960).

C From the dried seeds, 48% of a fixed oil, Zachun oil, has been obtained, whilst the seedcake contains 50% protein. The root, bark, fruit, leaves and seeds contain saponins (Kon and Weller, 1939). In the seeds, 6.7% of a tetraglycoside of diosgenin has been reported. The diosgenin content of the whole dried African plant was 5.6% (Hardman and Sofowora, 1972). In the other parts of the plant the genin consisted of two-thirds yamogenin and one-third diosgenin. *B. aegyptiaca* has been investigated as a source of steroidal sapogenins for the hemisynthesis of corticosteroids and hormones (Marker, 1947; Hardman and Sofowora, 1972).

P The root, bark, fruit pulp and seeds have been found to be lethal to fish and also to freshwater snails, which act as an intermediate host for bilharzia, and to the minute free stages of the parasite. The planting of *Balanites* alongside infested rivers to combat bilharzia propagation was therefore recommended by Archibald in 1933. Since then, other plants with stronger molluscicidal properties, such as *Polygonum senegalense* have been reported (Dossaji *et al.*, 1977). *Tephrosia* and *Jatropha curcas* (mentioned above) are also lethal to the molluscs. Antimicrobial properties were reported by Malcolm and Sofowora (1969). The saponoside is strongly haemolytic, with a toxicity to tadpoles similar to that of digitoxin, but less rapid in action. At a concentration of 10^{-6} of the saponoside in water, the tadpoles survive for more than 24 h. On frog's heart, a dose of 1 mg given subcutaneously produces no apparent digitalis effect; the isolated heart is stopped by a dilution of 10^{-3}. Thus a digitalis-like action exists but it is so weak as to be almost negligible (Caiment-Leblond, 1957).

In *Schwenkia americana* and in *Carica papaya*, described earlier, as well as in **Swartzia madagascariensis** Desv. fruit (Beauquesne, 1947), the presence of haemolytic saponosides has also been reported. In *Swartzia* these have been named Swartzia saponosides A and B, the genin of them being swartzigenin. The drug is ichthyotoxic and is strongly haemolytic (Beauquesne, 1947).

Group C_3: plants with an anti-anaemic action.

Although a few plants have been recommended for the treatment of anaemia in local West African medicine, their action has not been confirmed. No plants possessing a proven anti-anaemic effect seem to have been reported, nor has mention been made of folic acid or vitamin B_{12} contents in West African plants. Spinach, yeast, *Streptomyces griseus* and *S. aureofaciens* seem to remain the only few, generally known, vegetable sources of these compounds. In sickle cell anaemia, *Zanthoxylum zanthoxyloides* has been used, as mentioned under the description of this plant.

3

The nervous system

In this chapter are described plants having a direct action on the central nervous system (CNS) and those plants that are used in the treatment of mental and nervous diseases, including those acting via the cholinergic and adrenergic systems.

The CNS, which is best considered as a whole, controls all sensory and integrated motor activity. Plants which interefere with its function may be classified as follows:

I CNS stimulants
- Analeptics, stimulants
- Antidepressants, hallucinogens

II CNS depressants
- General anaesthetics, narcotic analgesics
- Analgesic-antipyretics
- Hypnotics, sedatives and tranquillizers
- Anticonvulsants and antiepileptics

III Peripherally acting depressants of the CNS
- Local anaesthetics (on sensory nerves)
- Neuromuscular blockers (curare action) (on motor nerves) and anticonvulsants

IV Those with cholinergic and adrenergic actions
- Depressants acting on both autonomic nervous system (ANS) and CNS terminals
- Antispasmodics acting mainly on sympathetic terminals
- Stimulants of the cholinergic and adrenergic systems

However, in practice no such clear classification can be made. Even a single plant constituent can act on both the CNS and the ANS. The plants are therefore grouped by the main resulting effect when both the CNS and ANS are involved. Also, most plants have many different constituents and these often have divergent actions (Table 3.1). In addition only a restricted number of the constituents may be known and pharmacological screening has in many cases been carried out on extracts containing several constituents in varying proportions or doses. An arbitrary decision may therefore have to be taken, in many cases based on our very partial knowledge about the plants, in considering the quantitatively and qualitatively most important constituent(s) of the plant and the most frequently reported effects.

I still consider it important to classify the plants by their clinical indications, in spite of the fact that the greatest amount of research is still to come, to keep in mind the aim, and to see rapidly and more clearly which of the far too numerous utilizations of each plant by local healers appear the most important and are confirmed by scientific observations. The enormous treasure of natural remedies should not be used in a haphazard way. A simple extraction might in some cases enable a healer to increase or isolate a fraction with a certain action or to eliminate a toxic constituent. The experience with the same plants in similar climates even in different continents should not be ignored (although the amount or quality of the constituents can vary and should be checked).

We saw in mentioning their mode of action in the Introduction (Chapter 1) that the activity of drugs acting via the nervous system can be based on their interference with the chemical mediators of nervous transmission (acetylcholine and catecholamines) at their receptor level. Anticholinergic action can be localized:

(1) at the level of the parasympathetic terminals, thus producing antispasmodic and antisecretory activities (used to prepare for anaesthesia) and mydriasis in ophthalmology (atropine action);
(2) at the level of the sympathetic and parasympathetic ganglia;
(3) at the level of the neuromuscular junctions (curare action).

On the other hand, the drugs can interfere with the catecholamines (noradrenaline and adrenaline) or the chemical intermediates in their synthesis (e.g. dopamine) in the post-ganglionic sympathetic nerves and their terminals. This action is called 'adrenergic' action. The catecholamines are also involved in central activity. In this field, the so-called 'antidepressant' drugs, for example, are supposed to act in correcting the basic biochemical trouble of the depression which could be the insufficiency of noradrenaline at the level of the encephalic synapses through degradation of cerebral catecholamines. The enzyme responsible for intracellular degradation of catecholamines is monoamine oxidase (MAO). The activity of this enzyme is blocked by a group of drugs called monoamine oxidase inhibitors (MAOIs), often found in the Rubiaceae (Table 3.2), which thereby inhibit the inactivation of noradrenaline. Unfortunately, these prevent not only degradation of cerebral catecholamines but also that of the catecholamines of the peripheral sympathetic system and of certain substrates. Thus, the biochemical properties of the MAOIs explain their long-lasting action, the numerous secondary effects observed and the diversity of the accidents produced. Their effects are very difficult to control and they are now seldom employed for psychiatric treatments. They cannot be taken together with alcohol, or with food including biogenic amines (especially tryptamine) or amino glucosidic antibiotics, in order to avoid attacks of hypertension, nausea and cephalgia (Lechat *et al.*, 1978; Goodman and Gilman, 1980).

Table 3.1. *More important constituents and divergency of pharmacology of West African Menispermaceae species*[a]

Plant	Part used	Constituents	Chemical group	Local use of plant	Pharmacology of constituents
Tricliseae					
Triclisia dictyophylla Diels syn. (*T. gilletii* (d. Wild.) Staner, *Tiliacora tricantha* Diels)	Leaves	Phaeanthine *N,N*-Dimethylphaeanthine Cocsuline Isotetrandrine Stebisimine Gilletine Trigilletimine	Bisbenzylisoquinoline alkaloids (Guha *et al.*, 1979)	Anaemia Oedema of legs	Muscular relaxation when introduced in the bloodstream, muscle–nerve transmission blocking ensues, hence muscular relaxation in moderate doses
		O-Methylmoschatoline	Oxoaporphine alkaloids		
		Tridictyophylline	Morphinan alkaloid		
Triclisia patens Oliv.	Leaves	Phaeanthine *N,N*-Dimethylphaeanthine Pycnamine Cocsuline	Bisbenzylisoquinoline alkaloids (Guha *et al.*, 1979)	—	Phaeanthine and dimethyl derivative have muscle relaxing and curarizing action strongly increased in methiodide
		Aromoline *O*-Methylmoschatoline	Oxoaporphine alkaloids		
Triclisia subcordata Oliv.		Phaeanthine Tricordatine Fancholine Tetrandrine Cocsuline	Bisbenzylisoquinoline alkaloids	—	Tetrandrine has antitumour action

Tiliacora dinklagei Engl. syn. (*Glossopholis dinklagei* (Engl.) Stapf)	Roots	Nortiliacorine A Funiferine Tiliacorinine Tiliageine Dinklacorine	Bisbenzylisoquinoline alkaloids (Guha *et al.*, 1979)	Anaemia Oedema of legs	Curare action of dimethiodide of funiferine and nortiliacorine
Tiliacora funifera (Miers) Oliv.	Roots	Funiferine Nortiliacorine A	Bisbenzylisoquinoline alkaloids	—	Antitumour (leukaemia)
	Leaves	Tiliafunimine Isotetrandrine Thalrugosine = Nortiliacorine A			
Epinetrum cordifolium Mangenot & Miège (in Ghana and Ivory Coast)	Root	Isochondrodendrine Cycleanine = Dimethyl-*o*, *o*-isochondrodendrine Norcycleanine Monomethyl-*o*-isochondrodendrine	Curare alkaloids (Debray *et al.*, 1967)	Anaemia Oedema of legs	Curare action Muscular relaxation Cycleanine also has anti-inflammatory and analgesic properties
Cocculeae					
Cocculus pendulus (J. & G. Forst.) Diels syn. (*C. leaeba* (Del.) DC., *Epibaterium pendulum* (J. & G. Forst.))	Leaves Stems	Cocsuline Penduline	Bisbenzylisoquinoline alkaloids	Biliousness Inermittent fever	Antitumour action
	Roots	Sangoline = Oxyacanthine	Aporphine alkaloids		Sympatholytic
		Palmatine Columbin	Bitter principles	Fruit; intoxicating drink	Stimulants
	Roots	Coccutrine Dihydroerysovine	Erythrinan alkaloids		Curare alkaloids

(*Table continued*)

Table 3.1. (*Continued*)

Plant	Part used	Constituents	Chemical group	Local use of plant	Pharmacology of constituents
Cissampelos owariensis Beauv ex DC. (*C. pareira*)	Roots	Chondrodendrine Isochondrodendrine	Bisbenzylisoquinoline alkaloids	Antipyretic Diuretic	Hyatin dimethiodide 2.5 times more potent than tubocurarine
Cissampelos mucronata Rich. (partly *C. pareira*)	Bark	Cycleanine Hyatinine Hyatine Dicentrine Dehydrodicentrine		Abortive Emmenagogue	
	Leaves	Cissampareine Cissamine			Cissampareine, anticancer
Stephania dinklagei (Enkl.) Diels syn. (*Cissampelos dinklagei* Engl.)	Roots	Dinklageine Stepharagine Corydine Norcorydine Steporphine *N*-Methylcorydine *N*-Methylglaucine	Isoquinoline and aporphine alkaloids	Given to barren women in menorrhagia Anthelmintic Sedative	Narcotic antitussive Antibiotic (TB and leprosy) From related species: rotundine, narcotic and hypnotic in Vietnam war
		Stepharine	Proaporphine alkaloids		
Rhigiocarya racemifera Miers syn. (*R. nervosa* (Miers) Chev.)	Roots	*O*-Methylflavinanthine Liriodenine Palmatine Menispermine	Morphine-like alkaloid	Sleeplessness Leafy twigs, seeds; aphrodisiac	Analgesic
		Magniflorine	Proaporphine alkaloid		
Kolobopetalum auriculatum Engl. syn. (*K. veitchianum* Diels)	Roots	*O*-Methylflavinanthine	Morphine-like alkaloid	Sleeplessness	Analgesic

Tinosporeae					
Chasmanthera dependens Hochst.	Root	Berberine In Ch. palmata: Columbamine Jateorhizine Palmatine	Berberis alkaloids Isoquinoline group	Bark tonic juice: sprains, bruises gonorrhoea	Bitter 'tonic', inhibits Leishmania tropica in concentrations of 1 : 80 000
	Leaves	Colombin Palmarin Chasmanthin	Bitter principles		
Jateorhiza macrantha (Hook.) Excell. & Mendonça syn. (*J. strigosa* Miers)	Root Leaves	Palmatine Columbamine Jateorhizine	Berberis alkaloids	Burns, snake bites	Bitter 'tonic' Hypotensive
	Root	Colombin Tinosporine	Bitter principles		
Tinospora bakis (Rich.) Miers	Root	Palmatine Tinosporine Colombin	Berberis alkaloids Bitter principles	Cholagogue Remittant fever Emmenagogue	Antipyretic Bitter 'tonic'
Penianthus zenkeri (Engl.) Diels syn. (Heptacyclum Engl.)	Stems, roots	Palmatine Jateorhizine Magnoflorine	Protoberberine alkaloids Aporphine alkaloids	Local infections Venereal diseases	See Duah *et al.* (1981)

[a] As can be seen from this table, the Menispermaceae produce not only curarizing bisbenzyl isoquinoline alkaloids (Guha *et al.*, 1979) (mostly asymmetric), which act on the neuromuscular junctions, but also some isoquinoline alkaloids with a morphine-like structure (bisbenzyl isoquinoline + phenanthrene) and aporphines and oxoaporphines with narcotic or analgesic activity. In addition they contain alkaloids of the berberine group (berberine, palmatine, chasmanthine, jateorhizine, columbamine) which have an isoquinoline nucleus and were called 'bitter principles'. The latter are mostly found in the Tinosporeae and find use as stimulants because of their bitter taste.

Table 3.2. *More important constituents and divergency of pharmacology of West African Rubiaceae species with an action on the nervous system*[a]

Plant	Part used	Active constituents	Pharmacological properties
Borreria verticillata (L.) Mey.	Aerial parts	Borrerine, borreverine (tetra-β-carboline nucleus)	Antibacterial, stimulant on rat uterus
	Root	Emetine (0.13%)?	
	Bark	Iridoids (chromogenic heterosides)	
Cinchona spp.	Bark	Quinine, quinidine and derivatives	Antipyretic, antiarrhythmic, protozoicidal
Coffea arabica L.	Seed	Caffeine (purine base)	Stimulant
Corynanthe pachyceras Schum. (*Pseudo-cinchona africana* Chev. ex Perrot)	Bark	Corynanthine, corynantheine, corynanthidine, corynantheidine, etc.	Mild local anaesthetic, hypotensive, sympatholytic, less toxic than yohimbine
Crossopteryx febrifuga (Afz. ex G. Don.) Benth.	Bark	Crossoptine = rhynchophylline	Antipyretic (?), local anaesthetic, protozoicidal
	Leaves	Glycoside, β-quinovine	Slight hypotensive, oxytocic
Feretia apodanthera Del.	Stem- and rootbark	Iridoids: feretoside, gardenoside, apodanthoside	Antineuralgic and CNS depressant (Bailleul *et al.*, 1979, 1980)
Leptactina densiflora Hook. f.	Bark and	Harmine derivatives	Antispasmodic to smooth muscles
Leptactina senegambica Hook. f.	leaves	Leptactinine	(Persinos and Quimby, 1967a)
Mitragyna inermis (Willd.) Ktze.	Bark and leaves	Rhynchophylline, rotundifoline	Hypotensive, local anaesthetic, protozoicidal

Mitragyna stipulosa (DC.) Ktze.	Bark and leaves	Mitraphylline, glycoside: β-quinovin	Stimulates intestinal and uterine contractions
Morinda lucida Benth.	Rootbark	Methylanthraquinones	Hypotensive, diuretic, stimulates ileum contractions
Morinda longiflora G. Don.	Leaves	Glycoside morindin	
Morinda geminata DC.			
Nauclea latifolia Sm.	Rootbark	Harmane derivatives, angustine, nauclefine, etc.; glucoalkaloids: cadambine, dehydrocadambine	Antipyretic ('African quinine' or Rio Nunez quinine), antidepressant
Nauclea diderrichii (de Wild. & Dur.) Merrill	Leaves		
Nauclea pobeguinii (Pob. ex Pellegr.) Petit	Root	Alkaloid and saponoside	Antipyretic, abdominal pains, oxytocic
Pauridiantha viridiflora (Schweinf. ex Hiern) Hepper	Bark	Harmane derivatives, pauridianthine, pauridianthinine (pyridine-harmanes)	Antidepressant, antipyretic, protozoicidal
Pausinystalia johimbe (K. Schum.) Pierre ex Beille	Bark	Yohimbine, mesoyohimbine, yohimbinine, etc.	Hypotensive, aphrodisiac, local anaesthetic, sympatholytic
Uncaria africana G. Don. syn. (*Uncaria talbotii* Wernh. partly)	Bark	Harmane derivatives, rhynchophylline in related *U. rhynchophylla*	Sedative, antispasmodic

[a]The Rubiaceae, a big, mainly tropical family, have constituents which act mainly on the ANS and the thermoregulating centres. Their components are quinolines, isoquinolines, purine bases and indole alkaloids including the harmala alkaloids. Many have antipyretic and protozoicidal actions, and often an antispasmodic effect on the striated muscles. Some have a stimulating action on the ileum and uterus. Local anaesthetic action is also frequent. Most of the harmala alkaloids have monoamine oxidase inhibiting (MAOI) properties and some have been used in sequels of encephalitis.

I CNS stimulants

Some plants producing central stimulation, like those containing xanthine derivatives, stimulate the CNS, but they are also cardiac and respiratory stimulants, produce diuresis and relax the smooth muscles.

Caffeine is mostly used only as a CNS stimulant, while theobromine and theophylline are used more often for their effects on the myocardium. These methylxanthines, especially theophylline, are competitive inhibitors of phosphodiesterase (the enzyme that inactivates cyclic AMP). Higher concentrations of cyclic AMP cause tissue glycolysis and may increase metabolic activity and this may explain the stimulant action (Burgen and Mitchell, 1972).

The most important region of stimulation by the strychnos alkaloids is the medulla. These alkaloids have an analeptic action, i.e. they are sometimes employed to overcome depression of the CNS due to overdoses of barbiturates, morphine and similar compounds by stimulating the centres in the medulla, but a sufficiently high dose of these drugs can produce generalized convulsions. Their use in electro-convulsive therapy has not proved successful and, owing to their toxicity, their therapeutic value is, on the whole, negligible (Burgen and Mitchell, 1972). Strychnine is used in investigations on the mode of action of convulsant drugs.

Hallucinogens have been classified as stimulants by Burgen and Mitchell (1972) although their effects are certainly not always of a stimulating nature: they produce a phase of stimulation of the muscles and mind. This is used in mental therapy as conflicts are revealed, resistance is overcome and introspection and insight increase. However, these effects can be accompanied by visual and/or other hallucinations and may be followed by depression. The effect varies with the individuals and may cause permanent psychological damage in some and there is the possibility of the development of serious dependence.

(a) Analeptics and convulsants

The *Cola* spp. have, like coffee, tea or maté a stimulant action due to the presence of xanthine derivatives (Burgen and Mitchell, 1972). *Ocimum canum* contains camphor: the analeptic action of this compound on the heart and respiratory system is well known. Xanthones are also found in *Anthocleista vogelii* (Loganiaceae). *Centella asiatica* (Umbelliferae) is reported to improve the mental ability and behaviour of mentally retarded children through glycosides of triterpenic acids. The reputation of *Strychnos* spp. as 'bitter tonics' may be due to a bitter taste with stimulation of the taste papillae and, as a reflex action, hypersecretion of saliva and gastric juices. Strychnine also has a strong convulsant and analeptic action and small doses can produce nervous and skeletal muscle stimulation but higher doses cause tetanus-like convulsions leading to death, (from spasmodic contractions of the thorax and diaphragm). The use of strychnine has now been limited mainly to investigations on the mode of action of convulsant drugs (Burgen and Mitchell, 1972). However, most of the West African *Strychnos* spp. also contain muscle relaxant alkaloids (see below). Convulsant and sedative action is also noted in *Afrormosia laxiflora* (Fabaceae). The plant is, however, very toxic.

Cola acuminata (Beauv.) Schott & Endl. syn. (*Sterculia acuminata* Beauv. & Oware, *C. pseudoacuminata* Engl.) (Fig. 3.1) STERCULIACEAE
Cola nitida (Vent.) Schott & Endl. syn. (*C. acuminata* var. *latifolia* Schum., *Sterculia nitida* Vent. Jard. Malm., *C. acuminata* Engl.)
Kola nut trees

L The trees begin to bear nuts when 5–6 years old and bear fully after about 10 years (average annual yield is about 60 kg per tree). The nuts are used as a masticatory and stimulant (Dalziel, 1937).

C Kola nuts contain purines, about 2.5% caffeine, 0.023% theobromine, 1.618% tannins and a considerable amount of fructose. The caffeine content is about the same in the red and the white seeds. In addition the nuts contain two phenolic substances kolatin and kolatein, catechols, (−)-epicatechol and kolanin, the latter mainly in young nuts. Kola red is an anthocyanin pigment, phlobaphen, which occurs through oxidation of the catechols (Michl and Haberler, 1954; Dublin, 1965); Goodman and Gilman, 1976, p. 359).

Fig. 3.1. *Cola acuminata* (Beauv.) Schott & Endl.

P Caffeine excites the CNS at several levels and is a mental, skeletal muscle, respiratory and cardiac stimulant. Theophylline and theobromine have similar action but are more diuretic than caffeine. Because of the presence of catechols the action of kola is more moderate and it also relaxes smooth muscle (Burgen and Mitchell, 1972, pp. 41, 42). In high doses kola can nevertheless be dangerous: it can produce over-excitement followed by depression as it conceals normal tiredness; smaller doses produce a passing exciting action on the nervous system and increase the blood pressure and the strength of the heartbeat. The seeds are exported to be used in the preparation of soft drinks and medicinally for their stimulating and sustaining effect in sport and intellectual work. For export the nuts are merely dried and put into strong bags (Russel, 1955).

Ocimum canum Sims syn. (*O. americanum* L., O. hispidulum Schum. & Thonn., *O. thymoides* Bak.) LABIATAE

Hoary or American basil

L The leaves are used in most areas of West Africa as an infusion in the treatment of fevers and dysentery and also to relieve toothache. The leaves are strongly flavoured and find use in flavouring and to repel mosquitos. In Guinea the leaves cooked with groundnuts are given before parturition and as an emmenagogue.

C The composition of the essential oil varies according to its origin. In East Africa the oil contains 16–25% camphor (Beckeley, 1936) whilst in Central Africa methyl-cinnamate predominates in the oil (Schimmel *et al.*, 1914); further citral and smaller amounts of other essential oils have been reported (Paris and Moyse, 1971). During the flowering period the amount of essential oil decreases in the leaves and increases in the flowers, these proportions are inversed in the fruiting period (Kerharo and Adam, 1974).

P Camphor is an excellent cardiac and respiratory analeptic and is also a general stimulant. The essential oil is antiseptic (Nickel, 1959) and is used as a pulmonary antiseptic and expectorant. It has been recommended as a mild analgesic and rubefacient for rheumatism and for pains in external use (Dalziel, 1937).

According to the *Flora of West Tropical Africa* (F.W.T.A.) the essential oil of **O. basilicum** (botanically similar to *O. canum*) can contain up to 75% of estragol (methylchavicol) and has sedative and antispasmodic effects (Kerharo and Adam, 1974).

Ocimum gratissimum L. (*O. viride* Willd., *O. guineense* Schum. & Thonn.) LABIATAE

Tea bush

L *O. gratissimum* is locally a familiar febrifuge and diaphoretic and is also regarded as a stomachic and laxative (Dalziel, 1937).

C The essential oil obtained from the plant differs entirely from that of *O. canum*. It contains mainly thymol (32–65%) and eugenol.

P The oil is used externally to keep mosquitos away but negative results have been reported. Thymol is antiseptic, antitussive and antispasmodic (Paris and Moyse, 1971, p. 282).

Anthocleista vogelii Planch. syn. (*A. kalbreyeri* Bak., *A. talbotti* Wern.)
LOGANIACEAE

L The seeds and bark are used in Nigeria for their antipyretic, tonic and purgative action (Dalziel, 1937).

C Three tetraoxygenated xanthones (1-hydroxy-3,7,8-trimethoxyxanthone, 1,7-dihydroxy-3,8-dimethoxyxanthone and swertioside) have been isolated from the leaves of *A. vogelii* collected in Zaire (Chapelle, 1974).

P The medicinal properties attributed to the plant are believed to be due to the synergistic activity of the xanthones (reported to have MAOI action by Suzuki *et al.* (1981), tetraoxygenated xanthones (which have anticonvulsant properties) and seco-iridoids (which are stimulants and stomachics) (Ghosal *et al.*, 1973; Chapelle, 1974).

Centella asiatica (L.) urb. *Hydrocotyle asiatica* L. UMBELLIFERAE
Indian pennywort

L An infusion of leaves and stems has been used in India for the treatment of leprosy and other skin diseases, and as a diuretic (Indian Pharmaceutical Codex). Large doses are said to have a narcotic effect.

C Asiaticoside, a glycoside of a genin composed of a pentacyclic triterpenic acid, was isolated from the Madagascar plant by Bontemps (1942). Bhattacharya and Lythgoe (1949) found no asiaticoside in the Sri Lankan plant, but reported the presence of a related compound, centelloside, and the triterpenic acids centoic and centellic acid (Boiteau *et al.*, 1949; Boiteau and Ratsimamanga, 1956; Oliver, 1960). Appa Rao *et al.* (1969) found in Indian plants two free terpenic acids, brahmic and isobrahmic acid, two saponins, brahmoside and brahminoside (tri- and tetraglycosides of brahmic acid) and also betulic acid and stigmasterol. The saponins were found to be different from asiaticoside found in the Madagascar plants. Dutta and Basu (1967) had isolated and identified asiatic acid in an Indian variety of *Centella asiatica* and the presence of asiaticoside, *meso*-inositol and oligosaccharide of centellose in this variety has also been reported. Finally, it has been shown that, depending on the habitat, the saponins can be of two types, the more common one containing asiaticoside and medacanoside, and the less common one showing the additional presence of arabinose in the saponins, thus forming brahmoside and brahminoside. Sapogenins and flavonoids were the same in both varieties (Rao and Seshadri, 1969).

P Asiaticoside had been found to be active in the treatment of leprosy (by dissolving the waxy coating of *Mycobacterium leprae* whilst an oxidized form, oxyasiaticoside, inhibited the growth of tubercle bacillus in vitro and in vivo (Boiteau *et al.*, 1949). In clinical trials, Appa Roa *et al.* (1969, 1973) studied the effect of the plant on the general mental ability of mentally retarded children and its anabolic effect on normal healthy adults. They found that in 30 mentally retarded children (free from epilepsy and other neurological conditions) a significant improvement in both general ability and behavioural pattern was obtained when the drug was administered for a period of 12 weeks. In 43 normal adults the mean levels of blood sugar, serum cholesterol, total protein and vital capacity were increased by the drug and the mean levels of blood urea and serum acid phosphatase were decreased (Appa Rao *et al.*, 1973).

Strychnos alkaloids. These are known to be analeptic and convulsant and are medullary stimulants. As analeptics they were used to overcome depressions due to overdoses of barbiturates or morphine, but this is now mostly abandoned because of their high toxicity. They act on the spinal cord by antagonizing or blocking postsynaptic inhibitions. Brucine has a similar action to strychnine but is fifty times less toxic. Picrotoxin, obtained from the Indian *Anamirta paniculata*, is a medullary stimulant in small doses and is used in preference to strychnine to counteract barbiturate and bromide poisoning. The West African *Dioscorea dumetorum* (Dioscoreaceae) has a picrotoxin-like convulsive action on the medulla.

Strychnos spp. LOGANIACEAE

L The local uses of the *Strychnos* spp. vary greatly. Some, like *S. usambarensis*, *S. camptoneura*, *S. splendens* and *S. angolensis*, seem not to be used in local medicine. The seeds of *S. densiflora* were used in trial by ordeal and the fruits of *S. aculeata* as a fish poison (Dalziel, 1937). The alkaloids of the latter species were studied by Mirand *et al.* (1979).

C Among the West African *Strychnos* spp. only a few were found to contain small quantities of alkaloids with convulsant effects, whereas muscle relaxing alkaloids seem to be present in more West African spp.

The first isolation of a convulsive alkaloid from African *Strychnos* spp. was achieved by continual pharmacological screening for convulsive and muscle relaxant effects in the East African *S. icaja* Baill. (Sandberg *et al.*, 1969a, b). This led also to the detection of 4-hydroxystrychnine in that species. From the rootbark of *S. aculeata* Sol., Sandberg *et al.* (1969b) isolated strychnofendlerine and *N*-acetylstrychnosplendine as well as traces of brucine. The muscle-relaxing effects seem to be limited to extracts of the seeds and the pericarp of the fruit and was later considered to be weak.

As a result of further screening, new tertiary indole alkaloids with a pronounced muscle-relaxant effect but producing clonic convulsions in high doses were found in other species (Sandberg and Kristianson, 1970; Sandberg *et al.*, 1971; Verpoorte and Bohlin, 1976; Rolfsen and Bohlin, 1978).

In 1975 Bouquet and Fournet (1975b) reported that the only African *Strychnos* spp. with over 0.5% of total alkaloids were *S. camptoneura* Gilg & Busse, *S. splendens* Gilg, *S. angolensis* Gilg and *S. usambarensis* Gilg.

P Verpoorte and Bohlin (1976) screened 11 African *Strychnos* species for muscle relaxant and convulsant effects. They reported strong muscle relaxant effects in *S. usambarensis*, *S. afzelli* Gilg, *S. barteri* Sol. and *S. longicaudata* Gilg whilst in *S. aculeata* Sol., *S. malacoclados* Wright and *S. spinosa* only a weak effect was noted.

Further details are given on five of the very many West African *Strychnos* spp.

Strychnos camptoneura Gilg & Busse syn. (*Scyphostrychnos talbotti* Moore, *S. psittaconyx* Duvign.) LOGANIACEAE

C Sandberg *et al.* (1971), in studying the muscle relaxant properties of stem- and rootbark collected in the Cameroons, found 11 alkaloids, the main ones being serpentine and alstonine. Koch *et al.* (1972) and Garnier *et al.* (1974) found five new alkaloids in the bark of specimens from the Ivory Coast; these were identified as retuline, camptine, camptoneurine, camptinine and *N*-oxyretuline.

P Verpoorte and Sandberg (1971) reported muscle relaxant activity in the stem- and rootbark of *S. camptoneura* Gilg & Busse.

Strychnos usambarensis Gilg syn. (*S. micans* Moore, *S. cooperi* Hutch. & Moss) LOGANIACEAE

C Oxindole and bis-indole alkaloids have been reported in *S. usambarensis*. Alkaloids of this type were formerly known in certain species of Rubiaceae and Apocynaceae but their presence has now been detected in Loganiaceae. The alkaloids found in the bark and leaves of *S. usambarensis* in the Congo have been identified as harmine, strychnofoline, 2-isostrychnofoline, strychnophylline and isostrychnophylline. In addition, the presence of 18,19-dihydro-usambarine, usambaridine and their dihydro-derivatives strychnobaridine, strychnopentamine and isomers has been reported (Angenot, 1978; Angenot *et al.*, 1970, 1973, 1975, 1978a, b). Usambarine, usambaridine and strychnambarine have been isolated from the leaves, the two former are indole alkaloids biogenetically derived from two tryptamines and one monoterpene (Koch *et al.*, 1973).

P The indole alkaloids showed both convulsant and curarizing muscle relaxant activity (Angenot *et al.*, 1970, 1975).

Strychnos dolichothyrsa (Gilg) Onochi & Hepper LOGANIACEAE

C In the stembark of this climber, the following alkaloids were found: bisnordihydrotoxiferine, bis-nor-*C*-curarine, bis-nordihydrotoxiferine-*N*-oxide, and bis-nordihydrotoxiferine-di-*N*-oxide (Verpoorte and Svendsen, 1976; Verpoorte, 1980, 1981). In the stems and leaves only traces of alkaloids were found (Bouquet, 1970).

P From the stembark of **Strychnos decussata**, a tertiary alkaloid has been isolated which has a pronounced muscle relaxant activity both in vivo and in vitro. The blockade effect of this alkaloid, decussine, is not antagonized by synstigamine (Bisset and Phillipson, 1970, 1973; Olaniji and Rolfsen, 1980; Rolfsen *et al.*, 1980a, b).

Strychnos spinosa Lam. syn. (*S. lokua* A. Rich., *S. laxa* Solered., *S. buettneri* Gilg, *S. spinosa* var *pubescens* Bak., *S. djalonensis* Chev.) LOGANIACEAE

L Locally this species is used mainly in colitis, entero-colitis and diarrhoea (roots and bark); the root powder is sold in the markets in Senegal as a stomachic (Oguakwa *et al.*, 1980).

C Lofgren and Kinsley (1942) found no alkaloids in the seeds, stems and leaves and the different parts of the plant were found to be non-toxic to mice and guinea pigs. More recently, Mathis and Duquenois (1963) found, by using two-dimensional thin-layer chromatography, 0.009% of alkaloids in the pericarp and 0.012% in the fruit plus seed. In 1980, Oguakwa *et al.* reported the presence of akagerine and 10-hydroxyakagerine in the leaves. Akagerine had already been reported in *S. panamensis* (Marini Bettolo *et al.*, 1970).

P The toxicity of the plant appears to be very low. Verpoorte and Bohlin (1976) report only weak muscle relaxant effect.

Strychnos dinklagei Gilg LOGANIACEAE

L The plant, which is found in Liberia, Ghana, Guinea and the Ivory Coast, is used in local medicine in the treatment of diseases of the mouth, the kidneys and the heart (Oguakwa *et al.*, 1980).

C Traces of alkaloids have been found in the leaves but they are more abundant in the bark. In the stembark, ellipticine has been found as the main alkaloid. It is the first time that a non-corynane, strychnine-type of alkaloid has been isolated from a *Strychnos* sp. (Michel *et al.*, 1980). Akagerine and derivatives were also found in *S. dinklagei* (Oguakwa *et al.*, 1980).

P Ellipticine had been shown to have slight convulsive properties (Sandberg *et al.*, 1971; Verpoorte and Svendsen, 1976). Fractionation led to the discovery of five alkaloids with a convulsive action. In addition to the already known akagerine, four new indole alkaloids were isolated including *O*-methylakagerine, which produced strychnine-like convulsions but which is less potent than akagerine (Verpoorte and Bohlin, 1976; Oguakwa *et al.*, 1980).

Dioscorea dumetorum (Kunth) Pax. syn. (*Helmia dumetorum* Kunth, *Dioscorea buchholziana* Engl.) DIOSCOREACEAE

Bitter yam or cluster yam

L The tuber is used as a food only in cases of scarcity and requires to be sliced and steeped long before use to eliminate the poisonous alkaloid dioscorine, which produces CNS paralysis. It is said to have caused death when eaten during famine in the Sudan. In Senegal, the tuber is sometimes used externally as a rubefacient.

C The dried tubers contain, besides dioscin, the genin of which is diosgenin, small quantities of other steroid sapogenins and a convulsant alkaloid, dihydrodioscorine (Bevan *et al.*, 1956; Bevan and Hirst, 1958). Nigerian yams also contain 83.3% of glucides and 9.9% of proteins. Diosgenin has been much used as a starting compound in the synthesis of hormones, corticosteroids, etc. (Oliver-Bever, 1972).

P The LD_{50} in mice of a water-soluble extract of the tubers from the Congo (containing 6.2% of dihydrodioscorine) is 15.5 mg/20 g, and the convulsant dose ED_{50} is 12.5 mg/20 g. In the cat, the extract produces a long-lasting hypotension when injected intravenously in doses of 100 mg/kg. The total extract produces a contraction of the smooth muscle fibres of the intestine both in vivo and in vitro (Bevan *et al.*, 1956). In small doses (30 mg/kg) in the cat or monkey, Schlag *et al.* (1959) have noted a desynchronization of the cortical electrical record lasting over 0.5 h. With higher doses (200 mg/kg) there are progessive convulsive impulses preceded each time by an increase in the arterial pressure and of the intestinal peristalsis which, according to the authors, indicates an excitant action of the drug on the cerebral cortex. In mice the LD_{50} of dioscorine is 65 mg/kg. Dioscorine produces first clonic then tonic convulsions and in concentrations of 10^{-5} M it reduces the response to acetylcholine of the isolated intestine of the cat and of the duodenum of the rabbit (Bevan *et al.*, 1956). Dioscin is less toxic; the LD_{50} is 100 mg/kg in mice. It has a picrotoxin-like convulsive action and the effect on the blood pressure of the cat and on the isolated ileum of the guinea pig is distinct although less pronounced than that of dioscorine (Broadbent and Schnieden, 1958; Correia da Silva *et al.*, 1962).

Afrormosia laxiflora (Benth. ex Bak.) Harms syn. (*Ormosia laxiflora* Benth. ex Bak.)
FABACEAE
False Dalbergia

L The local medicinal uses of the leaves, bark and roots of *Afrormosia* are based on their analgesic and antipyretic actions. The root is said to increase the intoxicating effect of palm wine, and to be slightly intoxicating if taken by itself. The plant was formerly used in arrow poisons and as an ingredient in a complex prescription taken to impart strength or stimulus 'when undertaking a journey or other enterprise' (Dalziel, 1937).

C The stembark of the Nigerian plant contains 6.7% of total bases including six alkaloids. The three main alkaloids are non-quaternary; the main constituent of these is *N*-methylcytisine (a quinolizidine derivative), a second is probably ammodendrine. Three minor alkaloids are also non-quaternary and a quaternary fraction is almost entirely composed of choline (Bevan and Ogun, 1964). In the heartwood of the related *A. elata*, afrormosine (a dimethoxyhydroxyflavone) has been reported, and in the bark catechuic tannins (Caiment-Leblond, 1957). The related *Ormosia dasycarpa* Jacks. and *O. coccinea* Jacks. contain two alkaloids, ormosine and ormosinine, which have a physiological action smaller than that of morphine (Caiment-Leblond, 1957).

P *N*-Methylcytisine is very toxic to mice, which can only tolerate a dose of 1 mg, producing long-lasting sedation. Higher doses (4 mg) produce severe ataxia and convulsions followed by death within 5 min. In anaesthetized rabbits doses of 2 mg produce a prolonged increase in blood pressure (Raymond-Hamet, 1954; Bevan and Ogun, 1964). The constituent appears to have a cytisine-like action, paralysing the ganglia after a short stimulation.

Tinctures of the rootbark and the isolated total alkaloids show a distinct hypertensive action when injected intravenously to the chloralosed dog (Kerharo and Bouquet, 1950; Caiment-Leblond, 1957). Tests on cytotoxicity were negative but a certain insecticidal effect was observed and the toxicity for mice (10 g/kg killed 40%) is high. The extracts are much less toxic to daphnia and fish (Kerharo and Bouquet, 1950).

For those plants with a stimulant action on the CNS already described in Chapter 2 see Table 3.3

(b) Antidepressants and hallucinogens

The drugs with these properties have been treated by Burgen and Mitchell (1972) under CNS stimulants. It should, however, be stressed that antidepressant and MAOI action is usually related to anticholinergic activity and it is not possible

Table 3.3. *Plants with stimulant action on the nervous system which are described in Chapter 2*

Plant Family	Active constituent	Part(s) used	Action on nervous system	References
Achyranthes aspera Amaranthaceae	Achyranthine	Seeds	Respiratory analeptic	CNS. Chapter 2 and Massion (1934)
Holarrhena floribunda Apocynaceae	Triacanthine		Respiratory analeptic	CNS. Chapter 2 and Quevauviller and Blanpin (1961)
Mostua hirsuta Loganiaceae	Alkaloid closely related to sempervirine	Roots	Convulsant, sedative and analgesic	CNS. Chapter 2 and Chevalier (1947)
Erythrophleum guineense Fabaceae	Cassaine	Bark, seeds	Convulsant, medulla stimulant, treatment of dental and facial neuralgia	Chapter 2
Physostigma venenosum Fabaceae	Physostigmine	Seeds	Stimulant in myasthenia gravis	Inhibits cholinesterase, see Chapter 2

to make a rigorous distinction between ANS and CNS activities. However, a number of these drugs may prove useful in the treatment of certain forms of mental illness. Unfortunately, these effects can be accompanied by secondary effects, by visual or other hallucinations, and may be followed by depression.

Securinine and phyllochrysine, alkaloids found in *Securinega virosa* and *Phyllanthus discoideus* (Euphorbiaceae), are mainly muscular stimulants which have been shown to be successful in re-establishing mobility in cases of paralysis and paresis and in mental patients. Indole alkylamines and MAOIs (β-carbolines) have been reported in *Pauridiantha viridiflora* and other Rubiaceae which also act on the ANS, e.g. *Nauclea diderichii* and *N. latifolia* and *Borreria verticillata*. All of these plants have been found to have a stimulant effect in post-encephalitic conditions. Antidepressant action is also attributed to *Justicia insularis* (Acanthaceae).

Stimulating to the CNS in small doses but narcotic and hallucinogenic in large doses are *Tabernaemontana crassa*, through ibogaine; *Cannabis sativa*, through tetrahydrocannabinols and *Myristica fragrans*, through myristicin (also a MAOI). *Datura metel* and *Datura* spp. containing hyoscyamine and hyoscine are sedative in psychomotor agitation like the other hallucinogens, and the alkaloids are used as pre-anaesthetics. They mainly act, however, via the ANS (anticholinergic with blockade of muscarinic receptors).

Securinega virosa (Roxb. ex Willd.) Baill. syn. (*Phyllanthus virosus* (Roxb. ex Willd.), *Fluggea microcarpa* Blume, *S. microcarpa* (Blume) Pax & Hoffm. ex Aubrev., *F. virosa* (Roxb. ex. Willd.) Baill.) EUPHORBIACEAE

L *S. virosa* is widely used in Senegalese local medicine. A decoction of the roots is given mainly for disorders of the liver, the gallbladder, the kidneys (including stones), the bladder and the genitals. It is also recommended in many tribes to Bilharzia patients (Kerharo and Adam, 1974). The Hakims in India use *S. virosa* as a cure for diabetes mellitus (Watt and Breyer-Brandwijk, 1962).

C Numerous alkaloids have been isolated from the leaves, rootbark and stembark of this and closely related species. Paris *et al.* (1955) isolated a crystalline alkaloid which they called fluggeine, together with choline. Fluggeine was later identified as hordenine by Iketubosin and Mathieson (1963), who also isolated norsecurinine from the Nigerian plants. Securinine, first isolated from *S. suffruticosa* in 1956 by Muraveva and Bankovski, was later also found in *S. virosa*. In addition, norsecurinine (1.6%), dihydrosecurinine (0.06%) and virosecurinine (1.14%) were isolated from the rootbark, viroallosecurinine from the leaves and virosine (dihydrosecurinine) from the roots (Kjaer and Friis, 1962). Considerable differences have been observed in the alkaloid content of the male and female plants, and in the different parts of the plants (Chatterjee and Bhattacharya, 1964; Saito *et al.*, 1964a, b; Satoda *et al.*, 1972). The presence of thioglycosides has been reported in several botanically allied species, e.g. in the Indian *Putranjeva roxburghii* Wall., glucocochlearin, glucoputranjevin and glucojeaputin were found, and later, glucoilcomin (Kjaer and Friis, 1962). The same glycosides of *S. virosa* have been found in the fresh leaves, stems and roots and some may also exist in other *Securinega* spp. as the hypoglycaemic action appears to be related to the presence of thioglycosides (Chatterjee and Roy, 1965; Oliver-Bever and Zahnd, 1979). In *S. virosa* leaves rutin has also been reported.

P Securinine was found to have an action similar to strychnine (Quevauviller *et al.*, 1967). In pharmacological and clinical trials, Tourova (1957) noted that the alkaloid stimulates the CNS, including the spinal cord. In clinical trials with the nitrate (200 cases), Tourova (1957) observed that a rapid re-establishment of motility was produced, and hence it seems indicated in paresis, paralysis in poliomyelitis and diphtheria, and in apoplectic paresis and paralysis. The use of securinine is also indicated for impotence, decrease of cardiac activity, functional amaurosis, asthenia consecutive to extenuating diseases and ocular nerve atrophy. The drug was administered either by mouth (10–20 drops daily of a 1:250 solution of securinine nitrate) or subcutaneously (1 ml of a 1:500 solution daily). With these doses, no secondary effects have been noticed but overdosage produces a painful tension in the nape of the neck, face muscles and other muscle groups. In China securinine is used in the treatment of facial paralysis and of neurasthenia (Xiao, 1983). It has also shown promising results in the treatment of multiple sclerosis in man (Chang, 1974). Chang compares it with strychnine with regard to toxicity and pharmacological activity and reports that although both are powerful stimulants of the CNS, the LD_{50} of securinine in rats following intravenous administration is 26-fold higher than that of strychnine (Chang, 1974).

Phyllanthus discoideus (Baill.) Müll. Arg. syn. (*P. discoides* Müll. Arg., *Cicca discoidea* Baill., *Fluggea klaineana* Pierre ex. Chev., *F. obovata* var. *luxurians* Beille)

EUPHORBIACEAE

L An extract of the bark of this small tree is used as a purgative and antipyretic and topically in ophthalmias (Oliver, 1960).

C From the rootbark the alkaloids phyllalbine, securinine, phyllochrysine (allo-securinine), phyllanthine (methoxysecurinine) and phyllanthidine have been isolated (Janot *et al.*, 1958; Parello *et al.*, 1963; Parello, 1966; Foussard Blanpin *et al.*, 1967). Earlier, securinine had been found in *Securinega suffruticosa* by Muraveva and Bankovski (1956). Bevan *et al.* (1964) have reported in Nigerian species 0.4%, 0.2% and 0.06% of securinine in the rootbark, stembark and leaves, respectively, as well as the minor alkaloid allosecurinine. The phyllanthine formerly reported has been shown to be identical with methoxysecurinine.

P Securinine was shown by Tourova (1957) to have a stimulating action on the CNS comparable to that of strychnine. It increases reflex activity and breathing as well as muscular tone and the blood pressure. It was used by Tourova (1957) in various neuropsychic complaints with satisfactory results (see *Securinega virosa*). It has been reported to be ten times less toxic than strychnine. The action of phyllochrysine, securinine and phyllalbine have also been investigated in France. Phyllochrysine and securinine are sympathomimetic and excite the CNS. Securinine, however, is more toxic, does not act on the medulla and produces more violent convulsions (Quevauviller *et al.*, 1965). Phyllochrysine does not have local anaesthetic, analgesic or antispasmodic activity. Phyllalbine chloride has a LD_{50} in mice of 10 mg/kg when given intravenously and of 45 mg/kg when given subcutaneously. It mainly stimulates the suprarenal glands, producing secretion of adrenaline, has peripheral sympathomimetic properties and slightly inhibits monoamine oxidase (MAO) (Quevauviller and Blanpin, 1959; Quevauviller *et al.*, 1965, 1967).

Pauridiantha viridiflora (Schweinf. ex Hiern) Hepper syn. (*Urophyllum viridiflorum* Schweinf. ex Hiern) RUBIACEAE

L The bark of a related species, *P. lyallii* (not found in West Africa), has long been used as a remedy for fever and malaria (Pousset *et al.*, 1974).

C From the bark of *P. viridiflora* growing in the Congo, the following have been isolated: harmane (0.08%); pauridianthine (0.007%) and its isomer pauridianthinine (of pyridine-harmane structure), as well as an anthraquinone and a glycoside, lyaloside (Pousset *et al.*, 1971; Bouquet and Fournet, 1975a; Levèque *et al.*, 1975). *P. canthiflora* Hook., another species found in West Africa, has been shown to contain only traces of alkaloids (Bouquet, 1970). Harmane and related alkaloids (β-carbolines) have been found in several other African species (*P. lyallii*, *P. callicarpoides*) (Pousset *et al.*, 1971, 1974). Harmine and related alkaloids were first detected in *Peganum harmala* L. (Rutaceae) from India (Henry, 1949). Harmane is 1-methyl-9*H*-pyrido-3,4,6-indole and harmine is its 7-methoxy derivative. The configuration of these alkaloids shows a close relationship with that of serotonin (5-hydroxytryptamine) and tryptamine.

P The antipyretic and protozoicidal actions of *P. lyallii* have been confirmed by Pousset *et al.* (1974) and the chemical composition of *P. viridiflora* could justify the same activities. Harmane and its derivatives have similar properties to quinine alkaloids. They are also protozoicidal and are also said to be coronary dilators and oxytocics.

The harmala alkaloids are able to elicit or exacerbate abnormal reactions such as are shown in schizophrenia or in ethanol intoxication and some of their effects are reminiscent of productive symptoms in these cases (Hofer *et al.*, 1950). Thus an isomer of harmaline (6-methoxyharmalane) is a powerful serotonin antagonist, and it is suggestive that the highest concentrations of serotonin have been found in the pineal glands of schizophrenics (McIsaac *et al.*, 1961; Wooley, 1962; Naranjo, 1967). MAO-active alkaloids can alter 5-hydroxytryptamine and noradrenaline metabolism in the brain, producing enhancement of the serotonin effect on body temperature and counteraction to reserpine (Pletscher *et al.*, 1959).

The harmala alkaloids have been used in sequels of encephalitis (Hill and Worster-Drought, 1929; Cooper and Gun, 1931; Naranjo, 1967). In large doses they cause tremors and clonic convulsions. Harmine, harmaline and harmol are MAOIs (Burger and Nara, 1965; Slotkin *et al.*, 1970); harmine and harmaline show a short duration of MAO inhibition as compared to the early MAOIs (hydrazine derivatives). Pretreatment with harmaline can therefore reduce undesired secondary effects of these in blocking the MAO molecule receptors (Rommelspacher, 1981).

The harmala alkaloids are cholinergic and antagonistic to benzodiazepines (which are anxiolytic, anticonvulsive, muscle-relaxing and sleep-inducing). They are a group of substances with a broad spectrum of activity differing from compound to compound and require further research (Holmstedt, 1967; Rommelspacher, 1981). The LD_{50} of harmaline given subcutaneously, is 120 mg/kg for rats and mice; that of harmine is 200 mg/kg. The human therapeutic dose of harmine, given perorally, is 300–400 mg (Usdin and Efron, 1976).

Nauclea latifolia Sm. syn. (*N. esculenta* (Afz. ex Sab.) Merrill, *Sarcocephalus esculentis* Afz. ex. Sab., *S. sassandrae* Chev., *S. sambucinus* Schum., *S. russeggeri* Kotschy ex Schweinf.) RUBIACEAE

African peach, guinea peach, doundaké

L In Nigerian local medicine, the fruit is sometimes dried and used in the treatment of piles and dysentery. Eaten in excess the fruit acts as an emetic. The bitter bark has been widely used locally, in the form of an infusion or decoction, as a tonic and antipyretic. It is called 'African quinine'. In Northern Nigeria, a cold infusion of the bark is taken as a diuretic and anthelmintic, and to regularize bowel functions. The stembark has been used as a haemostatic, *N. latifolia* is also a timber tree and frequent intoxication (headaches and nausea) of the workmen cutting up the trees may be attributed to an alkaloid, found in the leaves, which has a marked and cumulative cardio-inhibiting action (Caiment-Leblond, 1957). The root of *N. pobeguinii* is prescribed in local medicine in Senegal as a powder for abdominal pains and as an oxytocic in the form of a decoction (Kerharo and Adam, 1974).

C Different indolo-quinolizidine alkaloids and glyco-alkaloids have been isolated from the rootbark. The former have been identified and named angustine, angustoline, angustifoline, nauclefine and naucletine (Dimitrienko *et al.*, 1974; Hotellier *et al.*, 1975). The glyco-alkaloids have been identified as cadambine and 3-α-dihydro-cadambine. These two alkaloids have also been reported to be present in the leaves of **N. diderrichii** (de Wild. & Dur.) Merrill in Senegal (McLean and Murray, 1970). Patel and Rowson (1964) noted the presence of heterosides in the rootbark of *N. latifolia* and Hotellier *et al.* (1977) isolated a precursor from it called strictosamide, which is closely related to vincoside lactam. The simultaneous presence of indolo-quinolizidine alkaloids and of corresponding heterosides seems to indicate a biogenetic relationship in this family of Rubiaceae; similar observations had already been made concerning *N. diderrichii* and *Pauridiantha leyallii* Brem. (Levèque *et al.*, 1975). The leaves of *N. latifolia* yield 0.8% of alkaloids, including naufoline and angustine, and the same two glyco-alkaloids which had been reported in the rootbark (Hotellier *et al.*, 1979). Harmane, pyridine and indole-pyridine alkaloids have been isolated from *N. diderrichii* (de Wild. & Dur.) Merrill, which is also found in West Africa (McLean and Murray, 1970).

P In the guinea pig an intraperitoneal injection of an aqueous extract of the leaves and bark of *N. latifolia* collected in Nigeria equivalent to 6 g/kg produced a lowering of the rectal temperature of 2°C, which lasted for several hours, and in dogs an aqueous extract of the leaves was reported to have distinct hypothermic action and to produce a sudden decrease of the carotid pressure followed by the opposite effect and by renal vasoconstriction (Raymond-Hamet, 1937). A non-identified (possibly indolic) alkaloid isolated from the roots in Portuguese Guinea by Almeida *et al.* (1963)

produced an inhibiting effect on smooth muscles and was anticholinergic (Correia da Silva *et al.*, 1964).

Cardio-inhibiting and cardiotonic activity of extracts of the leaves and bark were reported by Patel and Rowson (1964). The leaves of *N. latifolia* have anticancer action against transplantable sarcoma 180 tumours and against Lewis lung carcinoma, producing a reduction of 43% and 53% respectively (Abbott *et al.*, 1966).

Justicia insularis Anders syn. (*Adhatoda diffusa* Benth., *J. galeopsis* Anders, *Siphonoglossa macleodiae* Moore) ACANTHACEAE

L The plant does not seem to be used medicinally in West Africa. In the Kumaon region (India), it is used as an antifatigue and stimulating plant (Ghosal *et al.*, 1979a, 1981).

C In India and Japan a number of *Justicia* spp. have been screened as they produce a large number of aryl-naphthalide lignans with antidepressant properties. Thus, *J. prostata* has been shown to contain the prostalidins A, B, C and tetrochinensin (4-aryl-2,3-naphthalidine lignan). In *J. hayatai* var. *decumbens* and *J. procumbens* var. *leucantha* (from Formosa and Japan) a number of aryl-naphthalide lignans (sesamin, asarin, sesamolin) and in *J. simplex* a new lignan, simplexolin, have been recorded (Ghosal *et al.*, 1979a, b, 1981).

P Pharmacological screening of these lignans revealed significant action on the CNS in animals. The prostalidins A–C produced a mild antidepressant action in albino mice and rats. The action was potentiated by carpacin, which itself showed only a weak sedative action. The combined active constituents have a low toxicity (Ghosal *et al.*, 1979b). Another biological activity reported for bicyclo-octane lignans is the reversal of sickling and crenation in erythrocytes by plant extracts containing similar constituents (Sofowora *et al.*, 1975).

Tabernaemontana crassa Benth. syn. (*T. durissima* Stapf, *Conopharyngia durissima* (Stapf) Stapf) APOCYNACEAE

Tabernaemontana pachysiphon Stapf var. *pachysiphon* syn. (*Conopharyngia pachysiphon* (Stapf) Stapf)

L A decoction of the leaves of *T. crassa* is taken in West Africa as a tonic, appetizer and aphrodisiac whilst the juice of the bark is used in the treatment of leprosy and for wound disinfection (Kerharo and Bouquet, 1950).

In Nigeria and Ghana, the rootbark of *T. pachysiphon* is used as an infusion in the treatment of manias (Ainslie, 1937; Irvine, 1961; Watt, 1967). The rootbark of some species is said to be strongly sedative (Watt, 1967).

C In the roots and bark of *T. crassa*, indole alkaloids, isovoacangine, conopharyngine, conodurine, conoduramine and coronaridine, and in the seeds, voacamidine, coronaridine-hydroxyindolenine, tabersonine and coronaridine have been reported (Dass *et al.*, 1967). Although coronaridine is closely related to ibogaine (Gorman *et al.*, 1960), the narcotic alkaloids found in *T. iboga*, such as ibogaine, ibogamine and tabernanthine (Tyler, 1966; Pope, 1969), could not be detected in *T. pachysiphon* or *T. crassa* (Taylor, 1957; Dickel *et al.*, 1958). By thin-layer chromatography, Patel *et al.* (1967) could detect in the bark of Nigerian *T. pachysiphon* c. (cultivated) coronaridine, conopharyngine and voacangine plus small amounts of a number of non-identified alkaloids. Conopharyngine and ibogaine were reported in **T. contorta** (Stapf) Stapf. All known *Tabernaemontana* spp. thus appear to be characterized by ibogamine-type alkaloids (Haller and Heckel, 1901).

P *T. crassa* Benth, when intraperitoneally injected (suspension of a crude extract) in rats, produced decreased motor activity and muscle relaxant effects: the pupil of the eye was dilated and there was blanching of the ears. Death occurred 30 min after injection of 250 mg/kg (Sandberg and Cronlund, 1982).

T. crassa and *T. pachysiphon* have been reported to be stimulants of the CNS and to be hallucinogenic in large doses (Marderosian, 1967). They increase and extend the hypertensive action of adrenaline and also have local anaesthetic activity (Schneider and Sigg, 1957; Raymond-Hamet and Vincent, 1960; Paris and Moyse, 1971, p. 81). Coronaridine hydrochloride, isolated from the roots of the Indian *T. heyania*, but also present in the above-mentioned species, has been shown to prevent pregnancy in rats when administered 1–5 days post coitum (Mehrotra and Kamboj, 1978).

Cannabis sativa L. var. *indica* Lam. CANNABINACEAE

L The plant is nowadays subject to government restrictions in most West African countries (Nigeria, Ghana, Senegal, etc.). Formerly, it was used as an antidote to snake poison and to treat malaria and blackwater fever (*Hager's Handbuch*, 1972, Vol. III, p. 652).

C The narcotic resin is obtained from the dried flowering or fruiting tops or from the green shoots (the quality varies with the district from which it comes). The resin exudes from the surface during growth and is collected by pressing the tops or flailing the stems. It contains cannabinol, cannabidiol and several isomeric tetrahydrocannabinols, which are the chief active principles. The two main psychomimetically active components are the (−)*trans*Δ^1- and (−)*trans*Δ^6-isomers (Kettenes-van den Bosch *et al.*, 1980). In addition, cannabigerol, cannabichromene and cannabitriol have been reported. The amount of exudate is low in temperate regions and high in warmer regions. Choline, trigonelline and a coumaric glycoside have also been found in the tops (Gaoni and Mechoulam, 1964; Schulz, 1964; Farnsworth, 1969; Turner *et al.*, 1980). Ten flavonoid glycosides have been isolated by column and paper chromatography. One was found to be the acyl derivative of apigenol, the others are *O*-glycosides and *C*-glycosides (vitexine, isovitexine and orientine) (Paris *et al.*, 1976).

P Cannabis has been used in veterinary medicine as a sedative in the treatment of equine colic (Merck Index, 1976, item 1748) and in man as an analgesic and hypnotic and in the treatment of depressive mental conditions (Paris and Moyse, 1967). The analgesic effect is considered to be a consequence of its general effect on the cerebral cortex (Kettenes-van den Bosch *et al.*, 1980). It is rarely prescribed even in the countries where its medicinal use is still authorized. It is known under many local names such as ganjah, bhang or charas in India, marihuana in North America, hashish (purified alcoholic extract) in North Africa, kif in Morocco, takrouri in

Tunisia, dagga in South Africa, etc. Its production and use has been forbidden in 74 countries since 1956 and the 1961 New York Convention (Paris and Moyse, 1967).

When ingested or (more frequently) inhaled as smoke, the drug first produces a state of euphoria, intellectual excitement and indifference to surroundings, then come illusions, loss of the notion of time and space, hallucinations, incoordination of movements and drowsiness but not complete unconsciousness. The psychomimetic effect is more rapid with smoking than with ingestion (*Hager's Handbuch*, 1972; Paton, 1975).

Toxic effects of prolonged use of *C. sativa* include lassitude, indifference, lack of productive activity, insomnia, headaches, nystagmus, increased susceptibility to infections, gastrointestinal disturbances, sexual impotence and personality changes. The slow and prolonged hypotensive action of the drug, and its interaction with catecholamines in the peripheral system, suggest the possibility of an interaction with the brain amines being responsible for the behavioural effects observed (Arora *et al.*, 1976). Kettenes-van den Bosch *et al.* (1980) write: 'Investigation as to the therapeutic potential of (−)*trans*Δ^1-tetrahydrocannabinol as an anticonvulsant, anti-emetic, antiglaucoma, anticancer and analgesic drug have started only recently and the results to date have not been convincing, adverse effects and the development of tolerance have been the limiting factors.' Hence they consider that further investigations are required.

The antibacterial activity of the essential oil of *C. sativa* was assessed on *Staphylococcus aureus*, *Streptococcus faecalis*, *Mycobacterium smegmatis*, *Pseudomonas fluorescens* and *Escherichia coli*. The oil was found to be active on Gram-positive bacteria and has been used against Gram-positive bacteria in cases of resistance against penicillin (Fournier *et al.*, 1978). The antibacterial agent appears to be cannabidiolic acid (Farnsworth, 1969).

Myristica fragrans Houtt. MYRISTICACEAE
Nutmeg

L *Myristica fragrans* has been introduced into various parts of West Africa as a spice. Essential oil of nutmeg is used externally for rheumatism and internally as a carminative (Oliver, 1960).

C The essential oil is associated in the nut with a solid fat. The oil contains pinene, camphene, borneol, geraniol and eugenol and in the last portions of the distillate, myristicin (methylene-dioxy-methoxyallylphenol) belonging to the phenylisopropylamines. In addition, elemicin and safrol have been reported (Gottlieb, 1979). The varying proportions of these substances explain the differences in pharmacological action of various samples of nutmeg oil (Shulgin *et al.* in Efron *et al.*, 1967, pp. 202–14).

P The oil is an aromatic stimulant and in high doses it has convulsant and oxytocic properties. Doses of between 0.2 and 1 g/kg induce dose-dependent light-to-deep sleep in young chicks. In man, the seeds and arils have in some cases hallucinogenic properties (Truitt, 1967; Weil, 1967) although this effect is contested by Shulgin (1966). The ingestion of about 5 g of the seed (about one large nutmeg) can lead after a few hours to a more or less severe physical collapse, weak pulse, hypothermia, clamminess of the extremities, giddiness, vertigo, nausea and a feeling of congestion and pressure in the chest or abdomen. For about 12 h there is an alternation of delirium and stupor, usually resolved by heavy sleep. For several days there may be headaches, dryness of the mouth, tachycardia and perhaps spells of dizziness (Weiss, 1960). Myristicin has been said to be effective in quietening hysteric or delirious patients. The LD_{50} in rats is less than 1 g/kg. Myristicin has been shown to be a MAOI in vitro and in vivo (Truitt, 1967). It may by degradation undergo transamination, producing 3,4,5-trimethoxyamphetamine; more rapid biodegradation of pure myristicin, in contrast to a slow release, might suggest a greater efficiency of the crude drug (Weil, 1965; Shulgin *et al.*, 1967; Truit, 1967; Forrest and Heacock, 1972).

Datura metel L. syn. (*D. fastuosa* L. var. *D. alba* (Nees) C. B. Cl.) SOLANACEAE
Datura stramonium L. including *D. tatula* L.
Datura innoxia Mill. syn. (*D. metel* Chev. Berh.)
Datura candida (Pers.) Safford syn. (*Brugmansia candida* Pers., *D. arborea* Ruiz & Pavon) (Fig. 3.2)

L In tropical West Africa, *Datura* spp. are used in native beer or in palm wine to add a stupefying or narcotic effect. Thus, a drink made from the seeds of *D. metel* is given as an intoxicant to Fulani youth to incite them in the Sharo contest or ordeal of manhood (Dalziel, 1937). A decoction of the seeds has been used for eye diseases (Pobéguin, 1912).

C The main alkaloids, present in all species, are the parasympathetic alkaloids atropine (($\pm$)- hyoscyamine), ($-$)-hyoscyamine and hyoscine (scopolamine). They are found mainly in the flowers and leaves, and, to a lesser extent, in the seeds. Norscopolamine, meteloidine, hydroxy-6-hyoscyamine and tiglic esters of dihydroxytropane have been reported as secondary alkaloids (Shah and Khanna, 1963, 1964, 1965a, b; Shah and Saoje, 1967). *D. metel* is the species richest in hyoscine, the leaves containing approximately 0.5% of total alkaloids of which three-quarters consists of hyoscine. The total alkaloid content of *D. stramonium* leaves is roughly the same, but of this over two-thirds is hyoscyamine/atropine. In *D. innoxia* leaves hyoscine predominates, whilst in the seeds it is the hyoscyamine/atropine fraction which predominates. It was shown that in young leaves of *D. metel* hyoscyamine is the main alkaloid, but in adult leaves it is hyoscine. In *D. stramonium*, however, the proportions are inverted and hyoscyamine predominates in adult leaves (Jentzch, 1953). Hyoscine is formed in the leaves by epoxidation of hyoscyamine. Long and intense exposure of the plants to light produces an increase in the hyoscine content. The amount of alkaloids present also varies with the origin of the plants and can be increased by various methods, such as deflowering, mutations, fertilizers, etc. (Paris and Cosson, 1965; Karnick and Saxena, 1970a, b). Balbaa *et al.* (1979) could increase the percentage of active constituents in *D. tatula* by more than 100% above the control level through the use of fertilizers.

P *Datura* spp. are very toxic and their alkaloids can produce delirium with vertigo and hallucinations. The three main *Datura* alkaloids have both peripheral and central actions. By local application to the eye, ($-$)-hyoscyamine and atropine cause a pronounced and long-lasting mydriasis due to paralysis of the circular muscle of the eye. They also paralyse the ciliary muscle. The mydriasis produced by hyoscine is of shorter duration but quicker in onset than that produced by atropine. These alkaloids are therefore used as eye drops to dilate the pupil and to paralyse accommodation. They antagonize the activity of the parasympathetic nerves innervating smooth muscles, the glands and the heart by blocking the action of acetylcholine at the post-ganglionic nerve endings (anticholinergic effect) and can be used in conditions where paralysis of the parasympathetic activity is desired, such as bronchial and intestinal spasms. They are constituents of many asthma powders and

Fig. 3.2. *Datura candida* (Pers.) Safford.

sea-sickness and anti-chronic bronchitis preparations. Hyoscine also has spasmolytic and peripheral antispasmodic action, but depresses the CNS, whilst atropine stimulates the CNS, and it is a useful sedative and hypnotic for patients with psychomotor agitation, delirium tremens, paralysis agitans and Parkinson's disease. It also finds a use, generally associated with morphine, as a pre-anaesthetic or for relieving withdrawal symptoms in morphine addiction. A subcutaneous injection of 1 mg in adults can induce stupor, confusion of mind and loss of will power, and is reported to have been used for 'brainwashing' (Fattorusso and Ritter, 1967; Karnick and Saxena, 1970b; Lechat *et al.*, 1978).

II Plants with a depressant action mainly via the CNS

The plants of this group often have a simultaneous activity in several sections. This applies particularly to those having analgesic, narcotic, sedative, hypnotic and antipyretic activity: each effect may be the predominating consequence of a general action on the cerebral cortex (Turner and Richens, 1978).

Narcotic analgesics cause unconsciousness and produce sleep through aboliton of the reflexes, including the sense of pain, by paralysing the nerve centres. They cause respiratory depression and a reduction in the motility of smooth muscles (causing constipation, spasm of the sphincter of Oddi and bronchoconstriction). Most narcotics have at first a short stimulating action on the nervous system but then cause a depression with dumbness and stupefaction. An example of a powerful analgesic is morphine. Minor analgesics can abolish the sensation of pain without producing loss of consciousness. Some may have, in addition to the analgesic action, antipyretic or anti-inflammatory activity.

Hypnotics produce sleep (without abolishing the reflexes). Sedatives and tranquillizers decrease watchfulness and calm down motor activity and agitation, and tranquillizers more particularly weaken exaggerated emotional reactions and attenuate restlessness. Even in strong doses they are not hypnotic but some relax skeletal muscles (Lechat *et al.*, 1978). These drugs are used especially in alleviation of the symptoms of schizophrenia and allied disorders and have also been called 'anxiolytics'.

Anticonvulsants lower the excitability of certain central neurones and thus are used to inhibit or diminish the excessive nerve impulses in epileptic convulsions. Different forms of epilepsy (petit and grand mal, psychomotor epilepsy) are caused by lesions of the psychomotor connections in the cortical regions of the brain. They are characterized by convulsions, loss of consciousness and changes in the electrocardiogram. In seizures the central inhibition is suppressed and abnormal nerve impulse activity occurs in small feedback loops (Burgen and Mitchell, 1972). Antiepileptics act by central inhibitory processes.

Antipyretics regulate the body temperature by reducing hyperthermia to normal values. A raised temperature induces peripheral vasodilatation and increased perspiration in an attempt to restore the temperature to normal. The exact mechanism by which antipyretics regulate this process is still not known. A number of plants originally studied for their antipyretic effects have ultimately been shown to act on the cause of the fever and have antimicrobial, antimalarial or anti-inflammatory activities. A few, however, like some *Holarrhena* and some *Funtumia* spp. containing alkaloids (see Chapters 2 and 5) and perhaps some containing palmatine and related alkaloids (see below), may have true antipyretic–analgesic action. Many plants in this group show several depressant activities simultaneously. Those described in Chapter 2 (The cardiovascular system) are listed in Table 3.4.

Rhigiocarya racemifera and *Kolobopetalum auriculatum* (Menispermaceae) are reported to have analgesic effects attributed to *o*-methylflavinanthine which has a structure similar to that of morphine. Two other Menispermaceous plants, *Jateorhiza macrantha* and *Tinospora bakis*, are reported to depress the CNS and to have antipyretic and hypotensive activity. *Khaya senegalensis* (Meliaceae) is a sedative, anticonvulsant and antipyretic(?). Sedative and analgesic action is shown by *Andira inermis* (Fabaceae) and sedative and spasmolytic effects alongside a respiratory excitant action (due to nupharine) in *Nymphaea lotus* (Nymphaeaceae). *Anogeissus leiocarpus* (Combretaceae) has CNS antidepressant activity. *Elaeocarpus sphaericus* and *Passiflora foetida* have been noted to possess, respectively anticonvulsant and hypnotic and anticonvulsant and sedative properties. Antispasmodic action was also reported in *Guiera senegalensis*. Spasmolytic and hypnotic action was found in *Alstonia boonei* (Apocynaceae), which also has cholinergic properties. A sedative effect on the CNS and a stimulant action on the medulla has been found in *Waltheria indica* (Sterculiaceae) and sedative and anticonvulsant actions have been reported for *Piper guineense* (Piperaceae). The essential oil of *Anacardium occidentale* has tranquillizing and antispasmodic properties.

Although in all these plants several chemical constituents have been identified, it is not always clear which components are responsible for the various activities.

Rhigiocarya racemifera Miers syn. (*R. nervosa* (Miers) Chev.)

MENISPERMACEAE

Kolobopetalum auriculatum Engl. syn. (*K. veitchianum* Diels)

L In Sierra Leone the root of *R. racemifera* is scraped and put in palm wine. Both plants are used for sleeplessness (Dalziel, 1937).

C From these two species, *o*-methyl-flavinanthine has been isolated. This compound has a structural formula very similar to that of morphine (Gyang *et al.*, 1964). In *R. racemifera*, the alkaloids liriodenine, palmatine, menispermine (*N*-methylisocorydine) and magnoflorine have also been found (Mehrotra and Kamboj, 1978; Dwuma Badu *et al.*, 1980).

P *o*-Methyl-flavinanthine has been reported to have a morphine-like inhibitory action on the peristaltic reflex of the guinea pig's isolated ileum and on the contractions of the guinea pig's ileum obtained by coaxial electrical stimulation (Gyang *et al.*, 1964; Gyang and Kosteilitz, 1966). The depressant effect was dose-dependent and the dose-response was more reproducible with the alkaloid than with morphine. The effect was not antagonized by nalorphine (Noamesi and Gyang, 1980). *o*-Methyl-flavinanthine has been shown to have an analgesic activity equivalent to one-fifth that of morphine. Reduction of the ketonic group to an alcohol may increase the analgesic action (Tackie *et al.*, 1974c).

Table 3.4. *Plants with depressant action on the ANS and CNS which are described in Chapter 2*

Plant Family	Active constituent	Part used	Action on nervous system	References
Rauvolfia vomitoria Apocynaceae	Reserpine	Rootbark	Sedative, tranquillizing	See Chapter 2
Argemone mexicana Papaveraceae	Protopine, berberine	Leaves, stems, seeds	Antispasmodic, sedative	See Chapter 2
Holarrhena floribunda Apocynaceae	Conessine	Root- and stembark	Sedative, tranquillizing, CNS depressant, antipyretic, analgesic	See Chapter 2 and Quevauviller and Blanpin (1960); Goutarel (1964)

Jateorhiza macrantha (Hook.) Exell & Mendonça MENISPERMACEAE

L This species is closely related to *J. palmata* (Calumba root) which is naturalized in Ghana and locally used as a bitter tonic and in the treatment of dysentery in Indian (Oliver-Bever, 1968).

C The root contains, in addition to colombin, related substances such as chasmanthin and palmarin. It also contains 2–3% of alkaloids of the berberine group: colombamine, jateorhizine and palmatine (Chopra *et al.*, 1956; Barton and Elad, 1956, 1962).

P The alkaloids colombamine, jateorhizine and palmatine depress the CNS, and when injected intravenously are hypotensive in the frog. Palmatine, the most active, acts mainly on the respiratory system and the blood pressure. The two others increase the intestinal tonus (see *Tinospora bakis*) (Paris and Beauquesne, 1938; Henry, 1949). Calumba root (*J. palmata*), used as a stimulant for patients with atonic dyspepsia, contains no tannins, so it can be prescribed with iron salts. However, its use has been mainly limited to veterinary medicine (Martindale, 1958) as it has occasionally been reported to produce toxic phenomena such as vomiting, paralysis of the CNS and depression of the respiratory centre (Paris and Moyse, 1967, p. 180).

Tinospora bakis (A. Rich.) Miers syn. (*Cocculus bakis* A. Rich.)
MENISPERMACEAE

L The bitter roots were sold in Senegal as a remedy for remittent and 'bilious' fevers and also as an emmenagogue and cholagogue. In India, the plant was in the Indian Pharmaceutical Codex as a bitter tonic and the root as an antidysenteric. The plant has been called 'Indian quinine' (Dalziel, 1937).

C The roots contain palmatine and 2–3% of columbin (Beauquesne, 1938). Different bitter heterosides (picroretin = picroretinoside, tinosporide and cordifolin) and the glycosides giloin and giloinin have been reported to be present in *T. cordifolia* in India (Paris and Beauquesne, 1938; Paris and Moyse, 1963).

P The root can produce toxic effects: vomiting and depression of the respiratory centre. Columbin, in small doses, increases the secretion of the bile and of the glands of the stomach and intestines; at higher doses it produces fatty degeneration of the liver (Biberfeld, 1910). Toxicity trials on the total alkaloids have shown that 5 mg/kg are not toxic for guinea pigs whilst 100 mg/kg produce death within 20 min (without convulsions). In the chloralized dog, an injection into the internal saphenous vein of 10 mg of total alkaloids produces immediate hypotension followed by recovery to normal within minutes. In experimental hyperthermia in guinea pigs, the temperature is lowered in a more spectacular way than with quinine sulphate. Palmatine shows a stronger antipyretic effect than the total alkaloids and it lowers the blood pressure and depresses the CNS (frog and mammals). It paralyses the respiratory centre even more than morphine (Beauquesne, 1938; Paris and Beauquesne, 1938). It is supposed that the antipyretic effect is due, like that of berberine, to paralysis of the peripheral vessels and to the resulting heat dispersion, and not to its toxicity towards microorganisms (Kerharo and Adam, 1974, p. 556). *T. cordifolia* has also been found to have hypoglycaemic and diuretic effects (Namjoshi, 1955).

Khaya senegalensis (Desr.) Juss. syn. (*Swietenia senegalensis* Desr.) MELIACEAE
Dry zone mahogany, Cail cedrat.

L In Nigeria, Senegal and Guinea the bark is used locally mainly as an antipyretic and tonic. Moreover, medicinal and veterinary uses of the bark as an anthelmintic (taenia), an emmenagogue (abortifacient) and as an emetic have also frequently been reported (Dalziel, 1937).

C The bark contains a bitter principle, first called 'calicedrin', which later proved to be a mixture of different components (Moyse-Mignon, 1942). On further investigation these were found to consist of several triterpenoids with a lactone or epoxide function and a furan ring. 6-Desoxy-3-destigloyl-swietenin and its acetates have been isolated from all parts of the plant. The bark, in addition, contains nimbosterol (β-sitosterol), 7-diacetyl-7-oxo-gedunin, methyl angolensate, methyl-6-hydroxyangolensate and 6-desoxy-3β-destigloyl-12β-diacetoxy-swietenin. In the rootbark, methyl-6-hydroxy-angolensate has been reported, and in the root, khayasine. In the heartwood, khayasine and derivatives, methyl-angolensate and 7-diacetyl-oxogedunin have been found. In the seeds, khivorine- and swietenine-derivatives have been found (Bevan *et al.*, 1963, 1965; Adesogan *et al.*, 1967; Adesogan, 1968; Adesogan and Taylor, 1968). Kerharo and Adam (1974) suggest the adoption of a new nomenclature based on a common nucleus, which can be named methyl-meliacate, more so as the composition varies in samples of different regions.

Similar bitter principles were found in **K. ivorensis** Chev. (*K. klainei* Pierre ex Pellegr., *K. caudata* Stapf ex Hutch. & Dalz.) (Adosogan & Taylor, 1970; Aspinal and Bhattacharjee, 1970). Further chemical investigation of stembark extracts of the two species by spectral, analytical and chromatographic methods led to identification of the sterols campestrol, stigmasterol and sitosterol and of the coumarins aesculetin and scopoletin. *K. ivorensis* was found to contain mainly scopoletin and scoparone and only traces of aesculetin and umbelliferone whilst in *K. senegalensis* scopoletin was the major coumarin next to aesculetin and only traces of scoparone were found. Scopoletin could also be isolated from the fruit (Adesina, 1983).

P Subcutaneous or intraperitoneal injections of calicedrin (0.05 g/kg) produce a distinct hypothermic effect (temperature reduction of 2–3°C) on experimental hyperthermia in guinea pigs (Moyse-Mignon, 1942; El Said *et al.*, 1968). In dogs, calicedrin produces slight hypertension. An antibiotic action of an aqueous extract of the stems against *Sarcina lutea* and *Staphylococcus aureus* was reported, and in a dilution of 1:10 000 calicedrin kills *Paramecia* within 20 min (Malcolm and Sofowora, 1969).

Crude hydroethanolic extracts of the stembark of *K. ivorensis* and *K. senegalensis* caused depression, sedation and reduced locomotor activity in mice. The coumarins scopoletine and scoparone have recently been found to have antipyretic, analgesic and anticonvulsant activities (Adesina and Ette, 1982; Adesina, 1983; Ojewole, 1983b).

Andira inermis (Wright) DC. FABACEAE
Dog almond or wormback

L The bark has been used in Northern Nigeria for trial by ordeal in the same way as

Erythrophleum. It is a dangerous poison in large doses, causing vomiting with drastic purgation, delirium and narcosis (Ainslie, 1937; Dalziel, 1937). In Senegal, the roots are used as an anthelmintic and in the treatment of mental diseases (Kerharo and Adam, 1974).

C The bark contains andirine, *N*-methyltyrosine (β-(*p*-hydroxyphenyl)-α-*N*-methylaminopropionic acid). Extraction of the heartwood produced an isoflavonoid as well as biochanine A (a trihydroxy-isoflavone), small amounts of fatty acids with long ramified chains and β-sitosterol (Cocker *et al.*, 1962, 1965). γ-Aminobutyric acid was found in extracts of leaves and stems (Durand *et al.*, 1962).

P The bark extracts are considered narcotic (Watt, 1967) and also have some insecticidal action (Heal and Rogers, 1950).

Nymphaea lotus L. syn. (*N. liberiensis* Chev. and other spp.) NYMPHAEACEAE

L A narcotic use of the rhizomes existed in ancient Egyptian and Mayan rituals (Oliver-Bever, 1961; Emboden, 1981). In West Africa, the rhizome is a food of scarcity. The seeds are used by the Hausas (Northern Nigeria) in eruptive fevers. In Sierra Leone, an eye lotion is prepared from the leaves, an infusion of stems and roots is used as an emollient and diuretic, and a decoction of the flowers as a narcotic and sedative. In the Ivory Coast a decoction of *Nymphaea* is taken for coughs and bronchitis (Oliver, 1960).

C A number of alkaloids have been recorded from the flowers and rhizomes of different *Nymphaea* spp. The chief alkaloids are nymphaeine, nymphaline, nupharine and α- and β-nupharidine. Quercitin has been reported to be present in the leaves (Chopra *et al.*, 1956; Hegnauer, 1962–68, Vol. V, p. 441). Delphaut and Balansard (1941) observed confusion concerning the reported constituents and their botanical origin. They studied 'nupharine' from *N. alba* L. (not growing in West Africa) and this was found to consist of nelombine, nupharidine, nymphaeine and α-nupharidine.

P The rhizomes of *N. alba* were tested on mice, eels and dogs, and in all cases there was a narcosis that terminated in somnolence (Delphaut and Balansard, 1941). Convulsions induced by strychnine could be counteracted by *N. alba* rhizomes, and these were characterized as antispasmodic and sedative by Delphaut and Balansard (1943). These workers found that nymphaeine is mainly localized in the rhizomes of *N. lotus*; its minimum effective and lethal doses for frogs are 30 and 50 mg/kg, respectively, and in mice and pigeon 60 and 80 mg/kg, respectively; warm-blooded animals die from central respiratory paralysis. Nymphaline, found in the flowers, acts as a cardiac glycoside. Nupharine, also found in the flowers, produces paralysis of the cerebrum when administered to frogs, mice, rats, guinea pigs and pigeons; it acts also as a respiratory excitant and causes death by respiratory poisoning. Two un-named alkaloids found in the flowers and roots show sedative action in small doses (Chopra *et al.*, 1956, p. 177). More recently the pharmacology of a number of *Nymphaea* alkaloids was studied by Dimitrov (1965), who reported sedative, spasmolytic and hypertensive properties. *N. tuberosa* is very active against *Mycobacterium smegmatis* (Su *et al.*, 1975).

Anogeissus leiocarpus (DC.) Guill. & Perr. syn. (*Conocarpus leiocarpus* DC., *A. schimperi* Hochst. ex Hutch. & Dalz., *A. leiocarpus* var. *Schimperi* (Hochst. ex Hutch. and Dalz.) Aubrev.) COMBRETACEAE

L A decoction of the leaves is used in Nigeria for ablutions in the treatment of skin diseases and itch and is considered to be an antidiarrhoetic. The powdered bark is applied to wounds and ulcers and in some regions an infusion of the bark is given as a febrifuge and the bark or seeds as a vermifuge (mainly for tapeworm in horses and donkeys) (Dalziel, 1937).

C The leaves, roots and bark contain 17% tannins. In the gum exuding from the trunk 20% uronic acids have been found. These produced, via hydrolysis, 12% (+)-xylose, 32% (−)-arabinose, 5% (+)-galactose, 2% (+)-mannose, 20% of oligosaccharides and traces of rhamnose, ribose and fucose (Aspinal and Christensen, 1961). The stembark of *A. latifolia* grown in India contains sitosterol, flavellagic acid, 3,3′,4-tri-*O*-methyl ellagic acid, quercitin, myrecitin and procyanidin along with gallotannins, shikimic acid, quinic acid and free sugars. It also contains alanine and phenylalanine (Bhakuni *et al.*, 1970).

P *A. leiocarpus* appears to contain a non-toxic CNS depressant principle (Fong *et al.*, 1972). The related *A. latifolia* Wall. has been found to have CNS depressant action to counteract an amphetamine hyperactivity test in mice (Bhakuni *et al.*, 1969b).

Elaeocarpus sphaericus Schum. syn. (*E. ganitrus* Roxb.) TILIACEAE

L In traditional Indian medicine the fruits of the plants were used in mental diseases, epilepsy, hypertension, asthma and liver diseases (Bhattacharya *et al.*, 1975b). The plant has been introduced in West Africa.

C A fixed oil has been obtained from the seeds of *E. serratus* L. (Chopra *et al.*, 1956).

P Bhattacharya *et al.* (1975b) noticed a prominent CNS-depressant effect of the water-soluble portion of the 90% ethanol extract of the fruit. The effect was characterized by potentiation of hexobarbitone hypnosis and morphine analgesia and by anticonvulsant and anti-amphetamine activities. In addition, cardiostimulant, smooth muscle relaxant and choleretic activities were reported. These effects were partly based on a direct musculotropic effect or were mediated through β-adrenoreceptor stimulation (Bhattacharya *et al.*, 1975b).

Passiflora foetida L. PASSIFLORACEAE
P. edulis Sims and related spp.
Stinking passion flower

L *P. edulis* and related spp. are cultivated in West Africa for their fruits. *P. foetida* has fruits which are edible when ripe, but before maturity the leaves and green fruit contain a cyanogenetic glucoside (Dalziel, 1937). A decoction of the leaves and roots is regarded in the Antilles as an emmenagogue and useful remedy in hysteria. The leaves were applied as a dressing for wounds (Sébire, 1899). In India, a decoction of the leaves is used in the treatment of biliousness and asthma (Chopra *et al.*, 1956).

C From the aerial parts of *P. incarnata*, Neu (1954, 1956) isolated indole derivatives and identified harmane. Traces of harmine and harmol were reported by Lutomsky and Wrocinski (1960) who, in addition, isolated flavonic derivatives by paper chromatography (Paris, 1963; Paris and Moyse, 1967, p. 458).

P *P. incarnata* was registered in the French Pharmacopoeia, 1937, for its sedative and antispasmodic properties. The extract of the plant shows antispasmodic activity on the rabbit intestine and is synergistic to papaverine and antagonistic to barium chloride and pilocarpine-induced effects. It decreases motility in mice and rats (Paris, 1963; Paris and Moyse, 1963).

Guiera senegalensis Gmell. COMBRETACEAE

L The dried pounded leaves are taken by women after childbirth to increase lactation and as a general tonic and blood restorer after any exhausting conditions. In Sokoto, the plant has the reputation of preventing leprosy and the leaves are applied externally for skin diseases. In Bornu, the roots are powdered, boiled and used as a remedy for diarrhoea and dysentery (Dalziel, 1937). Certain tribes in Senegal and Guinea use the leaves like those of certain species of *Combretum* for the treatment of colds, bronchitis and fever (Dalziel, 1937).

C For the first time, indole alkaloids (harmine and tetrahydroharmine) have been found in a member of the Combretaceae. They are present in the roots (Combier *et al.*, 1977). The ash of roots and leaves is rich in mineral elements. Mucilage, gallic and catechuic tannins, flavonoids, amino acids and alkaloids (0.2% in the roots and 0.15% in the leaves) have also been reported (Koumaré *et al.*, 1968).

P Plants collected in Senegal and Mali showed a depressive action on the CNS. In addition, anti-inflammatory and antitussive actions (mainly of the leaves) were noted; an antidiarrhoeic effect was particularly spectacular in rats infected with parasites. Kerharo *et al.* (1948) report the spectacular effects of the treatment with *G. senegalensis* in an epidemic of choleriform diarrhoea with gallbladder infection in Upper Volta.

Alstonia boonei de Wild. syn. (*A. congensis* Chev. & Aubrev.) APOCYNACEAE
Pattern wood, stool wood

L In Nigeria the bark of *A. boonei*, which is the most common variety of *Alstonia* in the country, is widely used as an antipyretic in the treatment of malaria, and sometimes, together with the leaves and roots, in external applications for rheumatic pains. Smearing the latex on Calabar swellings caused by filaria has also been recommended. In Ghana a decoction of the bark is given after childbirth to help the delivery of the placenta (Dalziel, 1937; Irvine, 1961).

C The bark of *A. boonei* contains echitamine (the main alkaloid), two echitamidine derivatives and a lactone boonein (Marini Bettolo *et al.*, 1983). The yield of total alkaloids varies with the location: in Ghana, it is 0.38–0.56%; in Nigeria, 0.18–0.31%; in the Cameroons, 0.11% (Goodson and Henry, 1925; Monseur and van Bever, 1955). The triterpenes β-amyrin and lupeol have been reported in the bark and ursolic acid has been found in the leaves of *A. boonei* (Kučera *et al.*, 1972, 1973). In the flowers of an Indian species, *A. scholaris*, a number of indole alkaloids have been reported, the major one being picrinine, which was found to possess a CNS depressant action in rats and mice (Dutta *et al.*, 1976).

P The reputation of *Alstonia* spp. as antimalarial agents was such that *A. scholaris* and *A. constricta* bark had formerly been included in the British Pharmacopoeia, 1914. The antimalarial action could not be confirmed in many tests carried out by numerous authors on birds, monkeys and human beings (Henry, 1949, p. 720). It was noted, however, by some that the bark of *A. scholaris* produced a fall of temperature in human patients and while this lasted the patients appeared relatively free from symptoms (Chopra *et al.*, 1938). This antipyretic action and the fact that these drugs were used in days when the diagnosis of malaria was not always accurate may account for doubtful results and earlier use (Henry, 1949, p. 720).

Echitamine was reported to lower carotid pressure and increase the renal output (Raymond-Hamet, 1934, 1941; Esdorn, 1961). More recently, Kučera *et al.* (1973) and Marquis (1975) observed that echitamine causes a fall of the blood pressure in hypertensive cats. Later, however, Marquis and Ojewole (1976) noticed that the hypotensive effect occurred only occasionally after a first intravenous injection of 6 mg/kg and thought that the diuretic action on saline-loaded dogs and cats may explain this hypotensive action. Ursolic acid, at first considered an inert compound, was found to act on the electrolytic balance. Doses of 3 mg produced a sodium retention equivalent to that of 3 μg DOCA (desoxy-cortisone) and a considerably higher potassium retention in adrenalectomized rats (Wenzel and Koff, 1956; Marquis and Ojewole, 1976). These authors also reported that echitamine in particular potentiated the barbiturate sleeping time of mice and rats and enhanced the lethality of strychnine. Echitamine contracts the isolated toad rectus abdominis preparation and its action was enhanced by increasing concentrations of acetylcholine and reversed by physostigmine on the isolated rat hemi-diaphragm. This action could be an undesired side-effect in native malaria treatment with decoctions or infusions of the bark although the plant has been said to have a relatively low toxicity. The ineffectiveness of echitamine treatment against a strain of *Plasmodium berghei* was noted (Marquis and Ojewole, 1976).

The spasmolytic and hypotensive actions of echitamine have recently been confirmed by Ojewole (1983a). The alkaloid also blocked the neuromuscular transmission in various muscle–nerve preparations examined. The author was able to show that the depressor effect is unlikely to be mediated via a cholinergic mechanism or histamine H_1-receptor stimulation.

In Indian *Alstonia* spp. several alkaloids with a CNS action have been reported, thus picrinine was said to potentiate hexobarbitone hypnosis and morphine analgesia (Dutta *et al.*, 1976), and alstovenine from *A. venenata* (not in West Africa) was in low doses a MAOI and in high doses a marked CNS stimulant. Venenatine, also from *A. venenata*, has, on the contrary, a reserpine-like action. It potentiates the hexobarbital sleeping time, antagonizes amphetamine toxicity, morphine analgesia and the anticonvulsant action of diphenylhydantoin, is synergistic with reserpine and can reduce the pressor response to tyramine but not to adrenaline (Bhattacharya *et al.*, 1975a).

Waltheria indica L. syn. (*W. americana* L.) STERCULIACEAE

L In northern Nigeria and in Togo a decoction of the root is frequently given to children to strengthen their resistance against fever, etc., and the Hausas also recommend a root decoction to produce immunity. In Togo and also in India the root is used in a cough medicine, and in Senegal it is used for the healing of wounds. In Cayor (Senegal) it is sometimes used as an antiepileptic.

C An original analysis of the plant only revealed mucilage, tannins and sugars (Dalziel, 1937). Later, unidentified alkaloids were reported to be present in the leaves and rootbark. Finally, three alkaloids were isolated from whole plants from the Ivory Coast. These were called adouétines x, y and z. These alkaloids are of a particular type, and were named 'basic peptides' by Goutarel. Of the four nitrogen atoms in the molecule only one is basic, the three others are present in peptide linkages (Pais *et al.*, 1963).

P Pharmacological studies have been carried out with the amidosulphonate of adouétine z. The LD_{50} in mice is 52.5 mg/kg. The drug behaves as a sedative of the CNS and a stimulant of the medulla. In the dog it produces hypertension slows down the heartbeat (compensatory reflex?) and has a relaxing action on the smooth muscle fibres of the intestine (Blanpin *et al.*, 1963).

Piper guineense Schum. & Thonn. syn. (*P. leonense* DC., *P. famechonii* DC.)
PIPERACEAE

West African black pepper or Ashanti pepper

L The black berries are a much-used spice. The pepper is used externally as a counter-irritant or in a stimulating ointment, and internally as a stomachic and carminative. The pulverized grains are useful as an insecticide.

C Chromatographic analysis of the fruits has revealed the amides piperine, *N*-isobutyloctadeca-*trans*-2-*trans*-4-dienamide, sylvatine, $\Delta\alpha,\beta$-dihydro-piperine, trichostachine and a new naturally occurring amide, $\Delta\alpha,\beta$-dihydro-piperlonguminine (Addae-Mensah *et al.*, 1977a, b). In the roots, piperine, trichostachine, and in the leaves, dihydrocubebin, a new naturally occurring lignan, have been reported (Dwuma Badu, 1975d). Earlier, 0.2% of a lignan derived from shikimic acid, aschantine and another lignan, which has been named yangambine, had been reported by Hänzel *et al.* (1966). An essential oil composed of terpenes (phellandrene, pinene, limonene has been obtained from the berries (1–2.4%) (Dwuma Badu *et al.*, 1975d, 1976a; Tackie *et al.*, 1975a; Raina *et al.*, 1976).

P Some of the constituents have been reported to have antimicrobial, anticonvulsant, antihypertensive, sedative, tranquillizing and insecticidal properties (Fong, 1972; Addae-Mensah *et al.*, 1977a, b). Small quantities act as a gastric stimulant and carminative, increase the flow of saliva and gastric juice, have diuretic and diaphoretic properties and act as a nervous stimulant. They also have bactericial and insecticidal action. In high doses, they are irritating to the skin and mucosae and can produce convulsions and haematuria (Paris and Moyse, 1967, Vol. II, p. 113). A derivative based on piperine, isolated from *P. nigrum* seeds is used in Chinese medicine as an antiepileptic (Xiao, 1983).

Anacardium occidentale L. ANACARDIACEAE

(See also Chapter 2.)

P An essential oil obtained by steam distillation from the leaves of *A. occidentale* produced in rats, in doses of 150–300 mg/kg of a 5% oil emulsion given intraperitoneally, a dose-related decrease of spontaneous motor activity and potentiated sodium pentobarbitone-induced hypnosis. Rota rod performance was decreased and further investigations suggested a CNS depressive action of tranquillizer type similar to, but lower than that of chlorpromazine. The essential oil possesses, however, an additional analgesic action (Carg and Casera, 1984).

Sedation, decreased spontaneous motor activity, loss of muscle tone, potentiation of barbitone sleeping time and ether anaesthesia were also seen with the xanthones from *Calophyllum inophyllum* Guttiferae (see Chapter 5).

III Peripherally acting depressants of the CNS

(a) Local anaesthetics

Local anaesthesia is a selective inhibition of conduction in the afferent or sensory nerves and endings resulting in the loss of the sensations of pain, pressure and temperature in localized areas of the body, especially the skin and mucous membranes. Local anaesthetics may act by preventing the liberation of acetylcholine from the preganglionic nerve endings thus blocking nerve conduction when applied locally. In low concentrations local anaesthetics mainly prevent the generation and production of nerve impulses. Their site of action is the cell membrane which they depolarize in changing the permeability to potassium and sodium ions. They can produce a depressant effect on the heart and a relaxation of the smooth muscles, which can be explained by the ganglioplegic action but can also be due to a direct stabilizing effect on the axonic membrane (Lechat *et al.*, 1978, pp. 582–3). Local anaesthetics are broken down in the liver to non-toxic constituents. Overdoses may lead to tremor, restlessness and convulsions (Burgen and Mitchell, 1972; Turner and Richens, 1978).

The local anaesthetic activity of cocaine from *Erythroxylum coca* (Erythroxylaceae) has long been known. Local anaesthetic action has also been recorded in *Cassia absus* (Caesalpiniaceae) through chaksine and isochaksine. The majority of the other plants with local anaesthetic action seem to act through the ANS. The local anaesthetic activity of *Jatropha podagrica* is attributed to tetramethylpyrazine and that of *Erythrophleum guineense* to cassaine. Indole alkaloids with local anaesthetic action have been reported in *Mitragyna* spp. (Rubiaceae) (mitraphylline), *Pausinystalia johimbe* (Rubiaceae) (yohimbine) and *Voacanga africana* (Apocynaceae) (voacangine) and a local anaesthetic steroid alkaloid has been noted in *Picralima nitida* (Apocynaceae).

Erythroxylum coca Lam. ERYTHROXYLACEAE

Cocaine plant

L Cultivated in West Africa (Hutchinson and Dalziel, 1954, p. 356), the astringent leaves are used locally in India as a stimulant and masticatory (Chopra *et al.*, 1956, p. 111).

C The leaves contain the alkaloid cocaine which is also present in the bark and seeds. In India, the leaves contain 0.4–0.8% of total alkaloids, largely cocaine (methylbenzoylecgonine), but also other pseudotropanol derivatives such as cinnamylcocaine, truxillines and tropacocaine (benzoylpseudotropanol), as well as some monocyclic *N*-methylpyrrolidine derivatives (Henry, 1949, pp. 93–104; Paris and Moyse, 1967, p. 283–4).

P Cocaine has pharmacological actions on the nervous and cardiovascular systems similar to those of other local anaesthetics but it blocks the uptake of catecholamines into nerve terminals and so has sympathomimetic properties. It produces surface anaesthesia on the eye and mydriasis. Despite vasoconstrictive properties it is readily absorbed from mucous membranes and is used for anaesthesia of respiratory passages (bronchoscopy) but more suitable drugs are now available. Cocaine stimulates the CNS and has been used as a stimulant in neurasthenia but must be given under strict medical control as it is habit-forming. It produces a short spell of intellectual stimulation and euphoria followed by depression. Large doses cause convulsions followed by central paralysis and finally by failure of respiration (Burgen and Mitchell, 1972; Turner and Richens, 1978).

When *E. coca* leaves (or powder) (5–10 g) are taken orally by human subjects cocaine is immediately detected in the blood by gas chromatographic mass spectrometry. It reaches peak concentrations after 40 min to 1 h and persists in the plasma for more than 7 h (Holmstedt *et al.*, 1979).

Cassia absus L. CAESALPINIACEAE

Four-leaved senna

L The seeds are used in West Africa as a fomentation in ophthalmias and are also used to treat ringworm infections. The leaves are used in Northern Nigeria and in Togo as a dressing for ulcers and swellings believed to be of venereal origin. In India the leaves are employed to treat asthma and the seeds are used for the treatment of ringworm and ophthalmias (Chopra *et al.*, 1956; Oliver, 1960).

C The seeds contain a fixed oil and a toxalbumin absin, similar to abrin from *Abrus precatorius*, as well as two alkaloids, chaksine and isochaksine. β-Sitosterol and β-sitosterol glucoside are found in the seed oil (Oliver, 1960; Qureshi *et al.*, 1964). Chrysophanol, aloe emodin, chaksine and isochaksine have been isolated from the roots and in addition to the two alkaloids, quercitin and rutin have been found in the leaves (Siddiqui and Ahmad, 1935; Oliver, 1960; Krishna Rao *et al.*, 1979).

P The pharmacology of chaksine and isochaksine has been extensively studied by Pradhan *et al.* (1953), Bukhari and Khan (1963) and Khan *et al.* (1963). Both alkaloids proved to have a local anaesthetic action on guinea pig skin when administered intradermally. The action is inferior to that of procaine, which proved to be 3.6 times more active than chaksine and 1.7 times more active than isochaksine. The anaesthetic action was confirmed in Man. By intradermal injection they produce histamine-like reactions. Chaksine and isochaksine also have distinct hypotensive and depressant effects on the parasympathetic nerve terminals of the bronchi, intestines and bladder (an action comparable to atropine) and also have a ganglioplegic and curariform action, isochaksine being generally somewhat less active than chaksine. Thus, both have a general depressive action on the CNS and the neuromuscular junctions (Pradhan *et al.*, 1953). Strong anti-5-hydroxytryptamine action has also been reported. The LD_{50} for chaksine given perorally to mice was around 70 mg/kg; in frogs it was 100 mg/kg. Chaksine and isochaksine also have an antibacterial action (Gupta and Chopra, 1953). Chaksine inhibits the growth of *Staphylococcus aureus* and of *Bacillus haemolyticus* at dilutions of 1 : 100 000 (Cheema and Priddle, 1965).

Jatropha podagrica Hook. EUPHORBIACEAE

L A native of Central America, this species is much cultivated in West Africa. The local medicinal uses in Ghana and Nigeria are as an antipyretic, diuretic, choleretic and purgative; stems and roots are used as chewing sticks (Irvine, 1961). *J. curcas* seed oil is used in local medicine in dropsy, sciatica, paralysis, worms and skin diseases (Oliver, 1960).

C Tetramethylpyrazine has been obtained from the stem of *J. podagrica*. This substance had formerly been reported to be present in fermented soya beans, cocoa beans and tobacco smoke (Odebiji, 1978). In *J. curcas* a toxalbumin (curcin) and small quantities of glycosides have been reported (Chapter 2, this volume).

P Tetramethylpyrazine demonstrated antibacterial activity (Odebiji, 1978). In anaesthetized cats it produced depressor effects, reduced the heart rate and blocked neuromuscular transmissions and appeared to have a spasmolytic activity on smooth muscles (Ojewole, 1980; Ojewole and Odebiji, 1980). Further studies confirmed blockage of adrenergic and cholinergic transmission by tetramethylpyrazine. The compound depressed and abolished the electrically evoked contractions of the chick oesophagus, rabbit duodenum and guinea pig vas deferens in vitro. It also inhibited the electrically induced contraction of the rat isolated hemi-diaphragm and of the cat's nictating membrane in vivo. Apart from its possible central effects, and those on the cardiac muscle and blood vessels, it could be suggested, from the results obtained in this study, that the hypotensive effect in experimental animals is likely to be contributed to by, or mediated via, its local anaesthetic (membrane stabilizing) activity. Through this action, the drug probably acts to block sympathetic and parasympathetic neurones and ganglia (Ojewole, 1981).

Tetramethylpyrazine has a number of other pharmacological actions. A main central effect was found to be tranquillization and sedation (Ojewole and Odebiji, 1984).

In China, tetramethylpyrazine originating from plants is used in the treatment of occlusive cerebral vessel diseases such as cerebral embolism (Xiao, 1983).

Plants having local anaesthetic action which are described in Chapter 2 (The cardiovascular system) are listed in Table 3.5.

(b) Neuromuscular blockers (curare action) and anticonvulsants

Plants which act on the neuromuscular junctions do so through the curare alkaloids. When introduced into the bloodstream the curare alkaloids act by interrupting the transmission of the nerve impulse at the neuromuscular junctions thus producing a profound and progressive paralysis of the voluntary movements. Continued administration leads to paralysis and finally death through paralysis of

Table 3.5. *Plants with local anaesthetic action which are described in Chapter 2*

Plant Family	Active constituent(s)	Part(s) used	Action on nervous system	References
Picralima nitida Apocynaceae	Akuammine, akuammidine	Stem- and rootbark	Local anaesthetic: stembark equivalent to cocaine, rootbark threefold that of cocaine hydrochloride	Adrenergic action: Chapter 2
Voacanga africana Apocynaceae	Voacangine	Stem- and rootbark	Local anaesthetic and analgesic	Adrenergic action: Chapter 2
Pausinystalia johimbe Rubiaceae	Yohimbine	Bark	Local anaesthetic	Adrenergic action: Chapter 2
Corynanthe pachyceras Rubiaceae	Corynanthine	Bark	Mild local anaesthetic	Adrenergic action: Chapter 2
Mitragyna stipulosa Rubiaceae	Mitraphylline, rhynchophylline	Rootbark	Local anaesthetic	CNS

the diaphragm. Many curare alkaloids have to be injected as they are not absorbed from the intestine and the injection produces a short-lasting reversible effect as the concentration of the alkaloids in the plasma is reduced by half every 13 min. A secondary effect of the curare alkaloids is the lowering of the arterial blood pressure, caused in some cases through the liberation of histamine and in others by ganglionic blockade. In small doses the curare alkaloids provide the muscular relaxation needed in abdominal and thoracic operations, in endoscopy and in painful spasmodic conditions found in tetanus and strychnine poisoning (Burgen and Mitchell, 1972; Lechat *et al.*, 1978). The relaxation spreads down from the head to the abdomen and members and affects the diaphragm last. However, on complete relaxation of the abdominal muscles already 50% of the respiratory muscles are involved. Hence the use of these blockers requires ventilation and precautions (Lechat *et al.*, 1978).

Curare-like compounds act as antagonists to acetylcholine by competing for the acetylcholine receptors, but other forms of curare can be acetylcholine-mimetic occupying the actual site of acetylcholine on the neuromuscular junctions and these forms are slowly destroyed by cholinesterase (Lechat *et al.*, 1978).

Some of the South American, Japanese and Indian species of the Menispermaceae are the chief sources of curare alkaloids. They have been used as arrow or hunting poisons and were prepared in pots or bamboo tubes according to the region, as opposed to *Strychnos* curare, which was prepared in calabashes and is more toxic. The main alkaloids of curare are asymmetric bis-benzylisoquinoline (or bis-coclaurine) alkaloids such as (+)-tubocurarine, stereoisomeric (−)-chondrodendrine, ((−)-bebeerine = buxine = pelosine) and also (+)- and (−)-isochondrodendrine, chondrofoline and oxyacanthine. Of all these, (+)-tubocurarine appears to be the most important and the South American *Chondrodendrons* remain its main source.

Most of the West African members of the Tricliseae and Cocculeae tribes contain curare alkaloids derived from bis-benzylisoquinoline, although not in large amounts. *Cissampelos owariensis*, *C. mucronata*, *Cocculus pendulus*, *Tiliacora dinklagei*, *Triclisia dictyophylla* and *Epinetrum cordifolium* have all been shown to have neuromuscular blocking action.

Erythrina alkaloids, with the exception of erythroidine, have neuromuscular blocking and CNS depressant and smooth muscle relaxant activities. They have the advantage of being active if taken orally but their action is short-lasting as they are tertiary bases and lose part of their activity through transformation into quaternary bases (which are 10–12 times less active). Their toxicity is high, however, and they have therefore now been mostly replaced by other drugs. West African *Erythrina* spp. with active constituents are *E. senegalensis*, *E. excelsa* and *E. sigmoideae*. Curariform activity is also reported in two further Fabaceae, *Mucuna pruriens* and *Desmodium gangeticum*.

Cissampelos mucronata A. Rich. syn. (*C. pareira* of F.T.A.) MENISPERMACEAE
Velvet leaf

Cissampelos owariensis Beauv. ex DC. syn. (*C. pareira* L. var. *owariensis* (Beauv. ex DC.) Oliv., *C. robertsonii* Exell.) (Fig. 3.3)
Pareira brava

L The roots of both species (which are often confused) are used locally as an emmenagogue, abortifacient, antipyretic and diuretic. A decoction of the leaves is used as a light purge. In India, the leaves are used as a local application for itch (Oliver-Bever, 1968). The roots of *C. owariensis* are known in commerce as *Pareira brava*.

C Flückiger and his co-workers isolated from the roots of *C. owariensis*, 0.5% of the alkaloid bebeerine, which was later separated into two stereoisomeric forms and a neutral crystalline substance, deyamitine (Oliver-Bever, 1968). The alkaloids of the West African *C. owariensis* were examined by Dwuma Badu *et al.* (1975a), who have reported dehydrodicentrine, dicentrine, cycleanine, insularine and isochondrodendrine. In Portuguese Africa, Ferreira *et al.* (1965) isolated from the roots of *C. mucronata*, isobebeerine ((+)-isochondrodendrine), hyatine ((±)-bebeerine) and hyatinine. In India, hyatine and hyatidine (±)-4′-*O*-methylbebeerine) and curine ((−)-bebeerine) were also found in the local *C. mucronata*, and in total eleven quaternary and five tertiary alkaloids were detected (Srivastava and Khare, 1964; Boissier *et al.*, 1965; Bhatnagar and Popli, 1967; Bhatnagar *et al.*, 1967; Roychoudhury, 1972).

P The curarizing activity of *C. mucronata* was tested on toad recto-abdominal striated muscle and on rat isolated nerve-diaphragm preparations of rats (Correia da Silva and Pavia, 1964). Some curarizing action was observed, which was greater in the methachloride and methiodide derivatives of hyatine. Hyatine methiodide was 2.5 times more active than (+)-tubocurarine. The activity of the hyatine derivatives on blood pressure and respiration was greatest in those with the highest curarizing

Fig. 3.3. *Cissampelos owariensis* Beauv. ex DC.

potency and decreased in parallel with the neuromuscular activity (Pradhan and De, 1959; Sen and Pradhan, 1963; Bhatnagar *et al.*, 1967, 1971; Basu, 1970). Another bis-benzyl isoquinoline alkaloid isolated from the bark of *C. mucronata*, cissampareine, was found to have significant and reproducible inhibitory action against human carcinoma cells of the nasopharynx in cell culture (Kupchan *et al.*, 1965).

Cocculus pendulus (J. & G. Forst.) Diels syn. (*C. leaeba* (Del) DC., *Epibaterium pendulum* J.R. & G. Forst.) MENISPERMACEAE

L The small red fruits are used by the Arabs to make an intoxicating drink. The roots of the plant are used in Nigeria as an antipyretic and in Senegal as a cholagogue and diuretic (Dalziel, 1937; Kerharo and Adam, 1974).

C The roots contain a bitter principle, colombin, and three alkaloids, pelosine (Chopra *et al.*, 1956), palmatine and sangoline (oxyacanthine) also found in *Berberis* (Beauquesne, 1938). Chemically, pelosine is believed to be identical with bebeerine or chondrodendrine, but slightly controversial opinions exist (Henry, 1949). Palmatine is a phenolic base belonging to the protoberberines whilst sangoline is a bis-benzylisoquinoline alkaloid. The stem and leaves contain, besides mineral elements such as potassium, sodium, magnesium, iron, aluminium, copper and zinc, two new bases, penduline of a biscoclaurine type and a bis-benzylisoquinoline alkaloid, an isomer of trilobine, cocsuline (Bhakuni *et al.*, 1970; Gupta *et al.*, 1970; Joshi *et al.*, 1974; Bhakuni and Joshi, 1975). In addition to the tertiary phenolic alkaloids, the rhizome of *Cocculus trilobus* has been found to contain cocculine and cocculidine (Wada *et al.*, 1967) and two erythrinan alkaloids, which were named coccutrine and dihydroerysovine (Ju-Ichi *et al.*, 1978). The Chinese drug Hang-Fang-Chi is obtained from *C. laurifolius*, *C. diversifolius* and *C. japonicus*, which have alkaloids related to those of *C. sarmentosus*, *C. trilobus* and *C. hirsutus*. An insecticidal alkaloid, cocculidine, has been obtained from *C. trilobus* (Oliver-Bever, 1968).

P Colombin and palmatine have strong stomachic and bitter tonic action, and *Cocculus* root has been used as a tonic in a similar way to *Chasmanthera* root. The root extract had a relaxant and antispasmodic effect on the rat ileum and stimulated the uterus of the albino rat. Toxicity was low; up to 40 mg/kg given intravenously and 100 mg/kg given orally produced no toxic effects in rabbits and albino rats. A solution of the alkaloids has a cardiotonic action which is five times that of the total extract and is comparable to that of 50–100 μg of digoxin (Das *et al.*, 1964). Sangoline produced long-lasting hypotension in dogs and, in doses of 0.1–0.2 g, clonic convulsions and death through paralysis of the respiratory centre in rabbits (Henry, 1949; Das *et al.*, 1964). Palmatine has similar paralysing action and is also strongly hypotensive. An extract of the stems and leaves of *Cocculus pendulus* has a distinct anticancer action. In sarcoma 180 it produces a 60% reduction, and in Lewin's lung carcinoma, 50% reduction, of the number of tumours compared to control animals (Abbot *et al.*, 1966). The extract also has an anticancer activity against human epidermal carcinoma of the nasopharynx in tissue culture (Bhakuni *et al.*, 1969b).

Tiliacora dinklagei Engl. syn. (*Glossopholis dinklagei* (Engl.) Stapf)
MENISPERMACEAE
Tiliacora funifera (Miers) Oliv. (Troupin includes *T. warnecki* Engl. ex Diels. and *T. johannis* Exell. in this species).

L Both plants have been used in local medicine in Ghana in the treatment of gastric fevers and menstrual irregularities (Tackie and Thomas, 1968).

C The major alkaloid obtained from the roots of *T. funifera* is the bis-benzyl-isoquinoline biphenyl base, funiferine. Isotetrandrine, thalrugosine (nortiliacorine A) and a new imino-bis-benzylisoquinoline alkaloid, tiliafunimine, have been isolated from the leaf extract (Tackie and Thomas, 1968; Tackie *et al.*, 1973b, c, 1975b; Ayim *et al.*, 1977; Dwuma Badu *et al.*, 1977). In *T. dinklagei* roots, in addition to nortiliacorine A and funiferine, tiliacorinine, tiliageine and dinklacorine (which later could be identified as a positional isomer of tiliacorine) have been found (Tackie *et al.*, 1974d, 1975b; Dwuma Badu *et al.*, 1976b, 1979).

P Funiferıne and nortiliacorine have weak antimalarial and antimicrobial activities, but their dimethiodides have a potent neuromuscular blockade action of the curare type (Tackie *et al.*, 1979). Using the isolated phrenic nerve diaphragm preparation of the rat, it has been shown that they are slightly less potent than (+)-tubocurarine in vitro but more potent than gallamine; this also applies to phaeanthine, a bis-benzylisoquinoline alkaloid from *Triclisia* (see below) (Boissier *et al.*, 1963; Ansa Asamoah and Gyang, 1967). Tackie *et al.* (1973b) noted that funiferine dimethiodide has a slightly higher activity than tubocurarine hydrochloride. Funiferine has been shown to have some activity in *P*-388 lymphocytic leukaemia and to inhibit the growth of the acid-fast bacillus *Mycobacterium smegmatis* (Geran *et al.*, 1972).

Triclisia dictyophylla Diels syn. (*T. gilletti* (de Wild.) Staner, *Tiliacora trichantha* Diels)
MENISPERMACEAE
Triclisia patens Oliv.
Triclisia subcordata Oliv. syn. (*Tiliacora subcordata* Chev.)

L The *Triclisia* have various local medicinal uses such as the treatment of oedema of the legs, anaemia, diarrhoea, 'joint pains' and in the case of *T. patens*, of malaria (Oliver, 1968).

C From leaf extracts of *T. dictyophylla*, the bis-benzylisoquinoline alkaloids cocsuline, isotetrandrine, stebisimine, gilletine (a menisarine-cocsuline type alkaloid), trigil-letimine, tricordatine and tridictyophylline have been isolated, as well as phaean-thine and *N*,*N*-dimethylphaeanthine (Dwuma Badu *et al.*, 1975b, Spiff *et al.*, 1981). In the leaves of *T. patens*, phaeanthine, *N*,*N*-dimethylphaeanthine, cocsuline, pycnamine and aromoline were found to be present, as well as an oxo-aporphine alkaloid, *O*-methylmoschatoline, which is also present in *T. dictyophylla* (Dwuma Badu *et al.*, 1975c). In *T. subcordata* the dimeric isoquinoline alkaloids fanchinoline, tricordatine, cocsuline, tetrandrine and phaeanthine have been reported (Kronlund *et al.*, 1970; Tackie *et al.*, 1973a, 1974a; Dwuma Badu *et al.*, 1975b, c, 1978; Sandberg, 1980; Spiff *et al.*, 1981).

P Phaeanthine has been shown to have one-ninth of the curarizing potency of

(+)-tubocurarine. The activity of the dimethiodide is greater and dimethylphaeanthine is also an effective skeletal muscle relaxant. Antitumour tests revealed a weak antitumour activity for phaeanthine, no activity for isotetrandrine and slight activity for fanchinoline, a 6-hydroxy analogue of tetrandrine. Methylation with diazomethane of fanchinoline produced tetrandrine with significant antitumour action (Guinaudeau *et al.*, 1975).

Epinetrum cordifolium Mangenot et Miège MENISPERMACEAE

L The local uses of this plant are the treatment of anaemia and of oedema of the legs (Debray, 1966; Debray *et al.*, 1966).

C In this species, which is also a member of the Tricliseae tribe, Debray *et al.* (1966) identified cycleanine (a dimethyl-*o*-*o*-isochondrodendrine), norcycleanine (a monomethylisochondrodendrine) and isochondrodendrine.

P The alkaloids of *E. cordifolium* have curare action, producing muscle relaxation. Cycleanine also has anti-inflammatory and analgesic properties (Debray, 1966; Debray *et al.*, 1966; Bouquet and Cavé, 1971; Guinaudeau *et al.*, 1975).

Erythrina senegalensis DC. FABACEAE
Erythrina vogelii Hook. f.
Erythrina excelsa Bak. syn. (*E. sereti* de Wild.)
Erythrina sigmoidea Hua syn. (*E. dybrowski* Hua, *E. erythrotricha* Harms)
Erythrina mildbraedii Harms syn. (*E. altissima* Chev.)

L In Northern Nigeria the bark of *Erythrina* spp. is chiefly used in the treatment of jaundice, as an infusion for gonorrhoea and as a diuretic for horses. In Ghana the bark is recommended as an emmenagogue, and in Guinea it is given to women after childbirth (Dalziel, 1937). In Senegal the bark is considered a remedy for dysentery and colitis (Kerharo and Adam, 1974).

C The first curarizing *Erythrina* alkaloid was obtained in 1937 by Folkers and Major from the seeds of *E. americana* (Folkers and Unna, 1939). It was named erythroidine and was subsequently shown to be a mixture of two isomers, α- and β-erythroidine (Folkers and Koniuszy, 1940; Folkers *et al.*, 1941). In a systematic investigation of 50 species of *Erythrina*, all were shown to contain alkaloids with curariform activity (Raven, 1974). Many of the species examined contained as their main alkaloids erysodine, erysovine (11-methoxyerysodine), erysotrine and erysopine. In addition, 11-hydroxy-erysodine and erythroline have been mentioned (over 30 alkaloids are known). A number of the alkaloids have phenolic groups, others such as erysothiopine and erysothiovine are in the form of sulpho-acetic esters (Paris and Moyse, 1969, p. 404; Hargreaves *et al.*, 1974) or of glycosides and are only liberated by hydrolysis. A variation in the alkaloid content of different samples of seeds of the same species collected at different times and locations has been noted (Barakat *et al.*, 1977). Alkaloids are present in the roots, stems, leaves, heartwood and flowers of most species but in smaller quantities than in the seeds (Krukoff, 1977; El Olemy *et al.*, 1978; Staunton, 1979). Games *et al.* (1974) indicated that in *E. senegalensis* erysodine represented 75% of the total alkaloids and oxo-erysodine 9%; in *E. excelsa*

48% of the total alkaloids is represented by erysodine and 35% by erysovine and in *E. sigmoidea* 43% of the total alkaloids is erysodine, 28% is erysovine and 21% is erythraline. The seeds from *E. mildbraedii* collected in Nigeria contain mainly erysodine, and to a lesser extent erythraline and erysopine (El Olemy *et al.*, 1978). In addition, secondary *Erythrina* alkaloids and in most species hypaphorine have also been reported (Games *et al.*, 1974).

P The curarizing action, based upon the grams of frog curarized per gram of seed, has been examined by Folkers and Unna (1939). This relationship was 20 000 for *E. senegalensis* and 44 000 for *E. sigmoidea*. Erythroidine and its more potent derivative, dihydroerythroidine, have been used as muscle relaxants in a number of clinical applications such as the control of convulsions, in the shock therapy of psychiatric patients and as an adjunct to general anaesthesia to produce the muscular relaxation desirable in some surgical operations (Folkers *et al.*, 1941; Bhattacharya *et al.*, 1971). The alkaloids have the advantage of being active if taken by mouth but are less active than curare. They are quaternary bases and their action is less lasting as they lose their activity through bio transformation to ternary bases. They have, however, now been replaced by other drugs because of their high toxicity. Neuromuscular blocking, smooth muscle relaxant, CNS depressant, hydrocholeretic and anticonvulsant effects are also described (Bhattacharya *et al.*, 1971; Ghosal *et al.*, 1972; El Olemy *et al.*, 1978). In general, it was reported that erysothiovine and erysothiopine, the sulpho-acetic esters of erysovine and erysopine, possess greater activity than the free alkaloids (El Olemy *et al.*, 1978).

Stephania dinklagei (Engl.) Diels syn. (*Cissampelos dinklagei* Engl.)

MENISPERMACEAE

L The roots are used by local West African healers as a sedative, an anthelmintic and an antimenorrhagic (Dalziel, 1937) and also in the treatment of infertility (Tackie *et al.*, 1974b).

C Analysis of the roots has revealed six alkaloids of the aporphine and isoquinoline group. Corydine (also found in several species of Papaveraceae and in the rootbark of some *Zanthoxylum* spp.) is the most abundant as 0.27% are found in the roots (Tackie *et al.*, 1974b). Dinklageine, norcorydine, stephanine, steporphine and stephalagine have also been reported and most of them identified by thin-layer chromatography (Paris and Le Men, 1955; Ray *et al.*, 1979a). Dinklageine shows many structural analogies with biscoclaurine (from *Cocculus laurifolius*) but has a higher melting point. In addition, *N*-methyl-glaucine and *N*-methylcorydine have been reported in *S. dinklagei* roots (Dwuma Badu et al., 1980).

P Another alkaloid fraction of *S. dinklagei* (possibly a mixture of isocorydine and decentrine) has been reported to be an excitant of the CNS, acting on the cerebral but not the medullary centres. On the ANS it has a peripheral and central sympathetic action (Quevauviller and Sarasin, 1967).

Corydine produces slight narcosis, slows down the heart and respiration rates and causes emesis in warm-blooded animals. As with other aporphines small doses have sedative and hypnotic action (Henry, 1949). *S. dinklagei* is said to be used as an antitussive (Tackie *et al.*, 1979). An infusion of the roots of *S. dinklagei* has distinct antispasmodic properties in the isolated rabbit intestine at concentrations of 1‰ and above. At 4‰, there is a drop of tone and arrest of peristalsis. In the intestine of the guinea pig contracted by acetylcholine or barium chloride, decontraction is immediate with a concentration of 5‰ (Paris and Le Men, 1955). Several Asiatic species of *Stephania* contain coclaurine and related alkaloids. Rotundine from *S. rotunda* was used as a narcotic and hypnotic in the Vietnam war. Rotundine and (+)-isocorydine are considered to be spasmolytic and analgesic in Chinese medicine (Xiao, 1983). From *S. cepharantha* Japanese workers isolated cepharanthine, used in Japan for the treatment and prophylaxis of tuberculosis and leprosy (Büchi, 1945). Cycleanine from *S. glabra* (Caucasus) in addition has anti-inflammatory properties (Khanna *et al.*, 1972) and epistaphine from the aerial parts of *S. hernandifolia* (India) was found to possess significant adrenergic neurone blocking activity, like guanethidine (Ray *et al.*, 1979b).

Desmodium gangeticum (L.) DC. var. *gangeticum* syn. (*Hedysarum gangeticum* L.) FABACEAE

L In India the plant is regarded as an antipyretic and anticatarrhal (Dalziel, 1937). The leaves are used in Liberia to bathe a child having convulsions (Harley, 1941).

C Twelve alkaloids consisting of carboxylated and decarboxylated tryptamines, β-carbolines and β-phenylethylamines have been isolated from different parts of *D. gangeticum* at different stages of its development (Ghosal and Bhattacharya, 1972). The aerial portion of the plant contained 2-indole-3-alkylamines, *N*-methyltetrahydroharmine and 6-methoxy-2-methyl-β-carboline. The alkaloid content in fresh material was three times that found in the dried plants (Ghosal and Bannerjee, 1968).

P The total alkaloid fraction of the stems and leaves of *D. gangeticum* collected in India exhibited curariform activity on the rectus muscle of the frog (Bhattacharya and Sanyal, 1976). The total alkaloid fraction also has an inhibitory effect on the isolated frog heart, and a relaxant effect on the smooth muscles of the rabbit and dog and on the isolated rat uterus. It is non-toxic and has a mild diuretic effect as well as an inhibiting action on respiration (Prema 1968; Ghosal and Bhattacharya, 1972). The aerial portions of the plant were reported to have anticholinesterase activity and to induce smooth muscle stimulant, CNS stimulant and depressor responses. The roots of the plant produce a nicotine-like effect on the dog intestine and on the carotid blood pressure. This has been attributed to the presence of catecholamine releasers, tertiary phenylethylamines and candicine. In addition, the aqueous extract of the roots has shown anti-inflammatory, antibacterial and antifungal activities.

Mucuna pruriens (L.) DC. var. *pruriens* FABACEAE

Cowhage, cow-itch

L The spicular hairs of the pods penetrate the skin, causing intense irritation. They have been used in Senegal as an anthelmintic prepared as an electuary with treacle or honey (Sébire, 1899).

C It has been shown that the hairs contain 5-hydroxytryptamine (serotonin) and that the itching produced by the hairs is due to the liberation of histamine in the epidermal layer of the skin (Broadbent, 1953; Bowden *et al.*, 1954). In the leaves, fruits and seeds, four indole-3-alkylamines (*N*,*N*-dimethyltryptamine, two derivatives and bufotenine), a 5-oxindole-3-alkylamine and a β-carboline have been reported as well as choline (Ghosal *et al.*, 1971). In the seeds a sterol, a fatty acid, two alkaloids, mucunine and mucunadine, and two water-soluble bases, prurienine and prurieninine had previously been found (Majumdar and Zalani, 1953) and L-DOPA has been obtained from the seeds (Bell *et al.*, 1971; Ghosal *et al.*, 1971).

P The indole bases have been shown to have an unspecified spasmolytic action on spasms induced in smooth muscles by acetylcholine, histamine or oxytocin. Two of them, 5-methoxy-*N*-*N*-dimethyltryptamine and 5-oxyindole-3-alkylamine, produce a block of a tubocurarine type and indole-3-alkylamine causes blockade of the striated muscle of the frog abdomen. In the anaesthetized dog, 5-oxindole-3-alkylamine and 5-methoxy-*N*, *N*-dimethyltryptamine cause severe depression of respiration and the blood pressure with spasm of the bronchi, which can lead to death. In the rat, to the contrary, they induce hyperactivity (Ghosal *et al.*, 1971). In view of the presence of L-DOPA and serotonin in the plant, its possible use in Parkinson's disease has been suggested (Bell and Jansen, 1971). The plant also has a hypoglycaemic and cholesterol lowering effect (Pant *et al.*, 1968).

IV Plants with cholinergic and adrenergic actions

Plants with a depressant or stimulant action on the nervous system that have been described in Chapter 2 (CV) are mentioned after those described below. The plants listed first are those acting on mixed receptor sites affecting both autonomically and centrally innervated organs and tissues.

Acetylcholine liberated from motor nerve terminals transmits the nerve impulse across a synaptic gap to the end plate region of skeletal muscle cells. Acetylcholine also acts as a neurotransmitter at Renshaw cells in the spinal cord, at autonomic ganglia and at motor terminals on smooth muscles.

Cholinergic drugs can act on muscarinic or nicotinic receptors. The action on muscarinic receptors resembles the effect of stimulating the parasympathetic nervous system and can be blocked by atropine. The effects on the nicotinic receptors are blocked by curare. Among the neurological complaints treated by anticholinergic plants is Parkinson's disease, which is characterized by involuntary movements and a depressive state. It is attributed to a disturbance in the functioning of the *Corpus striatum* in the brain controlling muscular tonicity. It was shown that acetylcholine stimulates the extrapyramidal nerve cells and tropane alkaloids with anticholinergic action were therefore used in the treatment of this condition. Further observations showed that the dopamine concentration was reduced in the *Corpus striatum* of patients with idiopathic and post-encephaletic Parkinsonism, numerous trials with L-DOPA were carried out and this was considered to be a drug of choice (Turner and Richens, 1978).

The use of tropane alkaloids with anticholinergic action in the treatment of Parkinsonism explains the use in neurology of several spp. of Solanaceae, e.g. *Datura* spp., with (−)-hyoscyamine (see encephalic stimulants); *Solanum nigrum* with solanine, which has anticholinesterase action and which is also analgesic and sedative to the CNS; and *Withania somnifera*, which also has sedative and narcotic action on the CNS. *Boerhavia diffusa* (Nyctagynaceae) is parasympatholytic and anticonvulsant and analgesic to the CNS. *Tetrapleura tetraptera* and *Securidaca longepedunculata* have sedative, tranquillizing and anticonvulsant properties which could be attributed to oleanolic acid glycoside. *Anthocleista procera* and *A. djalonensis* are hypotensive and also sedative and analgesic, probably through swertiamarin.

The aporphine alkaloids of many Annonaceae, e.g. *Uvariopsis guineensis*, *Enantia chlorantha*, *Isolona campanulata*, *Polyalthia oliveri*, *Annona muricata* and *Pachypodanthium staudtii*, show different degrees of antispasmodic activity, sometimes accompanied by sedative action on the CNS. Some of these plants have found use in the treatment of mental disease, e.g. *Enantia chlorantha*, with isocorydine, in paralysis agitans and *Polyalthia oliveri*, with oliveroline, has anti-Parkinson activity in mice. In *Isolona campanulata* isochondrodendrine and curine are found as well as aporphine alkaloids. In *Stephania dinklagei* (see above) isocorydine and other aporphine alkaloids have also been reported.

Antispasmodic action on the ANS was found in *Euphorbia hirta* used in the treatment of asthma; the active constituent is probably shikimic acid. *Uncaria africana* (Rubiaceae) is said to have antispasmodic and sedative action. *Alchornea cordifolia* contains a sympatholytic alkaloid which is used as a spasmolytic and the flowers of *Grewia bicolor* and spp. contain a sedative essential oil, farnesol. *Pentaclethra macrophylla* (Mimosaceae) and *Newbouldia laevis* (Bignoniaceae) are anticholinergic and have been used locally in the treatment of epileptic fits in children. *Lantana camara* (Verbenaceae) reported as an antispasmodic, has proved to be too toxic.

Stimulant action on the ANS was noted with *Cissus quadrangularis* (Vitaceae) (cholinergic), *Sida cordifolia* (Malvaceae) (adrenergic) and *Crateva religiosa* (Capparidaceae) and perhaps with *Borreria verticillata* (Rubiaceae) and *Cardiospermum halicacabum* (Sapindaceae).

(a) Depressants acting on both ANS and CNS terminals

Solanum nigrum L. syn. (*S. nodiflorum* Jacq., *S. guineense* (L.) Lam.)

SOLANACEAE

L The berries are used in India in the treatment of fever, diarrhoea and eye diseases (Chopra *et al.*, 1956). In Nigeria a decoction of the leaves is said to be diuretic and laxative and that of the young shoots is given in the treatment of psoriasis and skin disease (Dalziel, 1937).

C In the fruit, solanine has been found (especially in the green fruit). Hydrolysis of solanine produces glucose, rhamnose and solanidine. In addition, heterosides derived from spirosolane (the genin being solasodine) are present: solasonine, solamargine and solanigrine (Henry, 1949; Paris and Moyse, 1971).

P In the French Pharmaceutical Codex 1937, *S. nigrum* is used in 'Huile de Jusquiame' and a poultice of the leaves is used in France as an emollient and antineuralgic and has a slightly narcotic action (Paris and Moyse, 1971, p. 190). Solanine is reported to be a cytostatic and a cholinesterase inhibitor in vitro (Manske and Holmes, 1950–71, Vol. 10, p. 116). It is also considered to be analgesic for migraine and gastralgia, and a nervous sedative for paralysis agitans and for chronic pruritus in certain skin diseases (Denoël, 1958, p. 899). It is also said to have antimitotic action (Danneberg and Schmäl (1953) quoted in Paris and Moyse (1971). Solasodine is antagonistic to tachycardia provoked by adrenaline (Sollmann, 1957, p. 669).

Withania somnifera L. Dunal syn. (*Physalis somnifera* L.) SOLANACEAE

L In West African local medicine, both roots and leaves are used internally, and the freshly pounded leaves also externally, against fever, chills, rheumatism, colics, etc. The juice of the plant is said to be diuretic and emmenagogic. In local medicine in East Africa, the root is considered to have narcotic and antiepileptic actions (Pichi Sermoli, 1955). In India, the bruised leaves and ground root are used as a local application to painful swellings, carbuncles and ulcers as the root and leaves are considered to be sedative and the root has been included in the Indian Pharmacopoeia and Codex for its narcotic and sedative properties (Chopra *et al.*, 1956; Oliver, 1960).

C The roots contain choline, tropanol, pseudotropanol, 3-tigloyl-hydroxy-tropane, cuscohygrine, (±)-isopelletierine, anaferine and anhygrine. In the leaves, withaferine (5,6-epoxy-4β,22,27-trihydroxy-1-oxo-5β-ergosta-2,24-dien-26-oic acid δ-lactone) has been detected (Khanna *et al.*,1963; Schwarting *et al.*, 1963; El Olemy and Schwarting, 1965; Lavie *et al.*, 1965). Although the total alkaloid content appears high, the alkaloids resinify during extraction and the yield in pure substances is low. Schröter *et al.* (1966) isolated withasomnin (4-phenyl-1,5-trimethylenepyrazole) from the roots.

P Withasomnine showed distinct sedation in mice and induced sleep and narcosis in higher doses. The acetone-soluble fraction of the leaves, when given perorally, had a mild depressant effect (tranquillizer-sedative type in dogs, albino rats and mice). It exacerbated the convulsions produced by metrazol but protected against supramaximal electroshock seizures in rats. It had no analgesic activity in rats and produced hypothermia in mice. There was potentiation of barbiturate-, ethanol- and urethane-induced hypnosis in mice which could not be antagonized by lysergic acid diethylamide and dibenzylene. The two other fractions had no significant neuropharmacological actions and were devoid of irritant effect on mucous membranes (Prasad and Malhotra, 1968). The effect of the total extract on the smooth muscles, cardiovascular system, respiration and skeletal muscles had been studied earlier by Malhotra *et al.* (1960a, b). Other authors obtained no therapeutically useful sedation in animal tests. Watery, 50% methanol or absolute methanol extracts with and without pre-extraction in amounts up to 1 g/kg neither reduced the central activity nor affected rats in motility tests. Even with doses of 5 g/kg the different *Withania* extracts produced no sedation (Fontaine and Erdös, 1976). Withaferine has been reported to have antitumour action and to delay the onset of arthritis in rats.

It is also antibiotic towards Gram-positive organisms and certain fungi (Shohat, 1967).

Boerhavia diffusa L. syn. (partly *B. repens* var. *diffusa* (L.) Hook., partly *B. adscendens* Willd.) (Fig. 3.4) NYCTAGYNACEAE
Hog weed, Punarnava

L The plant is used in local medicine to treat convulsions and as a mild laxative and febrifuge. The roots and leaves are considered to have an expectorant action, to be emetic in large doses, and are of use in the treatment of asthma. The thick roots, softened by boiling, are applied as a poultice to draw abscesses and to encourage the extraction of guinea worms (Dalziel, 1937).

C An alkaloid, punarnavine (0.04% in the roots), has been extracted, as well as boerhavic acid, reducing sugars, potassium nitrate and tannins including phlobaphens (Oliver, 1960; Singh and Udupa, 1972a). Extracts from *B. diffusa* and *B. repens* are official in the Indian Pharmacopoeia as a diuretic (Chopra *et al.*, 1956). In addition a nucleoside, hypoxanthine-9-L-arabinofuranoside, has been isolated from the roots of *B. diffusa* (Ojewole and Adesina, 1983).

P Intravenous injection of punarnavine produces a distinct and lasting rise in the blood pressure of cats, with marked diuresis (Chopra *et al.*, 1956). Clinically, doses of 4–16 g of a liquid extract prepared from the fresh plant produced diuresis in 33 patients with oedema and ascites and seemed extremely active particularly in cases of cirrhosis of the liver and chronic peritonitis (Chopra and Ghosh, 1923). Mudgal (1975) confirmed the diuretic effect and also reported anti-inflammatory activity.

Fig. 3.4. *Boerhavia diffusa* L.

An active anticonvulsant principle was localized in the roots of *B. diffusa* (Adesina, 1979). Test solutions were prepared by methanol extraction of the roots, dissolution in water of the residue after evaporation, extraction with petroleum ether and lyophylization of the water-soluble portion. Intraperitoneal injections of this extract have shown anticonvulsant as well as analgesic properties on male albino mice with leptazol-induced convulsions. While the control animals died from convulsions within 2–3 min, a 1 g/kg dose of the extract gave 20% protection, and with 1.5–2.0 g/kg doses all animals were protected for over 10 min, 60% eventually dying (Adesina, 1979).

Further tests with the nucleoside isolated from the roots showed that, like inosine and adenosine, it relaxes the isolated coronary artery of the goat contracted with potassium chloride. The action, like that of inosine, is thought to be 'a direct vasodilator effect, not involving vascular adenosine receptors' (Ojewole and Adesina, 1983). (Also see Singh and Udupa, 1972b, c.)

Tetrapleura tetraptera (Schum. & Thonn.) Taub. syn. (*Adenanthera tetraptera* Schum. & Thonn., *T. thonningi* Benth.) (Fig. 3.5) MIMOSACEAE

L An infusion of the fruit is used in Nigeria as a tonic and stimulant. The bark is used in Ghana as a purgative and in Guinea and Senegal as an emetic. Elewuda mentions that the fruits, with other parts of the plant, are used as an anticonvulsant drink (Adesina and Sofowora, 1979).

Fig. 3.5. *Tetrapleura tetraptera* (Schum. & Thonn.) Taub.

C Bouquet (1972) reported the presence of saponosides and perhaps of tannins in the rootbark of the plant collected in the Congo, but found no flavonoids, quinones, cyanogenetic glycosides or steroids. Later Adesina and Sofowora (1979) isolated oleanic acid triglycoside and Ojewole (1983b) obtained scopoletin, a coumarin, from the fruit.

P Alcoholic as well as aqueous extracts of the fruits have been found to exhibit marked tranquillizing properties on male albino mice and to cause lowering of their body temperature. Oral doses of an alcoholic extract of the fruit sedated mice within 30–40 min after intraperitoneal injection of a convulsant drug (leptazol) and over 60% of the animals were protected (Adesina and Sofowora, 1979).

Anti-bronchoconstrictive and anti-arrhythmic effects of scopoletin were recently demonstrated in vivo and in vitro (Ojewole, 1983b). Scopoletin and tetramethyl-pyrazine (see *Jatropha podagrica*) were found to protect guinea pigs from, and to suppress ouabain-induced arrhythmias, increasing the functional refractory period of the myocardium in the same way as quinidine. They also relax acetylcholine-, 5-hydroxytryptamine- and histamine-induced contractions of the guinea pig isolated tracheal muscle (Ojewole, 1983a).

The inhibitory effects of scopoletin on electrically induced contractions, relaxations and twitches of cholinergically and adrenergically innervated muscle preparations are thought to be linked with the non-specific spasmolytic action of the coumarin, and also probably to be exerted via its local anaesthetic (membrane stabilizing) activity (Ojewole, 1983b, c).

Securidaca longepedunculata Fres. (Fig. 3.6) POLYGALACEAE
Violet tree, wild wisteria, Senega root tree

L Pieces of root covered with bark are sold in Hausa markets (Northern Nigeria) as a charm and a medicine. In small doses this is a drastic purgative and the powdered root causes violent sneezing. The root is used as a taenifuge and anthelmintic in French Guinea and Senegal. The seeds are rich in oil and are given medicinally for febrile and rheumatic conditions (Dalziel, 1937; Oliver, 1960; Kerharo, 1970).

C The roots have been shown to contain saponin (0.4%) and 4% methyl salicylate. A systematic examination of roots gathered in Angola indicated 27% lipids and 0.36% protides, tannins and steroids. Hydrolysis of the saponoside produced a steroid genin and glucose. Methyl salicylate was present as a monotropitoside similar to gaultherin; the sugars were glucose and xylose (Kerharo, 1970).

P The saponin was reported to have a certain ichthyotoxicity (Prista and Correia Alves, 1958). The LD_{50} lies around a concentration of 0.018% for small sweet water fish remaining for 24 h in the solution, and all fish died after 24 h in a concentration of 0.024% (Fraga de Azevedo and Medeiros, 1963). *Securidaca* is one of 24 plants used in treating convulsions in children. Oral administration of a decoction of the root produces a sedative effect and several hours of sleep. The active factor appears to be oleanolic acid glycoside; this is also found in *Tetrapleura tetraptera* which has the same sedative properties (Adesina and Sofowora, 1979; Sofowora, 1980).

Anthocleista procera Lepr. ex Bureau syn. (*A. frezoulsii* Chev., *A. nobilis* Lepr.)

LOGANIACEAE

Anthocleista djalonensis Chev. syn. (*A. kerstingii* Gilg ex Volkens, *A. procera* Chev.)
Anthocleista nobilis Don. syn. (*A. parviflova* Bak.)
Cabbage tree

L The seeds and bark of these species are used locally in Nigeria for their antipyretic, stomachic and purgative actions. Watt (1967) reported that the Abrous (Ghana) prepare a decoction of the leaves of *A. nobilis* Don. with lemon as a remedy for epilepsy (Irvine, 1961). In Casamance (Senegal) the decoctions are used as a diuretic (Kerharo and Adam, 1974).

C From the dried leaves of *A. procera*, 5% of a bitter monoterpenic heteroside, swertiamaroside or swertiamarin, has been obtained in the Ivory Coast and in Nigeria. It has been proved to be the precursor of an earlier-found indole alkaloid in the leaves and roots, gentianine or erythricine, which is formed from the former by treatment with ammonia (Canonica, 1962; Lavie *et al.*, 1963; Plat *et al.*, 1963; Koch *et al.*, 1964). Gentianine (1%) in crystal form has also been isolated from the dried leaves (Koch, 1965). Another crystallized compound, anthocleistine, a triterpenic

Fig. 3.6. *Securidaca longepedunculata* Fres.

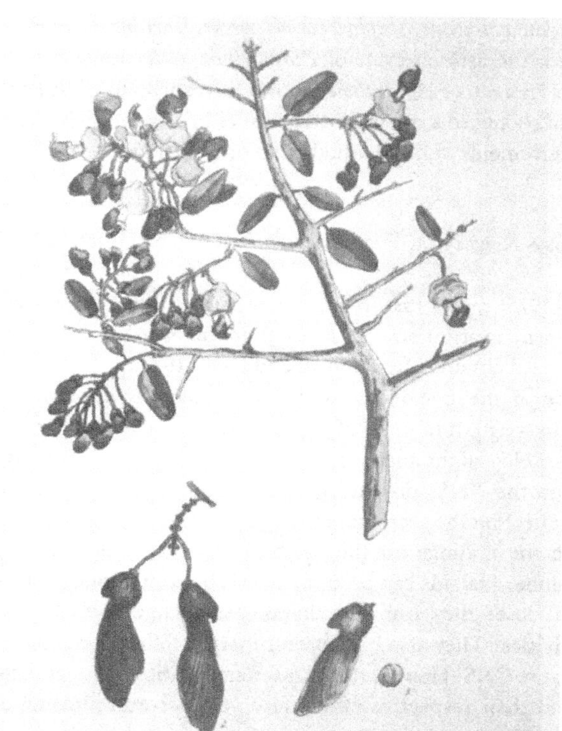

pentacyclic acid, was obtained by Taylor-Smith (1965) from the Nigerian *A. procera*. No swertiamarin was found in the leaves of *A. djalonensis* (Poisson *et al.*, 1972).

P The leaves of a non-West African species (*A. rhizophoroides*) which also contains swertiamarin were found to decrease nervous excitability and to have a cardio-moderating action (Poisson *et al.*, 1972). Gentianine (also found in the Gentianaceae) has low toxicity. The LD_{50} of extracted dried leaves in mice is 500 mg/kg subcutaneously and 0.5–1.3 g/kg perorally (Steinegger and Weibel, 1951; Koch, 1965). It is hypertensive in small doses but hypotensive in higher doses, resulting in a reversible decrease in the blood pressure, which with very high doses may become irreversible. It also has an inhibitory action on the isolated frog heart, and on the guinea pig ileum which has been stimulated by histamine or by acetylcholine (Steinegger and Weibel, 1951). Gentianine has a distinct analgesic action in the rat (Keng-Tao Liu *et al.* in Koch, 1965). One of the main interests of gentianine lies in its antihistamine and anti-inflammatory actions, which have been demonstrated on rats and guinea pigs. Thus, an intraperitoneal injection of 90 mg/kg prevents or considerably reduces the swelling of the rats' hindlegs when the injection is carried out prior to a subcutaneous injection of 0.1 ml of ovalbumin or formol, which in the control animals causes a very slow receding inflammation of the articulations. The injection of gentianine also protects the guinea pig against a histamine-containing aerosol. In the anti-inflammatory tests, gentianine is more effective than chloroquine and cortisone (Chen-Yu Sung *et al.*, 1958; Hsiu-Chuan Chi *et al.*, 1959). Gentianine does not influence the growth of yeast, *Candida albicans*, *Staphylococcus aureus*, *Bacillus coli* or *B. anthracis*, nor does it influence the asexual cycle of *Plasmodium gallinaceum* (Ghosal *et al.*, 1973). An aqueous extract of *A. djalonensis* was investigated by Adesogan and Olatunde (1974) and produced a rise in blood pressure in cats and an increase in tone and amplitude of movements of rabbit duodenum preparations.

Uvariopis guineensis Keay syn. (*Uvaria spectabilis* Chev. ex Hutch. & Dalz.)

ANNONACEAE

C Guinaudeau *et al.* (1975) report that from the rootbark of this tree, oxo-aporphines and open aporphines (uvariopsamine, uvariopsamine-*N*-oxide, uvariopsine and ushinsunine) have been obtained by Leboeuf and Cavé (1972a).

P The total alkaloids of the powdered root of *U. guineensis* have antispasmodic properties similar in strength to those of papaverine but longer lasting. They have little effect on the CNS (slight analgesic and antipyretic action) but have a sympatholytic action on the ANS, producing vasodilatation, and hypotension and a slight stimulating effect on the secretion of the gallbladder. These observations are in agreement with the findings for the rootbark (Quevauviller *et al.*, 1977). In general, the aporphine alkaloids can produce emesis by stimulation of the vomiting centre. In smaller doses they can have hypotensive, digestive, diuretic and/or antispasmodic activities. They also have been reported to have sedative and slight hypnotic action on the CNS. However, the different constituents vary considerably in their actions and their respective effects also vary according to the dose used (Guinaudeau *et al.*, 1975).

Enantia chlorantha Oliv. ANNONACEAE
African yellow wood

L Bark extracts are used in local medicine in different West African countries as an antipyretic and are also applied to ulcers and as a haemostatic to wounds (Dalziel, 1937).

C The main alkaloids from the root- and stembarks are quaternary protoberberines: palmatine, bebeerine, jatorrhizine and columbamine. Two oxyaporphines (*O*-methyl moschatoline and lysicamine) have also been isolated from the bark of the roots. In the leaves, flavonic heterosides and the alkaloids atherospermine and argintinine and the phenanthrene alkaloids (open aporphines) have been reported (Leboeuf and Cavé, 1972b; Hannonière *et al.*, 1975). From the leaves of **E. polycarpa**, Leboeuf and Cavé (1972b) isolated 1.8% of total alkaloids of which the the main alkaloids, L-(+)isocorydine (aporphine group), had been previously reported in species of *Stephania* and in the families Papaveraceae and Rutaceae but never in such high proportion.

P Palmatine has been shown to have an antipyretic effect as well as a depressive action on the arterial blood pressure and the nervous system (Beauquesne, 1938). Isocorydine, which is also present in several other species of Annonaceae, induces (like bulbocapnine) catalepsy. This led to its use in the treatment of diseases in which involuntary movement is a symptom as in the case in paralysis agitans and St Vitus dance (Raymond-Hamet, 1936; Henry, 1949, p. 313).

Isolona campanulata Engl. & Diels syn. (*I. leonensis* Sprague and Hutch., *I. soubreana* Chev. ex Hutch. & Dalz.) ANNONACEAE

L In the Ivory Coast, the bark of the roots and stem is given during pregnancy and for the treatment of bronchial infections, skin diseases, fever, haematuria and bilharzia (Oliver, 1960).

C The bark of the roots and trunk contains 0.2% and 0.3% of total alkaloids, respectively. Six aporphine alkaloids have been isolated and identified: anonaine, *nor*-nuciferine, oliveridine, oliverine, *N*-oxy-oliverine and liriodenine. Two of the alkaloids, oliverine and oliveridine, are present in major amounts. In addition, there are two neutral constituents, a triterpene (polycarpol) and an indole-terpene. In the stembark isochondrodendrine and curine have been reported. The species is thus characterized by virtue of its containing aporphine alkaloids and showing an analogy in its composition with some *Enantia* spp. (Leboeuf and Cavé, 1972b; Nicto *et al.*, 1976; Hockemiller *et al.*, 1977, 1978).

P In view of the alkaloids reported, an antispasmodic effect on the skeletal muscles and a depressive action on the ANS can be expected.

Polyalthia oliveri Engl. syn. (*P. acuminata* Oliv.) ANNONACEAE

L The Mendes tribe are said to use a decoction of the bark for blackwater fever. In Liberia, an infusion of the bark is used as an anthelmintic and in the Cameroons, a cold infusion of the bark is drunk for stomach troubles (Dalziel, 1937).

C Eleven aporphine alkaloids and some terpenes were found in the rootbark and leaves of this small tree. The main alkaloids of the rootbark are oliveridine and oliverine, whilst in the leaves mainly oliveroline has been reported (Hannonière *et al.*, 1974, 1975).

P Oliverine has been shown to have antihypotensive action whilst oliveridine is only hypotensive through relaxation of the vascular smooth muscles, comparable in this to the effect of papaverine. Oliveroline has antiparkinson activity in mice (distinct dopaminergic effect). This activity is weaker in oliveridine and non-existent in oliverine, thus indicating that methylation of one OH group (in position 7) decreases, and methylation of two OH groups (in positions 7 and 9) causes the disappearance of the activity of oliveroline (Quevauviller and Hannonière, 1977).

Annona muricata L. ANNONACEAE

Sour sop

Annona reticulata L.

Custard apple, bullocks heart

L Introduced and naturalized in many places, these species are natives of tropical America and the West Indies. A decoction of the leaves of *A. muricata* is used as a soothing and sudorific remedy for fever, whilst the pounded fresh leaves are applied to cicatrize wounds. In general, the leaves are considered a useful treatment for fever and dysentery (Dalziel, 1937). In India, the rootbark is given in ptomaine poisoning (Nadkarni, 1954). The leaves of *A. reticulata* are considered in popular medicine to be an insecticide and anthelmintic and the fruit an antidiarrhoeic. In Ghana, *A. arenaria* is used as a remedy for epilepsy (Watt and Breyer-Brandwijk, 1962).

C From the leaves and stems of *A. reticulata*, anonaine, roemerine, corydine, isocorydine and many other aporphine alkaloids have been isolated (Guinaudeau *et al.*, 1975). In addition L-dopamine, salsinol (1-methyl-6,7-dihydroxy-1,2,3,4-tetrahydroquinoline) and coclaurine have been identified by thin-layer and gas chromatography (Forgaes *et al.*, 1981). γ-Aminobutyric acid has been found in the fruit and stems of *A. muricata* (Durand *et al.*, 1962) and in the leaves also two phytosterol glycosides, ipuranol and anonol, have been reported to be present (Merck, 1976). From the roots, two alkaloids, muricine and muricinine, have been isolated (Meyer, 1941; Manske and Holmes, 1950–71, Vol. IV, p. 142). The main alkaloids which have been obtained from the roots of *A. muricata* from Guyana are (+)-reticuline and (+)-coclaurine. Three minor alkaloids of the isoquinoline type, coreximine, anomurine and anomuricine, have also been reported (Leboeuf *et al.*, 1980). Hydrocyanic acid has been found in variable amounts in the roots, leaves and mainly in the bark (Merck, 1976). From the leaves and stems of *A. squamosa*, a benzyl-tetrahydro-isoquinoline alkaloid, higemanine, has been isolated (Leboeuf *et al.*, 1981); it had been found earlier in *Nelumbo nucifera* seeds and is an effective adrenergic agonist.

P The constituents of *A. reticulata*. explain the cardiotonic activity of the plant which is inotropic and chronotropic and also spasmolytic (Forgaes *et al.*, 1981). An extract of the leaves and stems of *A. muricata* has a passing depressive effect on the blood pressure, which has been attributed to γ-aminobutyric acid (Durand *et al.*, 1962). Intraperitoneal administration of this acid (3 g/kg) has been shown to protect animals against the convulsive acid of brucine if it is given three days beforehand (Sorer and

Pylko, 1965). Intracerebroventricular administration of 25 μg (+)-coclaurine in mice suppresses or prevents the locomotor activity induced by dopaminergic stimulating agents (Watanabe *et al.*, 1981). An insecticidal principle, resistant to heat but not to saponification, has been detected in the seeds of *A. muricata*; it was effective against a great number of insects (Heal and Rogers, 1950). The screening of different extracts of the plant as to its anticancer action gave no significant results (Abbot *et al.*, 1966), although in related species, such as *A. squamosa*, aporphine alkaloids isolated from the bark (Bhakuni *et al.*, 1972) and constituents of the bark of *A. senegalensis* were found to have significant antineoplastic activity (Durodola, 1975; Adesogan and Durodola, 1976).

Pachypodanthium staudtii Engl. & Diels ANNONACEAE

L The bark was formerly used as an arrow poison and, in Liberia, medicinally. In the Congo, the bark is also considered to be analgesic and odontalgic, and a decoction is used in coughs and dyspnoea by certain tribes (Bevalot *et al.*, 1976).

C Seven isoquinoline alkaloids have been isolated from the bark: two tetrahydroprotoberberines, corypalmine and discretine; and oxo-aporphine, liriodenine; three aporphines, pachypodanthine (7-methoxy aporphine), pachystaudine and *nor*-pachystaudine (4-hydroxy-7-methoxy aporphine), and a new alkaloid, staudine, which proved to be the result of the combination of jateorrhizine (a protoberberine alkaloid) and 2,4,5-trimethoxysterine (Bevalot *et al.*, 1976, 1977; Cavé *et al.*, 1980). Earlier, a flavonic genin, dihydroxy-5,4′-trimethoxy 3,7,3′-flavone had been isolated from the stem- and rootbark by Cavé *et al.* (1973). An unusual neutral constituent, trimethoxy-2,4,5-styrene, which might be identical with the trimethoxystyrene obtained from *Piperomia lucida*, was found to be present in *Pachypodanthium confine* and in *P. staudtii* (Bevalot *et al.*, 1976).

P *P. staudtii* is said to have an excitant action on the nervous system; it produces peripheral vasodilatation and has a weak spasmolytic action of the papaverine type (Bevalot *et al.*, 1976).

(b) Antispasmodics acting mainly on sympathetic terminals

Euphorbia hirta L. syn. (*E. pilulifera* Chev.) (Fig. 3.7) EUPHORBIACEAE
Australian or Queensland asthma herb

L In the native medicine of Nigeria, the plant is used in an enema for constipation and in the treatment of dysentery. The latex is squeezed into the eye to cure eye troubles. Extracts of the dried entire plant are mentioned in the British Extra Pharmacopoeia (27th edn) for asthma and the treatment of coughs. In Konakry hospital, good results were achieved with a decoction of the fresh plants in cases of acute enteritis and dysentery (Dalziel, 1937).

C Analysis of the latex has revealed (−)-inositol, pyrogallic and catechuic tannins and the alkaloids xanthorhamnine (Oliver, 1960). Gupta and Garg (1966) found taxerol, friedelin, β-sitosterol, myricyl alcohol, ellagic acid and hentriacontane in extracts of the stem whilst Blanc *et al.* (1972) reported ellagic, gallic, chlorogenic and caffeic acids, kaempferol, quercitol, quercitrin (as genin of a heteroside), and a number of amino acids.

P Early pharmacological trials seemed to indicate the presence of two active principles, one with an antispasmodic action on smooth muscles and one with a spasmodic or histamine potentiating action (Hallet and Parkes, 1953). In 1978, these two pharmacologically active constituents were isolated and identified. The relaxing principle was identified as shikimic acid and the contracting principle as choline. *iso*-Inositol, glucose and sucrose were also reported (El Nagger *et al.*, 1978). An ethanolic extract of the above-ground portion of the plant produced relaxation of the guinea pig ileum. This extract, containing shikimic acid, was used in the treatment of asthma, hayfever, bronchitis and other respiratory conditions. It was shown to have an antispasmodic effect on the smooth muscles of mice and guinea pigs (Hellerman and Hasleton, 1950). A clinical trial on 53 patients with amoebic dysentery and later on 150 cases of dysentery in Upper Volta, the Ivory Coast and Madagascar showed 83% cure. There was no relapse for 5–12 months following the oral course. The substance was perfectly tolerated and was put on the market in France and the USA (Socambine) (Martin *et al.*, 1964; Ridet and Chartol, 1964) but its use has since been abandoned (Kerharo and Adam, 1974). The alcoholic extract of the whole plant had an anticancer action against Friend leukaemia virus in mice. It also showed hypoglycaemic action in albino rats and an antiprotozoal effect (Dhar *et al.*, 1968).

Fig. 3.7. *Euphorbia hirta* L.

The maximum dose of the total extract, given perorally, that was tolerated in mice was 1 g/kg. Purification of the total extract led to a new extract with a twenty times higher activity against *Entamoeba histolytica*. Further purification is in progress (Ndir and Pousset, 1981). Irritant and carcinogenic phorbol esters have been reported to be present in the stem latex (Ayensu, 1979).

Uncaria africana G. Don. var. **africana**; **Uncaria africana** var. **angolensis** Havil. syn. (*U. talbotti* Wernh., *U. angolensis* (Havil) Welw. ex Hutch. & Dalz.)

RUBIACEAE

L The bark of these spp. is chewed as a remedy for coughs (Lane Pool, in Dalziel, 1937) and has been used for the treatment of stomach pains and syphilis (Staner and Boutique, 1937).

C Some forty alkaloids have been identified in *Uncaria* spp. They are mainly of the hetero-yohimbine and corresponding oxindole types although pyridino-indoloquinolizidines (e.g. angustine) and harmane are also present. Rhynchophylline, rotundifoline and isorhynchophylline, etc. have been reported as well as gambirine and roxburghines (the latter in extremely limited quantities) as could be expected in view of the close relation with the genus *Mitragyna* (Merlini *et al.*, 1972; Phillipson *et al.*, 1973; Phillipson and Hemingway, 1973, 1975).

P Antispasmodic and sedative action is attributed to the 'hooked thorns' of the plant (Kariyone, 1971). The related *U. rhynchophylla* (not found in West Africa) is, according to the dose administered, reported to have either muscle-relaxant or muscle-stimulant activity (Harada and Ozaki, 1978, 1979).

Alchornea cordifolia (Schum. & Thonn.) Müll. Arg. syn. (*Schousboea cordifolia* Schum. & Thonn., *A. cordata* Benth.) EUPHORBIACEAE
Alchornea floribunda Müll. Arg.

L The root of *A. floribunda*, called *niando* in the Congo region (Zaire) is used by the Africans as a stimulating intoxicant and aphrodisiac. After reducing it to a powder, it is either mixed with food or macerated for several days in palm wine and consumed to provide energy for tribal festivities or warfare. It is said to provide a state of intense excitement followed by a deep, sometimes fatal, depression. It is further considered an excellent remedy for urinary, respiratory and intestinal disorders (Dalziel, 1937; Oliver, 1960).

C According to their origin (Ivory Coast, Guinea) and the time of conservation, the root and bark of *A. cordifolia* have been found to contain 0.03–0.26% of total alkaloids: the highest amounts were found in the most recent samples (Bennet, 1950). Thin-layer chromatography revealed two principal alkaloids, closely related to but not identical with yohimbine (Paris and Goutarel, 1958). Ferreira *et al.* (1963b) found in the roots of the plant in Portuguese Guinea 0.07% of alkaloids together with gentisic and anthranilic acids and suggest that gentisic acid could be a possible precursor in the biosynthesis of yohimbine. In addition, in the leaves and bark of this species, 10% and 11% of tannins, respectively, have been found (Bennet, 1950).

P The extract of the roots of *A. floribunda* has sympatholytic activity (Raymond-Hamet, 1933). It considerably increases the sensitivity of the nervous system towards adrenaline. In the dog, doses of 0.5 g/kg produce a slight hypotension, followed by a slight hypertension. Higher doses produce a gradual decrease of the carotid pressure, which returns only very slowly to its original value (Raymond-Hamet, 1954). A patent has been obtained for the use of the leaf alkaloid as a spasmolytic (Goutarel R. Brevet français 2087982 of 5.5.1970). Guedel (1955) obtained positive results in clinical experiments with root and leafy stem extracts in the treatment of icterus in Abidjan.

Grewia bicolor Juss. syn. (*G. salvifolia* Heyne ex Roth.) TILIACEAE
Grewia carpinifolia Juss. syn. (*Vinticena carpinifolia* (Juss.) Burret)
Grewia lasiodiscus K. Schum. syn. (*G. kerstingii* Burret, *Vinticena lasiodiscus* (Schum.) Burret, *V. kerstingii* (Burret) Burret)

L The Grewias have edible fruits (sometimes made into a fermented drink). The shoots of *G. carpinifolia* are given to sheep at lambing to help delivery (Irvine, 1930) and women put the roots in soup when approaching childbirth (Dalziel, 1937).

C The roots of all three species contain mucilage and catechuic tannins. In *G. carpinifolia* gallic tannins are also reported. The barks of all spp. were found to contain amines (aspartic acid and probably proline) but no alkaloids, quinones, saponosides or histamine-like substances were reported. The flowers, like those of other Tiliaceae, contain farnesol (Paris and Théallet, 1961).

P The bark extract of the *Grewia* spp. has a more or less distinct depressive action on the guinea pig ileum. This is contradictory to the effect observed on rat or rabbit duodenum (Paris and Théallet, 1961). The action appears to be directly muscular, and the authors suppose that the active principle belongs to the aminophenols. Binet *et al.* (1972) observed that farnesol has spasmolytic action on the smooth muscle fibres of the intestine as well as on those of the Oddi sphincter, while oxytocic action has been reported in a non-West African species, *G. elyseoi* (Paris, 1956). Further, farnesol has been found to have psychosedative action in cases of psychic over-excitement; in higher doses it influences psychomotor defence reactions. Farnesol has proved to be antagonistic to the excitant effect of caffeine and to potentiate the hypnotic effect of barbiturates without being hypnotic itself (Paris, 1956).

Pentaclethra macrophylla Benth. MIMOSACEAE
Oil bean tree, Congo acacia

L A decoction of the bark is used in Nigeria in a lotion for healing sores. In Sierra Leone, the bark has been reported to be an anthelmintic. The Ibos hold a baby suffering from adekuku (epileptic fits) in the smoke of the burning leaves (Irvine, 1961).

C The beans yield 30–36% of a non-drying oil, the seed kernels 44–45%. This oil is suitable for soap and candles and for lubrication (Oliver, 1958). In the shell of the nut the presence of the alkaloid paucine, as well as a fixed oil and resinous constituents, have been reported (Henry, 1949, p. 776). In the Ivory Coast, the nut has been found to contain 55.4% lipids, 28.5% of protides and 12.2% of glucides. The oil contains mainly glycerides of linoleic, oleic and lignoceric acids (Busson, 1965).

P Correia da Silva and co-workers reported that 0.5–1 ml of a 1 : 10 aqueous decoction of the bark causes violent and long-lasting contractions of the isolated guinea pig uterus (Correia da Silva *et al.*, 1960). Later they found that the same extract decreases smooth muscle tone in the guinea pig trachea and in rat, rabbit and pig blood vessels and that it antagonizes the effect of acetylcholine and histamine on the intestine of the guinea pig (Correia da Silva and Quiteria Paiva, 1970). The anticholinergic effect was confirmed by Sandberg and Cronlund (1982).

Newbouldia laevis (Beauv.) Seem. ex Bureau syn. (*Spathodea laevis* Beauv.) (Fig. 3.8)

BIGNONIACEAE

L In Ghana in 1891 Easmon found *N. laevis* bark effective in the treatment of malaria and dysentery and attributed the action to a tonic effect on involuntary muscles and mucous membranes. In Lagos, an infusion of the bark and rootbark is used against convulsions in children and the flowers and leaves are used in a liniment for skin

Fig. 3.8. *Newbouldia laevis* (Beauv.) Seem. ex Bureau.

diseases. The juice of the fresh leaves is applied to wounds (Dalziel, 1937; Watt and Breyer-Brandwijk, 1962).

C From the roots of trees growing in Portugese Guinea, four different alkaloids have been isolated. One of the bases was identified as harmane (Ferreira *et al.*, 1963a). Alkaloids have also been found in the bark but not the leaves of Congolese specimens. In the leaves and bark, no flavonoids, saponins, quinones, terpenes or steroids could be detected (Bouquet, 1972).

P The alkaloids obtained from the specimens from Guinea show an inhibiting action upon the isolated rabbit duodenum and guinea pig ileum. They antagonize the action of acetylcholine and histamine on the guinea pig ileum but do not act on the isolated uterus (Correia da Silva *et al.*, 1966).

Lantana camara L. syn. (*L. antidotalis* Thonn.) VERBENACEAE
Wild Sage, Bahama tea

L In Nigeria and Senegal, an infusion of the leaves is used as a treatment against coughs and colds. In Senegal, it is also given to asthma patients as it is said to relieve dyspnoea and suffocation. A mixed infusion with *Ocimum* is considered to have diaphoretic and antipyretic action.

C The leaves, stems and flowers were found to contain the triterpenes α-amyrine, β-sitosterol, lantaden B and a triterpenoid acid. In addition, a lactone was obtained from the hydro-alcoholic extract. The triterpenoid acid, later named lantaden A, has been identified as rehmannic acid (Louw, 1948, 1949). The essential oil found in the leaves is rich in caryophyllene, eugenol, α-phellandrene, dipentene, terpineol, geraniol, linalol, cineol, citral, furfural and phellandrone (Ahmed *et al.*, 1972). Sugars and lipids have also been reported to be present.

P The plant, if ingested, causes photosensitization in sheep (Seawright, 1963; Seawright and Allen, 1972). In an extensive bibliographical research, Watt and Breyer-Brandwijk (1962) showed that the icterogenous action of the plant is the basis of the intoxication. Intoxications in children accidentally eating the berries showed an icterogenous effect (Wolfson and Solomons, 1964) and rehmannic acid has been shown to be an icterogenous triterpenoid acting on the permeability of the liver cells, mainly by blocking the excretion of bile pigments (especially bilirubin and phylloerythrine), thus causing icterus and abnormal sensitization (Heckel *et al.*, 1960; Dhillon and Paul, 1971; Dwivedi *et al.*, 1971; Seawright and Allen, 1972). Lantanin, later identified as the isomeric triterpenes lantaden A and B, was mentioned in The US National Dispensary, 1926, as an antispasmodic, but has proved to be too toxic. Some of the constituents of the essential oil account for the antiseptic effect.

(c) Stimulants of the cholinergic and adrenergic systems

Cissus quadrangularis L. syn. (*Vitis quadrangularis* (L.) Wall. ex Wight & Arn.) (Fig. 3.9) VITACEAE
Edible stemmed vine, Vigne de Bakel (Senegal)

L The fresh leaves and pounded stems are applied to burns, wounds and also to saddle sores of horses, camels, etc. The stem is also used for gastrointestinal complaints or as a stomachic sometimes taken in the form of the succulent stem boiled and sugared. In Guinea the stems and leaves are given to cattle to induce milk and in Senegal a decoction of the stems and leaves is used as a friction and wash in pains with fever and in malaria (Dalziel, 1937).

C The plant was found to contain a steroid which can be separated into two fractions (Sen, 1966). Further, a water-soluble glycoside has been obtained from the plant, which on oral administration had no toxic effect in mice, rats or guinea pigs (2 mg/kg for 10 days). On intravenous administration, however, the animals showed convulsions and died with 5 min.

P Das and Sanyal (1964) noticed that an alcoholic extract of the plant (containing resins and sterols) acted upon the isolated intestine and the uterus of rabbits and albino rats in a manner comparable to that of acetylcholine. The effect was also observed in situ on the tracheal and intestinal muscles of the dog. The LD_{50} was 15.5 mg/kg in guinea pigs. The extract has a favourable effect on gastrointestinal evacuation and is recommended in cases of indigestion, dyspepsia and gastritis (Das and Sanyal, 1964). In dogs the glycoside fraction produced dose-dependent hypotension. The negative chronotropic effects on the myocardium can be overcome by 7.5 M calcium (Subbu, 1970). It is believed to act on the cell membrane by inhibiting the movement of Ca^{2+} into the cell substance (Subbu, 1971). Intramuscular administration of an extract of *C. quadrangularis* to rats and local use as an ointment in dogs was shown to reduce the convalescence time of experimental cortisone-treated fractures by 33% (cortisone has an anti-anabolic action and delays consolidation) (Udupa and Prasad,

Fig. 3.9. *Cissus quadrangularis* L.

1964; Prasad and Udupa, 1963). A potent anabolic steroid isolated from the plant has been shown to have a marked influence on the rate of fracture healing; it induces an early regeneration process of all connective tissues involved in the healing and a quicker mineralization of the callus (Udupa *et al.*, 1965). Indeed, after 6 weeks the bones recovered 90% of their original strength. Calcium-45 uptake studies indicated early completion of recalcification and earlier remodelling. This steroid fraction appears to have androgenic properties and produces an increase in body weight and the total weight of the testes in animals (Prasad and Udupa, 1963; Udupa and Singh, 1964; Udupa *et al.*, 1965; Prasad *et al.*, 1970).

The pathway to the site of action of the phytogenic steroid can be studied by tagging it with radioactive ^{14}C. The site of action is located in the rat by microautoradiography. The probable pathway seems to be through the anterior pituitary gland, then by the adrenals and testes. After some metabolism in the liver, the steroid reaches the osteogenic cells at the fracture site, where it seems to exert a stimulating effect on the healing of the fracture (Prasad and Udupa, 1972).

Sida cordifolia L. MALVACEAE

L The plant provides fibre and is a troublesome weed. In India the rootbark, pulverized and mixed with oil of sesame and milk, has been said to be effective in cases of facial paralysis and sciatica (Chopra *et al.*, 1956). Ephedrine has been found in this plant and has been listed in the Indian Pharmaceutical Codex (1953) for the relief of hay fever and asthma (Oliver, 1960).

C Recent analyses have revealed that ephedrine and ψ-ephedrine constitute the major alkaloids in the aerial parts (minor bases in the roots). From the aerial parts and the roots the following were also obtained: β-phenylethylamine, carboxylated tryptamines, quinazoline alkaloids, S(9)Nb-tryptophan methylester, hypaphorine, vasicinone, vasicine and vasicinol in varying amounts (Ghosal *et al.*, 1975). In **S. acuta** and **S. rhombifolia**, also found in West Africa, the main alkaloid in Sri Lanka proved to be cryptolepine, which was first found in *Cryptolepis* (Gunatilaka *et al.*, 1980). In *S. acuta* growing in India, the alkaloids cryptolepine and ephedrine were found in the roots as well as α-amyrin (Krishna Rao *et al.*, 1984).

P The favourable combination of three sympathomimetic amines and a potent bronchodilator principle (vasicinone) would, according to Ghosal *et al.* (1975), account for the major therapeutic uses in asthma, hay fever, etc. Vasicine is said to be a promising uterotonic abortifacient (it is mainly obtained from *Adhatoda vasica* (Gupta *et al.*, 1978).

Anticonvulsant and antipyretic activities of the plant (collected in India) have been observed by Dhar *et al.* (1968). In addition these authors note extensive antibacterial, antifungal and antiviral effects as well as antiprotozoal action on *Entamoeba histolytica* and anthelmintic action towards *Hymenolepis nana* and *Ascarides galli*. A hypoglycaemic effect and action on the smooth muscles and heart were also reported. Anticancer activity against human nasopharynx carcinoma (in tissue culture) and lymphoid leukaemia and sarcoma 180 in mice were revealed in CCNSC tests in the USA (Dhar *et al.*, 1968).

Crateva religiosa Forst. syn. (*C. adansonii* Oliv.) CAPPARIDACEAE

L The root is used in Nigeria as a febrifuge and the Yorubas apply the leaf as a mild counter-irritant for headache (Dalziel, 1937). In India, the stembark has been used as an antipyretic, stomachic, laxative and diuretic (Chopra *et al.*, 1956). In the Philippines, the juice of the bark is used to treat convulsions (Quisumbing, 1951).

C In India, the bark was found to contain a gum, a saponoside and tannins. From the air-dried powdered bark, the triterpenes lupeol, β-sitosterol and lupeol acetate have been isolated. The water-soluble portion contained traces of quaternary and tertiary bases and sugars. The tertiary bases were found to contain both sulphur and nitrogen (Bhandari and Bose, 1954; Chakravarti *et al.*, 1959; Kjaer and Thomsen, 1963; Smolenski *et al.*, 1972; Kondagbo and Delaveau, 1974). In the leaves and twigs nine flavonoids, mainly rutin, quercetin and isoquercetin have been reported (Hegnauer, Vol. 3, p. 362).

P The water-soluble fraction of the air-dried bark had spasmodic action, which was not blocked by atropine, on the uterus of the rat, guinea pig, rabbit, dog and humans. It was also observed to have cholinergic action on the isolated ileum of the guinea pig and on dog tracheal muscle preparations. Nicotinic action of the extract on the ganglia has also been noted by Deshpande (1973). Clinically, ingestion of a powder of the whole plant has been observed to improve the tone of the urinary bladder in 12 cases of post-operative prostatic enlargement and it could even remove smaller stones from the ureter and bladder and control various urinary tract infections. This action has been attributed to the cholinergic action of the drug on smooth muscles (Deshpande, 1973).

A petroleum ether extract and sterols isolated from the stembark significantly inhibited the acute inflammation induced by carrageenan and histamine in albino rats and inhibited the early and delayed phases of inflammatory changes in formaldehyde-induced arthritis (Ramjelal *et al.*, 1972). Total extracts had an inhibiting effect towards *Shigella dysenterica* and the leaves and stembark had (considerable) anticancer action on sarcoma 180 (Kerharo and Adam, 1974) and on lung cancer (leaves) (Abbot *et al.*, 1966).

Borreria verticillata (L.) Mey. syn. (*Spermacoce verticillata* L., *S. globosa* Schum. & Thonn.) RUBIACEAE

L A lotion of the plants is used in Senegambie for febrile children and in the treatment of leprosy, furuncles and paralysis (together with *Datura metel*). It is also said to be diuretic and abortive (Dalziel, 1937; Kerharo, 1968).

C Plants collected in the Ivory Coast and Senegal have been found to contain 0.20% of total alkaloids. Emetine or cephaline, reported in older publications (Moreira, 1963) could not be detected, but two other alkaloids have been reported to be present in the aerial parts. They are borreverine, with a tetrahydro-β-carboline nucleus, and borrerine (Pousset *et al.*, 1973). An essential oil is found in the aerial parts; from it, a sesquiterpenic lactone was isolated (Benjamin, 1979). Iridoids, daphyloside, asperuloside and feutoside have been obtained from the rootbark (Sanity *et al.*, 1981).

P Pharmacological screening in Brazil showed that this plant has a stimulating action on the uterus of the rat and on the duodenum of the rabbit. No actions on the blood pressure and respiration of the cat could be detected nor were any effects observed on the striated frog muscle and guinea pig intestine. Toxicity to mice and fish was nil (Barros *et al.*, 1970) and no insecticidal activity was found (Heal and Rogers, 1950). The antimicrobial activity of the essential oil was tested by Benjamin (1979) and revealed inhibition of *Escherichia coli* and *Staphylococcus aureus*.

Cardiospermum halicacabum L. syn. (*C. microcarpum* Kunth) SAPINDACEAE

L In Nigeria the leaves are sometimes rubbed on the skin for the treatment of skin eruptions, itch, etc. or applied as a poultice to swellings. The juice of the stem is dropped in the eye to treat ophthalmia. The leaf and root are used as a remedy for nervous diseases in many countries (e.g. Australia and South Africa) (Watt, 1967). After eating the seeds in quantity children may develop epileptiform convulsions.

C Stigmasterol, probably in the form of a glycoside, and quebrachitol have been isolated from the air-dried plant in India and proanthocyanidin and apigenin have been isolated from an alcoholic extract of the roots (Dass, 1966).

P The water-soluble fraction of a dried alcoholic extract of the seeds produced an initial depression followed by marked stimulation in the isolated frog heart preparation (Moti and Deshmanhar, 1972).

4

Anti-infective activity of higher plants

Ideally, the plants used for anti-infective therapy should be toxic to infectious organisms and devoid of toxicity for human beings. The aim is to obtain the highest possible favourable ratio between the dose toxic to man and that active against the agent of infection. Biochemical differences between the infective agent and the host should allow the finding of plant constituents that are selectively toxic to the infecting organism.

The plants used for anti-infective therapy can be divided into two groups:

I *Anti-infective higher plants*

These include:

(a) antibacterial plants;
(b) antifungal plants;
(c) antiviral plants;
(d) antiprotozoal plants;
(e) antimetazoal plants (anthelmintics).

II *Plants with insecticidal and molluscicidal activity* (parasitic hosts)

(a) insecticidal plants;
(b) molluscicidal plants.

The term antibiotics, which was first given to substances produced by fungi and bacteria that inhibit the vital processes of certain microorganisms other than the species producing them, has been extended by many authors to those constituents of higher plants which have similar effects in very low concentrations (e.g. Paris and Moyse, 1965). Those constituents which act on pathogenic fungi, viruses and protozoa are also included (Patel *et al.*, 1967) and the term has been used in this sense throughout the book.

The antibacterials inhibit the multiplication of bacteria and can be either bactericidal or bacteriostatic. The lowest concentration that completely inhibits the microorganisms after exposure in vitro for a specified period is referred to as the minimum inhibiting concentration (MIC). Conventional antibacterial drugs are known to inhibit cell division in the microorganisms by interfering with *p*-aminobenzoic acid, which is a co-factor in the synthesis of folic acid, others interfere with protein synthesis in the bacteria, others again attack the cytoplasmic membrane by dissociating its lipoproteic structure and another group of antibacterial drugs inhibit the building up of new cell walls during cell division by blocking the enzyme transpeptidase, which controls their synthesis.

The main endemic bacterial diseases in West Africa include leprosy, tuberculosis, occasionally cholera, bacillary dysentery, enteric fevers and undulant fever.

The demand for antifungal drugs is considerable in a warm, damp climate in which fungal diseases are rife. The diseases are mainly ringworm (*Tinea imbricata*, *T. cruris*, *T. pedis*, *T. unguium*), caused by e.g. *Epidermophyton floccosum*, *E. concentricum*, *Trichophyton mentagrophytes*, *T. rubrum* and *Nocardia minutissima*; pityriasis versicolor (Tinea versicolor), caused by e.g. *Cladiosporum mansoni* and *Malessesia furfur*; and aspergillosis (in lungs), caused by *Aspergillus fumigatus*. Many of the antimycotic drugs are used in topical application.

Virus diseases in West Africa include yellow fever, rabies, poliomyelitis and trachoma, but of course dengue, influenza and measles also occur. As opposed to the bacteria, which have their own reproductive system, viruses use certain syntheses of the cell itself for their reproduction. This explains why they can change the function of normal cells. No relevant information about the action of antiviral drugs has emerged from the pharmacological tests described in this section.

The main protozoal diseases in West Africa are malaria (parasite genus *Plasmodium*); African trypanosomiasis; leishmaniasis, including kala-azar (visceral form) and that caused by *Leishmania tropica* (oriental sore); and, rarely, amoebiasis (amoebic dysentery, parasite *Entamoeba histolytica*) and giardiasis (parasite *Giardia intestinalis* (Lamblia)).

Metazoal diseases in West Africa are caused by nematodes, trematodes and cestodes. The main infectious agents include (Manson-Bahr, 1952):

Nematodes acting on the intestine; hookworm (*Ancylostoma duodenale*), whipworm (*Trichuris trichiura*), pinworm (*Oxyuris vermicularis*), roundworm (*Ascaris lumbricoides*), trichinose (*Trichinella spiralis*)

Nematodes acting on tissues: guinea worm (*Wucheria bancrofti*) (filariasis), guinea worm (*Dracunculus medinensis*) (dracontiasis)

Nematodes acting on eyes: blinding worm (*Onchocerca volvulus*) (onchocerciasis), eye worm (*Loa Loa filaria*) (loa loa filariasis).

Trematodes: flukes (*Schistosoma mansoni*) (bilharzia)

Cestodes: adult and larval tapeworms (*Taenia solium*, *T. saginata*, *Echinococcus granulosis*, *Hymenolepis nana*)

Plants with anti-infective activity which have already been described in Chapter 2 for their action on the cardiovascular system are indicated hereunder by CV. Similarly, the plants already described in Chapter 3 for their effects on the nervous system are indicated by N.

The plants marked by an asterisk (*) in the enumeration of the constituents at the beginning of each action group are described in this text, the others can be found in the corresponding tables. Often several uses are indicated after each plant name, in which case a dagger (†) is placed after the use under which the description of the plant can be found.

So far, anti-infective experimentation with plants has been very superficial. Often, the testing amounts to little more than in vitro exposure of material to organisms which may or may not be pathogenic to man. Nevertheless, the tests carried out by many authors on non-pathogenic organisms are indicated here: some organisms formerly believed to be non-pathogenic have lately been found to be possibly pathogenic or may be pathogenic to cattle or domestic pets.

The organisms discussed below that are considered non-pathogenic are *Aspergillus flavus*, *Bacillus cereus*, *B. subtilis*, *B. mycoides*, *B. megaterium*, *Giardia muris* (rodents), *Micrococcus leirodeicticus*, *Mycobacterium smegmatis*, *Mycob. phlei. Penicillin chrysogenum*, *Saccharomyces cerevisiae*, *Sarcina lutea* and probably *Penicillin echyogenum*. *Paramecia*, a formerly much-used genus for testing, has been abandoned.

I Plants with antibiotic activity

A great variety of components belonging to different chemical groups were found to have antibiotic activities. Similar observations had been made by Lechat *et al.* (1978) and Brannon and Fuller (1973) concerning the antibiotics isolated from Streptomycetaceae and Thallophytae and were explained by them by the fact that the different antibiotics attack different sites in the pathogenic organisms

Comparison of the MIC of essential oils from plants in vitro tests with their tissue concentrations active in vivo showed that the mechanism of their anti-infectious action is different from that of the antibiotic type of action. Antibiotics require an in vivo concentration in the tissues that is equal to, or greater than, their minimum active concentration in vitro. In the case of essential oils, however, concentrations in the blood of about one-hundredth of their active concentration in vitro have been shown to heal patients with acute or chronic infections. Thus patients have been cured with essential oil arriving in the organism in doses quite insufficient to deal with the infectious agent in the laboratory, the in vitro activity of the essential oils being 100 times higher than that of antibiotics, whilst in vivo the actions are comparable. The results of tests on 268 clinical cases (Valnet *et al.*, 1978) suggest that essential oils have a general action on the organism of the patients. Valnet and co-workers call the microbiocidal use of the oils 'aromatotherapy' and their test-records 'aromatograms'. They believe that the action of the essential oils is based on a global action which modifies the general condition of the patient (action on neuro-endocrine functions?)

(a) Antibacterial plants

Active constituents

The plant constituents with antibacterial action are tentatively grouped according to their main chemical groups.

Phenols. In *Anacardium occidentale** the aromatic phenols cardol and anacardol (decarboxylated derivatives of anacardic acid) are not only bactericidal† and antifungal but also vermicidal and protozoicidal. The phenol chloropherin in *Chlorophora excelsa* is antibacterial and antifungal and in *Ocimum viride*, thymol (also vermicidal) and eugenol are antimicrobial.

Quinones. The napthoquinone plumbagin, found in *Plumbago zeylanica** and also in *Diospyros mespiliformis* and *Drosera indica**, is antibacterial,† antifungal, antiprotozoal and anthelmintic, and the benzoquinone from *Embelia schimperi** is slightly antibacterial but mainly anthelmintic.†

Acids. Citric acid in *Bryophyllum pinnatum**, is antibacterial and the fatty acids with a cyclopentene nucleus (chaulmoogric and gorlic acids) in *Caloncoba echinata** act on the lepra bacillus. The acidic phenols gallic acid (and ethylgallate) are reported to be antibacterial in *Bombax* spp. and antibacterial and anthelmintic in *Acacia farnesiana* and *Mangifera indica*.

Alkaloids. Berberine in *Argemone mexicana* (also found in *Chasmanthera dependens*) is said to be antimicrobial and antiprotozoal; sanguinarine acts as a lipolytic pro-drug and is antifungal. Cryptolepine in *Cryptolepis sanguinolenta** and *Sida* spp., canthine-6-one, chelerythrine and berberine in *Zanthoxylum zanthoxyloides** and solanine in *Solanum nodiflorum* are all antibacterial;† and this also applies to the indole alkaloids of *Strychnos afzeli* and funiferine from *Tiliacora funifera*.

Flavonoids. *Ageratum conyzoides** contains an antibacterial† flavone (the plant extract is also anthelmintic in vitro). *Combretum micranthum** and *C. racemosum** are antibacterial;† they contain flavonoids, alkaloids (combretines) and catechuic tannins. *Canscora decussta** (xanthones) inhibits *Mycobacterium tuberculosis** and is also antiviral, whilst *Psidium guaijava** has three flavonoids with a strong antibacterial† effect on *Mycob. tuberculosis*. The flavonoids of *Uvaria chamae** act on *Mycob. smegmatis**; they are also reported to be larvicidal.

Sulphur heterosides. Allicin in *Allium** spp. is antibacterial† and antifungal and this is also the case for the isothiocyanate glucosides in *Capparis decidua** (glucocapparin), *Lepidium sativum* (glucotropeolin), *Moringa oleifera** (rhamnosyl-oxybenzylisothiocyanate); and *Carica papaya** seeds (tropaeolin). Cleomin from *Ritchiea longipedicillata** is antibacterial and anthelmintic† (mainly). Hydrogen cyanide (HCN) found in *Acalypha wilkesiana** may account for the antibacterial action of this plant.

Terpenoids. Antibacterial activity is reported for *Borreria verticillata** (sesquiterpenic lactone), *Xylopia aethiopica** (diterpene, xylopic acid), *Azadirachta indica** (seeds, triterpenoids, also antiviral) and *Ekebergia senegalensis** (stembark, meliacins). The leaves of *Azadirachta* are insecticidal and those of *Ekebergia* are ichthyotoxic.

Proteolytic enzymes. Papain from *Carica papaya*, calotropine from *Calotropa procera* and bromelain from *Ananas comosus* are able to digest bacterial and parasitic cells and bromelain even digests worms.

Polyacetylenes and phenylheptatriene from *Eclipta prostrata* and *Bidens pilosa* (leaves) respectively, have a strong UV-mediated toxicity to bacteria and *Candida* and are also toxic to insects and larvae (Wat *et al.*, 1978, 1979, 1980).

Anacardium occidentale L. (Fig. 4.1) ANACARDIACEAE
Cashew nut tree

This tree has been described earlier (CV).

P Anacardic acid, which constitutes 39% of the 'cashew balsam' from the fleshy parts of the fruit (Gellerman and Schlenk, 1968) and is found also in crude extracts of cashew nut shells (Rahman *et al.*, 1978), is a mixture of 6-(n-C_{15}-alkyl) salicylic acids with side chains varying in degrees of unsaturation. Anacardic acid has been found to have moderate bactericidal activity against *Staphylococcus aureus* (Ogunlana and Ramstad, 1975). Investigation of certain decarboxylated derivatives of anacardic acid, cardol and anacardol, showed that these were not only bactericidal but also fungicidal, vermicidal, protozoicidal, parasiticidal and even anti-enzymatic (Eichbaum *et al.*, 1950; Jacquemain, 1959; Gulati *et al.*, 1964; Laurens and Paris, 1977). More recently anacardic acid, its acetate and the fully saturated analogues have been found to be very active against *Mycobacterium smegmatis* and moderately active against *Bacillus subtilis*. Good antifungal action against *Trichophyton mentagrophytes* and moderate action against *Saccharomyces cerevisiae* have also been reported (Adawadkar and El-Sohly, 1981). The antimicrobial activity of anacardic

Fig. 4.1. *Anacardium occidentale* L.

acids (which are also the main constituents of *Ginko biloba*) had been reported earlier by Gellerman *et al.* (1969). The anthelmintic action of the *Anacardium* nut-shell liquid had been tested before in ancylostomiasis, ascaridiasis and trichuriasis and had given satisfactory results (Eichbaum *et al.*, 1950). It has further been shown that the molluscicidal activity of crude extracts of nut shells requires both the unsaturated side chain and the carboxyl group of anacardic acid (Sullivan *et al.*, 1982).

The antibacterial activity of metallic complexes of anarcardic acid with mercury, zinc, copper, manganese and cobalt was the subject of tests by Chattopadhyaya and Khare (1970) against 17 test organisms. The strongest antimicrobial activity of anacardic acid proved to be against *Staphylococcus aureus*. The mercury complex was particularly active against *Staph. aureus*, *Streptococcus pyogenes*, *Escherichia coli* and *Bacillus pumilis*.

Another Anarcadiaceae, *Heeria insignis* (Del.) O. Ktze. has also been found to have antimicrobial activity; it gives positive reactions for tannins and saponins. In West Africa it is much employed as an anthelmintic and antidysenteric (Delaveau *et al.*, 1979).

Plumbago zeylanica L. PLUMBAGINACEAE

Ceylon leadwort

L The root is a vesicant and counter-irritant. Dried and pulverized it is added to a maize pap as a remedy for parasitic skin diseases and in Southern Nigeria the leaves are put in soup as a remedy for worms and for fever. In Ghana the root is administered as an enema to treat piles (Dalziel, 1937). In the Ivory Coast and Upper Volta it is used in the treatment of leprosy (Kerharo and Bouquet, 1950).

C The roots contain plumbagin or plumbagol, a 2-methyl-5-hydroxy-1,4-naphthoquinone, in amounts of 1.26% in the Ivory Coast (Paris and Moyse-Mignon, 1949; van der Vijver and Lötter, 1971). (Plumbagin is also found in *Diospyros* and *Drosera* spp.) The leaves and stems of *P. zeylanica* contain only very small amounts in addition to a fixed and a volatile oil (Watt and Breyer-Brandwijk, 1962).

P Plumbagin has vitamin K action and antibacterial properties. In India it is also said to stimulate the secretion of sweat, urine and bile and to have a stimulating action on the nervous system and on the muscular tissue (Bhatia and Lal, 1933). In a concentration of 1:50000, plumbagin has a marked antibiotic action on staphylococci and certain pathogenic fungi (*Coccidoides imminentis*, *Histoplasma capsulatum*, *Trichophyton ferrugineum*) and on parasitic protozoa (Carrara and Lorenzi, 1946; Van der Vijver and Lötter, 1971). Intravenous injections in patients with boils, anthrax or cystitis have been well tolerated and brought about a rapid recovery (Saint Rat *et al.*, 1946, 1948; Saint Rat and Luteraan, 1947; Blanchon *et al.*, 1948; Vichkanova *et al.*, 1973a). Experimental *Microsporum* infections in mice have been healed by local applications of 0.25–0.5% solutions in 40% alcohol or 1% emulsions of plumbagin (Vichkanova *et al.*, 1973a). In vitro, the growth of *Staphylococcus aureus*, *Streptococcus pyogenes* and *Pneumococcus* has been completely inhibited by solutions of plumbagin at concentrations of 1 : 100000; that of *Mycobacterium tuberculosis* at concentrations of 1:50000 and the growth of *Escherichia coli* and *Salmonella* at concentrations of 1:10000 (Skinner, 1955; Oliver, 1960; Vichkanova *et al.*, 1973b). Antispasmodic activity of plumbagin was also reported, by Bézanger-Beauquesne and Vanlerenberghe (1955), but it proved inactive in the treatment of infection by *Haemophilus pertussis*, Plumbagin (isolated from *P. capensis*) showed a potent antifeedant activity against larvae of the army worms (*Spondoptera exempta*) at 10 p.p.m. and of *S. littoralis* at 20 p.p.m., and caused their complete inhibition at a concentration of 12.5 μg/ml; it has proved to be antimicrobial in *Candida utilis* and *Saccharomyces cerevisiae* (Kubo *et al.*, 1980).

Drosera indica L. DROSERACEAE

L The plant is used in India as a powerful rubefacient and a maceration is applied topically to corns in Vietnam (Chopra *et al.*, 1956).

C The plant contains naphthoquinones, mainly plumbagin (see *Plumbago zeylanica*) (Bézanger-Beauquesne and Vanlerenberghe, 1955).

P *Drosera rotundifolia*, *D. coryifolia* and *D. intermedia* have been reported to prevent bronchospasms produced by acetylcholine or histamine and to decontract in vitro spasms of the intestine caused by acetylcholine or barium chloride. They are said to be antitussive and to prevent coughing induced by excitation of the larynx nerve in the rabbit (Paris and Moyse, 1967, pp. 227–9). The napthoquinones also have an antimicrobial action; a plumbagin solution inhibits the growth of *Staphylococci*, *Streptococci* and *Pneumococci* in concentrations of 1:50000 and has been used against whooping cough although practically no action against *Haemophilus pertussis* was noted (Bézanger and Vanlerenberghe, 1955) and is used in the treatment of other severe, persistent coughs. France was said to use about ten tons yearly in 1969 as an antitussive (Paris and Moyse, 1967, Vol. II, p. 229). Denoël (1958) reported that plumbagin is an isomer of phthiocol, a constituent of the tubercle bacillus, and some authors suppose that substitution of phthiocol explains the activity of *Drosera* on the respiratory tract. In fact certain plumbagol-sulphamide compounds showed an antituberculostatic activity in vitro. According to Vichkanova *et al.* (1973a) the antimicrobial spectrum of plumbagin includes Gram-positive and Gram-negative bacteria, influenza virus, pathogenic fungi and parasitic protozoa. However, they report it to be ineffective against *Giardia muris* (a protozoan parasite to rodents) and tuberculosis in mice when administered orally for 5 days. They successfully treated experimental *Microsporum* infections in guinea pigs by local application of 0.25–0.5% solutions in 40% alcohol or 1% emulsions of plumbagin.

Bryophyllum pinnatum (Lam.) Oken syn. (*B. calycinum* Salisb., *Cotelydon pinnata* Lam., *Kalanchoe pinnata* (Lam.) Pers.) CRASSULACEAE

Never die or Resurrection plant (from viviparous properties)

L The crushed leaves (or the juice squeezed out after heating) are mixed with sheabutter and oil and the mixture is applied to abscesses, swellings, ulcers and burns or used to rub the bodies of young children suffering from fever. The juice is also applied for earache and ophthalmia (Dalziel, 1937). In India, where the leaves are also used as an application on bruises, wounds, boils and insect bites, the leaves of an Indian species (*Kalanchoe laciniata*) are similarly used and are said to allay irritation and promote healing; the juice is considered styptic and is also administered in the treatment of bilious diarrhoea and lithiasis (Chopra *et al.*, 1956).

C The leaves contain bryophyllin, potassium malate and ascorbic, malic, isocitric and citric acids (Mehta and Bhat, 1952; Chopra *et al.*, 1956; Gaind and Gupta, 1969).

P Studies of the antimicrobial activity of the juice obtained from the heated leaves, using the agar diffusion method, showed an inhibition zone of more than 20 mm when used on test organisms of *Staphylococcus aureus*, *Bacillus subtilis*, *Escherichia coli* and *Pseudomonas aeruginosa*. It has also been noted that the heated leaf, when applied to inflamed areas, produces a soothing effect, keeping wounds clean and preventing them from going septic (Boakaiji Yiadom, 1977).

Caloncoba echinata (Oliv.) Gilg syn. (*Oncoba echinata* Oliv.) FLACOURTIACEAE
Caloncoba glauca (P. Beauv). Gilg syn. (*Ventenatia glauca* P. Beauv., *Oncoba glauca* (P. Beauv.) Hook. f., *C. dusennii*)
Caloncoba Welwitschii (Oliv.) Gilg
Gorli

L A lotion made from the plant is used by several native tribes in Guinea, Sierra Leone and Ghana for pustular eruptions of the skin. The seeds contained in the fruit capsule have long been known to contain chaulmoogric acid but the toxic ingredients of the acid cause nausea and vomiting and irritation of the mucous membrane of the stomach, and hydnocarpic acid is therefore preferred (Dalziel, 1937). The *Caloncoba* spp. are not used much in modern leprosy treatment.

C The three *Caloncoba* spp. have nearly the same constituents as chaulmoogra oil, which is obtained by squeezing out the fresh ripe seeds of *Taraktogenus kurzii* from Burma. However, the oil of the *Caloncoba* spp. is less appreciated than chaulmoogra or mainly hydnocarpus oil (from *Hydnocarpus* spp.) as the seeds are very small, extraction is laborious and the fat is less suitable for injection. The *Caloncoba* spp. contain in general 30–50% lipids of which 1–3% is non-saponifiable. The fat consists in West Africa of 60–80% chaulmoogric acid (against 50–70% hydnocarpic acid in *Hydnocarpus* spp. which have been acclimatized in Nigeria, the Cameroons, Guinea and the Ivory Coast); the remaining lipids consist in both cases of 8–15% gorlic acid and 10–12% of ordinary fats (oleic and palmitic acids) (Chevalier, 1928; Pelt, 1959; Paris and Moyse, 1963). Both chaulmoogric and hydnocarpic acids have saturated side chains.

P In 1967 there were some 10–12 million lepers in the world (Paris and Moyse, 1967). Chaulmoogra remains a classic medicine although often associated with sulphones, which are more effective in the malignant forms of the disease, and might be used as a suspension of diaminodiphenylsulphone in acetylchaulmoograte. The treatment is lengthy (250 ml/year is required) and should be combined with an adequate diet. Chaulmoogra oil is also used for certain skin complaints (lupus) and as a parasiticide in veterinary medicine (Pelt, 1959; Zenan and Podkorny, 1963).

Cryptolepis sanguinolenta (Lindl.) Schltr. ASCLEPIADACEAE

The hypotensive and antipyretic properties of this plant have already been briefly described (CV). In local medicine it is also reputed to be active in the treatment of urogenital infections and malaria. Hence Dwuma Badu *et al.* (1978) and Boakaiji Yiadom (1979) have carried out antibacterial tests. In those against urogenital pathogens the aqueous extracts of the roots showed antimicrobial activity against *Neisseria gonorrhoeae*, *Escherichia coli* and *Candida albicans* but not against *Pseudomonas aeruginosa*. With 20.0 and 10.0 g/l solutions the inhibition zones were approximately equivalent to those produced by a solution of 25 μg/ml of trihydrate of ampicillin used as a control, only the effect on *Candida albicans* was slightly weaker, especially with the smaller dose. The effect of 5.0 g/l was inferior to that of the control for all the test organisms. Cryptolepine has no plasmocidal effect (Boakaiji Yiadom, 1979).

Similar effects were reported with the hydrochloride of cryptolepine (an indoloquinoline alkaloid isolated from the roots) (Gellert *et al.*, 1951; Dwuma Badu *et al.*, 1978; Boakaiji Yiadom and Heman Ackah, 1979). Recently, Bamgbose and Noamesi (1981) have reported an inhibition by cryptolepine of carrageenan-induced oedema.

Zanthoxylum zanthoxyloides (Lam.) Watson syn. (*Fagara zanthoxyloides* Lam., *F. senegalensis* (DC.) Chev., *Z. senegalense* (DC.) Chev., *Z. polyganum* Schum.) RUTACEAE

Prickly ash, toothache bark, candlewood (see also CV)

L The root of this plant is much used as a chewing stick in Nigeria. Other common chewing sticks in the country are: *Vernonia amygdalina* root and stem, *Terminalia glaucescens* root, *Massularia acuminata* stem, *Garcinia kola* root. Chewing sticks are believed to have antimicrobial properties (Fadulu, 1975).

C Chromatographic and chemical purification of an ethanolic extract of the powdered root of *Zanthoxylum zanthoxyloides* yielded four compounds which showed antimicrobial activity: canthin-6-one, a tertiary phenolic alkaloid, two quaternary alkaloids with antimicrobial action (chelerythrine and berberine) and a compound whose structure is still to be determined (Odebiji and Sofowora, 1979).

P The antimicrobial effect of the buffered extracts of all these chewing sticks on the oral flora has been tested by the streak-plate method and it has been shown that all are to some extent active, whilst the controls showed a heavy increase in microorganisms although to differing degrees. The action against more than 20 organisms, including Gram-positive and Gram-negative bacteria as well as *Candida* spp. and protozoa (*Entamoebia gingivalis*) was examined. Canthine-6-one (also isolated from *Zanthoxylum elephantiasis*) has been shown to be consistently active against *Staphylococcus aureus*, *Klebsiella pneumoniae*, *Mycobacterium smegmatis* and *Candida albicans* (Mitscher *et al.*, 1972a, b). The rootbark also contains fagarol, which has been found to be identical with sesamine from *Sesamum indicum* (Karrer, 1958). The root of *Z. zanthoxyloides*, in addition to giving the biggest zone of inhibition of all the chewing sticks tested, in the seeded blood agar plate also preserved the colour of the blood in that zone (El Said *et al.*, 1971), but this effect was absent in the blood agar plates for the other plant extracts. Further investigation of this curious phenomenon showed that the extract was able to reverse sickling and crenation in erythrocytes in vitro (see CV) (Isaac-Sodeye, 1971; Sofowora *et al.*, 1975).

Ageratum conyzoides L. COMPOSITAE

L The common medicinal use of the plant in West Africa is for healing wounds, especially burns. For this treatment the juice of the bruised leaves is squeezed into the wound, which is then covered by a bruised but intact leaf. In Nigeria the patient's chest is also rubbed with the leaves of the plant as a treatment for pneumonia (Durodola, 1977).

C The whole plant contains an essential oil (0.16% of the dried plant) composed of phenols (traces of free eugenol), phenolic esters, coumarin and ageratochromone. From the Indian plant 6-dimethoxy-ageratochromene and ageratochromene have been isolated (Kasturi and Manithomas, 1967) and Rudloff (1969) has found ageratochromone to be the principal constituent (75%) of the essential oil of the leaves in the Indian plant. In the West African plant a new flavone, 5,6,7,8,3′,4′,5′-heptamethoxyflavone, has been identified in the stems and leaves as well as stigmasterol dotriaconthene, 7-methoxy-22-dimethylchromene and a new flavone, conyzorigun (Adesogan and Okunade, 1979).

P The untreated leaves of the plant have proved significantly superior to vaseline gauze as a wound dressing, and the juice has displayed antibacterial activity in vitro against *Staphylococcus aureus*. Different fractions of the plant extract have been tested against *Staph. aureus* and the greatest activity was observed with the petroleum ether fraction containing more than 90% of a new flavone, assumed to be 5-methoxynobilitin (Durodola, 1977). *A. conyzoides* has also shown an in vitro anthelmintic activity (Albert *et al.*, 1972).

Combretum micranthum G. Don syn. (*C. altum* Perr., *C. floribundum* Engl. & Diels, *C. raimbaultii* Heck.) COMBRETACEAE
Kinkeliba

L In Nigeria and Guinea a decoction of the root is considered an anthelmintic and is also used as a wash for sores. Diuretic and cholagogic properties are attributed to the leaves, which are given as an infusion for the treatment of bilious haematuric fevers accompanied by vomiting, and also for ordinary colic and nausea, to prevent vomiting. Hot decoctions of the leaf and root are applied as a vapour bath, as a wash for febrile patients and for lumbago. The pulverized dry fruits are mixed with oil to form an ointment for application to suppurating swellings, abcesses, etc.

C Flavonoids, heterosides of vitexine (8-C-D-glucopyranosyl apigenine-12) and its isomer saponaretine have been reported to be present in the leaves as well as quaternary amino bases comprising two major alkaloids (combretines A and B, stereoisomers of betonicine) and traces of a third base (Jentzch *et al.*, 1962; Ogan, 1972). In addition, choline has been found (also in the flower buds) as well as betaine and gallic acid, both free and esterified as catechols and catechuic tannins, together with organic acids and mineral salts (potassium nitrate) (Paris, 1942; Paris and Moyse-Mignon, 1956). Alkaloids have been reported in the bark (Popp *et al.*, 1968).

P The antidiuretic and anticholagogic action of the drug has long been known (Paris, 1942). Amongst the constituents, combretum-catechine has proved to be strongly diuretic and slightly hypotensive (Paris, 1942), whilst the choleretic action is believed to be related to combretoside (saponaretin heteroside) (Grégoire, 1953). The leaves and extracts of *C. micranthum* are reported to have an antibiotic action against *Staphylococci*, *Streptococci* and *Escherichia coli* (Mela, 1950). The bark of the tree growing in Nigeria has shown an antibiotic action against Gram-positive and Gram-negative organisms (Malcolm and Sofowora, 1969).

The young leaves of **C. racemosum** Beauv. (Fig. 4.2) are considered to be anthelmintic in Nigeria and are used by certain tribes in Senegal (Balant) against internal parasites (Dalziel, 1937; Kerharo and Adam, 1974).

Two other Combretaceae are known for their anthelmintic action: **Anogeissus leiocarpus** (DC.) Guill. and Perr. syn. (*A. schimperi* Hochst ex Hutch. & Dalz.), the root- and stembark or seeds of which are used mainly for tapeworm in horses and donkeys (see Table 4.5) and **Quisqualis indica** L. (rangoon creeper), commonly grown in the area, the seeds of which are used in India, its country of origin, as an anthelmintic (ascarides) (Chopra *et al.*, 1956), and the active principle of which is believed to be an alkaloid.

Fig. 4.2. *Combretum racemosum* Beauv.

The roots of **C. rhodanthum** Engl. and Diels (*C. comosum* of F.T.A.) act on Gram-positive and Gram-negative organisms, in particular on *Sarcina lutea*, *Staphylococcus aureus* and *Mycobacterium phlei* (Malcolm and Sofowora, 1969).

Canscora decussata (Roxb.) Roem. & Schult. syn. (*Pladera decussata* Roxb.)
GENTIANACEAE

L In India the fresh juice of the plant is prescribed for insanity, epilepsy and nervous debility (Chopra *et al.*, 1956).

C Triterpenes, alkaloids, mangiferin (flavonic heteroside) and a number of xanthones have been isolated from the roots (Chaudhury and Ghosal, 1971; Ghosal *et al.*, 1971).

P The alcoholic extract of the plant has been found to have CNS depressant action in mice, producing a fall in blood pressure, and to stimulate the isolated smooth muscles of the rabbit intestine and of the uteri of rats and guinea pigs. It enhances the effect of acetylcholine on skeletal muscles. The extract also possesses antiviral activity (Bhakuni *et al.*, 1969). Spermicidal action of the plant has also been reported (Madran, 1960). Rat sperm had been killed in 5 min by a concentration of 1 : 50000 of the alcoholic extract of the plant and a concentration of 1 : 20000 of the aqueous extract. Human sperm required higher concentrations (Madran, 1960).

A number of xanthone derivatives isolated from *C. decussata* have shown an inhibitory activity in vitro against *Mycobacterium tuberculosis*, equivalent to that of streptomycin. Toxicity to mice was low. Mangiferin, also isolated from the plant, has proved to have weak antitubercular properties (Ghosal and Chaudhury, 1975), and also to have CNS depressant and significant anti-inflammatory activity in rats (Shankaranarayan *et al.*, 1979).

Psidium guajava L. MYRTACEAE
Guava

L The fruits are used locally for an antidiarrhoeial.

C The leaves contain an essential oil rich in cineol, tannins, four triterpenic acids and ursolic and oleanolic acids (Arthur and Hui, 1954). In addition Khadem and Mohammed (1958) have isolated three flavonoids from the leaves: quercetin, its 3-L-4-arabinofuranoside (avicularin) and its 3-L-4-pyranoside with strong antibacterial action.

P Nickel (1959) had reported the antibacterial action of guava leaves on Gram-positive and Gram-negative organisms. In 1969, Malcolm and Sofowora confirmed the antibacterial action on Gram-positive organisms (*Sarcina lutea* and *Staphylococcus aureus*) and also noted action on *Mycobacterium phlei* by Nigerian plants. The flavone derivatives isolated by Khadem and Mohammed (1958) were reported to inhibit the growth of *Staph. aureus* in a dilution of 1 : 10000.

Uvaria chamae Beauv. syn. (*U. cylindrica* Schum. & Thonn., *U. cristata* R. Br. ex Oliv., *U. nigrescens* Engl. & Diels, *U. echinata* Chev.) ANNONACEAE

L The root is regarded in Nigeria and Ghana as a purgative and antipyretic. The rootbark is used in the treatment of dysentery and respiratory catarrhs. The juice of the leaves is sometimes applied to wounds and sores or an infusion is used as a lotion for injuries, swellings, ophthalmia, etc. (Dalziel, 1937).

C Tannins and alkaloids were reported to be present in the roots and stembark by Persinos and Quimby (1967). *C*-benzylated flavonoids have also been isolated, including a *C*-benzylated monoterpene, chamaenen, dihydrochalcones and flavanones (chamanetin 5-methyl ether and dichamanetin methyl ether) (Lasswell, 1977; Hufford and El Sohly, 1978; Hufford and Lasswell, 1978; El Sohly *et al.*, 1979; Leboeuf and Cavé, 1980).

P The MIC values of these flavonoids and certain of their derivatives against *Staphylococcus aureus*, *Bacillus subtilis* and *Mycobacterium smegmatis* compare favourably with those of streptomycin sulphate. No activity was observed against *Escherichia coli* and the fungi *Aspergillus niger* and *Saccharomyces cerevisiae*. The dihydrochalcones were slightly more active than the flavanones. A number of the *C*-benzylated flavonoids are also cytotoxic. A larvicidal compound in this Annonaceae was also found to be a benzylflavanone (Towers and Wat, 1979).

Allium sativum L. LILIACEAE

Garlic

Allium cepa L.

Onion

These plants have already been mentioned as possessing hypotensive and antidiabetic properties (see CV). As they also have major antibiotic properties, they deserve mention here. In fact they are very potent fungicidal and antibacterial agents, useful in the treatment of fungal and staphylococcal skin and alimentary tract diseases (Vohora *et al.*, 1973). The unusually strong action of the *Allium* spp. in candida infections is of great importance as there are few active antibiotics which can be used in dermatology cases (Tynecka and Gos, 1973).

The antibiotic effects have been attributed to the action of allicin (diallyldisulphide oxide) (which is contained in the juice of garlic and onion) on the growth and respiration of microorganisms such as *Candida albicans*, *Staphylococcus aureus* and *Escherichia coli*. *Candida* was the most sensitive of these organisms to allicin, whilst *E. coli* seemed to be more resistant than *Staph. aureus* (Kabelik, 1970) The –SO–S– grouping is essential for the bacterial action of allicin as it inhibits the –SH– enzymes whereas –S–S–, –S– and –SO– groupings were not effective (Willis, 1956). It has been observed that the permeability of bacterial cells to allicin is greatly influenced by the lipid content of the microorganisms (Small *et al.*, 1949). *E. coli* contains about 20% lipids in the cell wall as compared to only about 2% in *Staph. aureus*, which may account for the fact that allicin more easily penetrates *Staph. aureus* cells as low lipid content facilitates penetration. The optimum conditions for the action are obtained in combination with oxidizing agents, which improve the antibiotic effect. The therapeutic effect of allicin is weakened through reduction and in the presence of blood; it is inactivated by cysteine but can be reactivated by H_2O_2 and reduced by thiosulphates (Kabelik, 1970). The allicin content of the bulbs varies with origin, moment of vegetation and soil conditions. In storage the amount decreases, depending on the storage conditions and species. Besides allinin, from which allicin is formed, *Allium* spp. contain several other sulphoxides which are converted by allinase to allicin analogues.

Capparis decidua (Forsk.) Edgew. syn. (*C. aphylla* Hayne ex Roth, *Sodada decidua* Forsk.) CAPPARIDACEAE

L The root and rootbark are pungent and bitter and are given in Persia and India for intermittent fevers and rheumatism. They are also used for boils and swellings of the joints and for their anthelmintic action (Dalziel, 1937; Chopra *et al.*, 1956).

C In the non-saponifiable petroleum ether fraction of the rootbark of the Indian plant, Gaind and Juneja (1969) identified *n*-pentacosane, triacontanol and β-sitosterol, and in the alcoholic extract two alkaloids, one of them being (−)-stachydrine, were found. From the flowers and fruits, *n*-pentacosane, β-sitosterol, (−)-stachydrine, choline and phthalic acid could be isolated. Flowers and seeds yielded 0.4% and 0.6%, respectively, of a volatile material composed of an organic sulphur compound. Later, Juneja *et al.* (1970a, b) identified glucocapparin (an isothiocyanate glucoside) in the seeds.

P An aqueous extract of the flowers, fruit pulp and rootbark has been found to have anthelmintic properties on earthworms (Gaind *et al.*, 1969a). Whereas the alcoholic extracts of flowers, fruit-husks and seeds were highly active against *Staphylococcus aureus*, *Escherichia coli*, *Bacillus subtilis*, *Proteus vulgaris*, *B. megaterium* and *Vibrio cholerae* (*inaba*), none had antifungal properties. A steam-volatile organic sulphur containing compound from the seeds showed both high antibacterial and antifungal action particularly at dilutions of 50 μg/ml against *Aspergillus flavus*, *Penicillium echyogenum* and *Candida albicans* (Gaind *et al.*, 1969b, 1972) and of 25 μg/ml against *V. cholerae*. In the case of *V. cholerae ogawa* complete inhibition was obtained with 25 μg/ml within 5 h, in that of *V. cholerae inaba* within 13 h and in the case of *V. cholerae el tor* in 5 h when the dose was 60 μg/ml (Gaind *et al.*, 1972).

Moringa oleifera Lam. syn. (*M. pterygosperma* Gaertn.) (Fig. 4.3) MORINGACEAE
Horseradish tree

Some properties of this plant and of the sulphurated amino bases of the rootbark of *moringa* have been described earlier (see CV). As mentioned the roots contain three antibiotic constituents, pterygospermine, athomine and spirochine.

Pterygospermine is a condensation product of two benzylisothiocyanate molecules with one benzoquinone molecule. This substance has powerful antibiotic and antimycotic effects. The toxicity of pterygospermine in the mouse (LD_{50}), given perorally, is 350–400 mg/kg, whilst at higher doses all animals die from respiratory arrest. The purified or crystallized pterygospermine has a broad antibacterial spectrum, acting on Gram-positive and Gram-negative organisms including *Micrococcus pyogenes* var. *aureus*, *Bacillus subtilis*, *Escherichia coli*, *Aerobacter aerogenes*, *Salmonella typhi*, *Salm. enterides*, *Shigella dysentariae* and *Mycobacterium tuberculosis*. Pterygospermine further inhibits filamentous fungi, including plant parasites. Thiamine and glutamic acid antagonize its antibiotic action, whilst pyridoxine enhances it (Kurup and Narasimha-Rao, 1954; Watt and Breyer-Brandwijk, 1962; Kerharo, 1969). The antibacterial action of pterygospermine on *M. pyogenes* appears to be based on its interference with the glutamic acid metabolism of the micro-organism (Eilert *et al.*, 1981).

Athomine, the second antibiotic substance, is particularly active against the cholera vibrion, showing a degree of activity which is intermediate between that of chloramphenicol and streptomycin (Chatterjee in Watt and Breyer-Brandwijk, 1962). This substance is entirely non-toxic to the rabbit (Kurup and Narasimha-Rao, 1954; Gupta *et al.*, 1956; Das *et al.*, 1957a, b; Kurup *et al.*, 1957).

The third antibiotic in the roots is spirochine, a sulphurated amino base which also acts on the myocardium. When administered intramuscularly or in local application spirochine is prophylactic and antiseptic against wound infections even in patients with a marked existing infection. Its action against *Staphylococcus aureus* is observed at a dilution of 1:70000 in vitro (Chatterjee, 1951 in Watt and Breyer-Brandwijk, 1962, p. 782). It has been observed to promote epithelization and some analgesic and antipyretic activities have been attributed to spirochine as well.

In the seeds of *Moringa oleifera* another derivative of benzylisothiocyanate (a glycosidic mustard oil) has been reported (Das *et al.*, 1958). This is 4(4′-acetyl-

Fig. 4.3. *Moringa oleifera* Lam.

α-L-rhamnosyloxy)-benzylisothiocyanate, a compound which has proved to have bactericidal action in a concentration of 56 μmol/l against *B. subtilis* and of 40 μmol/l against *Mycob. phlei*. It has proved more effective than the currently used isothiocyanate and could be a welcome substitute (its pure crystals being non-volatile, water-soluble and easy to handle) (Das *et al.*, 1958; Kjaer *et al.*, 1979; Eilert *et al.*, 1980, 1981).

The seed oil is used for lubrication and in cosmetics and the seed itself finds use in the Sudan and elsewhere as a natural coagulant for water purification (Eilert *et al.*, 1980, 981).

Ekebergia senegalensis A. Juss. syn. (*Charia chevalieri* DC., *E. chevalieri* (DC.) Harms, *C. indeniensis* Chev., *E. indeniensis* (Chev.) Harms, *E. dahomensis* Chev.)
MELIACEAE

L Mainly a timber tree, according to Sébire, the bark is used locally in Senegal for epilepsy (Dalziel, 1937).

C From the roots a bitter principle related to calicedrin has been extracted by Paris and Moyse-Mignon (1939). This has been identified by Bevan *et al.* (1965) as a mixture of two meliacins, which were named ekebergalactones B and C.

P Paris and Moyse-Mignon (1939) noticed that the water extract of the leaves stupefied goldfish and concentrations of 3 g/l killed them. Gaudin and Vacherat (1938) also noticed the ichthyotoxic effect of the macerated leaves. The stembark gave positive results for *Staphylococcus aureus* and *Sarcina lutea* in short antimicrobial tests on a series of Nigerian plants on Gram-positive organisms (Malcolm and Sofowora, 1969).

Carica papaya L. (see also CV) (Fig. 4.4) CARICACEAE

The aglycone of tropaeolin, a benzylisothiocyanate found in the seeds of *C. papaya*, has been studied for its antibiotic and pharmacological activity and for its toxicity. Maximum antibiotic action has been achieved with freshly crushed seeds and the average concentration of 6 mg/g of seeds has proved sufficient for therapeutic purposes. An antifungal effect has been found useful in the treatment of fungal skin diseases and benzylisothiocyanate may be a useful bactericidal agent for intestinal and urinary infections, being active against a wide range of microorganisms and eliminated in the urine 3–6 h after administration (Georges and Pandelai, 1949). Further the toxicity of the seeds is low; the therapeutically effective single dose of 4–5 g of seeds (25–30 mg of benzylisothiocyanate) is believed to be innocuous (El Tayeb *et al.*, 1974). The seeds also have anthelmintic activity (Krishnakumari and Majumdar, 1960; Dar *et al.*, 1965) and are carminative. Ripe and unripe fruits (epicarp, endocarp and seeds) showed very significant antibacterial activity on *Staphylococcus aureus*, *Bacillus cereus*, *Escherichia coli*, *Pseudomonas aeruginosa* and *Shigella flexneri* (Georges and Pandelai, 1949). The MIC was 0.2–0.3 mg/ml for Gram-positive organisms and higher (1.5–4 mg/ml) for Gram-negative organisms. The purified bactericidal agent showed protein-like properties (Emeruwa, 1982). Papain, an enzyme present in the fruit, has proteolytic properties. It is reported to be an anticoagulant (Pillai *et al.*, 1957) and may contribute to the antibiotic action. The leaves contain the alkaloid carpaine, which has amoebicidal action (Burdick, 1971).

Borreria verticillata (L.) Mey. syn. (*Spermacoce verticillata* L., *S. globosa* Schum. & Thonn.)

RUBIACEAE

L The plant is mainly found in the wet season and is used as an anti-eczematic in parts of Nigeria. The juice obtained from the aerial parts is applied topically (Benjamin, 1979). In East Africa the plant is also used in the treatment of skin diseases (Kokwaro, 1976).

C Two major indole alkaloids, borrerine, which appears to be a condensation product of tryptamine with isoprene, and borreverine, which may be formed by the union of two borrerine molecules, have been isolated by Pousset *et al.* (1973) from the aerial parts of the plant. These authors have not, however, been able to confirm the presence of emetine in the roots as had been reported by Orazi (1946). Seven iridoids have been isolated from *B. verticillata* rootbark (Sainty *et al.*, 1981).

Fig. 4.4. *Carica papaya* L.

P An antibacterial action of borreverine towards *Staphylococcus aureus* and *Sarcina lutea* has been reported by Maynart *et al.* (1980). The air-dried aerial parts provide a brown oil after several extractions. This oil inhibits the growth of Gram-positive and Gram-negative bacteria, and has been separated by chromatography into a hydrocarbon and an oxygenated terpene fraction. Purification of the hydrocarbon fraction produces a sesquiterpene fraction containing gaiene, caryophyllene and cadinene. This fraction inhibits the growth of *Escherichia coli* but not of *Staph. aureus*. The oxygenated terpene fraction, violet with a pungent smell, inhibits both *E. coli* and *Staph. aureus*. Evidence for antibacterial action of a sesquiterpenic lactone has been provided (Benjamin, 1979). Research designed to reveal insecticidal action has given no significant results (Heal and Rogers, 1950).

Xylopia aethiopica (Dunal) A. Rich. syn. (*Annona aethiopica* Dun., *X. eminii* Chev.)
ANNONACEAE

Guinea of Ethiopian pepper, spice tree

L In Nigeria the fruits are used as a cough cure, a carminative and a stimulating additive to other remedies. A fluid extraction or a decoction of the fruit or bark is recommended in the treatment of bronchitis, dysentery and biliousness. An extract of the seeds is also used to eliminate roundworm. The fruit is used as a condiment (Dalziel, 1937).

C From the fruit of the Nigerian species xylopic acid, a new diterpenic acid (15 β-acetoxy(−)kaur-16ene-19-oic acid), three diterpenic alcohols, one of them identified as kauran-16-α-ol, 4 diterpenic acids, including kaurenoic and 15-oxo-kaurenoic acid, two acyclic compounds, fats, oils and an essential oil have been obtained and cuminal (isopropylbenzaldehyde) has been obtained from the seeds (Ekong and Ogan, 1968; Ekong *et al.*, 1969a; Ogan, 1971).

P An extract of the fruit has been found to have some antibiotic effect on *Sarcina lutea* and *Mycobacterium phlei* (Malcolm and Sofowora, 1969) and on *Staphylococcus aureus*, *Bacillus subtilis*, *Pseudomonas aeruginosa* and *Candida albicans*. It is, however, inactive against *Escherichia coli* (Boakiji Yiadom *et al.*, 1977). This action seems due mainly to xylopic acid and to a lesser extent to two other diterpene isolates. The effect of xylopic acid is comparable to that of chloramphenicol.

Azadirachta indica A. Juss syn. (*Melia azadirachta* L., *M. indica* (Juss.) Brandis) (Fig. 4.5)
MELIACEAE

Neem, margosa tree

(Considerable confusion between this tree and *Melia azedarach* is found in the literature and must be watched!)

L Introduced and now naturalized in West Africa and Nigeria, this tree is mainly ornamental. The seed oil (margosa oil or neem oil) is used in India for the treatment of skin diseases and as a hair tonic (Indian Pharmacopoeia). The dried flowers find use as a tonic and stomachic and an infusion or tincture of the dried bark as a tonic and antipyretic.

C In India the seed oil has been found to contain bitter constituents. Those reported have been mainly 1.1% nimbidin (containing sulphur) and 0.1% nimbin and 0.01% nimbinin (both free from sulphur), which also occur in the stembark (Chatterjee *et al.*, 1948). Meliacins found in the seeds include gedunin, 7-desacetylgedunin, desacetylnimbin and azedarachtin (Manske and Holmes, 1950–71, Vol. 5, p. 423).

From the flowers a flavonoid, nimbicetin, later found to be identical with kaempferol, has been isolated, together with a bitter substance and a pungent bitter essential oil. In the dried bark the same bitter components as in the seed oil have been found, and in the pericarp of the fruit a bitter principle, bakayanin, has been found (Narayanan and Iyer, 1967). From the Nigeria species El Said *et al.* (1968) isolated nimbin, nimbidol and other bitter tetra-nor-triterpenoids as well as a resin and a tannin. Ekong *et al.* (1969b) obtained two new meliacin cinnamates, nimbolin A and B, from the wood of the trunk besides a small quantity of nimbin, fraxinellose and gedunin. In the fresh leaves they found deacetylnimbine and a lactone, nimbolid.

Fig. 4.5. *Azadirachta indica* A. Juss.

P In India, a marked antibacterial action of the seed oil on Gram-positive and Gram-negative organisms and even against resistant strains of tuberculosis has been reported. The growth of all strains of *Mycobacterium tuberculosis* and *Micrococcus pyogenes* var. *aureus* was inhibited by a concentration of 1 : 800 000 and the growth of *Shigella typhosa*, *Salmonella paratyphi* and *Vibrio cholerae inaba* was notably inhibited by a concentration of 2 mg/l; that of *Klebsiella pneumonia* by concentrations of 5 mg/l (Murthy and Sirsi, 1957; 1958). Nimbin and nimbidin also inhibited potato virus X (Verma, 1974). The leaves appear to be insecticidal and are used in India to protect woollen fabrics and books against insects (Chopra *et al.*, 1956). The triterpenoids of the seeds have been reported to repel certain parasites of cultivated plants and to have antiviral as well as antibacterial properties (Rai and Sethi, 1973; Anisimov *et al.*, 1974; Kraus *et al.*, 1978).

Azedarachtin, together with warburganal from East African *Warburgia cyanensis*, is considered one of the strongest antifeedants of African army worms hitherto discovered (Kubo *et al.*, 1977).

An aqueous extract of the root has been reported to reduce sarcoma 180 (Abbot *et al.*, 1966). An aqueous extract (1 : 1000) of the leaves of the Nigerian tree was found to act on the isolated guinea pig intestine, to contain a histamine-like substance (Oladele Arigbabu and Don Pedro, 1971) and to cause, on intravenous injection in the dog, an initial increase in blood pressure followed by a prolonged decrease with accelerated breathing (Luscombe and Taha, 1974; Thompson and Anderson, 1978). The fresh leaf extract has also been shown to have hypoglycaemic and anti-hyperglycaemic effects in dogs. When injected intravenously it prevented adrenaline-induced, as well as glucose-induced, hyperglycaemia. This action lasted from 30 min to 4 h after administration (Satyanarayan Murty *et al.*, 1978). Anti-inflammatory and antipyretic properties of *Azadirachta* leaf extract have been reported in Nigeria by Okpanyi and Ezeukwu (1981).

Acalypha wilkesiana Müll. Arg. EUPHORBIACEAE

Cultivated (from tropical Asia)

L In Nigerian local medicine the plant is administered in the form of a tincture, a decoction or an infusion. The leaf juice, obtained by rubbing the leaves between the palms of the hands, is smeared on parts affected by *Pityriasis versicolor* or similar types of fungal skin infections. Alternatively, the decoction of the leaves may also be drunk by itself or in combination with other remedies.

C The active principle responsible for the antimycotic and antimicrobial action is being investigated by the agar diffusion method (Adesina *et al.*, 1980b).

P Five different extracts have been tested (the preparation of each extract including a number of chemical operations) against seven test organisms and marked anti-microbial properties (inhibition zone over 15 mm) have been displayed against *Bacillus subtilis*, *Staphylococcus aureus* and *B. cereus* by extract B and to a certain extent against *Escherichia coli*, *Klebsiella pneumoniae*, *Proteus vulgaris* and *Serrata marcescens*. (Adesina *et al.*, 1980b).

Eclipta prostrata (L.) L. syn. (*E. alba* (L.) Hassk., *Verbesina prostrata* L.)

COMPOSITAE

L In Nigeria the plant is used as a pot-herb and in Ghana it is used to combat constipation (Dalziel, 1937). In India and in Senegal (Wolof) the root is considered emetic and purgative and is applied externally as an antiseptic to ulcers and wounds (Chopra *et al.*, 1956; Kerharo, 1968). The leaves can also be used but are less active.

C The plant has been reported to contain the alkaloid nicotine (0.08%) (Pal and Narasimhan, 1943). The leaves have been found to contain wedelolactone and desmethylwedelolactone and its glucoside besides polyacetylenes and their thiophene derivatives (Govindachari *et al.*, 1956; Dhar *et al.*, 1968) and the presence of free triterpenoids in the leaves has also been observed (Kerharo and Adam, 1974).

P The alcoholic extract of the plant has been reported to have antiviral activity against Newcastle disease virus. The leaves and the polyacetylenes of *E. prostrata* were highly active when exposed to UV light (phototoxicity) to microorganisms in antibiotic tests against *Escherichia coli*, *Saccharomyces cerevisiae* and *Candida albicans* (Dhar *et al.*, 1968; Towers and Wat, 1979; Wat *et al.*, 1980).

For a more comprehensive list of plants with antibacterial action see Table 4.1.

(b) Antifungal plants

The fungicidal substances present in some plants appear to play an important role in the defence against invasion by cryptogamic parasites. Thus a high concentration of sanguinarine just beneath the epiderm of the roots of Berberidaceae and *Sanguinaria* has suggested defence against root-rot fungi (*Phymatotricum omnivorum*). Similarly, solanine accumulation, observed around wounds in potatoes, has been considered a defence against *Fusarium avenaceum* (dry rot disease) (McKee, 1955).

Active constituents

The plant constituents with antifungal properties are often the same constituents or similar to those that are active against other microorganisms. They include *flavonoids* like tetra- and trihydroflavones, glycoflavones and kaempferol as in *Cassia occidentalis**, *C. alata**, *C. tora**, *Euphorbia prostrata**, *E. thymifolia** and *Phyllanthus niruri**; *polyphenols*, like those of *Anacardium occidentale* (anacardic acid), *Chlorophora excelsa* (chloropherin) and *Cocos nucifera*; *naphthoquinones* like plumbagin in *Plumbago zeylanica** and *senevols* as in *Allium sativum**, *Lepidium sativum* and *Carica papaya*. *Triterpene glycosides* are also reported to inhibit fungal growth (Olsen, 1975). All of these have already been mentioned under the chemical groups of the antibacterial plants. These plants are mainly used for fungal skin diseases and are applied locally. Simeray *et al.* (1981) noted that in Europe antifungal plants are mostly found amongst Liliaceae, Scrophulariaceae (containing saponosides), Leguminosae, Compositae and Rhamnaceae.

Table 4.1. *Antibacterial plants*

The names are given here without the authorities; these are mentioned in the Botanical Index. The names of the infectious organisms are abbreviated as follows: *B.*, *Bacillus*; *Cor.*, *Corynebacterium*; *Ent.*, *Entamoeba*; *Ep.*, *Epidermophyton*; *Esch.*, *Escherichia*; *M.*, *Microsporum*; *Mycob.*, *Mycobacterium*; *Ps.*, *Pseudomonas*; *Rh.*, *Rhinosporidium*; *Sacch.*, *Saccharomyces*; *Salm.*, *Salmonella*; *Sh.*, *Shigella*; *Staph.*, *Staphylococcus*; *Strep.*, *Streptococcus*; *Tr.*, *Trichophyton*.

Plant	Part used	Active constituent(s)	Action/acts on	References
Acantospermum hispidum	Leaves	Essential oil	*B. pumilis*, *B. subtilis*, *Cor. B. diphtheriae*, *Salm. typhi*, *Sh. dysenteriae*	Jain and Kar (1971)
Achyranthes aspera	Seeds	Oleanolic glycoside	Leprosy	Gopalchari and Dhar (1958) Ojha *et al.* (1966)
Alafia multiflora	Latex	Alcohol-phenol, vanillic acid	*Staph. aureus*, *Strep. faecalis*, *Esch. coli*, *Ps. aeruginosa*	Balansard *et al.* (1980)
Ananas comosus	Juice	Bromelain (enzyme)	Antibacterial, anthelmintic	Shukla and Krishnamurti (1961a, b) Sakkawala *et al.* (1962)
Argemone mexicana	Leaves, stems	Berberine, sanguinarine (acts as lipolytic prodrug allowing greater intracellular penetration)	*Mycob. tuberculosis*, *Staph. aureus*, *Salm. typhi*, *Esch. coli*, *Sh. dysenteriae*	CV Lambin and Bernard (1953) Mitscher *et al.* (1978)
Bombax malabaricum, *B. buonopozense?*	Seeds	Gallic acid, ethylgallate	*Staph. aureus*, *Esch. coli*	Dhar and Munjal (1976)
Butyrospermum paradoxum subsp. *parkii*	Nut (fat)	Triterpenic alcohols 'in butter'	*Sarcina lutea*, *Staph. aureus*, *Mycob. phlei*	Malcolm and Sofowora (1969) Mital and Dove (1971)
Calotropis procera	Latex	Calotropain (enzyme)	*Micrococcus leirodeikticus*	Shukla and Krishnamurti (1961b) Sakkawala *et al.* (1962)
Cannabis sativa	Resin, leaves	Phenols (2.15% in leaves)	Gram-positive organisms, *B. megaterium*, *Staph. aureus*, *Esch. coli*	N Nickel (1959) Veliky and Genest (1972) Fournier *et al.* (1978)
Capsicum annuum	Fruit	Capsicidin (steroid saponin)	'Antibiotic', yeast (*Saccharomyces cerevisiae*)	Gal (1964, 1967)

Cardiospermum halicacabum	Seeds	Essential oil	Ophthalmias	N Modi and Deshmankar (1972)
Carica papaya	*Fruit, seeds*	*Protein*	*Staph. aureus, B. cereus, Esch. coli, Ps. aeruginosa, Sh. flexneri;* 0.2–0.3 mg/ml for Gram-positive organisms and 4 mg/ml for Gram-negative organisms, of purified extract	Shukla *et al.* (1973) Emeruwa (1982)
Cassia absus	Seeds	Chaksine, isochaksine	*B. haemolyticus, Staph. aureus*	Inayat Khan *et al.* (1963)
Centella asiatica	Leaves, stem	Asiaticoside	*Mycob. tuberculosis, B. leprae, Ent. histolytica*	Cheema and Priddle (1965) N Bhattacharya and Lythgoe (1949) Boiteau *et al.* (1949) Dhar *et al.* (1968)
Cnestis ferruginea	Roots, leaves	Squalene, myricyl alcohol, β-sitosterol, methyl-linolenate homologues	*Sarcina lutea, Staph. aureus*	Malcolm and Sofowora (1969) Boakaiji Yiadom and Konnig (1975)
Crateva religiosa	Stembark	Total extract	*Sh. dysenterica*	Olugbade *et al.* (1982)
Cyperus rotundus	Plant	Obturastyrene (cinnamylphenol)	*Staph. aureus*	N Towers and Wat (1979)
Diospyros mespiliformis	Rootbark	Plumbagin	*Staph. aureus, Cor. diphtheriae,* also antifungal	Khan *et al.* (1980b) CV Paris and Moyse-Mignon (1949) Goutam and Purohit (1973) Steffen and Peschel (1975)
Dracaena manni	Rootbark		*Sarcina lutea, Staph. aureus, Mycob. phlei*	N Malcolm and Sofowora (1969)
Ekebergia senegalensis	Stembark	Saponin? meliacin?	*Staph. aureus, Sarcina lutea,* also ichthyotoxic (leaves)	Sofowora and Olaniji (1975) N
Eucalyptus globulus	Leaves	Phenol acids, essential oil	Antitubercular (*Mycob. tuberculosum*)	Malcolm and Sofowora (1969) Low et al. (1974) Boukef *et al.* (1976)

(*Table continued*)

Table 4.1 (*Continued*)

Plant	Part used	Active constituent(s)	Action/acts on	References
Guiera senegalensis	Leaves	Gallic and catechuic tannins	Choleriform diarrhoea	Kerharo *et al.* (1948)
Khaya senegalensis	Bark	Meliacins	*Sarcina lutea*, *Staph. aureus*, *Paramecium coli*	N
Lasiosiphon kraussianus	Roots	Heteroside	*Bacillus leprae*	Tubery (1968)
Lawsonia inermis	Leaves	Lawsone, gallic acid	Bactericidal	CV Carrara and Lorenzini (1946) Bendz (1956) Abd El Malik *et al.* (1973) Khorrami (1979)
Mangifera indica	Kernel	Mangiferin, ethylgallate, phenylpropanoids	Antibiotic and anthelmintic	CV Singh and Bose (1961) Malcolm and Sofowora (1969)
Momordica charantia	Leaves	Aqueous extract	*Esch. coli*, *Staph. aureus*	Georges and Pandelai (1949)
Moringa oleifera	Roots	Athomine	Vibrio cholera	Sen Gupta *et al.* (1956)
		Pterygospermine	*B. subtilis*, *Staph. aureus*, *Mycob. phlei*, *Salm. typhosa*	CV Kurup and Narasimha-Rao (1954) Kurup *et al.* (1957)
Ocimum basilicum	Plant	Thymol, eugenol	Antimicrobial (Gram-positive bacteria and mycobacteria), antiparasitic	N Jain and Jain (1972) Jain *et al.* (1974) Valnet *et al.* (1978)
Ocimum canum	Plant	Camphor	Gram-positive bacteria and mycobacteria	N Balansard (1936) Jain and Jain (1972) Jain *et al.* (1974)
Piper guineense	Fruit, leaves	Piperine, amide-alkaloids (terpenes), dihydropiperine	*Mycob. smegmatis* in concentration of 100 μg/ml, also *Candida albicans*, *Mycob. smegmatis* and *Klebsiella pneumoniae* (100 μg/ml)	N Dwuma Badu *et al.* (1976) Addea Mensah *et al.* (1977)

Polygonum salicilifolium, P. senegalense	Leaves, roots, stems	Flavonoids	*Salm. typhi, Sh. dysenteriae, B. subtilis, Esch. Coli*	Abd El-Gawad and El Zait (1981)
Salvadora persica	Leaves, twigs	Organic sulphur compounds	(Tooth brush and chewing stick) Bactericidal	Faroqui and Srivasta (1968)
				Ray *et al.* (1975)
				Lewis and Elvin-Lewis (1977)
				Etkin (1981)
				Ezmirly *et al.* (1979)
	Oil	Trimethylamine derivatives		Githens (1949)
Sida acuta, S. cordifolia	Aerial parts	Cryptolepine and vasicine	Antibacterial, *Proteus vulgaris* (600 μg/ml)	Gunatilaka *et al.* (1980)
				Prakash *et al.* (1981)
Solanum nigrum (S. nodiflorum)	Leaf	Solanine	*Staph. aureus, Candida albicans*	N
Strychnos afzelii	Stembark	(Chewing stick) dimeric indole alkaloids	*Strep. mitis, S. mutans, S. sanguis*	Verpoorte *et al.* (1978, 1983)
Tabernaemontana glandulosa	Bark	Hydroxycoronaridine and hydroxyibogamine	Antibiotic	Patel *et al.* (1967)
				Aschenbach *et al.* (1980)
Terminalia glaucescens, T. avicennoides	Bark	Gallic tannins	Gram-positive organisms; *Sarcina lutea, Staph. aureus, Mycob. phlei*	Delaveau *et al.* (1979)
				Malcolm and Sofowora (1969)
Thalictrum rugosum	Leaves	Thaliadanine (alkaloid)	*Mycob. smegmatis*	CV
				Mitscher *et al.* (1972c)
				Wu *et al.* (1976)
				Liao *et al.* (1978)
Thevetia neriifolia	Leaves, fruits	Aucubigenol	Antibiotic	CV
				Paris and Etchepare (1966)
				Sticher and Meier (1978)
Tiliacora funifera		Funiferine	*Mycob. smegmatis*	Ayim *et al.* (1977)
Withania somnifera	Leaves	Withaferine (steroidal lactone)	Gram-positive and Gram-negative organisms, also antifungal	N
				Sethi *et al.* (1974)
				Merck's Index (1976 (9716))
				Ikram and Inamul-Haq (1980)

Cassia occidentalis L. CAESALPINIACEAE
Coffee senna, stinkweed
Cassia tora L.
Foetid cassia

L The seeds of *C. occidentalis* have been used as a coffee substitute. In local medicine the leaves are used as a purgative, febrifuge and diuretic and also as an eye lotion. The dried or fresh leaves of *C. tora* are used in Northern Nigeria in the treatment of ulcers, ringworm (Tinea cruris) and other parasitic skin conditions. The treatment may consist of application of the powdered dried leaves or of a lotion made from a decoction of fresh leaves.

C The seeds of *C. tora* and *C. occidentalis* have been found to contain rhein, aloe-emodin and chrysophanol as indicated by thin-layer chromatography studies (Tewari *et al.*, 1965; Shah and Shinde, 1969; Tewari and Rajbehari, 1972) and extracts of *C. occidentalis* furnished in addition two pigments, physcion and α-3-sitosterol and a xanthone (Ginde *et al.*, 1970). In the leaves kaempferol and probably a flavanone have been found by Anton and Duquenois (1968) and a major fungicidal principle, chrysophanic acid anthrone, was isolated by Acharya and Chatterjee (1974, 1975). The fresh seeds of *C. occidentalis* contain a phytotoxin which loses its toxicity on roasting so that the seeds can be used as a coffee substitute.

P Plant cell cultures in vitro of both species were antibacterial against Gram-negative organisms (Quadry and Zafar, 1978). Antibiotic activity of *C. occidentalis* seeds against *Staphylococcus aureus*, *Bacillus subtilis*, *B. proteus* and *Vibrio cholerae* and against the fungi *Aspergillus niger*, *A. flavus* and *Penicillium chrysogenum* has been reported by Gaind *et al.* (1966), who believe that this activity is due to the presence of a volatile oil contained in the seeds.

Alcoholic extracts of *C. tora* leaves showed antibacterial action against Gram-positive organisms and were also antifungal (Osborne and Harper, 1957; Shah *et al.*, 1968). The seeds were strongly antibiotic especially in patients with skin diseases and Anton and Duquenois (1968) attribute this effect to chrysophanol derivatives of kaempferol. *C. tora* was also found to have antiviral action against Newcastle disease virus and Vaccinia virus (Dhar *et al.*, 1968).

Cassia alata L. c. and naturalized CAESALPINIACEAE
Ringwormbush

L The juice of the fresh leaves is universally recognized by the local healers as a remedy for parasitic skin diseases, and is used in the treatment of many eruptive and pustular skin conditions by simply rubbing the crushed leaves either alone or mixed with lime juice or an oil, on the skin. In Ghana the juice of the fresh leaves is squeezed into the eye to cure eye troubles (Irvine, 1930; Dalziel, 1937).

C Anton and Duquenois (1968) reported that the leaves contain rhein and its glucoside, aloe-emodol, small quantities of free chrysophanol, free kaempferol and probably a flavanone. Chrysophanic acid is found in all parts of the plant, particularly in the fruit. The residue after extraction of the leaves with a chloroform–ethyl alcohol mixture produces by distillation an 'oil' containing sesquiterpenic and phenolic compounds (Benjamin and Lamiranka, 1981).

P The biological activity of *C. alata* leaf extracts is comparable to that of the medicinal *Senna* species, the extracts being rich in anthrones and anthraquinone. They are used as a purgative. The leaves were also used in the first half of this century by 'orthodox doctors' in West Africa in the form of an ointment made from a watery extract in combination with lanolin, etc. in the treatment of crural ringworm (dhobie itch), i.e. Tinea cruris (fungal skin disease caused by *Trichophyton mentagrophytes*, *Epidermophyton gloccosum*, etc.) (Dalziel, 1937), An alcoholic extract of flowers and leaves also had a certain antibiotic effect against Gram-positive organisms (Nickel, 1959). Antibacterial properties towards *Bacillus mycoides*, *B. subtilis*, *Staphylococcus aureus*, etc. had already been reported in 1949 by Anchel. The activity of the 'oil' prepared by Benjamin and Lamiranka (1981) on Gram-positive and Gram-negative organisms including *Pseudomonas* has also been reported by these authors. *Cassia* seeds are often discarded after the medicinal constituents have been extracted. However, they may be valuable as sources of amino acids and proteins which could be used to supplement diets, as suggested by Dale and Court (1981). Arginine and aspartic and glutamic acids seem to be the main constituents in most species.

Euphorbia prostrata Ait. EUPHORBIACEAE

Euphorbia thymifolia L. syn. (*E. burmannia* Gay partly, *E. aegyptiaca* Soiss. and *E. scordifolia* Jacq. of F.T.A. partly)

L *E. prostrata* is considered in Northern Nigeria to be a good anthelmintic for tapeworm and is used as a purge (Dalziel, 1937). The juice of *E. thymifolia* is used locally in India in the treatment of skin diseases including ringworm (*Tinea cruris*) and for snake bite.

C The leaves and stems of *E. thymifolia* contain 5,7,4-trihydroxyflavone-7-glycoside (Chopra *et al.*, 1956; Sankara Subramanian *et al.*, 1971) and Caiment-Leblond (1957) isolated three flavone-derivatives by paper chromatography, including one closely related to rutoside. *E. thymifolia* also proved to have insecticidal action (Heal and Rogers, 1950).

P The ether extracts of *E. thymifolia* and *E. prostrata* have been found to be highly effective against sarcoptic mange in sheep (Lal and Gupta, 1970), and to have powerful antifungal activity in vitro against *Trichophyton mentagrophytes* (Rao and Gupta, 1970). This antifungal activity was repeatedly found by Pal and Gupta (1971), who used petroleum ether and linseed oil extracts of the two plants and observed activity against *Tr. simii* and *Mycobacterium gypseum* as well.

For a more comprehensive list of plants with antifungal action see Table 4.2.

Table 4.2. *Antifungal plants*
See also Chaumont and Bourgeois (1978) and Chaumont and Senet (1978).

Plant	Part used	Active constituent(s)	Acts on/action	References
Alpinia speciosa c. *A. officinarum* and allied species	Rhizome	Flavonoids	*Trichophyton rubrum*, *Tr. mentagrophytes*, *Epidermophyton floccosum* and antibacterial	Chopra et al. (1956) Ray and Majumdar (1976)
Argemone mexicana	Leaves, stems, root	Berberine, chelerythrine	Antibacterial (Table 4.1), antiprotozoal (*Leishmania*, *Trypanosoma lewis*)	Adgina (1972)
	Seeds	Sanguinarine (toxic)	*Phymatotrichum microsporum*	Greathouse (1939)
Bidens pilosa	Leaves	Phenylheptatriyene (phototoxic polyacetylene, external use)	Anti-*Candida albicans*, wounds, ulcers, also toxic to certain insects and larvae, reacts with DNA and kills human fibroblast cells (concentration 10 ppm)	Morton (1962) Degener (1975) Wat *et al.* (1978, 1979) Tomassini and Mathos (1979)
Carica papaya	Seeds	Benzylisothiocyanate	Antifungal in skin diseases, antibacterial	El-Tayeb *et al.* (1974) Emeruwa (1982)
Chlorophora excelsa (iroko)	Wood (leaves)	Chlorophorin (phenol)	Fungicidal (*Rhinosporidium flavipes* and *Rh. lucifungus*) and antibiotic, acts also against termites	Arndt (1968) Lewis and Elvin-Lewis (1977)
Cocos nucifera	Nutshell	Phenols	Antifungal against 3 *Microsporum* and 4 *Trichophyton* spp. (100 μg/ml), and 1 *Epidermophyton* sp. (200 μg/ml), antitubercular	Gaind and Singla (1966) Malathi *et al.* (1959)

Desmodium gangeticum	Roots	Tannins in related species, indole alkaloids	Antifungal, antibacterial	N Ghosal and Bannerjee (1968)
Haematoxylon campechianum	Wood	Ethylgallate	Mycobacterium	Little *et al.* (1953)
Hygrophyla auriculata syn. (*Asteracantha longifolia*)	Plant	Steroid and triterpene glycosides	*Trichophyton rubrum Tr. mentagrophytes, Mycobacterium gypseum, Candida albicans*, also acts on *N. gonorrhoea*	Venkataraman and Radnakrishnan (1972)
Hymenocallis littoralis	Plant	Lycorine	Antifungal, antitubercular	Verpoorte *et al.* (1983)
Lepidium sativum c. & nat.	Plant	Glucotropeolin essential oil with senevols	Fungicidal and bactericidal	Chaumont and Bourgeois (1978) Chaumont and Senet (1978) Chaumont *et al.* (1978) Ieven *et al.* (1978) Githens (1949) Watt and Breyer-Brandwijk (1962)
Phyllanthus niruri	Roots	Glycoflavones (kaempferol 4′- and eryodictyol 7-rhamnopyrosid	Ringworm (tinea cruris), ulcers, scabies, jaundice	N Chauhan *et al.* (1977)
Piper nigrum	Seeds	Essential oil	Fungicidal	Chaurasia and Kher (1978)
Ricinus communis	Oil	Undecylenic acid prepared from oil	Fungistatic and antibacterial *Tinea* spp., moniliases	Raina *et al.* (1976) Oliver (1960) Martindale (1958, p. 1331) Paulose *et al.* (1964)

(c) Antiviral plants

Viruses possess an outer coat, or capsid, of protein or lipoprotein and an inner core of nucleic acid, either DNA or RNA, but not both. The viruses have an exclusively intracellular mode of reproduction and like all living entities share a common mechanism of replication at the molecular level. Hereditary traits of the virus can be transferred from one type of cell to another by purified fragments of DNA. All viruses exist in two basic forms, the metabolically inert transmissible virion and the vegetative virus, an active constituent of the host cell. In situ fragmentation of the RNA in poliovirus has been reported by van den Berghe and Boeije (1973). Cohen *et al.* (1964) had observed that extracts of *Melissa officinalis* (lemon balm plant) had antiviral activity against several viruses (Newcastle disease, Vaccinia, Herpes simplex and Semliki forest viruses) in eggs and in chick embryo fibroblast monolayers. They considered the active principle to be a tannin, which attaches itself to the virus and thus prevents its adsorption to the host cell receptors. The antiviral activity was prevented by injection of gelatin into the eggs following injection of the *Melissa* extracts. The authors explain this by dissociation of the weak complex formed by the virus and the active principle (Kučera *et al.*, 1965). Lysogeny is attributed to the presence of bacteriophages in the bacterial genome.

The immunological character of some viruses, e.g. the influenza viruses, varies a great deal, so it is difficult to prepare the appropriate vaccine in advance of an outbreak of the diseases caused by these viruses, and a vast amount of research has been done on the development of drugs which will prevent or cure new outbreaks (Beveridge, 1977).

Certain viruses (arbo or toga viruses) are transmitted to Man and animals by arthropods (mosquitos, flies, ticks) and they multiply in the arthropod as well as in the vertebrate host. Their virion (RNA) is able to act as a mRNA in the host cell (Beveridge, 1981).

Antiviral constituents of higher plants

Poliomyelitis virus. Antagonistic action on this virus is claimed for two Apocynaceae, *Plumeria rubra** and *Allamanda cathartica*, and is attributed to a bitter glycoside, plumeried and a lactoflavone, fulvoplumierin. Antipoliomyelitis virus action has also been reported for *Anagallis arvensis* cultivated in the area (*A. pumila* widespread in West Africa, might be examined). Saponosides are believed to be responsible for this action.

Vaccinia virus is antagonized by three members of the Caesalpiniaceae, *Caesalpinia bonduc**; *Cassia fistula** and *Caesalpinia pulcherrima*. The latter also has an antiviral action on the Influenza virus.

Influenza virus is antagonized by *Canavallia ensiformis* (through xanthones), *Drosera indica* (through plumbagin) and by *Gossypium** spp. (through gossypol).

Measles virus and *coxsackie virus.* Lycorine from *Hymenocallis littoralis* has an antiviral effect on these viruses. For further details and references see the text and Table 4.3.

Arbo-viruses, with a lipoidal envelope, are destroyed by salts of mono- and poly-unsaturated acids from the roots of *Olax latifolia* (e.g. *Aedes aegypti*, which transmits dengue fever).

Herpes simplex. Action on Herpes simplex has been obtained with gossypol from *Gossypium* spp., with eugeniin from *Syzygium guineense* and with saponosides from *Anagallis arvensis*.

Newcastle disease (fowl pest) provides a convenient test-condition for antiviral activity and positive anti-Newcastle virus action has been noted in *Anagallis arvensis*, *Cassia fistula*, *Cocculus pendulus* and *Milletia congolensis*.

Plumeria rubra L. (Fig. 4.6) APOCYNACEAE
P. alba L.
Frangipani

L Use as a medicine is avoided in West Africa as the latex is a drastic purgative (Dalziel, 1937). In tropical America and East Asia it is sometimes used as a purgative and abortifacient and to treat venereal diseases (*Hager's Handbuch*, 1977).

C In the roots, plumericin, isoplumericin, their dihydro-derivatives and fulvoplumierin have been found (*Hager's Handbuch*, 1977). Rao and Anjaneyulu (1967) and Rao *et al.* (1967) reported the presence in *P. rubra* bark of fulvoplumierin, lupeol, β-sitosterol and the bitter glucoside, plumeried (up to 2% in the bark), which is also present in the leaves (these constituents had already been isolated from Indian *P. acutifolia* by Rangaswani *et al.* in 1961). In the flowers kaempferol and quercetin

Fig. 4.6. *Plumeria rubra* L.

glycoside have been found, and the latex also contains lupeol and its acetate and esters with fatty acids (Rangaswani *et al.*, 1961; Rao and Anjaneyulu, 1967). Fulvoplumierin is a lactoflavone; plumeried, also called agoniadin, is a pseudo-indican glucoside (Malvan *et al.*, 1974).

P The bark of *P. acutifolia* is very toxic, producing vomiting and mydriasis and slowing down the heart (*Hager's Handbuch*, 1977, p. 783). Extracts of the stem have been found to produce significant surface anaesthesia on the cornea of the rabbit (Chak and Patnaik, 1972). In the guinea pig all extracts produced, by infiltration, an anaesthesia lasting over 1 h. In mice all extracts prevented pain induced by phenylquinone, but none of the extracts showed analgesic action in mice if pain was induced by a hot plate or pinching of the tails. Spasmolytic activity was not specific (Chak and Patnaik, 1972). Plumeried was found to be mainly purgative (in doses of 0.2–0.3 g in man) (Watt and Breyer-Brandwijk, 1962; Harrison *et al.*, 1973). Antibiotic activity was reported for plumericin (*Hager's Handbuch*, 1977) and fulvoplumierin has been shown to be bacteriostatic (tuberculostatic) in doses of 1–5 μg. Extracts of *P. rubra* have also been shown to be highly active against the polio virus (van den Berghe *et al.*, 1978). *P. rubra* var. *alba* contains a lower percentage of antibiotic constituents than *P. rubra* L. (Cochran and Lucas, 1958–59).

Cassia fistula L. (Fig. 4.7) CAESALPINIACEAE

Indian laburnum (cultivated in West Africa)

In India, the rootbark, seeds and leaves are considered laxative, the fruit is cathartic and is applied in the treatment of rheumatism and for snake bite. The roots are used as an astringent, tonic and febrifuge and the juice of the leaves in the treatment of skin diseases.

C The leaves of *C. fistula* have been found to contain free rhein, rhein glucoside and sennoside A and B (Kaji *et al.*, 1968). A butanol extract of the powdered stembark contained tannins whilst the benzene extract yielded lupeol, β-sitosterin and hexacosanol (Sen and Shukla, 1968). From the alcoholic extract of the pods a new anthraquinone (fistulic acid) was obtained and identified as 1,4-dihydroxy-6,7-dimethoxy-2-methylanthraquinone-3-carboxylic acid (Agrawal *et al.*, 1972). In an acetone extract of the flowers kaempferol and a proanthocyanidin (leucopelargonidin tetramer) have been found and in one of the bark a leucopelargonidin trimer has been found (Narayanan and Seshadri, 1972).

P Lillykutty and Santhakumari (1969) studied the antibacterial action of the leaves, stembark and fruit pulp of *C. fistula* and found the fruit pulp to be the most potent. The ether-soluble fraction of the fruit pods had the most potent activity against *Staphylococcus aureus*, *Staph. albus*, *Bacillus megaterum*, *Shigella flexneri*, *Sh. Shiga* and *Salmonella paratyphi* A and B, and in vitro 1 g of this extract was more potent than 1 μg of chloramphenicol. This activity was attributed to the presence of rhein. The acetone extract of root- and stembark was found to have antifungal action against *Trichophyton tonsurans*, *Tr. rubrum* and *T. megnini* (*rosaceum*). In vitro 100 μg of the rootbark proved more potent than 16 μg of griseofulvin (Lillykutty and Santhakumari, 1969). The activities might be due to the presence of flavonoids. The pods and stembark of *C. fistula* have also been found to have antiviral activity against Newcastle disease virus and Vaccinia virus (Cutting *et al.*, 1965) and in addition extracts of *C. fistula* also have a hypoglycaemic action in rats (Dhar *et al.*, 1968). On screening various extracts of 620 Indian plants for their antiviral activity (Balbar *et al.*, 1970) found that *C. fistula* extracts were amongst those which were most effective against Vaccinia and Newcastle disease viruses. They inhibited the cytopathy of the host viruses and also their replication in the cells of the host and were more effective when given before or along with the virus. The virus-inhibitory activity of the plant extracts is believed to be associated with some interferon-like factor.

Caesalpinia bonduc (L.) Roxb. syn. (*C. crista* L., *C. bonducella* (L.) Fleming, *Guilandina bonduc* L.) CAESALPINIACEAE
Bonduc or nicker nut

L A decoction of the root is used in Guinea as an antipyretic, and a gargle for sore throats is prepared from the boiled leaves. In India, the dried seeds are reputed to be

Fig. 4.7. *Cassia fistula* L.

antipyretic, styptic, tonic and anthelmintic (*Hager's Handbuch*, 1967–80, Vol. III, p. 561). In Nigeria, a decoction of the rootbark is used as a rubefacient and in cases of dyspepsia, and the seed oil is considered to be a remedy for convulsions in Nigeria and in South Africa (Watt, 1967).

C The seed kernel was known to contain, besides 20–28% of albuminoids, 35% of starch and 5–6% of sugars, a bitter glucoside, bonducin or guilandinin, a saponin, a phytosterin (sitosterol) and an oil (20%). Neogi *et al.* (1958) separated the bitter constituent into four fractions, A, B, C and D. Further analysis of the seeds revealed three bitter principles, α-, β- and γ-caesalpin. These are diterpenic derivatives with three aromatic rings and one lactone ring. Later, fourth and fifth principles, δ- and ϵ-caesalpin were reported and their formulas established (Canonica *et al.*, 1963, 1964, 1966; Qudrat-I-Khuda and Efran, 1964; Balmain *et al.*, 1967). The wood contains two dyes, brasilin and brasilein, which are related to haematoxylin.

P The alcoholic extract of four fractions produced slight hypotension in the dog and depressed the heart of the frog. Fractions A and B were antibiotic, fraction B also antipyretic and diuretic, and fraction D anthelmintic (Watt and Breyer-Brandwijk, 1962). Alcoholic extracts of the roots and stems were found to have an antiviral action against *Vaccinia virus* (Dhar *et al.*, 1968). An antimalarial action could not be detected (Spencer *et al.*, 1947a). In India, the powdered seeds were found to have an anti-oestrogenic activity in rabbits and mice, and an anti-fertility action was noted in mice and rats (Bhide *et al.*, 1976). Aqueous extracts of the roots, stems and leaves showed a distinct action on sarcoma 180 (Abbot *et al.*, 1966).

Gossypium spp., **Gossypium hirsutum, G. barbadense,** etc. MALVACEAE
Cotton plant

L A number of *Gossypium* spp. and hybrids are widespread in tropical and subtropical West Africa. Many species have been cultivated and are naturalized and some are now subspontaneous. Kano (Nigeria) has been a cotton market since the ninth century. The seed floss provides the cotton and the seed oil is used locally as a source of oil for cooking and cattle food. The husks are sometimes used as a fuel. Medicinally, a cold infusion of the leaves with lime juice is said to give relief in dysentery; the root is believed to be emmenagogic and oxytocic, the active part being the rootbark. The leaves and crushed seed-kernels are applied to sores or as a poultice to bruises and swellings and the lint is used as a dressing for wounds (Dalziel, 1937).

C Absorbent cotton wool is obtained directly from the seed floss as well as from card strips and comber waste in cotton mills. The kernel oil is composed of 47.8% linoleic acid, 23.4% palmitic acid and 22.9% oleic acid plus small amounts of myristic and myristoleic acids and other fatty acids. The residual cake is rich in proteins (97%)

and contains most of the essential amino acids, together with flavonoid pigments (glucosides of quercetol, kaempferol, isoquercetrin, etc.) while the pigment glands of the seeds contain a toxic polyphenol, gossypol, which is destroyed by heating (thus rendering the oil suitable for consumption) (Aizikov and Kurmukov, 1973). The rootbark contains a yellow oil which consists of furfuraldehyde, a phenolic acid (probably 2,3-dihydroxy-benzoic acid), betaine, a phytosterol and a resin (with 1–2% gossypol).

P Gossypol has been demonstrated to have antiviral activity and apogossypol formed by deformylation of gossypol retains the potent antiviral activity but is at least 10 times less toxic for human Ep cell cultures (Abbot *et al.*, 1966). Influenza virus is inactivated by treatment with gossypol resulting in a 96–100% protection rate. Both gossypol and apogossypol inactivate the enveloped virus para influenza 3 and herpes simplex in vitro (Dorsett *et al.*, 1975). The non-enveloped polio virus is not affected by either agent. Incubation of infected cells with gossypol or apogossypol does not alter subsequent plaque formation, indicating that the antiviral effect does not occur within the cell (Vichkanova and Shipulina, 1972). From the roots of *G. hirsutum* and *G. barbadense*, 6-methoxygossypol and 6,6′-dimethoxygossypol (dimeric sesquiterpenes) have been isolated, respectively. They have proved to be active in vitro against *Penicillium* and *Cladisporum* spp. and frequently inhibit *Aspergillus fumigatus* (lung aspergillosis) (Stipanovic *et al.*, 1975). When instilled three times daily into the conjunctival sac of rabbits suffering from experimental herpetic keratitis, a solution of 0.005–0.1% of gossypol could eradicate the infection (Maichuk *et al.*, 1972; Dorsett *et al.*, 1975). Gossypol had antiherpetic action in infected mice on both oral and subcutaneous administration. It was, however, more effective against dermotropic than against keratogenic strains of the herpes virus. Further, a strong viricidal action against Influenza A_2 virus (Frunze strain) in vitro and in vivo (mice) has been reported for gossypol (Vichkanova and Gorunyunova, 1972).

Gossypol is a nerve and cellular poison and hence before feeding cattle with oil-cakes the pigment pockets should be eliminated in a pre-press process. Fed to pigs in amounts of 0.02–0.03% of the total ration, gossypol causes liver congestion and oedema especially in the lungs and body cavities, while the grazing of sows in harvested cotton fields leads to abortion. A horse weighing 450 kg cannot tolerate more than 450 g of oil-cake per day and dies when the dose is as much as 1.4 kg. Small amounts over a long period cause a bad general condition and anaemia in calves and sheep; larger amounts cause haemorrhage, inflammation, haematuria, muscular weakness, respiration difficulties and paralysis. Abortion and blindness are also reported (Steyn in Watt and Breyer-Brandwijk, 1962). The LD_{100} for gossypol given intraperitoneally to rats is 20 mg/kg, death occurring between the thirteenth and twentieth day after the injection.

For a more comprehensive list of plants with antiviral action see Table 4.3.

Table 4.3. *Antiviral plants*

Plant	Part used	Active constituent(s)	Acts on/action	References
Allamanda cathartica	Bark	Plumericin, isoplumericin	Antiviral to polio virus	CV Pai *et al.* (1970) Wacker and Eilmes (1975)
Anagallis arvensis	Plant	Oleanane triterpenes, curcubitacins	Polio virus, Newcastle disease and herpes viruses	Bhakuni *et al.* (1969) Amoros *et al.* (1977, 1979) Yamada *et al.* (1978) Bezanger-Beauquesne *et al.* (1980)
Caesalpinia bonduc	Root, stem	α- and β-caesalpins, δ-caesalpin	Anti-vaccinia virus, anthelmintic	Dhar *et al.* (1968) Watt and Breyer-Brandwijk (1962)
Caesalpinia pulcherrima	Plant	Alcoholic extract	Influenza and vaccinia virus, also anti Gram-negative bacteria and *Candida* spp.	Watt and Breyer-Brandwijk (1962) Cochran *et al.* (1966) Dhar *et al.* (1968) Ieven *et al.* (1979)
Canavalia ensiformis	Seeds	Xanthones, canavanin (2-amino-44-guanidinooxybutyric acid), canavalin	Influenza virus and bactericidal, also insecticidal	Kitagawa and Tomiyana (1930) Farley (1944) Heal and Rogers (1950) Burger (1960, p. 1103) Thomson *et al.* (1964) Muelenaere (1965) Olson and Liener (1967)

Catharanthus roseus	Roots, leaves	α-Acylindolic acid = perivine	A_2 influenza virus	Farnsworth *et al.* (1968)
Citrus spp.	Rind	Hesperidin	Vesicular stomatitus virus	Kaleyra (1975) Wacker and Eilmes (1975) Hoizey *et al.* (1978)
Cocculus pendulus and related *C. indicum*	Aerial parts	Cocculidine, 1% picrotoxin	Newcastle disease virus	Bhakuni *et al.* (1969)
Desmodium gangeticum	Roots	Aqueous extract, flavonoids	Antiviral, antibacterial	N Ghosal and Banerjee (1968) Prema (1968)
Hymenocallis littoralis c.	Plant	Lycorine	Antiprotozoal, moderate plaque inhibition of measles and coxsackie viruses	Berghe, van den *et al.* (1978) Ieven *et al.* (1978)
Maesopsis eminii	Stems	Saponoside	Antiviral	Soulimov *et al.* (1975) Chaumont *et al.* (1978)
Milletia barteri	Extract of stems	Saponin	Strongly viricidal to Newcastle disease	Balbar *et al.* (1970) Soulimov *et al.* (1975)
Olax latifolia	Roots	Mono- and poly-unsaturated acids (salts)	Acts on arboviruses with lipoidal envelopes	Jenkin (1973) Soulimov *et al.* (1975)
Porterandia cladantha	Root, stem	Saponosides	Viricidal, abscess, furunculosis, bronchitis	Balbaa *et al.* (1970) Soulimov *et al.* (1975)
Syzygium spp. and related *S. guineense?* (*S. aromatica*, not in West Africa)	Buds	Eugeniin	Acts on herpes simplex virus in concentration of 10 μg/ml	Takechi and Tanaka (1981)

(d) Antiprotozoal plants

General tests for antiprotozoals were formerly carried out on paramecia, which are non-pathogenic to man, but these proved to be of no value. The malaria parasites, of which the most malignant is *Plasmodium falciparum*, have two main phases: intracorporal, in Man, and extracorporal or exogenous in the mosquito (*Anopheles*). The intracorporal phase has three distinct cycles: asexual (schizogony), sexual (sporogony) and an extra-erythrocytic cycle in the liver. Thus Man is an intermediary; the mosquito is the final host in which the sexual cycle is completed. Trypanosomiasis is also conveyed by an animal vector, the tsetse fly (*Glossina morsitans*). Similarly, in leishmania tropica the intermediate host is *Phlebotomus papatassi* and visceral leishmaniasis is transmitted through the sandfly *Phlebotomus argentipes*. In bilharzia the vector is the freshwater snail. In all these cases, prophylactic action against insects and molluscs is an essential part of any campaign against the disease.

Active constituents

The activity of chemical constituents (e.g. alkaloids) against protozoa such as *Trypanosoma lewisi* is determined by the method of Hopp *et al.* (1976) in terms of the immobility produced on the organism and is expressed as the average percentage of non-motile organisms from four microscopic fields. Phenol and pentamidine are comparatively tested as controls. In looking at the effects registered in this chapter, it appears that in their activity against protozoa, the constituents seem to be chemically less varied than in the preceding groups. Many are *alkaloids*.

Plasmodium spp. Apart from the antimalarial action of the alkaloid quinine from *Cinchona* spp., which has been known for many years, brucine (or bruceolides?) from *Brucea antidysenterica*, funiferine from *Tiliacora funifera*, rhynchophylline from *Crossopteryx febrifuga* and meliacins from *Trichilia roka* are said to act against *Plasmodium berghi*.

Activity against *Leishmania tropica* has been observed with the alkaloids echitamine from *Alstonia boonei* and berberine (also anti-amoebic) from *Argemone mexicana*. The latter species also contains sanguinarine, which is trypanocidal. Berberine is also found in *Chasmanthera dependens*, which has both anti-leishmania and trypanocidal action. At one time berberine was also believed to be antimalarial but Chopra *et al.* (1938) could show that berberine acts on the spleen, increasing its volume and rhythmic contractions, and that the parasites thus liberated in the peripheral circulation are then able to respond to quinine action. The diagnosis of latent malaria as carried out by certain specialists is based on this property. Harmane derivatives in certain plants like *Acacia nilotica*, *Newbouldia laevis* and *Pauridiantha lyalli* are also antiprotozoal.

The motility of *Trypanosoma lewisi* can be inhibited by harmine, harmol, harmane chloride, palmatine iodide and, most of all, sanguinarine derivatives (nitrate and sulphate). These compounds caused 70–100% immobility of the test organisms after 1–24 h. The most active is sanguinarine from *Argemone mexicana*; it causes 100% non-motility even at dilutions of 1:40000 and in vitro had eighty times the activity of pentamidine on *Trypanosoma lewisi*. Palmatine from *Zanthoxylum zanthoxyloides*,

Chasmanthera dependens and *Cocculus* and *Tinospora* spp., at dilutions of 1:8000 led to 95% non-motility after 1 h, but at dilutions of 1:16 000 only 70% non-motility was achieved after 24 h (Hopp *et al.*, 1976; Al-Shamma and Mitscher, 1979; Al-Shamma *et al.*, 1981). (The antiprotozoal action of *Zanthoxylum zanthoxyloides* and of *Anacardium*, *Plumbago*, *Drosera* etc. have been mentioned under Antibacterial plants.)

Conessine from *Holarrhena floribunda** bark has proved to be very active against *Entamoeba histolytica* in intestinal and hepatic amoebiasis, preventing multiplication of the amoeba. It was found to produce neurological side-effects and therefore its clinical use was suspended. (Conessine is also active against *Trichomonas vaginalis* and is now mostly used in the topical treatment of trichomoniasis.) Other antiprotozoal alkaloids have been found in *Alchornea cordifolia* (alchornine) and *Mitragyna inermis* (rhynchophylline). Anti-amoebic effects have been observed with carpaine from *Carica papaya* leaves, an ethanolic extract of *Phyllanthus niruri** and borrerine (closely related to emetine) from *Borreria verticillata*. Antiprotozoal activity is also associated with some terpenoid and meliacin constituents. Thus *Curcuma domestica** rhizomes and the aerial parts of *Euphorbia tirucalli** have also been reported as active anti-amoebics, owing to the presence of sesquiterpenic ketones (turmerones) and terpenoids, respectively, and the anti-amoebic effect of *Euphorbia hirta** has been attributed to euphosterol.

Paramecia are mentioned as responding to the meliacins from *Khaya senegalensis*, *Crossopteryx febrifuga** and *Trichilia roka** and to pseudocedrelin from *Pseudocedrela kotschyi**, whilst *Melia azedarach* does not seem to be active against paramecia (Spencer *et al.*, 1947a). On the other hand, the triterpenic saponins in *Paullinia pinnata** are said to be active paramecidals. It is interesting to observe from this list that the majority of the plants with antiprotozoal action are plants which show a depressive action on the central nervous system.

Chasmanthera dependens Hochst. MENISPERMACEAE

L The leaves and juice are used locally as a dressing for fractures and mixed with shea butter as an embrocation for pain, sprains, etc. An extract of the bark is given in the treatment of venereal disease and as a tonic for physical and nervous debility (Dalziel, 1937).

C The alkaloid berberine has been found in the roots. The related species *C. palmata* from Mozambique contains bitter principles without nitrogen such as colombin, chasmanthin, palmarin and 2–3% of alkaloids of the berberine group, namely, colombamine, jateorhizine and palmatine (Oliver-Bever, 1968).

P *C. dependens* is used as a bitter tonic. Berberine sulphate is used in India in the treatment of oriental sores; it inhibits *Leishmania tropica* in concentrations of 1:80 000. The alkaloid is also a uterine stimulant (Oliver-Bever, 1968).

Curcuma domestica Valeton syn. (*C. longa* L.) ZINGIBERACEAE
Turmeric

L The plant is cultivated in West Africa in the forest areas and often grown in native compounds as a dye plant for leather, palm fibre, etc. It is used amongst Moslems as an anthelmintic, an eye wash and an ointment for skin diseases (Dalziel, 1937). In India turmeric is the main constituent of curry powder and is used in local medicine (fresh juice) as an anthelmintic and antiparasitic in the treatment of many skin diseases. It is also said to be an ant repellent (Chopra *et al.*, 1956). In Vietnam turmeric is taken as an anthelmintic and diaphoretic.

C A colouring matter has been isolated from the rhizomes. It contains the curcumines belonging to the dicinnamoyl-methane group and an aromatic oil, turmeric oil (0.24% in the fresh rhizomes), composed of 25% terpene-carbon derivatives and 65% of sesquiterpenic ketones (turmerones) (Paris and Moyse, 1967, p. 78).

P Sodium curcuminate administered intravenously to dogs in a dose of 24 mg/kg doubles the rate of bile flow without any appreciable disturbance of the blood pressure and respiration. It has in addition a powerful antibacterial action against *Micrococcus pyogenes* var. *aureus* (Ramprasad and Sirsi, 1956). Basu (1971) compared the antibacterial action of *C. domestica* to that of penicillin on Gram-positive organisms and that of streptomycin on Gram-negative organisms, and showed that *C. domestica* was less active than these antibiotics. Antiprotozoal activity against *Entamoeba histolytica* has been reported for the alcoholic extracts of the rhizomes (Dhar *et al.*, 1968).

Euphorbia tirucalli L. EUPHORBIACEAE

L In India *E. tirucalli* latex is used as a fish poison. It is applied medicinally to warts and used in the treatment of rheumatism and neuralgia (Chopra *et al.*, 1956). In contact with the eye it is a dangerous poison, producing inflammation of the cornea and iris (Crowder and Sexton, 1964). Ingestion of 'remedies' containing the latex can produce ulceration of the gastrointestinal mucous membrane (Crowder, 1964 in Kerharo, 1968).

C The latex contains 75–82% of a resin and 14–15% of a rubber. From the latex three terpenic alcohols have been isolated: taraxasterol, tirucallol and euphol (Haines and Warren, 1949, 1950; McDonald *et al.*, 1949). From the fresh latex, isoeuphorol has been obtained, which, when the latex is dried out, is replaced in a few months by euphorone, a ketone. The stem of *E. tirucalli* has been found to contain hentriacontane, hentriacontanol, β-sitosterol, taraxerol, 3,3′-di-*o*-methylellagic acid, ellagic acid and also a glucoside which on hydrolysis produces kaempferol and glucose (Gupta and Mahadevan, 1967).

P Alcoholic extracts of the aerial parts of *E. tirucalli* were shown to possess antiprotozoal activity against *Entamoeba histolytica* (Dhar *et al.*, 1968; Rao and Gupta, 1970). The protozoal effect of *Euphorbia hirta* might be due to euphosterol, also reported to be present in this plant (Watt and Breyer-Brandwijk, 1962). A fraction of lyophilized extract, at a dilution of 1:200, had after 48 h the same anti-amoebic effect as emetine hydrochloride at a dilution of 1:80000 after 24 h (Ndir and Pousset, 1981).

The alcoholic and aqueous extracts of the stems were observed to reduce adenocarcinoma and sarcoma to a considerable extent (Abbot *et al.*, 1966).

Holarrhena floribunda (Don.) Dur. & Schinz var. *floribunda* syn. (*H. africana* DC., *H. wulfsbergii* Stapf, *Rondeletia floribunda* Don.) APOCYNACEAE

Conessi or kurchi bark

For local uses and chemical constituents see CV.

P *Holarrhena* alkaloids and berberine hydrochloride were studied on intestinal amoebiasis in rats and on hepatic amoebiasis in hamsters and their activity was compared to that of emetine hydrochloride. All the alkaloidal fractions of *Holarrhena* effectively reduced the infection in rats and hamsters. Kurchamine appeared to be one of the most promising active compounds. Berberine showed marked effectiveness in both types of infection (Bertho, 1944). Berberine is also remarkably toxic against *Leishmania tropica*, inhibiting the growth of *Leishmania* in vitro at a 1 : 80 000 dilution. A 2% solution injected at the base of the lesion caused healing after 2–3 administrations (Chopra *et al.*, 1932).

Clinical tests with conessine on patients with intestinal and hepatic amoebiasis have been found to give results comparable to those obtained with emetine. Use of conessine must, however, be closely supervised, as in some cases it can produce neurological troubles (vertigo, sleeplessness, agitation, anxiety and delirium) (Crosnier *et al.*, 1948; Tanguy *et al.*, 1948; Alain *et al.*, 1949; Siguier *et al.*, 1949). For this reason the administration of conessine in cases of dysentery has been discontinued and the substance is now limited to the external treatment of *Trichomonas vaginalis* and urethritis (Godet, 1950).

Phyllanthus niruri L. EUPHORBIACEAE

L Used locally for dysentery and intestinal spasms in Ghana. Sometimes used as a bitter stomachic and for constipation, the leaves of the plant are also chewed to cure hiccups. An infusion is sometimes used for genito-urinary disorders (Dalziel, 1937).

C In the aerial parts, three crystalline lignans including phyllanthine, hypophyllanthine and quercitin have been found (Krishnamurti and Seshadri, 1946). Five flavonoids have been identified: quercitin, astralgin, quercitrin, isoquercitrin and rutin (Nara *et al.*, 1977). Four leucodelphinidine alkaloids were separated from the leaves and stems, one of them being an enanthiomorph of securinine (Stanislas *et al.*, 1967; Rouffiac and Parello, 1969).

P Extracts of *P. niruri* (by 50% ethanol) have extensive antibacterial, antifungal and antiviral action. They also act on *Entamoeba histolytica* (protozoa) but not on the malaria parasite *Plasmodium berghei*, and they are anthelmintic to *Hymenolepsis nana*. The extracts have depressive action on the isolated guinea pig ileum and have an anticancer action on Friend leukaemia virus (Dhar *et al.*, 1968). Aqueous extracts of the leaves have been found to have an oral hypoglycaemic effect comparable to that of tolbutamide (Ramakrishnan, 1969).

The maximum dose of an extract of the whole plant tolerated by the mouse is 1 g/kg, given orally (Dhar *et al.*, 1968).

Pseudocedrela kotschyi (Schweinf.) Harms syn. (*Cedrela kotschyi* Schweinf., *P. chevalieri* DC.) MELIACEAE
Dry zone cedar

L The rootbark is used in Nigeria and Togo as a febrifuge and also in the treatment of dysentery and rheumatism (Dalziel, 1937).

C The bark contains 1% of a bitter non-nitrogenous principle, pseudocedrelin, which is related to calicedrin. Pseudocedrelin is an unsaturated phenolic lactone with OH and OCH_3 groups. In the wood oil, Ekong and Olagbeni (1967), Ekong *et al.* (1968, 1969b) and Taylor (1979) found meliacins (limonoids), including 7-desacetoxy-7-oxogedunin and the pseudrelones A, B and C. Each of these consists of a single limonoid alcohol which is esterified with various acids. In the bark 8.5% of mineral substances, 14.6% of tannins, a saponin and 1% of lipids were also reported.

P Pseudocedrelin in a dilution of 1/10 000 kills paramecia in 20 min and the bark (3 g/l) is toxic to goldfish inducing death after 1 h, whilst greater dilutions have a stupefying action. In guinea pigs with experimentally induced hyperthermia, 0.05 g/kg pseudocedrelin given subcutaneously or intraperitoneally produces a lowering of the temperature of 2–3°C compared with control animals.

Carapa procera DC. syn. (*C. guineensis* Sweet ex Juss., *C. touloucouna* Guill. et Perr., *C. gummiflua* DC., *C. velutina* DC., *C. microcarpa* Chev.) MELIACEAE
Crabwood, Monkey cola

L The bark exudes a gum resin, is bitter, and is said to have antipyretic and tonic properties. It is used in Nigeria in the treatment of fevers and as a general tonic. It is also given to women during pregnancy. In addition, it is a component of native cough mixtures and anthelmintic remedies, and is applied externally as a dressing for sores. In the Congo area, the bark, cut up and mixed with palm wine, is used as a quinine substitute.

C From the plant collected in the Casamance, a bitter principle named tulukinin, which is closely related to calicedrin, was isolated together with 19.5% mineral compounds and 12% tannins. This bitter principle was also found in the seed oil (Moyse-Mignon, 1942). Later, Bevan *et al.* (1963) obtained from the Nigerian plants a crystalline substance analogous to the triterpenoid of the limonin group obtained from *Cedrela odorata*. It was called carapin and its structure was identified as a bicyclononalid (Bevan *et al.*, 1963, 1965).

P The toxicity of the rootbark extract in mice was found to be 1.28 g/kg. It appeared to be inactive as an antimalarial agent (Popp *et al.*, 1968).

Trichilia roka (Forsk.) Chiov. syn. (*T. emetica* Vahl, *Eleaja roka* Forsk.)
MELIACEAE
Roka tree

L An extract or decoction of the root is used in West African local medicine as a purge and emetic, and the pounded bark and seeds are applied in the form of an ointment in parasitic skin diseases (ringworm, etc.). In Senegal, *T. roka* has also been reported to be used in skin diseases and to act as a tonic, to stimulate bronchial secretion and to have an antiepileptic effect, whilst the related *T. prieuriana* A. Juss. is considered in Senegal to have antipyretic, purgative and anti-arthritic actions (Kerharo and Adam, 1974). Also in India *Trichilia* spp. are used for skin diseases (Chopra *et al.*, 1956).

C In the rootbark of *T. roka*, a bitter principle related to calicedrin has been found (Paris and Moyse-Mignon, 1939) as well as resin and a tannin. In the Nigerian species of *T. prieuriana*, a meliacine called prieurianine was isolated by Bevan *et al.* (1965). The seeds of *Trichilia* spp. yield 49–60% of fat which is used in candle-making. It consists of palmitic, linoleic and oleic acids (Watt and Breyer-Brandwijk, 1962). The analysis of the bark of *T. heudelotti* has revealed an orange bitter principle, sterols and 10.2% of pyrocatechuic acid (Planche, 1949).

P The seeds are purgative and emetic and the toxicity of the seed-cake for cattle has been repeatedly reported (Watt and Breyer-Brandwijk, 1962).

Crossopteryx febrifuga (Afzel. ex Don) Benth. syn. (*Rondeletia febrifuga* Afzel. ex Don, *C. kotchyana* Fenzl, *R. africans* Winterb.) RUBIACEAE

L In local medicine the roots and bark are used in the treatment of fever, and a decoction of the roots is used in the treatment of coughs and gastrointestinal complaints (Dalziel, 1937). In Zambia the extracts have been used in trials by ordeal (Marwick, 1963).

C A glycoside and an alkaloid have been isolated from the bark. The glycoside has been identified as β-quinovin and the alkaloid has been called crossoptine (Blaise, 1932). Raymond-Hamet (1940) considered that the characteristics of the alkaloid described by Blaise are those of mitrinermine, later identified with rhynchophylline. The presence of an alkaloid in the Nigerian species has been confirmed by Persinos *et al.* (1964). A phytosterol and a phlobaphene have also been found in the bark (Caiment-Leblond, 1957).

P The alkaloid isolated by Blaise shows no toxicity upon ingestion in the dog at doses up to 14 mg/kg; it has a slight hypotensive effect and produces vasoconstriction in the kidneys. It does not appear to modify the hypertensive action of adrenaline and seems pharmacodynamically different from quinine and yohimbine (Blaise, 1932). A survey for supposed antimalarial activity gave negative results (Spencer *et al.*, 1947b).

Paullinia pinnata L. (Fig. 4.8) SAPINDACEAE

A flavotannin extracted from the leaves of this plant also has a cardiotonic effect (see Chapter 2, CV).

C The rootbark of *P. pinnata* collected in Brazil is said to contain a bitter principle, timboin, and an oily substance, timbol, which act as a violent nerve poison producing paralysis preceded by convulsions (Watt and Breyer-Brandwijk, 1962). An alkaloid, ichthyonine, a resin and a pigment have also been reported, but the botanical identity of the plant appears uncertain (Kerharo *et al.*, 1960, 1961, 1962).

P A saponoside with a triterpenic aglycone, present in the leaves and twigs of the plant in West Africa, could account for its toxicity (Boiteau *et al.*, 1964) (the plant had caused a number of deaths when it had been used as an abortifacient). Kerharo *et al.* (1960, 1961, 1962) think that the saponin is probably also the responsible constituent for the toxicity of *Paullinia* towards protozoa (*Paramecia* were killed in 1 h by a concentration of 1 : 500).

For a more comprehensive list of plants with antiprotozoal action, see Table 4.4.

Fig. 4.8. *Paullinia pinnata* L.

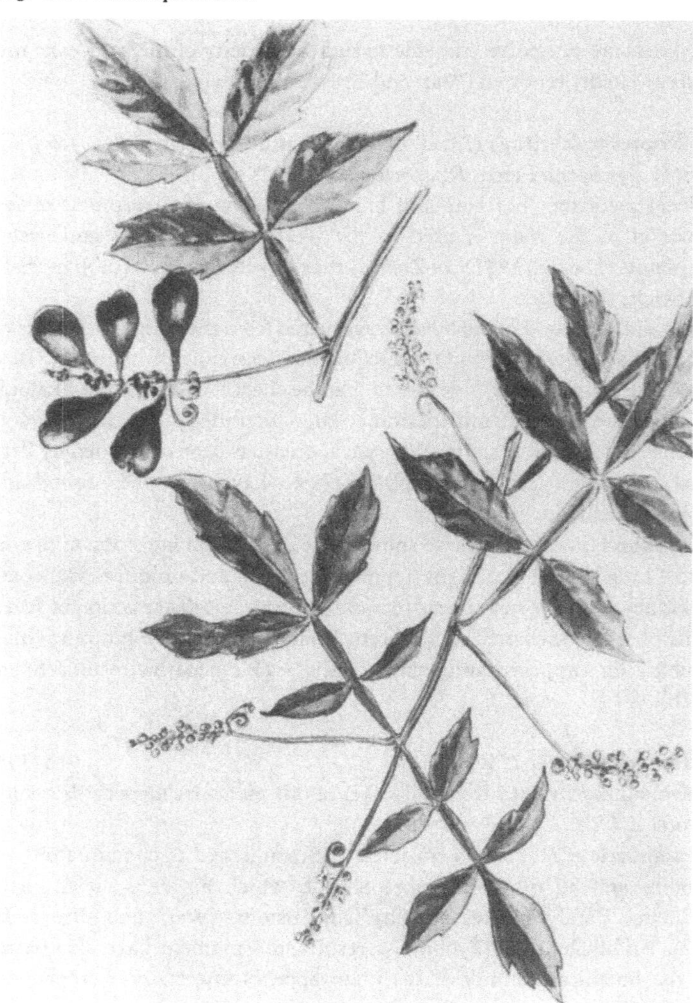

Table 4.4. *Antiprotozoal plants*

Plant	Part used	Active constituent(s)	Acts on/action	References
Acacia nilotica	Leaves	Tryptamine, tetrahydroharmane	*Entamoeba hystolytica*	Bhakuni *et al.* (1969)
Albizia lebbeck	Pods	Saponins (genins are triterpenoids, echinocystic acid, etc.)	Antiprotozoal	Dhar *et al.* (1968) Kaleyra (1975) Tripathi and Dasgupta (1974)
Alchornea cordifolia	Root	Alchornine (alkaloid)	Antiprotozoal	Bouquet (1972) N Ferreira *et al.* (1963a) Hart *et al.* (1970) Ogunlana and Ramstad (1975)
Alstonia boonei	Bark	Echitamine, plumeried?	Filariasis (Calabar swellings)	N
Argemone mexicana	Leaves, stems	Berberine	*Leishmania tropica*, *Trypanosoma lewisi* and antibacterial	CV Lambin and Bernard (1953)
Bixa orellana	Seed-coat	Wax-like substance	Paralyses intestinal parasites	Freise (1935)
Brucea antidysenterica, *B. guineensis*		Bruceolides (bruceantin, bruceantinol, dehydrobruceins)	*Plasmodium berghi*, also antifungal	Ngo Van Thu *et al.* (1979) Phillipson and Darwish (1981)
Carica papaya	Leaves	Carpaine (alkaloid)	Strongly amoebicidal, also inhibits *Mycob. tuberculosis*	CV Webb (1948b) Burdick (1971)
Chasmanthera dependens	Root	Berberine	*Leishmania tropica*, trypanocidal	Das Gupta and Dikshit (1929) Chopra *et al.* (1932) Henry (1949, p. 345) Mitscher *et al.* (1972a)
Cinchona spp.	Bark	Quinine	Plasmodium, anti-amoebic	CV
Euphorbia hirta	Plant	Euphorbon (triterpene)	Anti-amoebic	Kaustiva (1958) Kerharo and Bouquet (1950) Debaille and Petard (1953) Martin *et al.* (1964) Ridet and Chartol (1964)

(*Table continued*)

Table 4.4. (*Continued*)

Plant	Part used	Active constituent(s)	Acts on/action	References
Khaya senegalensis	Bark	Meliacin	*Paramecia* (*Balantidium coli*)	N Malcolm and Sofowora (1969)
Mitragyna inermis, *M. stipulosa*	Rootbark	Rhynchophylline	Antiprotozoal	CV
Newbouldia laevis	Bark	Harmane derivatives, harmine, harmol	Protozoicidal and anthelmintic	N Ferreira *et al.* (1963b) Ross *et al.* (1980) Al Shamma *et al.* (1981)
Nymphea lotus	Rhizomes	Nymphaeine	*Mycob. tuberculosis*, *Mycob. smegmatis*	Su *et al.* (1973, 1975)
Paullinia pinnata	Bark leaves	Triterpenic saponins	*Paramecia* (in vitro)	N Kerharo *et al.* (1960, 1962) Delaveau *et al.* (1979)
Pauridiantha lyalli (not in West Africa)	Bark	Harmane derivatives, harmine, harmol	Antiprotozoal	N Pousset *et al.* (1974) Levêque *et al.* (1975) Ross *et al.* (1980)
Phyllanthus niruri L.	Plant	Flavonoids	*Entamoeba histolytica*, also anthelmintic and antibacterial	Dhar *et al.* (1968) Nara *et al.* (1977)
Tiliacora funifera	Leaves	Funiferine	Antimalarial	N
Trichilia roka	Rootbark	Meliacins, catechuic tannins	Antimalarial	N Ogunlana and Ramstad (1975) Delaveau *et al.* (1979)
Urginea indica	Bulb	Scillarenin	Antiprotozoal, inhibits rhinovirus	CV Dhar *et al.* (1968) Sato and Muro (1974)

(e) Antimetazoal plants (anthelmintics)

Some anthelmintics act by paralysing worms (such as tapeworm), which then may have to be expelled by a purge. However, some drugs are also purgative themselves. Other anthelmintics can destroy the parasite through lysis, for example those containing proteolytic enzymes like bromelain from *Ananas comosus*, calotropain from *Calotropis procera* and papain from *Carica papaya* can digest worms. In other cases drugs (like antimony in the case of trematodes) act through inhibition of phospho-fructokinase, thus reducing the energy required by the worms to maintain their attachment to the walls of the mesenteric vessels and forcing them to emigrate into intra-hepatic portal veins, where they are destroyed by phagocytosis. Certain filaricides also cause the filaria embryos to emigrate to the liver, where they are subject to lysis by the cells of the reticulo-histiocytal system (Lechat *et al.*, 1978).

Active constituents

Many constituents with antibacterial activity are also anthelmintic. Apart from the vermicidal action already mentioned under antibacterial constituents, plumbagin (naphthoquinone) also appears to be active against cestodes; thus, *Diospyros mollis* also acts on *Hymenolepis nana*. Other plants containing gallic acid and ethylgallate such as *Acacia arabica*, *A. farnesiana* and *A. nilotica* also act on cestodes. This applies also to *Combretum racemosum* and another Combretaceae, *Anogeissus leiocarpus*, which contain gallic acid and tannins.

Alkaloids are often the active agents. Echitamine from *Alstonia boonei* acts against loa loa filariasis (Calabar swellings). The anthelmintic action of *Spigelia anthelmia* is attributed to spigeline, that of *Andira inermis* to berberine and/or andirine, that of *Punica granatum* to pelletierine tannate (acts on cestodes), that of *Quisqualis indica* to an alkaloid contained in the seeds and that of *Hunteria umbellata* to hunteramine or related alkaloids.

Flavonoids found in *Citrus acida* and *Albizia lebbeck* have antinematodal properties. In *Bixa orellana* the wax-like substances from the seed coat have been said to paralyse intestinal parasites.

Terpenoids. Many plants contain glucosides with terpenoids or resins as a genin and for many of these plants the action may be due to the genin. Amongst them we find some well-known anthelmintic plants such as *Chenopodium ambrosioides** (ascaridole, monocyclic terpene) and *Cucurbita maxima** and *C. pepo** (fruits), which have glucosides of tetracyclic triterpenes (cucurbitacines). These and ascaridole also force cestodes to emigrate but do not destroy them and their administration must therefore be followed up by a purge. Triterpenes are also found in *Melia azedarach* roots (act on cestodes and ascarides) and *Santalum album* (santalenes, tricyclic sesquiterpenes) both plants also being insecticides. Tetracyclic sesquiterpenes are reported in *Euphorbia* resins (euphol and euphorbol). Pentacyclic triterpenes derive from α- and β-amyrine, and oleanolic acid has been found as a constituent of the stembark of *Opilia celtidifolia**, which stimulates the activity of *Taenia pisiformis*. Several anthelmintics owe their action to resins or glucoside resin-compounds. *Albizia lebbeck*, *Embelia schimperi**, *Mallotus philippinensis* and *Phytolacca dodecandra* have been mentioned as such.

Mustard-oil heterosides. Plants with antibiotic properties that contain mustard-oil heterosides, such as *Allium* spp., *Cleome gynandra* and *Ritchiea longipedicilata*,* are also anthelmintic, and an active benzylisothiocyanate derivative (pterygospermine) is present in *Moringa oleifera* as well as triterpenoids.

The plants containing proteolytic enzymes have been mentioned above and under Antibacterial plants. In *Harrisonia abyssinica**, warburganal is considered to be the active agent.

Chenopodium ambrosioides L. CHENOPODIACEAE

American or Indian wormseed, sweet pigweed

L Cultivated and naturalized as a domestic anthelmintic. The pounded leaves are also applied to sores (Dalziel, 1937).

C The fresh aerial parts, harvested when the plant is flowering, yield 0.2–0.3% of an essential oil from the leaves, 0.5–1% of the oil from the flowering tops and over 1% from the fruits. The oil contains 20–30% of terpenoids (*p*-cymene, limonene, terpene) and 60–80% of ascaridol (peroxide of a terpenoid), which is mainly abundant in the fruit (Paris and Moyse, 1967, p. 131).

P Ascaridol is toxic to cold-blooded animals. It kills and paralyses *Ascaris* and hookworms (*Ankylostoma*) and to a lesser extent oxyurides and cestodes. Its administration must be followed by a saline or oily purge. In Man it may induce vertigo, vomiting and headaches, and is more often used in veterinary medicine (Paris and Moyse, 1967, p. 132; Kerharo and Adam, 1974).

Cucurbita maxima Duchesne **Cucurbita pepo** L. CUCURBITACEAE

Pumpkin, squash gourd

L The fruit (pulp) and leafy shoots of these plants are eaten as a vegetable and in soup whilst the gourds are used as domestic utensils. The seed kernels of both species are used in India and in some parts of Europe as an anthelmintic for tapeworm and as a diuretic, and the scraped pulp is applied as a poultice to burns, boils and swellings or even as a cooling application for headache or neuralgia (Dalziel, 1937).

C The pulp of the fruit of both species has been found to contain highly nutritive proteins, carbohydrates and minerals and also contains an amino acid, cucurbitine (amino-3-carboxy-3-pyrrolidine), believed to be responsible for the anthelmintic effect (Schabort, 1978). The seed oil of *C. maxima* is composed of 20.5% palmitic, 28% oleic, 8.5% stearic and 43% linoleic acids (Tewari and Srivasta, 1968) and that of *C. pepo* contains the same fatty acids but in slightly different proportions. In some bitter varieties of *C. pepo* (var. *ovifera*) the cucurbitacines B, D, E and I (toxic tetracyclic terpenes) have been found. Some of these which act as purgatives are also present in other Cucurbitaceae, particularly cucurbitacine E.

P The aqueous, ethereal and alcoholic extracts of the seeds of both plants have been found to have anthelmintic properties in vitro and in vivo. In vitro they significantly affect trematodes (*Fasciolopsis buski*) but they seem to be virtually inactive on nematodes and on *Ascaris lumbricoides* (Srivasta and Singh, 1967) and they are also inactive on cestodes (*Rallietina casticillus* in fowl) (Lahon *et al.*, 1978). The extracts have some cardiotonic action. The active principle is believed to be cucurbitine, which paralyses earthworms (*Lumbricoides terrestris*) within 117 min after application of a 2.9×10^{-3} M solution. This is an effect comparable to that of piperazine, a standard anthelmintic drug (Bailenger and Sequin, 1966). Gonzalez *et al.* (1974) noted that cucurbitine had a contractile action on the isolated rabbit intestine, which was counteracted by atropine but not by papaverine. Similarly cucurbitine paralyses *Taenia* spp. but its administration must be followed after 4–5 h by a saline purge to expel the parasite. Oral administration of a preparation of a stable, concentrated and deproteinized extract of the fresh seeds of *C. maxima* was, according to Junod (1964), not only well tolerated but in 80 patients with *Taenia saginata* produced 84.6% success. When the extract was given by duodenal or gastric tube, the treatment was successful in all of 54 cases (Gonzalez *et al.*, 1974; Bizyulyavichyus, 1969). The seeds of *C. pepo* are also reported to be an excellent anthelmintic especially against *Taenia* and *Botriocephalus*; in adults, doses of 50–60 g of fresh seeds are non-poisonous. Two fractions (not only cucurbitine) are thought by Bailenger and Sequin (1966) to be responsible.

Furthermore, it has been noted that the crude aqueous extract of the ripe fruit of *C. pepo* inhibited in vitro virulent strains of *Mycobacterium tuberculosis* in 1 : 10 000 dilution and retarded over 50% at a 1 : 100 000 dilution. This efficiency was confirmed in vivo in mice. Peposin (obtained from the acetone extract) inhibited the growth of *Mycob. tuberculosis* in a dilution of 1:50 000 for a period of 3 weeks (Gangadharan and Sirsi, 1955).

Opilia celtidifolia (Guill. ex Perr.) End. ex Walp. syn. (*Groutia celtidifolia* (Guill. & Perr.) Endl. ex Walp., *O. amentaceae* of Chev.; of Aubrev.) OPILIACEAE

L An extract of the bark of *Opilia* is used as an anthelmintic while its leaves are used in the treatment of sleeping sickness and as a diuretic.

C A methanol extract of the stembark yielded four saponins, the sugar moiety of all of these being arabinose and mannose. The aglycones of two of them have been identified as aleonalic acid whilst the aglycone of a third was found to be hederagin (Haerdi, 1964; Bouquet, 1972; Shihata *et al.*, 1977).

P Pharmacological studies by Shihata *et al.* (1977) revealed that intravenous injection of the saponin fraction of the stembark into anaesthetized dogs in doses of 20 mg/kg caused an increase in respiratory rate and a fall in blood pressure, which started to increase slightly after 30 min but did not return to its normal level. There was no effect on the renal circulation. The non-pregnant rat uterus was stimulated by doses of 40–100 mg/50 ml bath whilst the pregnant uterus did not respond. Doses higher than 10 mg inhibited intestinal motility but did not affect the intestinal response to acetylcholine. On the isolated rabbit heart 5–25 mg produced severe reduction in coronary outflow.

A study on the anthelmintic action of the saponins on intestinal worms isolated from dogs has shown slight stimulation of motility in *Toxocara leonani* by high doses (100–150 mg/50 ml bath). *Taenia pisiformis* was more easily affected: 50–150 mg/50 ml produced distinct stimulation of motility (Shihata *et al.*, 1977).

Embelia schimperi Vatke syn. (*E. abyssinica* Bak.) MYRSINACEAE

L In Uganda the leaf is used as a foodstuff; in different parts of Africa the berries are used as an anthelmintic against *Taenia* (an overdose is said to be often fatal). Related Indian species like *E. ribes* are also anthelmintic and mainly used against ascarides.

C The berries of the African species contain 6–7.5% embelin (2.5-dihydroxy-3-undecylbenzoquinone) whilst those of the Indian species contain 2.5–3% of embelin. In addition, quercitol, fatty ingredients and an alkaloid, christembine, are reported to be present (Chopra *et al.*, 1956; Kapoor *et al.*, 1975). The African *Embelia* is said to contain a toxalbumin (Watt and Breyer-Brandwijk, 1962, p. 786).

P In Africa embelin has been used in doses of 0.2–0.4 g as a taeniacide. In India embelin is said to have no effect on *Taenia* and hookworm but to be very effective in the treatment of ascarid infections. Clinical studies on 40 children infected by ascarides have shown a positive effect in 80% of the cases using an alcoholic and aqueous percolation of the berries whilst an aqueous extract cured 55% in both cases, eliminating ova from patients' stools. The worms were also expelled from the stools and no purging was required. No evidence of toxicity was noted during or after the treatment (Guru and Mishra, 1966). Perhaps the same results might be obtained in Africa if the toxalbumin could be eliminated. Aqueous extracts of the berries have proved to be antibacterial against *Staphylococcus aureus* and *Escherichia coli* in India (Chopra *et al.*, 1956) and have also been found to reduce fertility (Arora *et al.*, 1971). Gupta *et al.* (1976) have reported that the anthelmintic properties of embelin were greatly improved by using analogues obtained by chemical substitution such as isobutyl-embelin or *n*-hexylaminoembelin whilst di-imines were inactive. In concentrations of $1–3 \times 10^{-3}$ the analogues were active (contact period 30 min) on the parasites tested: *Paramphistomum cervi*, *Trichuris ovis*, *Oesophagostomum columbianum*, *Dipylidium caninum* (flukes, roundworms and tapeworms).

Ritchiea longipedicillata Gilg CAPPARIDACEAE

L In Nigeria, an extract of *R. reflexa* is used to treat guinea worm infection, the roots being used to treat earache (Dalziel, 1937). The leaves of the related *R. capparoides* (Andr.) Britten are reputed in Senegal to be antivenomous and antifilarial, and the roots plus leaves are used externally in the treatment of snake bites and of guinea worms infection. Roots and twigs are used as a plaster on the cervical (enlarged lymph nodes) 'ganglia' of patients suffering from *Gambiense trypanosomiasis* (Kerharo and Adam, 1974, p. 323).

C Cleomin has been isolated from the rootbark and identified as *S*(−)-ethyl-5-methyl-2-oxazolidine-thione, which is obtained by enzymatic hydrolysis of glucocleomin, a mustard-oil glycoside; it has also been found in *Gynandropsis gynandra*

(L.) Briq. (Misra and Sikhibhushan Dutt, 1937; Ahmed *et al.*, 1972; Oguakwa *et al.*, 1981).

P Mustard oils, including the oils of *Allium* spp., and senevols are reputed to have antiseptic, rubefacient and anthelmintic properties; this seems to justify the local uses of the plant.

Harrisonia abyssinica Oliv. syn. (*H. occidentalis* Engl.) SIMAROUBACEAE

L In Ghana the rootbark is boiled and drunk with palm wine as a laxative (Irvine, 1930). Watt and Breyer-Brandwijk (1962) report that the Teita and Jaluo tribes use the root as a remedy for bubonic plague, and the Sukuma administer it as an anthelmintic against oxyures and ascarides. The Nyamwezi swallow the smoke from the burning rootbark for ancylostomiasis until 'the smoke passes through the intestine'. The leaf is applied to abscesses and carbuncles.

C From *Harrisonia* (root?) cantine-6-one, harrisonin, obacunin, obacunoic acid, warburganal and muzigadial have been isolated (Mitscher *et al.*, 1972b; Kubo *et al.*, 1976, 1977).

P Githens (1949) has reported the action of *Harrisonia* roots against roundworm. In experimental malaria the roots proved ineffective (Spencer *et al.*, 1947a). Warburganal is a potent antifungal, antiyeast agent and a potent antifeeding agent against army worms (*Spodoptera littoralis* and *S. exempta*) (Kubo *et al.*, 1977; Vigneron, 1978), thus confirming local anthelmintic uses. Ether extracts of the root are reported to have an antimicrobial effect against *Neisseria gonorrhoea* and *Trichophyton mentagrophytes* (Uiso, 1979).

For a more comprehensive list of plants with antimetazoal action see Table 4.5.

Fig. 4.9. *Quisqualis indica* L.

Table 4.5. *Antimetazoal plants*

Plant	Part used	Active constituent(s)	Acts on/action	References
Acacia farnesiana	Aerial parts	Ethylgallate, gallic acid	Taenia, also antibacterial	Little *et al.* (1953) Hussain *et al.* (1979)
Acacia nilotica	Juice	Ethylgallate, flavonoids	Taenia	Camp and Norrel (1966) Attia *et al.* (1972)
Albizia lebbeck	Bark	Saponin	*Ascaris lumbricoides* in vitro	Sannie *et al.* (1963) Kaleyra (1975)
Alpinia galanga and spp.	Rhizome	Flavonoid	*Ascaris lumbricoides* in vitro	Mitsui *et al.* (1976) Ray and Majumdar (1976)
Andira inermis	Rootbark	*n*-Methyltyrosine, berberine	Anthelmintic (also insecticide)	N Heal and Rogers (1950)
Anogeissus leiocarpus	Root- and stembark	17% tannins	Taenia (veterinary)	N Dalziel (1937)
Carica papaya	Seeds, latex	Carpasemin (benzylthiourea), proteolytic enzyme, papain (ascaridol) digests worms	Anthelmintic and amoebicidal	CV Krishnakumari and Majumdar (1960) Shukla and Krishnamurti (1961b) Dar *et al.* (1965)
Carissa edulis	Twigs	Quebrachytol, cardio-glycosides	Taenifuge, anthelmintic and antiparasitic	*Hager's* Handbuch (1967–80, Vol. 3, p. 716) Bhaduri *et al.* (1968)
Citrus acida, *C. medica*	Rind	Hesperidin (flavonoid)	*A. lumbricoides* in vitro	CV Kaleyra (1975)
Diospyros spp.	Fruit	Plumbagin	*Hymenolepis nana*, also insecticidal	CV Sen *et al.* (1974) Farnsworth and Cordell (1976) Khan *et al.* (1980a)

Gynandropsis gynandra (Cleome gynandra)	Leaves	Glucocapparine (methyl senevol glucoside)	Anthelmintic	Ahmed *et al*. (1972) Misra and Sikhibushan Dutt (1937) Kjaer and Thomson (1963)
Hunteria umbellata syn. (*Polyadoa umbellata*) (Erin tree)	Bark	Alcoholic extract of bark	Paralyses and kills *Ascaris*, also smooth muscle depressant	Onuaguluchi (1964, 1966)
Mallotus oppositfolius var. *pubescens*	Leaves	Rottlerin	Locally used to expel tapeworms	Dalziel (1937)
Mallotus philippinensis c. ('Kamala' cultivated in French-speaking West Africa)	Hairs from fruit	Rottlerin, 'kamala'	Taenifuge in Indian Pharmacopoeia and British Veterinary Codex	Oliver (1960) Chopra *et al*. (1956)
Morinda geminata	Rootbark	Morindin = methylanthraquinone glycoside	*A. lumbricoides*	CV Nguyen Ba Tuoc (1953) Adesogan (1973)
Punica granatum c.	Bark	Pelletierin and isopelletierin, tannates, friedelin	Taenia, *Ascaris* (toxic), *Ankylostoma*	British Pharmacopoeia (1948)
Quisqualis indica (Fig. 4.9)	Fruit	Sesquiterpene (santonin-like), quisqualic acid	Cestodes, trematodes, thread- and hookworm	Guerrero *et al*. (1924) Henry (1949, p. 782) Xaio (1983)
Securidaca longepedunculata	Roots	Saponosides	Molluscicidal, intestinal parasites, *Tinea cruris* (fungus)	CV Fraga de Azevedo and de Medeiros (1963) Gaudin and Vacherat (1938) Odebiji (1978) Prista and Correia Alves (1958)
Spigelia anthelmia	Roots, leaves (fresh)	Spigeline, spigeleine (alkaloids)	Anthelmintic (toxic!), high doses cause spasms of eye and face muscles and even convulsions	British Pharmaceutical Codex (1923) US National Dispensary (1926)

II Plants with insecticidal or molluscicidal activity

In tropical climates especially, a number of infective diseases are conveyed by insects or molluscs, and the need for prophylactic action has already been mentioned in the discussion on protozoal and viral diseases. Some nematode infections are also transmitted by animal vectors; thus river blindness (onchocerciasis) is transmitted by the buffalo gnat, *Simulium damnosum*, and in filariasis caused by *Manzonella ozzardi* transmission is done by the midge *Culicoides furens* (Manson-Bahr, 1952). Insofar as the combating of these animal vectors can be based on the use of plants, they will be dealt with here.

(a) Insecticidal plants

Plant–insect relationships are numerous and often contradictory. Plants can provide a home and food for insects, and can live in partial symbiosis with some insects, for instance ants. Thus in Nigeria the spindle-shaped swellings of the branchlets of *Barteria nigritiana* Hook. (Passifloraceae) are occupied by stinging ants which defend the plant against larvae, caterpillars, beetles, and in fact, as I have found out to my considerable discomfort, against any living creature approaching the plant. Many plants attract insects, through bright colour and the smell of the flowers, to ensure pollination (Fraenkel, 1959). On the other hand, plants can also trap and 'digest' insects. An example is the insectivorous *Drosera*.

In insects, basic food requirements seem to be very similar and yet most insects feed preferentially only on a few closely related plant species. Fraenkel (1959) assumes that plant substances which are of secondary importance to the metabolism of the plants, such as glycosides, saponins, tannins, alkaloids and essential oils, may repel most insects or other animals but attract those few that feed on the particular plant species, which may have its particular smell or taste. Tests have shown that isolated active substances (glyco-alkaloids, mustard-oil glycosides, essential oils) even induced feeding when incorporated into neutral media (filter paper or agar jelly) or when applied to leaves commonly not accepted by the insects. Euw and Reichstein (1968) isolated 0.1 mg of aristolochic acid from swallowtail butterflies (the larvae of which feed exclusively on plants of the Aristolochia family). This acid protects the butterflies against vertebrate predators (Oliver-Bever, 1970). Like most insects, mites and microorganisms (bacteria and fungi) appear to be affected by the secondary plant products, and a number of plants have constituents with insecticidal action.

Most of the constituents of insecticidal plants are highly toxic for several cold-blooded animals and often their toxic properties are tested in fish. They are virtually non-poisonous to mammals (including Man) when given orally, but when intravenously injected some can produce respiratory paralysis and death by asphyxia (Heftmann, 1975).

A number of Fabaceae are insecticidal. The best known, which are widely used commercially, belong to the genera *Derris* and *Lonchocarpus*, and contain 3–20% of rotenone. Unfortunately, the West African species of *Derris** and *Lonchocarpus** appear to have less active components than the American or Indian species, but some other Fabaceae, such as *Entada africana**, *Mundulea sericea** and *Tephrosia vogelii**, which are found in West Africa, are very efficient.

All these insecticidal Fabaceae have roughly the same chemical constituents and their efficiency is often measured in relation to the amount of rotenone present.

Rotenone has an isoflavone nucleus combined with a furan and pyran ring. It is said to belong to the isoflavones (Paris and Moyse, 1967, p. 387), or to the phenylpropanoid flavonoids (Towers and Wat, 1979). Substances chemically related to rotenone, also found in these Fabaceae, are called rotenoids. Thus, degueline is an isomer with a 2-dehydrobenzopyran nucleus; tephrosine or toxicarol have phenolic hydroxyl groups. Derric acid constitutes the common half of the molecule of rotenone and these rotenoids. The oral lethal toxicity of rotenone is low for mammals: 3 g for the rabbit, 0.6 g for the rat and 0.06 g for the guinea pig. In insects and fish it acts on ingestion or contract and is a potent inhibitor of mitochondrial oxidation (Towers and Wat, 1979); it is active in dilutions down to 1 : 20 000 000. At this concentration it can kill goldfish in 2.5 h. As a stomach poison in silkworms, rotenone is 30 times as toxic as lead arsenate. It is 15 times as toxic as nicotine when used as a contact insecticide against bean aphis and is 25 times as toxic as potassium cyanide to goldish (Watt and Breyer-Brandwijk, 1962).

Saponosides have been reported to be present in other Leguminosae with insecticidal properties, such as *Dichrostachys glomerata* (also molluscicidal against *Bulinus globulus*), *Pentaclethra macrophylla* and *Tetrapleura tetraptera*.

Triterpenoids seem to be responsible for the insecticidal activity of *Melia azedarach**, *Annona* spp*. and *Santalum album**. The insecticidal action of *Sesamum indicum* has been attributed to a lignan (sesamin) and that of *Duranta repens* cultivated in West Africa, to an alkaloid. *Vernonia pauciflora*, with sesquiterpenic lactones (vernolide and hydroxyvernolide), is considered useful for killing termites.

Larvicidal properties have been found in *Thevetia neriifolia* (aucubine, flavonic heteroside), *Uvaria chamae* (chamaenetin, benzylflavanone) and *Spilanthus uliginosa* (spilanthol). The leaves of *Melia azedarach* are insect repelling through their meliacins and a repellant action is found in many essential oils. The oils of *Ocimum basilicum* and *O. canum* are reputed to repel ants and moths. (For details on *U. chamae*, *O. basilicum* and *O. canum* see Antibacterial plants).

Derris microphylla Miq. Jacq. FABACEAE

Derris spp. have been introduced in certain parts of West Africa (Congo, the Ivory Coast and Nigeria). They were originally used in China and India and were introduced into the USA and Europe towards 1930 (Paris and Moyse, 1967, Vol. II, p. 388). Commercial Derris is mainly obtained from Southern Asia (*D. elliptica* Benth. and *D. malaccensis* Prain). The roots of these contain 4–20% of rotenone. Seasonal variations of the rotenone content have been reported (Nandy and Gupta, 1968). Rotenone is colourless and odourless, and thousands of tons of rotenoids are used against agricultural parasites and to relieve domestic animals of insects. In its commercial formulation, rotenone is often associated with pyrethrins (from certain plants of the genus *Chrysanthemum*, referred to as Pyrethrum, not present in West Africa), which act more rapidly but are less stable and more expensive to produce (Paris and Moyse, 1967, Vol. II, p. 389). Prenylated flavonoids (derricin, derridin, lonchocarpin, etc.) are almost ubiquitous in *Derris* and *Lonchocarpus* spp. and may be considered as biogenetic precursors of the rotenoids (Della Monache *et al.*, 1978).

Lonchocarpus sericeus Poir. H.B. & K. syn. (*Robinia sericea* Poir.) FABACEAE

L The bark is employed as a stomachic and laxative and in Nigeria for convulsions and backache. It is applied locally for parasitic skin conditions and eruptions. In America this species has been listed, together with other *Lonchocarpus* spp. as an insecticide (Dalziel, 1937).

C From the roots, seeds and leaves of this tree, lonchocarpine has been isolated in the Congo (Castagne, 1938). It was identified as a 5-hydroxy-2,2-dimethyl-3-chromen-6-yl-strylycetone. From the leaves, sterols (including β-sitosterol), prenylated flavonoid precursors, *p*-coumaric acid, quercitin, rutin and hyperosid have been isolated (Kerharo and Adam, 1974; Della Monache *et al.*, 1978). The main commercial sources for rotenone are the American *L. nicou* DC. and *L. urucu* Phillips & Smith. In the Nigerian *L. laxiflorus* Guill. & Perr., Pelter and Amenechi (1969) reported an isoflavonoid and a pterocarpanoid.

Entada africana Guill. et Perr. syn. (*E. ubaguiensis* de Wild., *E. sudanica* Schweinf., *Entadopsis sudanica* (Schweinf.) Gilb. & Boutique) MIMOSACEAE

L This tree has many local uses. It yields an inferior quality gum and the leaves are used as cattle fodder. The bark of root and stem yields a long fibre used for cordage, commonly for roof-binding, the tying of grass matting, etc. In Northern Nigeria and Northern Ghana, an infusion of the leaves or of the bark is taken as a tonic and stomachic. The leaves also constitute a good wound dressing, preventing suppuration (Dalziel, 1937).

C Rotenone has been reported to be present in the plant (Gaudin and Vacherat, 1938) and saponosides have been found in the bark and leaves (Githens, 1949).

P An infusion of the leaves at a concentration of 1 : 1000 kills *Carassius auratus* (goldfish) in 12 h but was not toxic even in doses of 5 g/kg to the guinea pig (Gaudin and Vacherat, 1938).

Mundulea sericea (Willd.) Chev. syn. (*M. suberosa* (DC.) Benth.) FABACEAE

L The bark is known to be poisonous and both bark and seeds are used as a fish poison in Nigeria, Ghana, India, Sri Lanka, Tanzania and Zimbabwe. It is said to kill and not merely to stupefy. Both bark and roots have been suggested as an insecticide (Dalziel, 1937).

C Worsley (1936) has isolated rotenone, deguelin, tephrosin and some alkaloids and glycosides from the bark. Chopra *et al.* (1941) confirmed the presence of rotenone. Mundelone or mundulone of isoflavonoid structure has been isolated from the bark as well as a rotenoid, munduserone, and a flavonoid, sericetin (Finch and Ollis,

1960). In addition, the flavanones lupinifolin, lupinofolin, mundulin, mundulinol and a chalcone, sericone, have been reported by van Zyl *et al.* (1979). The plant is also said to contain hydrocyanic acid (Watt and Breyer-Brandwijk, 1962).

P In India the root, bark, leaves, stem and seeds are all said to be toxic and several authors consider the plant to be an extremely efficient fish poison capable, even, of killing small crocodiles (Chopra *et al.*, 1941). According to Worsley (1937) it is considerably more toxic to insects than *Tephrosia vogelli* Hook. The bark is said to be equivalent in toxicity to the root of *Derris elliptica*, despite a lower percentage of rotenone but the toxicity of plants grown under different conditions varies and the smooth bark from the closed forest region is much more toxic than the rough corky bark of the savannah areas. The leaves, root and bark have proved to be efficient against *Chrysanthemum* aphids (Watt and Breyer-Brandwijk, 1962, p. 636).

Tephrosia vogelli Hook. FABACEAE
Fish poison bean

L The shrub was formerly cultivated throughout West Africa, and used to stupefy fish (this has been prohibited in Southern Nigeria). If the pounded leaf is thrown into a creek the fish are temporarily paralysed and can be lifted out of the water. Most commonly the leaves are used alone but in some districts people also use them with the pods or use the pod only. A very dilute solution is sufficient to produce paralysis and death in fish.

C Two main constituents, deguelin and tephrosin (hydroxydeguelin) have been found not only in the leaves but also in the roots, fruit capsules and seeds. The leaves also contain a volatile oil, tephrosal, which is responsible for their pungent odour (Hanriot, 1907; Castagne, 1938). Furthermore, rotenone, dehydroxydeguelin, rutin and tetrahydroxy-3,6,7,4′-methoxy-5-flavone (which was called vogeletin) have been reported to be present in *T. vogelii* (Vleggaar *et al.*, 1978) and rotenoid extraction from the leaves was carried out by Barnes and Freyre (1966a, b, 1967).

P By simple contact, fish kept in a 2% dispersion of the leaves have died within 3 h and those kept in a 1% dispersion in 12 h. On the other hand, guinea pigs ingested as much as 5 g/kg without adverse effects. When injected intravenously, however, a leaf extract corresponding to 0.01 g/kg of tephrosin has killed dogs in 5 min whilst the injection of 0.01 g/kg of pure tephrosin produced death through respiratory arrest only the day after, preceded by accelerated breathing and convulsive effects alternating with paralysis (Hanriot, 1907). Concentrations of tephrosin as low as 1:50000000 paralyse fish and a delayed fatal effect may occur. The *Tephrosia* rotenoids are also toxic to insects, batrachians, worms and snails. Although deguelin is about half as efficient a piscicide as rotenone (it kills fish in a dilution of 1:20000000 at 27°C in 4.5 h whilst rotenone in the same circumstances kills fish in 2.5 h (Barnes and Freyre, 1966b, 1967)), in *Tephrosia* it is the leaves which contain the active rotenoids rather than the roots as in *Derris* and *Lonchocarpus* spp. After 22 years of storage the dried leaves of *Tephrosia* were still able to kill goldfish and larvae of worms (Watt and Breyer-Brandwijk, 1962).

Spilanthes uliginosa Sw. syn. (*S. acmella* Chev, *S. oleraceae* Jacq.) COMPOSITAE
Bresil cress, para cress

L The flowerheads of *S. uliginosa* have a pungent taste and cause salivation. They are chewed in Nigeria and the Cameroons to relieve toothache and are used in local application as a haemostatic and analgesic (Dalziel, 1937). In India, they are also used for sore throats and gums and in paralysis of the tongue (Chopra *et al.*, 1956).

C The flower tops of both species contain spilanthol, which is an unsaturated amide and may be identical with affinine (from *Erigeron affinis*) (*N*-isobutyldecatriene-2,6,8-amide), a sterol and a non-reducing sugar (Paris and Moyse, 1971, p. 458).

P Extracts of the plants have a depressant action on the guinea pig ileum and on the blood pressure of cats and dogs. In mice the maximum dose tolerated intraperitoneally is 100 mg/kg (Dhar *et al.*, 1968). Spilanthol has been said to have local anaesthetic action (Chopra *et al.*, 1956). It also has a larvicidal action. Extracts of the flowerheads in a soapy suspension and spilanthol kill *Anopheles* larvae in a dilution of 1:100000 (Kerharo and Adam, 1974) and the whole plant has insecticidal properties towards cockroaches and bedbugs (Heal and Rogers, 1950).

A 50% ethanol extract of the plant is reported to have antibacterial action against *Staphylococcus aureus*, *Salmonella typhi*, *Escherichia coli*, *Mycobacterium tuberculosis* and *Agrobacterium tumefaciens*. Antifungal activity was found against *Candida albicans*, *Trichophyton mentagrophytes* and *Aspergillus niger*. Antiviral action was seen in Raniket disease and against *Vaccinia* virus, and antiprotozoal action against *Entamoeba histolytica*. Finally, the plant has an anthelmintic effect towards *Hymenolepis nana* (Dhar *et al.*, 1968).

Melia azedarach L. (Fig. 4.10) MELIACEAE
Persian lilac, bead tree

L Cultivated in Nigeria, its bark has been used as an anthelmintic. In India, the country of origin, the juice of the leaves is used internally as an anthelmintic, antilithic, diuretic and emmenagogue (Chopra *et al.*, 1956). The seeds are prescribed for rheumatism and the oil is considered to be similar in its properties to Neem oil.

C Schulte *et al.* (1979) isolated triterpenoids, steroids and aromatic compounds from the air-dried roots. These compounds are 24-methylene-cycloartanone, cylco-eucalenone, 4-stigmastene-3-one-2,4-campestene-3-one, 4-methylene-cycloartanol, triaconthanol, cycloeucanol, β-sitosterol, β-sitosterol-D-glucoside, vanillic aldehyde, transcumanic aldehyde and vanillic acid (Kraus and Bokel, 1981).

In the fruit and heartwood they found the same cyclonastone derivatives as in the roots, plus two additional substances: 21, 23, 24, 25-diepoxy-tirucall-7-en-21-ol and a protolimonoid (Nath, 1954; Schulte *et al.*, 1979). An alkaloid paraisine (ocaziridine) was found in the leaves. In the rootbark of the Nigerian species the limonoids gedunin and 7-deacetoxy-7-oxogedunin, the nimbolins A and B (meliacin cinnamates), melianins A and B, as well as the products of decomposition (fraxinellose and azedainic acid) were reported, besides 24-methylene-cycloartanone, cycloeucalenone and cycloeucalenol (El Said *et al.*, 1968; Ekong *et al.*, 1969b; Okogun *et al.*, 1975). In Israel, the protolimonoids melianon, melianol, melianodiol and meliantiol as well as an apo-eupol derivative were found by Lavie *et al.* (1967) and Lavie and Levy (1969) in the roots and bark.

P Aqueous extracts of the fruits have been found to produce dyspnoea, tremor, convulsions and death in rabbits. In cats, dogs and sheep they produced paralysis and narcosis (Murthy and Sirsi, 1957, 1958). However, Ekong *et al.* (1968) found no components toxic to mice from Indian *M. azedarach*. The LD_{50} in mice was 1.04 g/kg for ether and chloroform extracts and 1.5 g/kg for a triterpene mixture.

The insecticidal properties of the leaves, stems and bark have been clearly demonstrated on crickets, cockroaches and *Aedes aegypti* (Heal and Rogers, 1950) and make *Melia* a very superior insecticide. It has also been noted that plants sprayed with triterpenoids of the fruit are not visited by insects (Henry, 1949, p. 781; Bézanger-Beauquesne *et al.*, 1981). The repellant substance found in *Melia* leaves (3.5%) has been named meliatin. It is very similar in its properties to cail-cedrin (Chauvin, 1946). Anthelmintic properties have also been reported for *Melia* by

Fig. 4.10. *Melia azedarach* L.

Chopra *et al.* (1938) and *Cortex azedarach* was included in the US Dispensary as an anthelmintic against tapeworm and ascarides, and as an antiparasitic.

The bark is said to act as an antispasmodic and tonic. Some antimicrobial activity of extracts of the leaves, bark and seeds has been reported by Nickel (1959), mainly on Gram-negative bacteria. The extracts were inactive against *Paramecia* (Spencer *et al.*, 1947a). In anticancer tests the extracts were only slightly active against sarcoma 180 and adenosarcoma 755 (Abbot *et al.*, 1966).

Annona muricata L. ANNONACEAE
Soursop
Annona reticulata L.
Custard apple
The leaves of these introduced spp. have been reported to have anthelmintic activities in their country of origin (Nadkarni, 1954; Watt and Breyer-Brandwijk, 1962) and an aqueous solution of the seeds of *A. reticulata* kills most wood bugs (*Lecanium*). Insecticidal activity has also been reported for the seeds, stems and roots of *A. muricata* and *A. reticulata* (Steenis-Kruseman, 1953). *A. muricata* has a potent insecticidal effect against *Macrosiphonella sanbornia*. Against *Aphis ruminis* the toxicity of the roots of *A. muricata* is inferior to that of *Derris elliptica* (Heal and Rogers, 1950; Kerharo and Adam, 1974). The insecticidal constituent of the leaves and roots is resistant to heat, but not to saponification, and may be related to fatty acids of high molecular weight (Hegnauer, 1962–68, Vol. III, p. 120). Both *Annona* spp. contain hydrocyanic acid and *A. reticulata* contains 0.12% of anonaine.

Annona senegalensis Pers. ANNONACEAE

L The people of Northern Nigeria use the leaves of this small tree, mixed with cereal cakes, or in the form of a large bolus mixed with soda and bran, as an anthelmintic for horses. The bark and root have been similarly used. The leaves are also applied as a dressing mixed with the latex of *Calotropis procera* and of *Euphorbia balasamifera* in the treatment of epizootic lymphaginitis in horses (Dalziel, 1937).

C The leaves of *A. senegalensis* have been found to contain rutin, quercetin and quercitrin (Mackie and Ghatge, 1958) and in the stembark, positive reactions for alkaloids, saponins and tannins have been obtained (Persinos and Quimby, 1967). The leaf contains a hard and a soft wax, both of which contain higher saturated fatty acids. From the hard wax, primary alcohols and palmitone (hentriacontanone), and from the soft wax a yellow sesquiterpenic oil has been obtained (Mackie and Misra, 1956).

P The soft leaf wax has proved to be effective against *Sclerostoma* larvae from horse faeces; this action has been attributed to the sesquiterpenes. The twigs of *A. senegalensis* produced 100% mortality of *Oncopeltus fasciatus* in concentrations of 500 μg/ml (Jacobson *et al.*, 1975).

Annona squamosa L. ANNONACEAE
Sweet sop, sugar apple

L In Northern America and in Gambia the leaves of the sweet sop are used as an insecticide and to eliminate bed-bugs (Dalziel, 1937). In Indonesia the seeds are used against head-lice (Steenis-Kruseman, 1953).

C From the bark, roots, seeds and stems of *A. squamosa* aporphine alkaloids (anonaine, roemerine, norcorydine, corydine, norisocorydine and glaucine) have been isolated (Bhakuni *et al.*, 1972).

P Anonaine has been found to possess antimicrobial properties against *Staphylococcus aureus*, *Klebsiella pneumoniae*, *Mycobacterium smegmatis* and *Candida albicans* at the 100 μg/ml level (Chen *et al.*, 1974). Corydine is reported to have anticancer activity (Bhakuni *et al.*, 1972). An ether extract of the seeds of *A. squamosa* has been found to act as a stomach poison in the larvae and eggs of *Bombyx mori* and to be moderately toxic to *Musa nebulo* and *Triboleum castanum* adults. When petroleum ether is added to the extract an insoluble resin is formed and this treatment has been found to increase the toxicity against *M. nebulo* by a factor of six (Mukerjee and Govind, 1958).

Santalum album L. SANTALACEAE
Sandalwood

L A native of India introduced in parts of West Africa (Nigeria and the Ivory Coast), this tree is used in India for its scented heartwood, yielding the fragrant sandalwood oil.

C From the bark a triterpenoid has been identified as urs-12-en-3β-yl-palmitate (Shankaranarayana *et al.*, 1980). The bark is a waste material after sandalwood oil extraction.

P The benzene extract of the bark has insect growth-inhibiting and antireproductive (chemosterilant) properties. Local application of the triterpenoid in micro-doses to freshly formed pupae of economically harmful forest insects (*Atteva fabriciella*, *Eligma narcissus*, *Eupterote geminata*, etc.) produced morphologically defective adults with crumpled wings and shorter abdomen, suggesting a growth inhibition. It has been observed that feeding a glucose solution containing a minute quantity of the compound to freshly emerged moths prevented them from mating and laying eggs, indicating an antireproductive (chemosterilant) effect (Shankaranarayana *et al.*, 1980). These authors write: 'Such chemosterilant compounds of late, are becoming popular as "third generation pesticides" in controlling forest pests without the bad side effects (like toxicity, environmental pollution etc.) generally possessed by common organic pesticides.'

For a more comprehensive list of plants with insecticidal and molluscicidal action see Table 4.6.

Table 4.6. *Insecticidal and molluscicidal plants*

Plant	Part used	Active constituent(s)	Acts on/action	References
Acacia nilotica	Fruit, stembark	Tannins	Molluscicidal, *Bulinus truncatus*, *Biomphalaria pfeifferi*	Hussein Ayoub (1983, 1984)
Afrormosia laxiflora	Rootbark	*N*-Methylcytisine	Insecticidal	N
Balanites aegyptiaca	Stembark	Saponosides, the genins are dios- and yamogenins	Molluscicidal	CV
	Seeds	Furocoumarin	Insect-feeding deterrent	Hardman and Sofowora (1972) Abdullah *et al*. (1978) Seida *et al*. (1981) Tomassini and Mathos (1979)
Bidens pilosa	Leaves	Phenylheptatriene	Toxic to insects and larvae	Degener (1975)
Dichrostachys glomerata	Roots	Saponosides	Molluscicidal to *Bulinus globulus*	Bouquet (1972) Adewunmi and Sofowora (1980)
Duranta repens c.	Fruit juice	Isoquinoline (alkaloid analogue to narcotine)	Insecticidal lethal to anopheline and culcicine larvae	Manson (1939) Chopra *et al*. (1941) Yousef (1973)
	Leaves	Flavonoids		Makboul and Abdel-Baki (1981)
Euphorbia hirta	Plant	Euphorbon (triterpene, quercetol)	Insecticidal anti-amoebic antibacterial	Heal and Rogers (1950) Debaille and Petard (1953) Ridet and Chartol (1964) Gupta and Garg (1966) Dhar *et al*. (1968) Ndir and Pousset (1981) Pousset (1981)
Hymenocardia acida	Roots	Hymenocardine (alkaloid)	Insecticidal	Heal and Rogers (1950) Pais *et al*. (1976)
Milletia ferruginea (Fig. 4.11)	Seeds	Rotenone, saponins	Insecticidal	Paris and Moyse (1967, Vol. II, p. 390)
Momordica charantia	Leaves	Momordicin	Insecticidal, bacteriostatic	Heal and Rogers (1950) Watt and Breyer-Brandwijk (1962)

Nicotiana tabacum	Juice	Nicotine, nornicotine, anabasine	Insecticide, guinea worm (toxic!)	Henry (1949, p. 49) Kerharo and Bouquet (1950)
Ocimum basilicum	Leaves	Methylchavicol, eugenol	Chases ants, anthelmintic	Heal and Rogers (1950)
O. canum	Leaves	Camphor	Chases moths	Nickel (1959)
Pentaclethra macrophylla	Root	Saponosides	Insecticidal	Jain and Jain (1972) N
Piper guineense	Fruit	Piperine, dihydropiperine, dihydropiperlonguminine, etc., dihydrocubebin	Insecticidal and antibacterial (see Table 4.1)	Soulimov *et al.* (1975) Paris and Moyse (1967, Vol. II, p. 114)
Pseudocedrela kotschyi	Bark	Pseudocedrelin (phenolic lactone)	Kills goldfish	Addea-Mensah *et al.* (1977a)
Quassia africana	Stem wood	Quassin	Insecticidal, expels threadworms	Moyse-Mignon (1942) Heal and Rogers (1950)
Securidaca longepedunculata	Roots	Triterpenic saponosides	Molluscicidal	Oliver (1960) Fraga de Azeveda and de Medeiros (1963)
Sesamum indicum	Plant	Sesamine (lignan)	Insecticide synergic to pyrethrum	Kerharo (1968) Webb (1948) Ramaswany and Sirsi (1957) Paris and Moyse (1971, Vol. III, p. 251)
Spilanthus uliginosa	Flower-heads	Spilanthol = affinine (*N*-isobutyl-decatriene-2,6,8-amide)	Larvicidal (anopheles, kills cockroaches and bedbugs)	Heal and Rogers (1950) Dhar *et al.* (1968)
Tetrapleura tetraptera	Fruit	Saponosides (oleanic acid triglycoside?)	Insecticidal	Adesina *et al.* (1980a)
Thevetia neriifolia	Fruit, leaves	Aucubine (iridoid heteroside)	Larvicidal and antibacterial	Sofowora (1980) CV
Vernonia pauciflora	Twigs (leafy)	Sesquiterpenic lactones (vernolide and hydroxyvernolide)	Termites	Sticher and Meier (1978) Kerharo and Adam (1974)

[a]Piperine is more toxic than pyrethrin to the housefly and 0.05% of piperine + 0.01% of pyrethrin is more toxic than a 0.1% solution of pyrethrin (Harvill and Hartzell in Kerharo and Adam, 1974, p. 637).

(b) **Molluscicidal plants**

As already mentioned, snails can play an important part in the transmission of diseases, especially those caused by *Schistosoma* spp. in their various forms. At present over 200 million people in about 70 tropical and subtropical countries are affected by endemic schistosomiasis (Hostettmann *et al.*, 1982). The developmental cycle in the snail (mostly *Bulinus* spp.) takes about 6 weeks. In the infected snail the parasites are very prolific, a single miracidium can produce in optimal conditions 100 000–250 000 cercaria, these emerging at the rate of 50–1000 daily (none emerging on dark days).

Active constituents

A number of plant extracts lethal to schistosome-transmitting snails were submitted by Mozley (1939, 1952) to comparative tests. The fruits of *Balanites aegyptiaca* Del., *Swartzia madagascariensis** Desv. and *Sapindus saponaria* L. (grown

Fig. 4.11. *Milletia aboensis* (Hook.) Bak.

in West Africa, introduced from tropical America) were found to be amongst the most promising molluscicides. Later, Lemma (1970) reported from Ethiopia the strong molluscicidal properties of the dried berries of *Phytolacca dodecandra* L'Hérit.*, which also occurs in West Africa and which is widely used in Ethiopia as a soap substitute. The active constituents in all these plants were found to be haemolytic saponosides. Another plant, *Croton macrostachys* Hochst. ex Del., with active molluscicidal properties, was reported from the Sudan (Amin *et al.*, 1972). The active factors of this were also saponosides (El-Kheir and Salih, 1979).

In 1977 Dossaji *et al.*, performed tests with *Polygonum senegalense**, a powerful molluscicide in Kenya, where it is a common weed along rivers, lakes, etc. The active components are pseudo-cyanogenetic glycosides. In West Africa, the molluscicidal properties of *Jatropha gossypiifolia** and *J. curcas** fruits were examined by Adewunmi and Marquis (1980) and the methanol-related extract of the fruit of *J. gossypiifolia* was found to be particularly potent. Hydrocyanic acid and steroid saponins are present in the plant.

In *Securidaca longepedunculata* (Polygalaceae) the molluscicidal action of the roots is due to triterpenic saponins (see Table 4.6). In testing the molluscicidal activity of 24 saponins from various plants against *Biomphalaria glabrata*, Hostettmann *et al.* (1978, 1982) reported that at concentrations of 32 p.p.m. and even of 16 p.p.m. monodesmosidic triterpenoid saponins and saponins of the spirostanol series are potent molluscicides. Bidesmosidic saponins and the aglycones (e.g. hederagin and oleanolic acid) were found to be inactive. In *Acacia nilotica* galloyl tannins bind the protein of snails, inhibiting their enzymes (Hussein Ayoub, 1984).

Swartzia madagascariensis Desv. CAESALPINIACEAE

L The fruits are used in parts of West Africa, the Zambesi valley, Madagascar and Portuguese East Africa as a fish poison, (although fish poisons are forbiddeen in Nigeria). In Northern Nigeria the roots are occasionally added to the crushed fruits (Dalziel, 1937; Githens, 1949). In Senegal the roots are used in the treatment of leprosy and syphilis and the fruits as an abortifacient (Kerharo and Adam, 1974). The heartwood is termite and borer proof (Dalziel, 1937). In Bechuanaland the leaves are considered a valuable means of combating the schistosoma-carrying snail and are said to have insecticidal properties, especially against termites (Schultes, 1979).

C From the heartwood, pterocarpanoid constituents have been isolated (Harper *et al.*, 1969). The fruit valves were found to contain a yellow flavonoid pigment which gave, on hydrolysis, swartziol (=kaempferol), rhamnose and glucose. In addition, triterpenoid saponosides and catechuic tannins have been reported. The seeds also contain saponosides and are strongly haemolytic (Beauquesne, 1947; Paris and Bézanger-Beauquesne, 1956). Harborne *et al.* (1975) confirm the presence of isoflavones.

P The fruits have been shown experimentally to possess ichthyotoxic properties and to kill fish in concentrations of 0.1% within 2 h, whereas they are toxic to guinea pigs only at higher concentrations (5 g/kg given perorally) (Gaudin and Vacherat, 1938; Beauquesne, 1947). They are also lethal to the snails transmitting bilharzia at higher concentrations (schistosomiasis).

Phytolacca dodecandra l'Hérit. syn. (*P. abyssinica* Hoffm.) PHYTOLACCACEAE
Endod, soapberry tree

L In some parts of West Africa the young shoots are used as a potherb and in the Ivory Coast as a substitute for spinach. On the other hand in East Africa and in Madagascar the berries and the juice of the fresh leaves are said to have caused death and the leaves are said to be the most dangerous of the Chagga abortifacients (a cheekful of leaves and young shoots is chewed and the juice swallowed, abortion commencing about 10 h later) (Watt and Breyer-Brandwijk, 1962). In Southern Nigeria the ashes of the burnt plant are used for making soap and in Somalia, Ethiopia and Uganda an infusion of the seeds or berries is used as a soap for washing clothes (Dalziel, 1937; Watt and Breyer-Brandwijk, 1962). A higher mortality of molluscs along the rivers in Ethiopia where the inhabitants washed their clothes with the berries induced Lemma to study the molluscicidal properties of the plant (Lemma, 1970).

C Investigations into the nature of the active principle showed that the active part was a saponoside, endod, the genin of which could be traced back to oleanolic acid by several authors. A freeze-dried aqueous extract was actively molluscicidal at concentrations of 1.25–2.5 p.p.m. (Jewers, Tropical Products Institute London, 1968 quoted in Lemma (1970)). A steroid saponin has been found in the roots (Ahmed *et al.*, 1949).

P After 24 h of exposure endod killed all species of snails tested in comparative experiments with other molluscicides (*N*-tritylmorpholine, copper sulphate, ethanolamine of niclosamide). The test animals were killed in all cases by concentrations of less than 30 p.p.m. Whole and powdered berries kept at room temperature (22°C) for 4 years displayed no change in their molluscicidal potency during six-monthly tests, whilst the crushed fresh berries, and a solution prepared from them, lost their potency within a few days. The potency of the crude berries remained stable over a wide range of pH values and temperatures, under ultra-violet irradiation and in various concentrations of river-bed mud. The low solubility of the saponin from the whole berries also appears to be a notable feature. For mammals and birds, the LD_{50} is above 2 g/kg. The berries do not harm germination, growth rate or the morphology of vegetation. They have no insecticidal or larvicidal action and their toxicity to fish is similar to that of the above-mentioned molluscicides. In a concentration of 4 p.p.m., leeches are killed in 6 h. The *Cercariae* and miracidia of *Schistosoma haematobium* are killed in 10 min by 1000 p.p.m., in 1 h by 100 p.p.m. and in 2 h by 50 p.p.m. Thus *P. dodecandra* seems to be a cheap and effective means of eliminating *Schistosoma* in certain areas (Lemma, 1970). Only short staminate plants produce berries and the haemolytic activity of the samples runs parallel with the molluscicidal effect. Although in Ethiopia the unripe berries are always used for washing clothes (highest saponin level), sometimes the over-ripe berries have also shown a marked capacity to kill snails (Lugt, 1980). Endod has a weak ovicidal effect: it is not ovicidal at the concentration at which it kills adult snails. This difficulty can, however, be overcome by repeated treatment as has been demonstrated in the field. The molluscicidal potency (CL_{90}) for eggs of *Bulinus truncatus sericinus* is of the order of 50–100 p.p.m. in 6 h (Lemma, 1970). The berries show approximately equal potency against different species of snail in a 24-h exposure at 30 p.p.m., whereas in a 6-h exposure, *Physa acuta* was the least and *Biomphalaria pfeifferi rupelli* the most resistant species of snail examined (Lemma, 1970).

Croton macrostachyus Hochst. ex Del. syn. (*C. guerzesiensis* Beille)

EUPHORBIACEAE

L In East Africa, the bark of this tree is reported to be used as a cathartic (Githens, 1949).

C The testa, constituting 40% of the weight of the kernels, possess no molluscicidal activity. Only alcoholic and aqueous extracts of the kernel were biologically active; chromatographic screening showed that the seeds contain alkaloids, amino acids and sterols as well as triterpenes or their saponins. Biological testing indicated that the zone corresponding to the sterols and triterpenes was the only active one. Finally, the molluscicidal action was found to be due to at least two saponins and the activity was greatly increased by freeing the saponins from the other constituents of the seeds. Activity was raised from 2 p.p.m. for the alcoholic extract to 0.06 p.p.m. for the purified saponins (El-Kheir and Salih, 1979).

P The crude ground seeds of this tree showed high molluscicidal activity compared to plants previously investigated (Amin *et al.*, 1972).

Polygonum senegalense Meisn. syn. (*P. glabrum* of F.T.A.) POLYGONACEAE
Polygonum senegalense forma **albotomentosum** R. Grah.

L These plants and related species are 'pounded with native natron and rubbed into the limbs as a remedy for rheumatic and other swellings, and applied to syphilitic sores' (Dalziel, 1937).

C The flavonoids luteolin, quercitin, luteolin-7-*O*-glucoside and quercitin-3-*O*-galactoside have been isolated and identified in *P. senegalense* (Abd El-Gawad and El Zait, 1981). As aqueous extracts of the plants release cyanate ions on alkaline hydrolysis, the molluscicidal compounds may be chemically related to pseudo-cyanogenic glycosides or azoxyglycosides such as cycasin and macrozamin (Dossaji *et al.*, 1977). In 1978, Maradufu and Ouma considered 2,4-dihydroxy-3,6-dimethoxy-chalcone to be responsible for the molluscicidal activity.

P Cut leaves of both above-mentioned varieties proved to be potent molluscicides. Tests were carried out with aqueous extracts of *P. senegalense* forma *senegalense* on the snails *Biomphalaria pfeifferi* and *Lymnaea natalensis*. Isolation of the highly water-soluble active principle increased the molluscicidal effects to 70% and 60% mortality after 1 h for *Biomphalaria* and *Lymnaea*, respectively, with concentrations of 25 p.p.m. and 100% mortality for both species after 24 h. The activce constituent was subsequently identified as quercetin-3(2'')-galloylglycoside (Dossaji and Kubo, 1980). The fact that the plant grows near rivers, lakes, etc. should facilitate its use in the control of snails (Dossaji *et al.*, 1977).

Jatropha gossypiifolia L. EUPHORBIACEAE
Wild cassada, red fig-nut flower
Jatropha curcas L.
Barbados nut

L The viscid sap is used to cure sores on the tongues of babies (?thrush) (Irvine, 1961) and in India the leaves of *J. gossypiifolia* are applied to boils and carbuncles, eczema and rashes (Chopra *et al.*, 1956). The rootbark of *J. curcas*, dried and pulverized, is applied as a dressing for sores (Irvine, 1930). The seeds are a (dangerous) purgative and are given as a remedy against venereal diseases (Dalziel, 1937).

C *J. curcas* contains a non-drying fixed oil consisting mainly of glycerides of stearic, palmitic, myristic, oleanic and curcanoleic acids (the latter belongs to the same group as ricinoleic and crotonic acids) (Watt and Breyer-Brandwijk, 1962). In the Indian plant, vitexin and isovitexin (flavonoids) were found (Sankara, 1971). The seeds contain a mucilage and a toxalbumin, curcin, with seven protein groups (Mourgue *et al.*, 1961; Hufford and Oguntimein, 1978). In the fruit, roots and bark, cyanic acid, and in the bark a steroid saponoside have been found (Watt and Breyer-Brandwijk, 1962).

P Molluscicidal activity against *Bulinus globulus* was shown by all *Jatropha* spp. tested. The methanolic extract of the fruit of *J. gossypiifolia* was the most potent. Its mean lethal concentration (LC_{50}) was 11.55–16.24 p.p.m. The LC_{50} of the seed and that of the rootbark of *J. curcas* were 120.55 and 125 p.p.m., respectively, and the LC_{50} of the seeds and that of the stem of *J. podagrica* were 130.73 and 125.08 p.p.m., respectively. Thus these were ten times less potent than the extract of *J. gossypiifolia*. The potency remained stable after UV radiation and over a wide range of pH values, but was reduced in the presence of minerals and faecal impurities (Adewunmi and Marquis, 1980). The molluscicidal properties of *Jatropha* were also examined by Amin *et al.* (1972) and the antibacterial and pharmacological properties of *J. podagrica* were studied by Odebiji (1980), Ojewole and Odebiji (1980) and Ojewole (1981).

5

Hormones of the adrenal cortex

I Introduction: the action of plants on hormone secretion in Man

A number of plant constituents can act as a substitute for natural hormones in cases of hormone deficiency. Some of these constituents have been found to be chemically identical with the natural hormone (e.g. oestrone in the kernels of date and oil palm). Other plant components have a structural similarity to the hormone (e.g. *Funtumia* and *Holarrhena* alkaloids and corticosteroids). Others again have an entirely different chemical structure (e.g. coumestans and isoflavones with oestrogenic action). All these constituents can act as a substitute for hormones and replace them in their biological functions.

Plant components often have an indirect action on the secretion of certain hormones by stimulating or inhibiting other areas like the hypothalamus and the pituitary gland which can control the function of most other glands (e.g. gonadotrophic, thyrotrophic and corticotrophic action). (For more details see Plants acting on sex hormones.) However, great precaution should be taken in using the hypothalamus–pituitary axis as a narrow interrelationship exists and often hormonal or pharmacological actions other than those desired can be obtained (Bianchi, 1962; Goodman and Gilman, 1976). Thus stimulation of lactation has been obtained with dried thyroid gland or thyroxine (Robinson, 1947; Naish, 1954), no doubt through indirect action via the pituitary gland and an impairment of the pituitary adrenal response to acute stress is observed in alloxan diabetes (Kraus, 1949).

Plant constituents capable of influencing hormone-controlled metabolic actions may also act by removing hormone-inactivating compounds such as enzymes (thus plant constituents can inhibit insulinase in hypoglycaemic plants or certain enzymes in antifertility plants). Also, certain plant substances can antagonize hormone-controlled action (thiocyanates can inhibit the secretions of the thyroid gland). As the future will no doubt reveal, there are many more ways in which the secretion of the hormones can be stimulated or inhibited.

Those plants already described in Chapter 2 as acting on the cardiovascular system are indicated hereunder by CV in both tables and text. Similarly, the plants already described as acting on the nervous system (Chapter 3) are indicated by *N* and those with an anti-infectious (antibiotic or antiparasitic action (Chapter 4)) are indicated by I.

The plants marked by an asterisk (*) in the enumeration of the constituents are described in this text, the others can be found in the corresponding tables.

II Plants acting like hormones of the adrenal cortex

The internal fasciculated zone of the adrenal cortex produces cortisol, hydrocortisone and corticosterone under the control of the pituitary corticotrophin adrenocorticotrophic hormone (ACTH) which is in turn controlled by the corticotrophin-releasing factor (CRF) of the hypothalamus. The external glomerulated zone of the cortex produces aldosterone. Cortisol is mainly used for its anti-inflammatory action and its effect on the glucides, whilst corticosterone and aldosterone mainly act on sodium retention. Cortisol can also act on the electrolyte balance; it is the prototype of the glucocorticoids or glucocorticosteroids. Besides their anti-inflammatory and immuno-depressive actions these substances can act on diuresis, intercranial hypertension and asthma. Plant equivalents of these compounds should thus be examined for each action separately. The role played by the pituitary-hypothalamus axis makes this examination still more complicated. For example, inflammatory action can be controlled not only by corticosteroids but also by other anti-inflammatory agents which often have analgesic and antipyretic properties as well (e.g. salicylates and phenylbutazone). In many cases it is difficult to decide whether the action of a constituent of a given plant depends on its action on the adrenal–pituitary axis if no indication has been given by the authors. In a few rare cases the general action of a plant on both the adrenals and the pituitary has been reported. Two examples of this are for the actions of *Funtumia africana** and *Holarrhena floribunda**, where the anti-inflammatory action is mentioned as part of the many other effects of the different constituents. The anti-inflammatory constituents in *Funtumia* are glucofuntamine and funtumine and those in *Holarrhena* are holamine and holaphylline. They are all steroid alkaloids.

For the reasons mentioned above the other plants with anti-inflammatory action will be described under the heading Anti-inflammatory plants.

Two non West-African plants known to act on the adrenal cortex are *Panax ginseng* and *Eleutherococcus*. Equivalent plants have not been reported in the area.

The stimulating action of ginseng on glucocorticoid production is believed to be of neurogenic origin. Ginseng appears to facilitate the adaptation of the adrenocorticol function to the need of the organism under changing conditions. Its effect is also manifested after hypophysectomy so it seems to act upon the peripheral site of the stress mechanism. Anti-inflammatory and anti-exudative as well as cardiovascular effects of the drug have also been reported (Brekhman and Dardymon, 1969; Chul Kim *et al.*, 1970).

Funtumia africana (Benth.) Stapf syn. (*Kicksia africana* Benth., *K. latifolia* Stapf, *F. latifolia* (Stapf) Schlechter, *K. zenkeri* Schum.) APOCYNACEAE

False rubber tree

L In Ghana the roots are mixed with palm wine and given to patients suffering from incontinence. The dried pulverized leaves are applied as a dressing to burns (Dalziel, 1937). The latex of *F. elastica* Stapf has been used as a source of rubber. Its bitter bark is used as a remedy for haemorrhoids.

C The leaves of *F. africana* contain up to 4% of total alkaloids. The main alkaloid, funtumine, is 3-α-aminopregnane-20-one, funtumidine is 20α-hydroxy-3-α-aminopregnane. The presence of glucofuntamine has also been reported (Quevauviller and Blanpin, 1960).

P Funtumine has hypotensive, antipyretic and local anaesthetic properties and the *Funtumia* alkaloids are similar in action and constitution to those of a related Apocynaceae, **Holarrhena floribunda** (see CV) (Fig. 5.1). This similarity is also confirmed in the hormonal activity of these alkaloids. Funtumine and holamine (3α-amino-5-pregnen-20-one) were observed to antagonize the effects of oestrogens. Holaphyllamine (3β-aminopregnen-20-one) enhanced the activity of oestrogens and testosterone in female and male animals. Funtumine, funtumidine, glucoholamine, holaphyllamine and holaphylline (3β-methylamino-5-pregnen-20-one) showed anti-gonadotrophin action as well as corticotrophic activity. Funtumine antagonized the release of corticotrophin like cortisone. All of the above-mentioned steroid alkaloids had anti-inflammatory properties, holamine and holaphylline being the most effective, followed by glucofuntamine and funtumine. The steroids increased liver glycogen and decreased the weight of the thymus. Holamine and holaphylline cause sodium retention and all three holarrhena alkaloids act as diuretics in rats. Holamine was found to decrease the protein and water content of the liver indicating an anti-anabolic action, and funtumine and funtumidine lowered the serum cholesterol and raised the phospholipid levels and may have anti-atherogenic properties. The six steroids appear to share the neurotrophic effects of steroids; all potentiated the narcotic action of pentobarbital and their toxic effect appears to be confined to the CNS. On intravenous injection in mice their LD_{50} values vary from 28 to 31 mg/kg, holamine being slightly less toxic (the LD_{50} is 37 mg/kg). All of them caused depletion of adrenal ascorbic acid (Blanpin and Quevauviller, 1960a, b; Quevauviller and Blanpin, 1960).

Funtumidine, holaphyllamine and to a lesser extent holamine have been found on subcutaneous injection of 5 mg/kg to exert antipyretic effects in rabbits made hyperthermic by injection of bacterial vaccine. An analgesic effect to thermal or mechanical stimuli in mice has been observed only with funtumidine (25 mg/kg given intraperitoneally). Glucofuntamine was more effective as a local anaesthetic on rabbit skin than cocaine; funtumidine and the holarrhena alkaloids were less effective. Funtumidine had the most consistent tranquillizing effect and was also hypotensive in rabbits (0.2 mg/kg). Funtumine had a spasmolytic action in particular against acetylcholine (Blanpin and Quevauviller, 1960a, b).

Corticosteroids are frequently prepared by hemisynthesis from plant sources such as diosgenin (see *Dioscora* spp. (Chapters 2 and 3)). Many other sources are found in West Africa, for example sarmentogenin (see *Strophanthus sarmentosus* (Chapter 2)), diosgenin plus yamogenin from the fruit of *Balanites aegyptiaca* (Chapters 2 and 4) (Hardman and Sofowora, 1971) and solasodine (from *Solanum torvum*). Another West African source might be costugenin (3% in *Costus afer* (Iwu, 1982)). Some of these constituents have oestrogenic activity themselves.

Fig. 5.1. *Holarrhena floribunda* (Don) Dür. & Schinz.

Anti-inflammatory plants

The inflammatory reaction

Inflammation, whatever its origin (which may be infectious, chemical or physical), generally occurs in three consecutive phases:

(1) an increase in capillary permeability with hyperaemia and oedemas;
(2) cellular infiltration (phagocytosis by polynuclear cells and interference of lymphocytes);
(3) proliferation of fibroblasts and synthesis of collagen fibres and mucopolysaccharides, forming new conjunctive tissue.

If the new conjunctive tissue is attacked in turn by phagocytes and lymphocytes, a chronic inflammation results. Rheumatoid arthritis is a classic example of these conditions (Lechat *et al.*, 1978).

Some of the anti-inflammatory drugs are more active in the initial stages of the inflammation (e.g. salicylates and *Curcuma*). Others tend to act in later stages (glucocorticosteroids and *Phytolacca*) whilst others again are active in both stages (*Commiphora*, *Crateva*, *Terminalia* and *Withania*).

Several pharmacological tests have been devised to measure anti-inflammatory activity, most being based on experiments with inflammation in rats. These are:

(1) carrageenan and kaolin-induced hind-paw oedema (or rat pedal oedema), formalin-induced arthritis of ankle joint in rats; yeast-induced paw oedema, croton oil-induced granuloma, cotton-pellet granuloma (Winter *et al.*, 1962; Benitz and Hall, 1963; van Arman *et al.*, 1965);
(2) adjuvant-arthritis, where injection of *Mycobacterium butyricum* in one hind-paw produces inflammation in the other paws and granulations in the ears;
(3) graft-versus-host reaction in chicks (against lymphocytes of hens), which is used to test immune reaction.

It has been noticed that drugs which block inflammatory- and arthritis-like syndromes in animals are also effective against rheumatic diseases in Man. Inflammation in patients with rheumatoid arthritis implies the combination of an antigen (gamma-globulin) with an antibody (the rheumatoid factor) and a complement, resulting in phagocytosis by leucocytes and release of lysosomal enzymes. These enzymes damage cartilage and other tissues and enhance the inflammation (Woodbury and Fingl, in Goodman and Gilman, 1975). Prostaglandins are also formed by leucocytes during phagocytosis. Local injection of prostaglandin E_1 or E_2 causes definite vasodilatation and hyperaemia and increases the permeability of cell membranes, and it is believed that inhibition of prostaglandin synthesis is one effect of anti-inflammatory drugs (Ferreira and Vane, 1974; Awouters *et al.*, 1978; Oriowo, 1982). Other effects are stabilization of the lysosome membranes by prevention of loss of enzymes from the lysosomal envelope, uncoupling of oxidative phosphorylation (Whitehouse, 1965; Whitehouse *et al.*, 1967) and inhibition of the synthesis of mucopolysaccharides (which constitute the fundamental substance of conjunctive tissue and cartilage (Paulus and Whitehouse, 1973; Lechat *et al.*, 1978). Various mechanisms interfering with antigen-antibody aggregation have also been considered, including inhibition of antigen-induced release of histamine.

Interference with prostaglandin synthesis has been observed in the case of *Zanthoxylum zanthoxyloides* and *Terminalia ivorensis*. Immuno-depressive action has been reported for *Allium* and *Withania somnifera*, and *Solanum* spp.

Chemical mediators in inflammation processes are histamine, 5-hydroxytryptamine (serotonin), vitexin and bradykinin (Prabhakar *et al.*, 1981; Saxena *et al.*, 1982). Antihistamine action has been reported for *Anthocleista*, *Arnebia*, *Callophyllum*, *Citrus*, *Crateva*, *Cryptolepis*, *Curcuma* and *Ipomoea*.

In some cases the anti-inflammatory action seems to be independent of the pituitary–adrenal axis (e.g. that of *Cyperus rotundus* and *Commiphora*) whilst in other cases the drugs are said to have a direct action on the adrenal cortex (e.g. curcumin (Chandra and Gupta, 1972) and glycyrrhetic acid (Gibson, 1978)).

As the plant triterpenoids often turned out to be active anti-inflammatory constituents, their anti-inflammatory and anticonvulsant properties were evaluated in rodents by Chaturvedi *et al.* (1974, 1976). For a number of natural plant triterpenoids the protection afforded against carrageenan-induced rat-paw oedema ranged from 9–48% when they were given intraperitoneally in doses of 40 mg/kg. Good correlation has been observed between the anti-inflammatory and antiproteolytic properties of these plant products. The latter activity was demonstrated by in vitro inhibition of trypsin-induced hydrolysis of bovine serum-albumin and casein. All the tested triterpenoids except friedelinoxime and acetylmethylursulate provided 10–40% protection against pentylene-tetrazol-induced convulsions in mice.

Active constituents of anti-inflammatory plants in West Africa

In West Africa the most effective anti-inflammatory plants seem to be those which have the following active constituents.

Steroid or triterpene glycosides. Steroids or their heterosides are found in *Commiphora indica**, *Costus afer**, *Cyperus rotundus**, *Leptadenia pyrotechnica*, *Solanum torvum** and other *Solanum* spp. and *Withania somnifera**. The triterpenic glycosides involved are the glycyrrhetic heterosides in *Lonchocarpus cyanescens** and *Terminalia ivorensis** or the oleanolic heterosides found in *Boerhavia diffusa*, *Gymnema sylvestra*, *Securidaca longepedunculata*, *Tetrapleura tetraptera* and probably *Phytolacca dodecandra**. Triterpenoids are also present in *Alstonia boonei* (bark), which is applied topically in the treatment of rheumatic pains.

Terpenes (and their heterosides). These are found in *Azadirachta indica*, *Atractylis gummifera* c., *Crateva religiosa*, *Xylopia aethiopica* and *Vernonia colorata*.

Alkaloids and amides. These seem to be responsible for the anti-inflammatory action in *Anthocleista procera*, *Capsicum frutescens**, *C. Annuum**, *Cryptolepis sanguinolenta** and *Zanthoxylum zanthoxyloides**. Pungent amides are also found in *Piper guineense* and *P. umbellatum* (chavicine and piperine) as well as in *Capsicum* spp. The latter may also act through their flavonosides.

Flavonosides. Flavonosides, of which a number contain coumarins, are found in *Afraegle paniculata**, *Arnebia hispidissima**, *Calophyllum inophyllum**, *Canscora decussata*, *Citrus nobilis*, *Dalbergia sissoo* and *Hibiscus vitifolius*. Few of the above-mentioned plants belong to the same family; there are, however, three Solanaceae and three Rutaceae, of which two have anti-inflammatory constituents belonging to the same chemical group.

Sulphur heterosides. Many of the plants used in local medicine in the treatment of rheumatic diseases are found in this group. Their constituents or scission products often contain pungent mustard oils or amides which act as counter-irritants or rubefacients, diverting hyperaemia by irritation of the skin or intestine, and these have been used, mostly locally, in the treatment of arthritis, lumbago, rheumatism bronchitis, congestion of the lungs, etc. In West Africa they are found in Capparidaceae, namely glucocapparin in *Capparis decidua*, *Crateva religiosa* and *Gynandropsis gynandra*; tropaeolin in *Carica papaya* seeds, glucotropaeolin (spirochine) in *Moringa oleifera* and alliin in *Allium sativa* and *A. cepa*.

Other constituents. Other active constituents are reported to be tetrahydrocannabinol in *Cannabis sativa* (the resin of which has also been found to contain coumarin glycosides), the dyes haematoxylin and brasilin in *Haematoxylum campechianum*, tertiary phenylethylamines in *Desmodium gangeticum* and curcumine in *Curcuma domestica**. The active fraction in *Salvadora persica** could be β-sitosterol; in *Ipomoea* spp.* it has not yet been identified.

Many of these plants also have an antibiotic action, which in some cases may contribute to their effectiveness.

A number of these plants (those marked with an asterisk (*)) are described in more detail to give a better understanding of their possible therapeutic interest and modes of action. The others are listed in Table 5.1 (p. 211).

Plants with steroid or triterpene glycosides as active constituents.

Commiphora africana (Rich.) Engl. syn. (*Heudelotia africana* Rich., *Balsamodendron africanum* (Rich.) Arn.) BURSERACEAE

African bdellium or African myrrh

L In Niger country a maceration of the stembark of *C. africana* is given perorally in the treatment of rheumatic diseases (Adjanohoun, 1980). In West Africa the gum-resin is boiled for treatment of inflammation of the eyes by holding the face over the steaming pot. For scorpion-bite the bark is applied after it has been chewed with natron (Dalziel, 1937).

C The gum-resin contains 70% alcohol-soluble resin and 30% water soluble gum. The resin contains 7–9% of essential oil (Dalziel, 1937; Kerharo, 1968) and is composed of free terpenoids and terpenoid glycosides and the gums are composed of polyholosides (Boiteau *et al.*, 1964).

In the Indian *Commiphora mukul* Hook. ex Stocks the essential oil has been found to contain 4–6% of myrcene, 11% of dimyrcene and some polymyrcene. The petroleum ether extract of the gum-resin has yielded sesamin, cholesterol and a few other steroids (Indian Council, 1976, p. 271). A number of steroids have been isolated and identified and the diterpenoid constituents cembrene A and mubulol as well as some fatty tetrols have been reported (Patil *et al.*, 1972).

P The oleo-resin fraction from *C. mukul* has shown significant anti-arthritic and anti-inflammatory activity (minimum effective dose 12.5 mg/100 g in albino rats). This activity has been localized in the acidic fraction of the oleo-resin and has been shown to occur even in adrenalectomized animals (Santhakumari *et al.*, 1964).

The aqueous extract of the oleo-gum-resin of *C. mukul* had suppressive action on carrageenan-induced acute rat paw oedema and in the granuloma pouch test as well. In adjuvant arthritis the secondary lesions were very effectively suppressed without any significant action on the primary phase. Side-effects were negligible as compared to those occurring in beta methasone-treated animals (Gujral *et al.*, 1960; Satyavati *et al.*, 1969).

A steroidal compound isolated from the petroleum ether extract of *C. mukul* showed a dose-dependent anti-inflammatory activity on rat paw oedema which was much more potent than that of the resin fraction. The steroid fraction had a pronounced effect on primary and secondary inflammation induced by Freund's adjuvant; it was less effective than hydrocortisone acetate in the primary phase but more effective in reducing the severity of secondary lesions (Arora *et al.*, 1972). Furthermore, *C. mukul* was also found to lower the serum cholesterol in hypercholesterolaemic rabbits and to protect the animals against cholesterol-induced atherosclerosis (Satyavati *et al.*, 1969; Nityanand *et al.*, 1973). Long-term experimental studies of its effectiveness as a hypolipaedemic agent gave satisfactory results and showed that the effect could be attributed to (a) an increase in the rate of removal/excretion of cholesterol, (b) a decrease in the input/synthesis of cholesterol and (c) mobilization of cholesterol from tissues (Indian Council, 1976, pp. 272–5).

In view of the interesting results obtained with Indian myrrh, African myrrh might be examined chemically and pharmacologically for similar properties.

Costus afer Ker-Gawl. syn. (*C. obliterans* Schum., *C. anomocalyx* Schum, *C. insularis* Chev., *C. lucanuscianus* Chev.) (Fig. 5.2) ZINGIBERACEAE
Ginger lily

L *C. afer* is widely used as a cough medicine, either as a decoction of the stems or the pounded fruit, or by chewing the succulent stem itself. The boiled root is applied to cuts and sores and a soothing fomentation for rheumatic pains is prepared with the boiled leaves (Dalziel, 1937, p. 472).

C The abundant juice of the leaves (69.7%) (Odutola and Ekong, 1968) contains 0.4% oxalate, furan derivatives and starches.

Thin-layer chromatography of extracts of the tubers with petroleum ether and chloroform yielded three compounds which were identical with lanosterol, tigonenin and diosgenin. Iwu (1982) could isolate from the chloroform extract 3% costugenin, the most abundant sapogenin (closely related to sarmentogenin), 1.5% stigmasterol and 0.8% diosgenin. Similar sterols had been reported to be present in *C. speciosus* (Bhattacharya *et al.*, 1973; Gupta *et al.*, 1980, 1981).

P In clinical trials, 25 patients, 17 of whom were suffering from rheumatoid arthritis and 5 from osteo-arthrosis, received, in groups of seven, differential solvent extracts (prepared according to traditional methods by native doctors) of *C. afer* (30 ml doses twice daily) for four days and, after an interruption of three days, the same treatment for ten days. Two further groups of seven patients received the same treatment with extracts of *Lonchocarpus cyanescens* and *Terminalia ivorensis* (see below for these plants). All the patients suffering from rheumatoid arthritis were relieved of their symptoms (Iwu and Anyanwu, 1982a, b). Three of the patients who received *Lonchocarpus* and one each of those receiving *Costus* and *Terminalia* reported complete recovery. Only two of five patients suffering from osteo-arthrosis showed some improvement (no results with *Costus*).

In pharmacological tests with the same three plants the extracts reduced carrageenan-induced oedema in the rat paw, checked diarrhoea due to arachidonic acid and castor-oil (Awouters *et al.*, 1978) and ameliorated all signs associated with adjuvant-induced polyarthritis in rats. The extracts were well tolerated in daily doses of 100–300 mg/kg except for the chloroform extract of *C. afer*, which caused the death of four out of ten experimental animals at that dose regimen.

Further chemical and pharmacological tests are planned by the authors (Iwu and Anyanwu, 1982b).

Fig. 5.2. *Costus afer* Ker-Gawl.

Cyperus rotundus L. CYPERACEAE

Nutgrass

L The rhizomes are slightly fragrant and the essential oil they yield is used in Asia as a perfume for clothes and to repel insects. In Nigeria the plant is used as cattle fodder and the tuberous rhizomes as a cough medicine for children. In the Congo Brazzaville the pulp of the roots is used in frictions for oedema and rheumatism (Dalziel, 1937; Bouquet, 1969). In India the tubers are reputed to be diuretic, emmanagogic and anthelmintic and are used for treating disorders of the digestive tract (Chopra *et al.*, 1956, p. 88; Hegnauer, 1964, Vol. III, p. 285).

C The tubers contain a fatty oil, which is chiefly made up of glycerides of oleic, palmitic and linolic acids with small quantities of essential oils. The crude volatile oil has been noted to contain about 40% of a sesquiterpenic ketone, α-cyperone (McQuillin, 1951). In tubers collected in India, the essential oil fraction is reported to be composed of pinene, traces of cineol, sesquiterpenoids, monoterpenic and aliphatic alcohols and β-sitosterol (Kalsi *et al.*, 1969). The sesquiterpenoids were identified by Kapadia *et al.* (1967). In *C. esculentus* cholesterol has also been found (Abu-Mustafa *et al.*, 1960).

P The anti-inflammatory action has been studied in India on oedema induced in the rat paw by carrageenan or by cotton-pellet implantation and was first attributed to the petroleum ether extract of the roots. Then a triterpenoid obtained from this extract by chromatographic separation was shown to possess an anti-inflammatory activity which was eight times greater than that of cortisone; the fraction also had antipyretic properties (in pyrexia induced by brewer's yeast) plus an analgesic action. On intraperitoneal administration the LD_{50} of the extract was 50 mg/kg; the ED_{50} was 1.6 mg/kg (Gupta *et al.*, 1970). Later, β-sitosterol isolated from tubers grown in India was found to be a powerful agent against inflammation in the above-mentioned tests (Bach, 1978). When it was administered intraperitoneally the effect was similar to that of hydrocortisone and oxyphenbutazone and the substance was also effective against carrageenan-induced oedema when given perorally. Its action proved to be independent of the pituitary–adrenal axis and to be similar to that of acetylsalicyclic acid. β-sitosterol showed a broad safety margin as in intraperitoneal administration the LD_{50} was more than 3 g/kg in mice and the minimum ulcerogenic dose was 600 mg/kg in rats (Gupta *et al.*, 1980). The essential oil was reported to have oestrogenic activity which could be attributed to cyperene I, a hydrocarbon fraction (Indira *et al.*, 1956).

Solanum torvum Sw. including *S. torvum* var. *compactum* Wright syn. (*S. mannii* Wright including var. *compactum* Wright) SOLANACEAE

L The small orange-red berries are eaten cooked or sometimes raw. In Sierra Leone a decoction of the fruit is used as a cough medicine for children (Dalziel, 1937, p. 435).

C The fruits contain sitosterol D-glucoside and 0.1% of the glucoalkaloid solasonine (solasodine-glycoside), from which solasodine is obtained. Solasodine is used as a starting product in the hemisynthesis of cortisone and sex hormones (Chopra *et al.*, 1956, p. 230). The glycoalkaloid contents vary considerably during growth (Paris and Moyse, 1971, Vol. III, p. 148). Furthermore, *S. torvum* yields a rare sterol, first reported from **S. cerasiferum** Dun. syn. (*S. xanthocarpum*) (Sayed and Kanga, 1936), which has been shown to have the structure of (22R)22 hydroxy-6-oxo-4α-methyl-5α-stigma-7-en-3β-yl benzoate (Beisler and Sato, 1971). *S. torvum* contains as much as 0.04% of carpestrol (Bhattacharya *et al.*, 1980).

P The steroidal alkaloid solasodine has been shown to cause thymolysis in rats and to have antiphlogistic properties in experimental arthritis in rats and in experimental burns in rabbit ears. Investigation of the immunomodulating properties of solasodine (isolated from *S. nigrum*) and of withanolide D from *Withania somnifera* showed that both substances had an immunodepressive action in vitro (Bär and Hänsel, 1982).

Carpestrol has also been reported to produce a dose-dependent inhibition of carrageenan-induced paw oedema in albino mice. As compared to withaferine A and hydrocortisone it showed the highest potency, being active in doses of 0.9 mg/kg when given intraperitoneally, whilst the LD_{50} for mice of carpestrol given intraperitoneally is 500 mg ($\pm$8 mg)/kg and that of withaferine A is 110 mg ($\pm$5 mg)/kg. Carpestrol has certain structural similarities to hydrocortisone and withaferine A, both of which also have anti-inflammatory activity (Bhattacharya *et al.*, 1980).

Withania somnifera L. SOLANACEAE

Winter cherry

The roots of this plant, already reported to have sedative and antibiotic properties (N and I), have been used for centuries in folk medicine to treat rheumatism, ulcers and skin diseases (Menssen and Stapel, 1973). The roots contain withaferine A and several other steroidal lactones and withanolides, and these were also isolated as minor constituents of the leaves (Abraham *et al.*, 1975). Withaferine A is also a tumor inhibitor (Kupchan *et al.*, 1965).

Withaferine A and withanolide D (isolated by Menssen and Stapel, 1973) have been shown to be active against inflammation. Six intraperitoneal doses of withaferine A of 25 mg/kg every second day in one series of tests, and twelve doses of 12.5 mg/kg every second day in a second series of tests, have been shown to delay the onset of adjuvant arthritis in rats and strongly to inhibit swelling and inflammation of the diseased area and of the secondary lesions (Fuegner, 1973; Roshchin and Geraschenko, 1973). Withaferine A has also been observed to inhibit the xenogenic graft-versus-host reaction in chicks to a great extent (Fuegner, 1973; Bär and Hänsel, 1982). The mode of action of withaferine appeared to be similar to that of prednisolone and azathioprin (purine antagonist). The similarity of their structures and actions to those of glucocorticosteroids had already suggested that withaferine and withanolide D might have, besides their antiproliferative effect, a complex influence on inflammation and immune responses (Bär and Hänsel, 1982). Shohat *et al.* (1978) also observed that in concentrations of 1 μg/ml withaferine A and withanolide E showed immuno-depressive action in cultures stimulated by 1 μg/ml and 0.3 μg/ml phytohaemaglutinin and in those without mitagenic stimulation. The functional activity of normal human T lymphocytes as assessed by local xenogenic graft via host reaction was also affected by these two steroidal lactones. Apparently, withanolide E had a specific effect on T lymphocytes whereas withaferine A affects both T and B lymphocytes. The authors believe that rheumatism, asthma and certain skin diseases all have an immuno-pharmacological basis.

Lonchocarpus cyanescens (Schum. & Thonn.) Benth. syn. (*Robinia cyanescens* Schum. & Thonn., *Philenoptera cyanescens* (Schum. & Thonn.) Roberty)

FABACEAE

West African wild indigo, Indigo vine

L The leaves are applied as a poultice to ulcers of the foot, etc. or as a dressing for skin diseases. The bark and root are a remedy for jaundice and are used as a general tonic (Dalziel, 1937). A decoction of the bark with native natron is, in certain districts, a treatment for abdominal troubles, with flatulence in horses. The leaf decoction is also used for the treatment of venereal disease and semen insufficiency (Ainslie, 1937).

C The fresh leaves of *L. cyanescens* contain 0.08–0.3% of indigo, yielding 43% of indigotin. The dried leaves can yield as much as 56% of indigotin (Dalziel, 1937). From the chloroform extract of the roots, glycyrrhetinic acid, rotenone and lonchoterpene have been isolated, and from the methyl alcohol extract oleanolic acid and ursolic acid have been isolated. Four unidentified minor compounds have also been reported from these extracts (Iwu and Ohiri, 1980). In the related Nigerian **Lonchocarpus laxiflorus** Guill. & Perr., isoflavans (laxiflorin and lonchoflavan) and pterocarpans have been reported (Pelter and Amenechi, 1969). Glycyrrhetinic acid (also known as glycyrrhetic acid) is a pentacyclic terpene. It was first obtained by hydrolysis of glycyrrhizin (or glycyrrhizic acid) from *Glycyrrhiza glabra*, which yielded two molecules of glucoronic acid and one of glycyrrhetic acid (Bombardelli *et al.*, 1979).

P Chloroform and methyl alcohol extracts of *L. cyanescens* roots reduced carrageenan-induced rat paw oedema and adjuvant (*Mycobacterium butyricum*-induced) polyarthritis in rats. Daily doses of 100 and 200 mg/kg of the extracts partially inhibited primary and secondary lesions of the rat hind paw and also reduced both body weight and arthritic symptoms. In combination with phenylbutazone both extracts alleviated all symptoms of polyarthritis in rats (Iwu and Ohiri, 1980). Glycyrrhetic acid and some derivatives have been found to have an anti-arthritic activity similar to that of hydrocortisone (Kraus, 1960; Parmar *et al.*, 1964; Tangri *et al.*, 1965). (Linnaeus (1707–1778) himself mentions the use of Glycyrrhiza for rheumatism).

Numerous papers deal with the biochemical basis of the anti-inflammatory properties of glycyrrhizin and were summarized in a comprehensive review by Gibson (1978). It is generally concluded that glycyrrhizin has an aldosterone effect in the body, causing retention of sodium ions and a plasma depletion of potassium ions. Glycyrrhizin appears to stimulate the adrenal cortex directly increasing the production of mineral corticoids, glucocorticoids and adrenal androgens. In addition, it inhibits the inactivation of corticoids in the liver and kidneys. These two effects result in a continuous and elevated plasma level of corticoids (Matsuda *et al.*,

1962). Glycyrrhizin has also been found to enhance the immuno-depressive action of cortisone and to inhibit the action of cortisone on the thymus and on liver glycogen deposition (Kumigai, 1969; Kumigai *et al.*, 1967a, b, c).

Glycyrrhizin and glycyrrhetic acid are also active against coughs and peptic ulcers and have given good results in the treatment of rheumatic thrombophlebitis (Kerharo and Adam, 1974, p. 440). The effect of glycyrrhizin in the treatment of peptic ulcer cannot be completely explained by the anti-inflammatory action, since the deglycyrrhinated drug is of considerable value in the treatment of ulcers (Brodgen *et al.*, 1974; Gibson, 1978). The anti-arthritic and anti-inflammatory effects have also been attributed to a reduction in the activities of serum-glutamin-oxaloacetic acid transaminase and serum-glutamic-pyruvic transaminase (Parmar *et al.*, 1964), and to an uncoupling of oxidative phosphorylation (Whitehouse *et al.*, 1967).

Another West African plant which produces glycyrrhizin, and which is used for its commercial extraction, is *Abrus precatorius* (5–10% in the leaves, 1.5% in the roots).

Terminalia ivorensis A. Chev. COMBRETACEAE
Satin wood, shingle wood

L Powdered bark and bark infusions are much used in local medicine for the dressing of wounds. They are also used in the treatment of arthritic conditions and piles and as a diuretic (Ainslie, 1937; Dalziel, 1937). Anti-inflammatory properties have also been reported for the related *T. avicennoides* (see CV).

C From the chloroform and methanol extracts of the stembark, terminolic acid, ellagic acid, sericic acid, quercetin, β-glycerrhetinic acid and 2–8 hydroxy 18α-glycyrrhetinic acid have been isolated (Bombardelli *et al.*, 1979; Iwu and Anyanwu, 1982b). The wood also contains β-sitosterol, terminolic acid and tri- and tetra-methyl ellagic acid and laxiflorin and sitosteryl palmitate were reported to be present in the stembark of most Nigerian *Terminalia* spp. (Ekong and Idemudia, 1967; Idemudia and Ekong, 1970).

P Extracts of the stembark with different solvents (chloroform and methanol giving the most active extracts) inhibited carrageenan-induced pedal oedema and adjuvant-induced polyarthritis (by Rosenthale method) in the rat. Daily administration of 100–300 mg/kg of the extracts reduced primary and secondary lesions of the rat's hind paws with concomitant reduction of body weight and reduction of the arthritic symptoms. The chloroform extract showed an activity comparable to that of phenylbutazone and in combination-therapy with this (100 mg/kg of each) has an activity almost equivalent to that of indomethacin. The extracts also effectively checked the diarrhoeas produced by arachidonic acid and castor oil. This action is related to the ability of anti-inflammatory drugs to interfere in the synthesis of the prostaglandins E_2 and F_2; this in turn reduces the inflammation associated with the prostaglandins (Awouters *et al.*, 1978; Iwu and Anyanwu, 1982a). The authors conclude that the results lend support to the use of *Terminalia ivorensis* in the treatment of arthritis and other inflammatory conditions.

The antiphlogistic action of β-glycyrrhetic acid is well known (see *Lonchocarpus*, above) and the anti-inflammatory action of *Terminalia ivorensis* could, in the view of Iwu and Anyanwu (1982a), be due at least in part to this acid and to the presence of other oleanane derivatives (terminolic acid and lonchoterpene).

Phytolacca dodecandra L'Herit syn. (*P. abystinica* Hoffm.?)
PHYTOLACCACEAE

Endod, soapberry

This plant has anti-inflammatory as well as molluscicidal properties (I).

C The constituents have been examined in detail and it was found that up to ten saponins (phytolaccosides) were distributed in all tissues of *Phytolacca* spp. (*P. americana* and *P. esculenta*). Polyphenols have been isolated from the seeds (Woo and Kang, 1978). The saponins contained glucose as a sugar component except for phytolaccoside A, which contained D-xylose. The genins were found to be: phytolaccagenic acid, phytolaccagenin, jaligonic acid, and esculentic acid in *P. americana* and *P. esculenta* (Woo *et al.*, 1976). In *P. dodecandra* the presence of the genins oleanolic acid and bayogenin had been reported by Powell and Whalley (1969).

P A water-insoluble saponin fraction, in doses of 30–50 mg/kg in rats, gave a 50% inhibition of carrageenan-induced paw oedema. The LD_{50} of the crude saponins was 181 mg/kg in mice and 208 mg/kg in rats (Woo *et al.*, 1976). Investigation of the anti-inflammatory activity of the saponins and of phytolaccagenin from the roots of *P. americana* in rats and mice showed that oral administration required six times the dose needed in intraperitoneal injection. The anti-exudative and anti-granulomatous properties were eight times higher than those of hydrocortisone, but in higher doses (160 mg/kg) the saponin produced severe thymolysis in rats (Woo and Shin, 1976).

Phytolaccoside B (1) and Phytolaccoside E (2) administered intravenously to rats inhibited exudate formation after sponge pellet and carrageenan-induced oedemas. The haemolytic activity of (1) was greater than that of (2). The LD_{50} doses for (1) in mice and rats were respectively, 4.5 and 10.8 mg/kg, and for (2) they were 23.6 and 42.3 mg/kg, respectively. Toxic and anti-inflammatory effects of the phytolaccosides were less than those of aescin (Shin *et al.*, 1979).

Plants with alkaloids and amides as active constituents.

A Menispermaceaous alkaloid, tetrandine (in *Tiliacora* spp.), showed anti-inflammatory action when given intramuscularly but was not active when given perorally. This suggests that it probably acts by stimulation of the adrenals (Yamahara *et al.*, 1974).

Capsicum frutescens L., *C. frutescens* var. *minimum* SOLANACEAE

Cayenne pepper, African pepper, paprika, *piment enragé*

C. annuum L.

Capsicum pepper, *piment doux*

L The fruits of these plants are popular spices and are used medicinally as rubefacients and counter-irritants. The small-fruited varieties are the chillies; when crushed and powdered, the fruits produce the condiment known as Cayenne pepper, which is sold commercially. *C. annuum* has bigger, less pungent fruits than *C. frutescens*.

C The pungency is due to the presence of a volatile phenolic compound, capsaicin, which is closely related to vanillin as it is a vanillylamide of 8-nonene-6 carboxylic acid (it has been called capsicin by some authors). *C. annuum* contains 0.1%, *C. frutescens* about 0.5% of capsaicin, but the figure varies–it may exceed 1% in both species and be as little as a tenth of this in some varieties. Furthermore, the presence of a steroid saponin, capsicidin, has been reported (Gal, 1964, 1967). The fruits are a source of vitamin C (7.3 mg/kg and 12 mg/kg for *C. frutescens* and *C. annuum*, respectively) and are also relatively rich in vitamin A and mineral elements. Fifteen to thirty per cent of the seed content has been extracted in the form of an oil rich in triolein.

P The counter-irritant and carminative actions are due to capsaicin, which, according to Molnar (1965), acts as a toxic stimulant of the receptors involved in the circulatory and respiratory reflexes. In small doses it increases intestinal peristalsis and the production of gastric acid but higher doses inhibit these effects.

Capsaicin is mainly used in local applications as a revulsive in rheumatism, lumbago, neuralgia, respiratory tract diseases, chilblains, etc., in the form of impregnated cotton-wool (Paris and Moyse (1971) Vol. III, p. 200). Capsicidin has a definite antibiotic action on certain microorganisms and on *Saccharomyces cerevisiae* (Gal, 1964).

Cryptolepis sanguinolenta (Lindl.) Schltr. PERIPLOCACEAE

P Cryptolepine, an alkaloid obtained from the roots, among other pharmacological properties (see CV and I), inhibited carrageenan-induced oedema of the rat hind paw in doses of 1, 5, 10 and 20 mg/kg. The effect was dose-related but was much less potent than that of aspirin or indomethacin. In doses of 1–20 μg/ml in the bath fluid cryptolepine did not inhibit prostaglandin synthesis in the isolated lizard lung, but it antagonized prostaglandin E_2 and not 5-hydroxytryptamine on a perfused isolated stomach strip of the rat. It also antagonized the action of histamine and of acetylcholine on the isolated guinea pig ileum. These observations seem to indicate that the inhibitory effect is probably receptor-mediated. Cryptolepine has been shown to be only slightly less potent than phentolamine in blocking adrenoreceptor stimulation (Noamesi and Bamgbose, 1980; Bamgbose and Noamesi, 1981). Further tests revealed that cryptolepine has preferential presynaptic α-adrenoreceptor blocking action (Noamesi and Bamgbose, 1982).

Zanthoxylum zanthoxyloides (Lam.) Watson syn. (*Fagara zanthoxyloides* Lam., *F. senegalensis* Chev.) RUTACEAE

Prickly ash, candlewood, toothache bark

P The rootbark of this plant, which also has cardiovascular and anti-infectious properties (see Chapters 2 and 3), contains, amongst other constituents, fagaramide, which has been shown to reduce inflammation. Fagaramide has been shown to reduce carrageenan-induced paw oedema in rats but is approximately twenty times less potent than indomethacin. The acute inflammatory reaction is characterized by a sequential release of mediators: histamine, 5-hydroxytryptamine, bradykinin and prostaglandin (Di Rosa *et al.*, 1971). The prostaglandin phase of acute experimental inflammation is usually assumed to cover events taking place from about one hour after carrageenan injection. This phase was modified by fagaramide and it was noted that in vitro fagaramide inhibited prostaglandin synthesis in a dose-dependent manner but had no effect on prostaglandin E induced potentiation of carrageenan-oedema in indomethacin-treated rats (Oriowo, 1982). Oriowo (1982) suggests that at least part of the anti-inflammatory effect of fagaramide is attributable to inhibition of prostaglandin synthesis.

Plants with flavonosides as active constituents.

Afraegle paniculata (Schum. & Thonn.) Engl. syn. (*Citrus paniculata* Schum. & Thonn., *Balsamocitrus paniculata* (Schum. & Thonn.) Swingle, *A. barteri* Hook.)
RUTACEAE

L In Lagos an *agbo* infusion is made of the fruit and leaves and sometimes the bark of *A. paniculata*, to be taken both internally and as a wash in rheumatic conditions (Dalziel, 1937). Sometimes *Clausena anisata* is added to the infusion (Irvine, 1961).

C The coumarins scoparone, imperatorin and xanthoxyletin have been isolated from the stembark (Adesina and Etté, 1982). The fruit contains imperatorin, γ-sitosterol and another coumarin, xanthotoxin (Quartey, 1963). In addition, free aliphatic acids and a triacid triglyceride have been isolated from the bark and fruits (Adjangba *et al.*, 1974).

The dried seeds yield 46% lipids; besides 34% oleic acid, stearic, palmitic, linoleic, linolenic and palmitoleic acids have been reported (Busson, 1965), and a mucilage of the fruit produced through partial hydrolysis 6-O-β-D-glucuronosyl D-galactose (Torto, 1961). The leaves contain 47.3% of glucides, 27.1% of protides and mineral elements with a predominance of calcium.

P The three coumarins xanthotoxin, xanthotoxol and imperatorin isolated from *A. paniculata* were tested for their anti-inflammatory properties following oral administration of 100 mg/kg to mice. Xanthotoxin had some activity but the other two had none. The oral LD_{50} of xanthotoxin was more than 4000 mg/kg in rats and over 1000 mg/kg in mice (Adjangba *et al.*, 1975).

Scoparone was found to have anticonvulsant properties while the other two coumarins found in the stembark proved to be mildly sedative and anticonvulsant (coumarin itself is a mild sedative). The structurally related compounds scopoletin and angelicin have been reported to possess anticonvulsant properties (Adesina and Etté, 1982).

Arnebia hispidissima (Sieber and Lehm) DC. syn. (*Lithospermum hispidissimum* Sieber ex Lehm, *A. asperrima* (Del.) Hutch. & Dalz., *Anchusa asperrima* Del.)
BORAGINACEAE

L A red or purple dye is obtained from the root.

C Vitexin (8β-D-glucopyranosyl-apigenin) was isolated from the flowers of *A. hispidissima* and *Ochrocarpus longifolius* L.

P Vitexin showed potent hypotensive, anti-inflammatory and non-specific antispasmodic properties. The hypotensive effect was attributed to its ganglion-blocking properties and the anti-inflammatory effects to its antihistamine, anti-bradykinin and anti-serotonin properties (Prabhakar *et al.*, 1981).

Vitexin has also been reported to be present in the leaves of *Lophira lanceolata* van Tiegh. ex Keay (Jacquemain, 1971), which are used in Senegal as an antitussive (whilst the stembark is said to cure oedemas) (Kerharo and Adam, 1974, p. 614). Vitexin heterosides have also been noted in the leaves of *Combretum micranthum* (Jentsch *et al.*, 1962).

Calophyllum inophyllum L. GUTTIFERAE
Mesua ferrea L.
Both introduced into West Africa

L In its country of origin (South India) the oil of the seeds of *Calophyllum* is specifically used to treat skin diseases and is also applied topically in cases of rheumatism. The gum and the juice are said to be purgative and the bark astringent in patients with internal haemorrhage. The leaves are used as a fish poison.

C From the leaves of *C. inophyllum*, friedelin and triterpenes of the friedelin group, namely canophyllal, canophyllol and canophyllic acid (Govindachari *et al.*, 1967), and from the heartwood xanthones (mesuaxanthone B and calophyllin B) have been isolated (Govindachari *et al.*, 1968). A number of 4-phenylcoumarin derivatives have successfully been isolated from the seeds (Mitra, 1957a; Farnsworth and Cordell, 1976, p. 424); they include calophyllolide, inophyllolide and calophyllic acid. In plants collected in India all three of these compounds were present, while in Indochina and Tahiti only calophyllic acid has been found (Polonsky, 1957). Furthermore, inophyllic acid (Mitra, 1957b) and two more 4-phenylcoumarins, ponnalide (Adinarayana and Seshadri, 1965) and calaustrin (Bhusan *et al.*, 1975) have been reported.

P The pharmacological properties of calophyllolide have been examined by Arora *et al.* (1962), who reported that in the isolated perfused rabbit heart it decreased the amplitude of contraction but increased coronary flow. It suppressed ventricular ectopic tachycardia resulting from acute myocardial infarction in dogs as effectively as quinidine and had vasoconstrictive action on the peripheral vessels. Its effectiveness as an anticoagulant puts it between dicoumarol and tromexan for this purpose (Arora *et al.*, 1962). Also, Saxena *et al.* (1982) found calophyllolide to be effective in reducing the increased capillary permeability induced in mice by various chemical mediators involved in inflammation processes, namely histamine, 5-hydroxytryptamine and bradykinin. Pretreatment of the mice with calophyllolide afforded significant protection against the inflammation induced by these mediators. When evaluated against carrageenan-induced rat pedal oedema, at 40 mg/kg given intraperitoneally, calophyllide and inophyllide reduce oedema by 60.7% and 29.8%, respectively. In this study hydrocortisone given intraperitoneally at a dose of 10 mg/kg reduced the inflammation by 44% (Chaturvedi *et al.*, 1974). The safety margin of calophyllolide is very similar to that of oxyphenbutazone (21.4 mg/kg and 25 mg/kg, respectively) (Bhalla *et al.*, 1980), which would warrant clinical studies of different inflammatory diseases with this compound (Saxena *et al.*, 1982).

The xanthones of *C. inophyllum* and *M. ferrea*, namely dehydrocycloguanandin, calophyllin B, jacareubin, 6-deoxyjacareubin and mesuaxanthone, produced in mice and rats various degrees of CNS depression (sedation, decreased spontaneous motor activity, loss of muscle tone, potentiation of barbitone sleeping time and ether anaesthesia). None had analgesic, antipyretic or anticonvulsant activities (Chaturvedi *et al.*, 1974). The xanthones displayed, both by intraperitoneal and oral routes, anti-inflammatory activity in rats in carrageenan-induced hind paw oedema, cotton-pellet granuloma and granuloma pouch techniques both in normal and in adrenalectomized rats. The xanthones did not have any mast cell membrane effect and did not alter the prothrombin time in rats (Gopalakrishnan *et al.*, 1980).

Plants with other active constituents.

Curcuma domestica Valeton syn. (*C. longa* L.) ZINGIBERACEAE
Turmeric

P Next to the anti-infectious and choleretic activities of the root already mentioned (I), the volatile oil obtained from the rhizome of *C. longa* has been noted to have significant anti-inflammatory properties in rats, comparable to those of hydrocortisone acetate and phenylbutazone. In oral doses of 0.5–1 mg/kg, curcumin inhibited carrageenan-induced oedema in rats and mice as well as formalin-induced acute oedema in mice and subacute arthritis and cotton pellet-induced granuloma formation in rats (Arora *et al.*, 1971; Srimal and Dhawan, 1973). Curcuma oil also strongly reduced the histamine content of the skin of the rat (by 50% as compared to a 42% reduction by hydrocortisone). This antihistamine effect is believed to explain the protective action of the volatile oil (0.01 ml/kg) in early inflammation lesions in adjuvant arthritis in rats. The inhibition of the late arthritis changes may be due to the activation of the adreno–hypophyseal axis (Chandra and Gupta, 1972). The volatile oil was also found to act on the proteases responsible for the acute inflammatory process as it was found to inhibit trypsin as well as hyaluronidases (Tripathi *et al.*, 1973).

In addition to the anti-inflammatory properties, aqueous extracts of the rhizome of *C. domestica* had a 100% antifertility action in albino rats at a dose of 100 mg/kg body weight. The petroleum ether extract had an 80% antifertility effect and also caused resorption of the implants (Garg, 1971). Furthermore, sodium curcuminate has long been known as an active choleretic (Ramprasad and Sirsi, 1956). Curcumol and curdione are reported to be active in the early stage of cervical cancer (Xaio, 1983).

Salvadora persica L. SALVADORACEAE

Salt bush or toothbrush tree

L The twigs and roots are used as a toothbrush. In Senegal the roots are mainly used as a diuretic and are sometimes used associated with kinkeliba leaves for blackwater fever, rheumatism and venereal diseases (Kerharo, 1968). In the Eastern Sudan the bark is pulverized and made into a paste with water, which is applied to the head in cases of serious febrile diseases. Dried leaves are taken for flatulent dyspepsias (Dalziel, 1937).

C The leaves and bark contain the alkaloid trimethylamine and the leaves and seeds yield about 45% of a fatty oil (composed of ±20% lauric acid, ±55% myristic acid and ±19% of palmitic acid (Gunde and Hilditch, 1939)), which is used for making candles and which contains a small proportion of volatile oil (mainly benzyl-mustard-oil) (Wehmer, 1935). From an extract of the roots, β-sitosterol and elemental and monoclinic sulphur were isolated (Ezmirly *et al.*, 1979) and from the leaves polyphenols (quercetin and caffeic and ferulic acids) were isolated. In the rootbark, the presence of small amounts of tannins and saponins has also been reported (Farooqi and Srivastava, 1969).

P Extracts of *S. persica* root have shown antibacterial, weak anti-inflammatory and mild hypoglycaemic properties (Ezmirly *et al.*, 1979).

Ipomoea purpurea (L.) Roth. CONVOLVULACEAE

Ipomoea pes-caprae (L.) Sweet subsp. *brasiliensis* (L.) Oostr. syn. *Convolvulus brasiliensis* L., *I. pes-caprae* Roth.) (Fig. 5.3)

Goat's foot convolvulus

Fig. 5.3. *Ipomea pes-caprae* (L.) Sweet.

L *I. pes-caprae* is a good sandbinder near the sea. The starchy root is a common Indian drug used for rheumatism, dropsy and colic (Dalziel, 1937) and the juice is given as a diuretic in cases of oedema whilst the bruised leaves are applied to the swelling (Chopra *et al.*, 1956, p. 142).

C The whole plant contains 7–8% of total resin, 0.05% of an essential oil, triaconthane, pentatriaconthane, a sterol, behenic acid and melissic, butyric and myristic acids (Cwalina and Jenkins, 1938; Dawalkar and Dawalkar, 1960). The roots contain glycorrhetins similar to those of other *Ipomoea* resins (Hegnauer, 1962–68, Vol. III, p. 551; Shellard, 1962).

P In Indonesia the plant is used against inflammation and cancer. An antihistamine action was reported for this plant by Wasuwat (1970), who noted that a volatile fraction of 6×10^{-5} M antagonized the contractile effect of histamine (2×10^{-6} M) on the guinea pig ileum in a similar fashion to benadryl or antistine (1×10^{-6} M). It also antagonized the contractile effect of jellyfish as effectively as did these antihistamines. *I. purpurea* (L.) Roth, which is naturalized in West Africa, produced 60% inhibition, in doses of 100 mg/kg, in carrageenan-induced rat paw oedema when the dose was repeated at 150- and 30-min intervals before carrageenan administration, when tested on Indonesian plants (Benoit *et al.*, 1976).

Table 5.1. *Additional anti-inflammatory plants*

The abbreviations used are as follows: adj. arthr., adjuvant arthritis; carr., carrageenan; cotton p. gran., cotton pellet induced granuloma; crot. i., croton oil induced; equiv. = equivalent; form., formalin; i.r.p.o., induced rat paw oedema; i.p., intraperitoneally; i.v., intravenously; p.o., perorally; s.c., subcutaneously

Plant Family	Part used	Active constituent(s)	Action/acts on	Pharmacological tests	Comments	References
Allium sativum, *A. cepa* Liliaceae	Bulb	Alliin Allicin	Rheumatism, hypoglycaemic	Allicin immunodepressive, also hypotensive and antibiotic		Ayoub and Svendsen (1981) Xaio (1983)
Alstonia boonei syn. (*A. congensis*) Apocynaceae	Bark	Triterpenoids	External application in rheumatism			Dalziel (1937) Esdorn (1961) *Hager's Handbuch* (1967–80, Vol. II, p. 1237)
Anthocleista procera Loganiaceae (also see Chapter 2)	Leaves	Gentianine (partly formed from precursor swertiamaroside)	Anti-histamine, anti-inflammatory, also analgesic	90 mg/kg given i.p. protects guinea pig against histamine aerosol and prevents albumin- or formol-i.r.p.o. gentianine = immunopotentiating	Preventive protection greater than that of chloroquine or cortisone	Chen-Yu *et al.* (1959) Keng-Tao Liu *et al.* (1959) Lavie and Taylor-Smith (1963)
Atractylis aristata Compositae	Rhizomes	Diterpene carboxyl-atractyloside	Anti-inflammatory	Related species Inhibit carr. i.r.p.o.		Bombardelli *et al.* (1979)
Azadirachta indica Meliaceae	Seed oil	Nimbidin (triterpenoid)	Anti-arthritic, anti-inflammatory also antipyretic and anti-ulcer action	40 mg/kg inhibits carr. and kaolin i.r.p.o. form. arthr. and croton oil gran. in rats. Low oral toxicity (13 g/kg) in mice. Important anti-gastric ulcer effect	Equiv. to 100 mg/kg phenylbutazone and to prednisolone	Pillay *et al.* (1978) Okpaniyi and Ezeukwu (1981) Pillay and Santhakumari (1981)

(*Table continued*)

Table 5.1. (*Continued*)

Plant Family	Part used	Active constituent(s)	Action/acts on	Pharmacological tests	Comments	References
Boerhavia diffusa *B. punarnava* (name given to white-flowered variety in India) Nyctaginaceae	Roots	Flavones (quercetin, quercitrin), oleanolic glycoside, β-sitosterol (also contains an alkaloid)	Anti-inflammatory, also in internal inflammations, oedemas and asthma (diuretic)	4 mg/100 g of aqueous and of acetone extract inhibits carr. i.r.p.o. and carr. i. gran. pouch arthritis	In arthritic animals aqueous extract inhibits serum aminotransferase like hydrocortisone	Chopra *et al.* (1956) Subramanian and Ramakrishnan (1965) Subramayan *et al.* (1965) Bhalla *et al.* (1968; 1971) Misra and Tiwari (1971) Singh and Udupa (1972a, b, c) Srivasta *et al.* (1972) Mudgal (1975)
Cannabis sativa Cannabinaceae	Resin	Δ^9-tetrahydrocannabinol	Anti-inflammatory	Various rat paw oedema models	Like hydrocortisone and aspirin	Sofia *et al.* (1974) Biswas *et al.* (1975a)
Canscora decussata Gentianaceae	Plant	Mangiferin (xanthone heteroside)	Anti-inflammatory, anti-arthritic, depresses CNS			Ghosal *et al.* (1971) Ghosal and Biswas (1979) Shankaranarayan *et al.* (1979)
Capparis decidua Capparidaceae	Seeds	Glucocapparine	Rheumatism, revulsive			Gaind and Junega (1970) Ayoub and Svendsen (1981)
Carica papaya Caricaceae	Seeds, latex	Tropaeaoline, papain	Anti-inflammatory			Rigaud *et al.* (1956) Yarington and Bestler (1964)
Citrus nobilis Rutaceae	Rootbark	Suberosin, xanthyletin (crenulatin, suberenol) (coumarins)	Anti-inflammatory and antihistamine In Traditional Medicine, antirheumatic	Also antipyretic		Paris and Delaveau (1977) Reisch *et al.* (1980)
Crateva religiosa Capparidaceae	Bark, seeds	Lupeol (triterpene), glucocapparine?	Anti-inflammatory revulsive	Inhibits carr. and histamine induced inflammation in rats, and early and delayed inflammatory lesions in form. i.r.p.o.		Chakravarti *et al.* (1959) Ramjelal *et al.* (1972) Biswas *et al.* (1975b)

Dalbergia sissoo c. Fabaceae	Wood, stem	Dalbergichromene neoflavonoids	Anti-inflammatory, anti-arthritic			Mukerjee *et al.* (1971) For Indian spp.: Kishore and Tripathi (1966) Singh and Chaturvedi (1966) Tripathi and Kishore (1967)
Desmodium gangeticum Fabaceae	Roots	Aqueous extracts, tertiary phenylethylamines, also also hypaphorine alkaloid	Anti-inflammatory, also antimicrobial	Non-toxic in toxicity tests		Hye and Gafur (1975) Ghosal and Banerjee (1971) Prema (1968) Ghosal and Bhattacharya (1972)
Gymnema sylvestre Asclepiadaceae	Leaves	Potassium salts of gymnemic acids (gymnagenin = hexa-hydroxy Δ^{12}-oleanene	43% inhibition in carr. r.p.o.	p.o. in 2 doses of 100 mg/kg[a] (plants collected in India)		Rao *et al.* (1972) Benoit *et al.* (1976)
Gynandropsis gynandra Capparidaceae	Leaves	Glucocapparine	Counter-irritant	Arthritis		Dhar *et al.* (1968)
Haematoxylum campechianum (introduced) Caesalpiniaceae	Wood	Brasilin, haematoxylin	Both substances anti-inflammatory	10 mg/kg perorally in carr. i.r.p.o. and fertility egg test	Brasilin, in doses of 100 mg/kg in the r.p.o. and fertility egg tests, is more active than berberine hydrochloride In the cotton p. gran. test activity of brasilin is equivalent to that of berberine hydrochloride	Hikino *et al.* (1977)
Hibiscus vitifolius Malvaceae	Petals	Gossypin (bioflavonoid)	Antiflammatory			Parmar *et al.* (1978)

(*Table continued*)

Table 5.1. (*Continued*)

Plant Family	Part used	Active constituent(s)	Action/acts on	Pharmacological tests	Comments	References
Leptadenia pyrotechnica Asclepiadaceae		2 triterpenoids, steroid glycosides (5-tri-triaconten-17-one, sitosterol, stigmasterol, 18-pentatriacontanol)	Gout, rheumatism			Iyer *et al.* (1974) Manavalan and Mithal (1980)
Moringa oleifera syn. (*M. pterygosperma*) Moringaceae	Roots, seed oil	Glucotropaeoline (benzylisothiocyanate derivatives)	External counter-irritant (oil)			Chopra *et al.* (1956) Dalziel (1937) Kerharo and Adam (1974, p. 599)
Securidaca longepedunculata Polygalaceae	Roots	4% methylsalicylate saponin (oleanolic glycoside)	Revulsive, antitussive	Also anticonvulsive		Dalziel (1937) Kerharo (1968) Odebiji (1978)
Tetrapleura tetraptera Mimosaceae	Bark, Roots	Saponosides, oleanolic triglycoside	Lumbago, antitussive	Also anticonvulsive		Odutola and Ekong (1968) Adesina and Sofowora (1979)
Vernonia colorata Compositae	Flowers, bark	Vernolid, sesquiterpenic lactone, glycosides	Inhibition of carr. i.r.p.o. in flowers 11%, in bark 19%	Vernolid is also cytostatic (100 mg/kg) given twice[a] (plants from Nigeria)		Satoda and Yoshi (1962) Benoit *et al.* (1976)
Xylopia aethiopica Dun. Annonaceae	Fruit	Kauren-diterpenes	Anti-inflammatory			Ekong and Ogan (1968) Ekong *et al.* (1969) Boakaiji Yiadom *et al.* (1977)

[a]Second dose given 2 h after first dose.

6

Sex hormones and thyroid hormones

I Sex hormones

(a) Female sex hormones

Oestrogens are the hormones concerned with the maturation of the female genital tract, the development of the secondary sexual characteristics and bone formation. The main oestrogens, oestradiol and oestrone, are secreted by the theca interna and appear in the fluid of the Graafian Follicles. Oestrogens are also secreted by the placenta and to a much lesser degree by the adrenal cortex. They are rapidly metabolized in the liver.

Progesterone is the hormone concerned with the maintenance of pregnancy. It is secreted by the corpus luteum and acts only on tissues formerly sensitized by oestrogens. It inhibits ovulation during pregnancy, depresses the action of oestrogens and produces further development of the breasts. It also plays a role in the development of the placenta and depresses uterine contractility.

The production of steroid hormones is controlled by the pituitary gonadotrophins (follicle-stimulating and luteinizing hormones). In large doses oestrogens can depress the gonadotrophic and lactogenic activities of the anterior pituitary gland. Because of their rapid metabolism by the liver, oral administration of the natural oestrogens is less effective than parenteral administration, with the possible exception of oestriol, which is claimed to be as potent when given orally as when given by injection.

The oestrogens are used to treat menopausal disturbances and cases of dysmenorrhoea and menorrhagia. Progesterone is used chiefly in the treatment of functional uterine haemorrhagia. Both oestrogens and progesterone are used in the contraceptive pill.

Plants with oestrogenic constituents

The first plant constituents capable of replacing oestrone in hormone deficiency were found to be chemically identical with the animal hormone and were reported as early as 1933 (Butenandt and Jacobi, 1933; Skarzynski, 1933). It was, however, not until thirty years later that the oestrogenic substances in plants were investigated further. A review of the steroid oestrogens in higher plants was published by Heftman in 1967.

Other plant substances, although chemically not identical to them, were subsequently found to be able to replace the biological functions of the hormones (Shutt, 1976). Bennets and his coworkers (1951) noted the infertility observed in sheep which grazed exclusively on clover for a certain length of time in Australia. A compound isolated from clover was found to be oestrogenic (Biggers and Curnow, 1954) and was identified as genisteine, an isoflavine. Failure of sperm transport was later found to be the cause of the infertility of the ewes (Lightfoot *et al.*, 1967). Subsequently, other isoflavones with oestrogenic activity were found in forage and other plants (Shutt, 1976; Adams, 1977; Livingstone, 1978). Also reported to be present in plants was another group of chemical compounds, the coumestans (skeletal structure 6H benzofuran 3-2-*c*-benzopyran-6-one), which have a higher oestrogenic activity than the isoflavones. A striking similarity of the skeletal structures of the isoflavones and coumestans with that of the synthetic oestrogen stilboestrol was noted by Farnsworth *et al.* (1975b). Of the constituents tested by these authors, the natural steroid oestrogens were found to be the most active, followed by the coumestans and then the isoflavones. It was further shown that daidzin (in **Pueraria thunbergiana** (Sieb. & Zucc.) Benth.) was the most active isoflavonoid derivative in the mouse uterine weight assay. Biochanin (in **Dalbergia sissoo** and spp. and in **Lupinus tassilicus** Maire) and genisteine (in **Trifolium baccarini** Chiov.) were of lesser but approximately equal activity (Farnsworth *et al.*, 1975a).

In search of plants with fertility regulating action, Farnsworth *et al.* (1975b) have listed the plants containing oestrone and oestriol. Some of these are found in West Africa: the date palm *Phoenix dactylifera*; the oil palm *Elaeis guineensis*; rice, *Oryza sativa* L.; wheat, *Triticum aestivum* L. and vegetable beans, *Phaeseolus vulgaris* L.

Phoenix dactylifera L. PALMAE

In Northern Nigeria dates, together with bran and the seeds of *Sterculia tomentosa*, are given to young heifers long before they are mature in order to cause them to become prolific breeders (Dalziel, 1937, p. 509). The pollen grains of the date palm, which is dioecious, are reported to act like gonadotrophins (El Ridi, 1960). Oestrone has been reported to be present in the kernels (Hassan and El Waffa, 1947; Heftman *et al.*, 1966; Paris and Moyse, 1967, Vol. II, p. 10), which also contain 5% protides and 8% lipids.

Elaeis guineensis Jacq. PALMAE

Disparity in results obtained by different investigators when analysing palm kernels for oestrone content can be explained by the fact that different varieties of oil palm produce fruits with a varying thickness of mesocarp and size of kernel and extremes of being shell-less and kernel-less. The oestrone content of certain varieties could be confirmed (Butenandt and Jacobi, 1933; Bradbury and White, 1951).

Triticum aestivum L. syn. (*Triticum sativum* Lam. (wheat) GRAMINEAE
Oryza sativa L. (rice)

The seeds of these and other cereals mainly contain starches but the embryo (germ) at the base of the grain contains 10–20% lipids compared to the 1–2% in the albumen. The lipid fraction contains the sterols and the tocopherols (vitamin E). Thus wheatgerm-oil contains 85% of glycerides of non-saturated fatty acids, sterols and vitamins (mainly vitamin E). The oil is administered in cases of sterility in women and is also used in veterinary medicine in epizootic abortions.

It is interesting to note that vitamin E (α-tocopherol mainly) is found in the germs of cereals and also in different oils (palm oil, castor-oil (1%) and olive oil (Langlois, 1941)) and that deficiency of tocopherols is characterized by gestation troubles in females and testicular modifications in males (vitamin E is known to be a fertility factor (Bacharach, 1940)). *Oryza sativa* c. and *Avena sativa* c., both containing oestrone (Farnsworth *et al.*, 1975b), have been shown to induce ovulation (Heftman, 1967; Paris and Moyse, 1967, Vol. II, p. 26).

Phaseolus vulgaris L. FABACEAE

The vegetable bean, also cultivated in some parts of West Africa, contains oestrone, oestriol and 17α-oestradiol (Kopcewicz, 1971). It contains more proteins (20–25%) than the cereals (8–15%) and only 1–7% lipids, including sterols and vitamin E, but has 61.6% glucides (Paris and Moyse, 1971, Vol. III, p. 201).

Arachis hypogeae L. FABACEAE

Another very important food plant, the groundnut, already mentioned in Chapter 2 (CV) contains an oestrogenic factor and a goitrogenic compound. The plant is treated in more detail below under Thyroid hormones.

Groundnuts contain an average of 42.8% of lipids and 26.2% of proteins and small quantities of vitamins, including vitamin E (Adrian and Jacquot, 1968).

Medicago sativa L. FABACEAE

Also cultivated as a food plant, *M. sativa* contains coumestrol and is rich in vitamins; mainly vitamin E is reported (Paris and Moyse, 1971, Vol. III, p. 391). Vitamin E is considered to be an antisterility factor (Bacharach, 1940). The activity of these oestrogenic food plants seems to be mostly concentrated in the germs and kernels. Other sources of oestrogens in West Africa might be the following plants.

Cyperus rotundus L. CYPERACEAE

The essential oil from the tubers shows oestrogenic activity. Fractionation of the oil produces an active hydrocarbon, cyperene I, which has a slightly less potent oestrogenic action than the oil and which also has an antispasmodic action on the uterus (Indira *et al.*, 1956a, b; Abu-Mustafa *et al.*, 1960) (see also Anti-inflammatory plants (Chapter 5)).

Holarrhena floribunda (Don.) Dur. & Schinz. APOCYNACEAE

A root decoction of Holarrhena is prescribed for sterility (Kerharo and Adam, 1974). The roots contain many steroid alkaloids. One of these, holophyllamine has significant oestrogenic activity (see more details under Plants acting like hormones of the adrenal cortex (Chapter 5)).

Funtumia africana (Benth.) Stapf APOCYNACEAE

The main alkaloid of the leaves, funtumine, is a 3α derivative of allopregnane with a ketone function at C-20. It is suggested that it might be of use in the hemisynthesis of hormones. Funtumine itself antagonizes the effects of oestrogens (see under Plants acting like hormones of the adrenal cortex (Chapter 5)).

It is only in the last few years that pharmacological and clinical details of the phyto-oestrogens have been mentioned. Most of the scientific data have been obtained in connection with the development of new methods of fertility regulation (Farnsworth *et al.*, 1975a, b, 1983; Farnsworth and Waller, 1982). This application of the sex hormones currently represents by far the biggest practical demand. I have limited the other applications of the hormones to a few examples only.

Phytoandrogens are hardly mentioned in the literature. An androgenic steroid with an anabolic action in the healing of fractures has been found in *Cissus quadrangularis* (see Chapter 3).

(b) Plants in birth control

In order to be able to decide which of the very many species of plants should be tested for their effectiveness in birth control it has been necessary to consider the mechanisms by which the constituents exert their effect and to classify them according to the anatomical sites involved. This has greatly contributed to a better understanding of the plant oestrogens. Farnsworth *et al.* (1975a, b) have therefore enumerated the different antifertility mechanisms in laboratory animals. These are described briefly below.

The oestrogens are of interest in studies on fertility regulation because they can act as contraceptives by inhibiting the mid-cycle surge of pituitary gonadotrophin that is associated with ovulation.

Antifertility mechanisms

The organs on which antifertility agents may act in females are the hypothalamus, the anterior pituitary, the ovary, the oviduct, the uterus and the vagina. An antifertility agent can be classified by its action on each of these organs.

The pituitary. The functioning of this organ is under the close control of the hypothalamus via the follicle-stimulating and luteinizing-hormone releasing factors

(Goodman and Gilman, 1980, pp. 1389–92). Therefore, antifertility action at this level should include: (a) disruption of the normal hormonal functions of the hypothalamus and/or pituitary, e.g. by oestrogenic steroids and (b) interruption of the neural pathways to the hypothalamus that control the liberation of gonadotrophin-releasing factors.

Since the hypothalamus receives contributions from other areas of the brain, substances having CNS depressant activity and/or effects on neuro-hormonal transmission could be expected to alter gonadotrophin transmission. And indeed pentobarbital, morphine, atropine, tranquillizers (reserpine), anaesthetics and adrenergic (as well as cholinergic) blocking agents have been shown to block the ovulatory surge of luteinizing hormone in laboratory animals by having inhibitory effects on the hypothalamus (Smith, 1963).

Studies on laboratory animals such as the rat, mouse, hamster and guinea pig provide much information but the reproductive cycles of these animals exhibit species-specific differences as do those of men and primates, and different results with a given compound are often obtained.

Interference with gonadotrophin secretion may have post-ovulatory antifertility effects, but the antifertility usefulness of drugs having such effects is questionable not only because they have other pharmacological actions but also because animals (e.g. rats) sometimes become coitus-induced ovulators following blockade of their gonadotrophin surge (Farnsworth *et al.*, 1975a).

In the guinea pig the pituitary can be removed after the third day of gestation without affecting the pregnancy and hypophysectomized women induced to ovulate by consecutive administration of follicle-stimulating hormone and human chorionic gonadotrophin have become pregnant without further gonadotrophin replacement therapy (Farnsworth *et al.*, 1975a).

The ovary. Substances having antifertility properties may exert their effects at the ovarian level by inhibiting ovulation and/or steroidogenesis. Oestrogen administration early in the luteal phase has been shown, in monkeys and women, to decrease progesterone secretion and to hasten the onset of menstruation (Knobil, 1973). The lowering of post-ovulatory plasma progesterone levels may be at least one mechanism by which post-coitally administered high doses of oestrogen exert their antifertility effect in women.

The oviduct. Since successful implantation depends on the correct timing in the menstrual cycle of the arrival of the blastocyst in the uterus, disturbances of tubal transport may be accompanied by failure of implantation. Accelerated transport of ova results in a reduction in fertility, either through expulsion of the fertilized ova from the reproductive tract or through degeneration of fertilized ova that arrive (too early) in the non-receptive uterus. Either oestrogenic or anti-oestrogenic compounds may play a role in this inhibition. Anti-oestrogenic compounds are compounds which inhibit the effects of standard oestrogens such as oestrone, oestriol and oestradiol. Androgens and progestogens may show this activity and the weak plant oestrogens coumestrol and genisteine have also been shown to have this inhibitory effect (Folman and Pope, 1966). The anti-oestrogenicity of a compound can be demonstrated by the blocking of a step in the reproductive cycle that requires oestrogen.

The uterus. Antifertility agents that prevent ovulation and/or fertilization are called contraceptive agents whilst those that act after implantation has taken place are usually called abortifacients. The term 'interceptives' refers to compounds that act after fertilization has occurred to prevent implantation from taking place. Their earlier or later administration might render some of them contraceptive or abortifacient, respectively.

Inhibition of implantation has been observed with a number of compounds and it appears possible that non-physiological treatment of the endometrium may impair fertility in the human. It has been suggested that an excess amount of oestrogen present in the uterus shortly after ovulation (in mammals) may prematurely sensitize the uterus so that the latter is in a non-receptive state at the time of arrival of the blastocyst. Other substances, instilled locally in the uterus, can impair fertility by affecting the endometrium or by causing physical obstruction of the lumen.

An abortifacient type of antifertility effect can be produced by compounds that stimulate uterine contractility. The hormone oxytocin can be used close to term to induce labour but is effective as an abortifacient if used earlier in a pregnancy. Prostaglandins, however, are now being widely used for the latter purpose. They appear to promote myometrial contractility, acting on both contraction and stretching of the myometrial wall. This produces a decrease in the levels of oestradiol and progesterone and the resulting endogenous stimulatory mechanism may be sufficient to complete abortion. If it is not, additional prostaglandin or oxytocin may be used.

Cervix or vagina. Antifertility activity exerted at the level of the cervix or the vagina can be based on the production of a cervical mucus that is 'hostile' to sperm penetration. Some of the spermicidal preparations are based on a trypsin-like protease (acrosin). Oestrogenic compounds produce vaginal keratinization (Farnsworth *et al.*, 1975a).

Anovulation has also been observed in a number of cases as a result of inhibition of hypothalamic-hypophysial function, for example in Rhesus monkeys treated with reserpine (from *Rauvolfia vomitoria*) (Bianchi, 1962). But, as mentioned earlier, the suitability of using plants that act in this non-specific way is questionable because of the many different hormones that are controlled by this axis, for example the gonadotrophins (oestrogenic, luteinizing, prolactin, androgenic), thyrotrophin, and corticotrophin, and the changes in target organ function that result from effects on these hormones.

It has been reported that contraceptive steroids produce changes in metabolic processes, for example deterioration in glucose-tolerance tests and an increase in lipoproteins (Briggs, 1976; Briggs and Briggs, 1981). The authors further note alterations in plasma proteins and a certain action on the pituitary: basal prolactin and plasma growth hormone are enhanced, and the concentration of thyrotrophin remains unchanged but adrenocorticotrophin is suppressed. Also, the pattern of plasma androgen is altered. The levels of plasma cortisol and to a lesser extent of plasma aldosterone are increased.

Plants having antifertility effects in females

Most of these plants have either anti-implantation, uterine stimulant or abortifacient action. Bioassays which can be used to evaluate the alleged fertility regulating effects of the plants include in vitro evaluation using the rat uterus and confirmation in the 29-day pregnant rabbit for utero-evacuant activity. A third bioassay involves testing for anti-implantation effects, initially in rats, with confirmation in hamsters.

The active principles of some Fabaceae, such as **Sophora occidentalis** L., *Sesbania bispinosa* (Jacq.) Wight and **Cajanus cajan** (L.) Mill., were found to be isomeric quinolizidine alkaloids; they had some value as ecbolics but were ineffective for inducing abortion during the early stages of pregnancy (Farnsworth *et al.*, 1975b). The activity of a principle of another Fabaceae, **Pisum sativum** L. (the common garden pea), *m*-xylohydroquinone (2,6-dimethylhydroquinone) has been studied extensively in women in India. However it proved to be only about 60% effective in preventing conception in the groups studied and was abandoned as a potentially useful fertility regulator (Sanyal, 1956, 1958; Sanyal and Rana, 1959; Furuya and Galston, 1965).

From numerous reviews on plants as sources of antifertility agents (Casey, 1960; Chaudhury, 1966; Vohora *et al.*, 1969; Brondegaard, 1973; Barnes *et al.*, 1975; Prakash and Mathur, 1979 and many others), Farnsworth *et cl.* (1975a, b) concluded that for the estimation of the potential value of new antifertility agents results from animal experiments were difficult to interpret, and a WHO meeting was convened in Mexico City in 1976 in order to develop a programme for a more systematic evaluation of antifertility plants. Priorities were set on the study of male contraceptives and 'morning after' and anti-implantation agents for use in the female.

In January 1978, six collaborating centres in different parts of the world were selected to contribute to the collection, testing and subsequent identification of antifertility agents from plants identified by an extensive computer analysis of published information (Soejarto *et al.*, 1978). In further research in the course of the fulfilment of this worldwide programme, Bingel and Farnsworth (1980) pointed out that plants tending to interfere with implantation in the female (Naqvi and Warren, 1971; Morris and van Wagenen, 1973; Mishra *et al.*, 1979) and those which may interfere with sperm formation and/or maturation in the male (Setty *et al.*, 1976; Banerji *et al.*, 1979) were those of particular interest to the antifertility programme. They found, however, that there are other factors to be considered.

Abortion or interception has frequently been found to be caused by cytotoxic plant constituents (Hartwell, 1976; Pakrashi *et al.*, 1977). Many cytotoxic agents have been found to be anti-spermatogenic, for example, vinblastine and vincristine from *Catharanthus roseus* (Fern, 1963) and cardenolides from *Calotropis procera*, (Atal and Sethi, 1962; Vilar, 1974; Atal, 1980). The use of such drugs entails the danger that neoplastic agents may induce chromosomal aberrations and thus not only affect the sperm number but also cause other adverse reactions. Therefore the usefulness of cytotoxic agents for interrupting pregnancy is limited by their potential toxicity to the maternal organism and by the possibility of their producing teratogenic effects in the surviving fetuses, when given in marginal doses or at marginal times (Morris, 1970, in Bingel and Farnsworth, 1980). In the case of *Abrus precatorius*, abortive action seems to be limited to a steroidal oil and the teratogenic action to an aqueous extract of the protein fractions. Separation of these two effects may be possible by extraction.

Cyanogenic agents have also been noted to influence reproduction and neonatal development in rats (Dlusi *et al.*, 1979) and rabbits (Eshiett *et al.*, 1980). Also, a correlation between the spermicidal activity and the haemolytic index of certain plant saponins has been reported (Ad Elbary and Nour, 1979).

Kamboj and Dhawan (1982) endeavoured to reduce the list of Indian plants used for contraceptive purposes in females. They reviewed the literature on these plants published in India together with the unpublished data of the Central Drug Research Institute and gave most attention to the identification of plants with interceptive properties. They concluded that in many cases the evaluation showed results which varied from no activity to 100% activity for the same plant. Lack of proper botanical authentication and inconsistent results in repeat tests or lack of facilities were indicated as possible reasons. (They also suggested modification to the feeding schedule of the plants to be screened.)

Of the plants examined, Kamboj and Dhawan (1982) excluded follow-up of *Hibiscus rosa-sinensis* and *Plumbago zeylanica* on the basis of preliminary toxicity studies. They recommended, in order of priority, follow-up of *Achyranthes aspera* (stembark), *Sapindus trifoliatus* (seed) and *Abrus precatorius* (seed). Because of similarities in the climates of parts of India and tropical West Africa, a certain number of these plants also occur in the area discussed in this book, and a selection of them (and of those indicated by Farnsworth *et al.* (1983)) is listed in Table 6.1. It should be noted that the African plants will have to be examined for analogy in chemistry and action with the Indian varieties.

Plants having antifertility effects in males

Plants used for regulating male fertility have to interfere either with sperm production and maturation (Parkhurst and Stolzenberg, 1975) or with sperm storage or with their transport in the female genital tract. Ideally, these plants should not exert any effects on non-reproductive systems and enzymes. With these objectives in mind, Bingel and Farnsworth (1980) selected from plants found worldwide a number of promising and interesting species (see also Farnsworth and Waller, 1982). Of these, those species which are found in tropical West Africa are listed in Table 6.2. Of course it will have to be checked whether the plants growing in West Africa have the same constituents and properties as the same species grown elsewhere.

As well as being included in the tables, a few antifertility plants which are of particular interest to West Africa have been described in more detail.

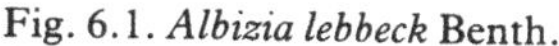

Fig. 6.1. *Albizia lebbeck* Benth.

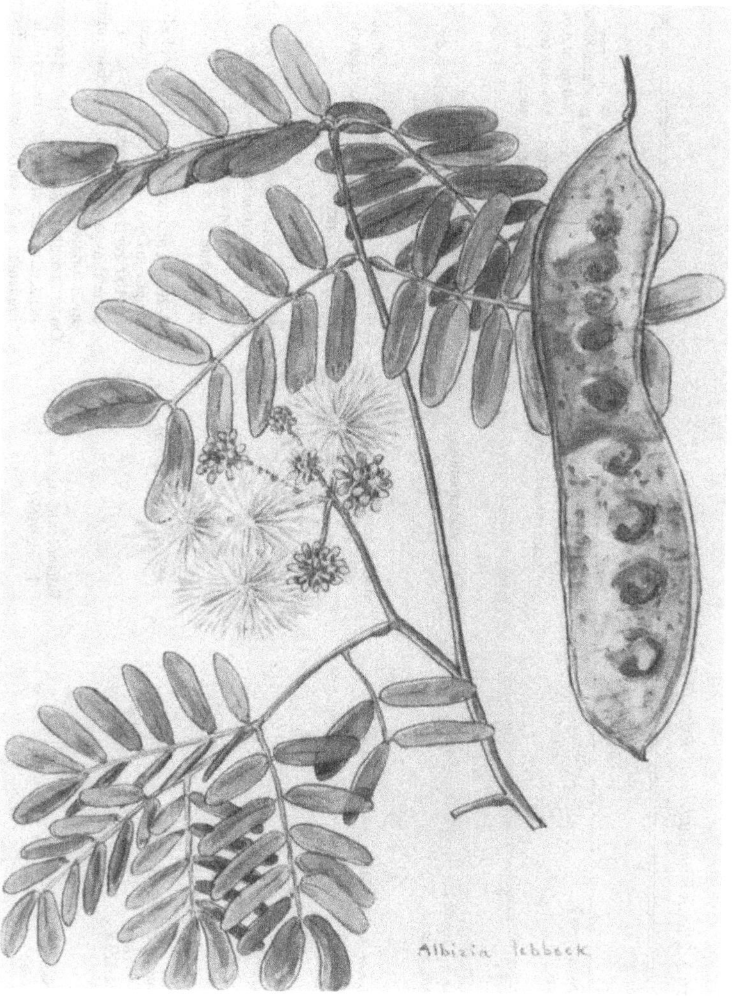

Table 6.1. *Plants with antifertility action in females*

Plant Family	Part used	Active constituent(s)	Observed activity	References
Abrus precatorius L. Fabaceae	Seeds, roots	Aqueous extract of seeds (proteins), steroidal oil, ethanol extract of root	See text. 150 mg of steroidal oil produces 80% sterility in rats. Oral doses of 300 mg of root extract per kg body weight 100% interception. LD_{50} in mice is 2 mg/kg. Oxytocic	Desai and Rupawala (1966) Desai and Sirsi (1966) Dijkman *et al.* (1966) Desai and Rupawala (1967) Gupta *et al.* (1968) Agrawal *et al.* (1969, 1970)
Achyranthes aspera L. Amaranthaceae	Bark	Oleanic acid saponin	Abortive and anti-implantation activity in mice using 50 mg benzene extract per kg body weight on Days 1–5 post coitum	CV Gopolachari and Dhar (1958) Hariharan and Ranjaswani (1970) Kamboj and Dhawan (1982)
Ananas comosus (L.) Merr. syn. (*A. sativa* Schult.) Bromeliaceae	Leaves	Stigma-5ene-3β7α-diol, stigma-5ene-3β7β-diol, βsitosterol	Anti-fertility activity in mice, abortive 50 ml unripe fruit-juice on Day 1–7 post coitum causes 60% interception	I Shukla and Krishnamurti (1961) Sakkawala *et al.* (1962) Bhaduri *et al.* (1968a) Pakrashi *et al.* (1975a, b)
Carica papaya L. Caricaceae	Fruit, (unripe) latex	Papain?	Ethanol extract at doses of 100 mg/kg exerted 60% anti-implantation activity, higher doses hardly any. Early abortifacient effect. Oxytocic antiovulatory	CV I Chandrasekar *et al.* (1961) Garg and Garg (1974) Kapoor *et al.* (1974)
Curcuma domestica Valeton syn. (*C. longa* L.) c. Zingiberaceae	Rhizome	Essential oil, curcumin	100–200 mg petroleum ether extract per kg body weight on Days 1–7 post coitum produces 80–100% anti-implantation. May act on adrenal-hypophysial axis. Essential oil inhibits hyalorunidase	I Arora *et al.* (1971) Chandra and Gupta (1972) Garg (1974) Garg *et al.* (1978)
Hibiscus rosa-sinensis Malvaceae	Flower	Ethanol extract, hibiscetin??	On tenth day of pregnancy rats having received benzene extract showed 80% reduction in implantation. No early abortifacient effect was noted. Ethanol extract is said to alter sex ratio of the pups	Batta and Santhakumari (1970) Singh *et al.* (1982)

Plant	Part	Constituent	Activity	Reference[a]
Hyptis suavolens (L.) Poit. Labiatae	Leaves	Essential oil?	Alcohol extract of leaves shows anti-implantation activity. Further investigation advised	Garg *et al.* (1978)
Oldenlandia affinis (Roem. & Schult.) DC. syn. (*O. decumbens*) (Hochst.) Hiern. *Hedyotis affinis* (Roem. & Schult.) Rubiaceae	Whole plant, aerial parts	Polypeptide (molecular weight 4000), serotonin	Polypeptide isolated from aqueous decoctions of aerial parts of the plant produced, in vitro, contractions of pregnant uterus. Action absent or doubtful in situ	Bingel and Farnsworth (1980) Gran (1973a, b, c, d, e) Bingel and Farnsworth (1980)
Plumbago zeylanica L. Plumbaginaceae	Whole plant, roots	Plumbagin (2-methyl-4-hydroxy-1-4-naphthoquinone)	Abortive in vitro, no abortion in pregnant rabbits or guinea pigs. In rats 75–100% mortality after 3–4 doses. Oxytocic in vitro	I Bhatia and Lal (1933) Ko (1933) Premakumari *et al.* (1977) Bingel and Farnsworth (1980) Santhakumari and Sujantham (1980)
Rhoe spathaceae (Sw.) Stearn syn. (*R. discolor*) Hance Commelinaceae	Plant		Antifertility and anti-nidatory. Stimulative action on mouse uterus	Weniger *et al.* (1980) Weniger *et al.* (1982)
Sapindus trifoliatus (L.) Sapindaceae	Seed, fruit-pulp	Saponin	An ethanol extract administered 1–7 days post coitum in rats produced 80% interception in 100 mg/kg doses and 100% interception with 500 mg/kg doses. A methanol fraction of the ethanol extract acted orally in doses of 25 mg/kg when given on Days 4 and 5 of pregnancy	CV Bodhankar *et al.* (1974) Garg *et al.* (1978)

[a] Where a plant has been discussed in another chapter, reference is made to that chapter. Thus CV refers to Chapter 2 (Cardiovascular plants) and I refers to Chapter 4 (Anti-infection therapy).

Table 6.2. *Plants with antifertility action in males*

Plant Family	Part used	Active constituent(s)	Observed activity	References[a]
Aeschynomene indica L. Fabaceae	Whole plant	Saponins	Spermicidal[b] at a concentration of 2% (instantaneous immobilization) for human and rat semen in vitro	Setty *et al.* (1976, 1977)
Albizia lebbeck Benth. syn. (*Mimosa lebbeck* L.) (naturalized) (Fig. 6.1) Mimosaceae	Seedpod, (root), (wood)	Echinocystic and oleanic acid saponins, lebbekanin E (heteroside of acacic acid)	Spermicidal[b] at a concentration opf 2%; according to test used, the minimum effective doses of pod and root are, respectively, 0.125–0.6% and 0.125–1.0%, for human spermatozoa in vitro An abortifacient glycoside, albitocin, found in *Albizia* spp. has shown high toxicity	I Dhar *et al.* (1968) Setty *et al.* (1976) Tripathi and Dasgupta (1974) Varshney *et al.* (1979) Lipton (1967)
Anagallis arvensis L. Primulaceae	Whole plant	Anagalligenin saponin (triterpenoid)	Spermicidal[b] to human sperm	I Banerji *et al.* (1978b, 1979b) Yamada *et al.* (1978) Bezanger-Beauquesne *et al.* (1980)
Azadirachta indica Meliaceae	(Seeds), leaves, water extract	Sodium nimbinate (triterpene)	1 ml of extract daily given orally, for one month produced in mice an 80% reduction in pregnancies. The males regain 100% fertility 45 days after stopping of treatment. Activity has been attributed to loss of libido in treated animals	I Sharma and Saksena (1959) Ekong *et al.* (1969) Garg *et al.* (1970) Luscombe and Taha (1974) Thompson and Anderson (1978) Deshpande *et al.* (1980) Farnsworth and Waller (1982)
Blighia sapida Koenig (Akee apple) Sapindaceae	Fruit	Hederagin and oleanolic saponins	At a concentration of 2% spermicical[b] against human semen in vitro	Banerji *et al.* (1979b)
Calotropis procera (Ait.) R.Br. Asclepiadaceae	Ethanol extract of flowers	Cardiac heterosides (cytotoxic)	In gerbils 20 mg/animal orally every second day for 30 days caused testicular necrosis and severe degenerative changes in the sexual organs and also liver damage. Fractionation of the active extract to separate the toxic factor is needed	CV I Atal and Sethi (1962) Crout *et al.* (1964) Hartwell (1976) Garg (1979) Farnsworth and Walker (1982)

Carica papaya L. Paw paw. Widely cultivated Caricaceae	Seeds	Oleanolic glycoside? (76–78% of oleic acid in oil of seeds)	20 mg/rat given perorally in aqueous suspension for 8 weeks to males reduced the ability to fertilize females by 40% although the weight and histology of the genitals, and spermatogenesis, seemed unaffected	Garg and Garg (1971) Bodhankar *et al*. (1974) Das (1980) Farnsworth and Waller (1982)
Catharanthus roseus (L.) Don syn. (*Vinca rosea L*., *Lochnera rosea* Reichb.) Apocynaceae	(Roots), leaves	Vinblastine, vincristine (alkaloids)	Intraperitoneal injection of the total leaf alkaloids produced degenerative changes in the spermatogenic elements of the testes in animals and oligospermia in men	I Joshi and Ambaye (1968) Bustos-Obregon and Feito (1974) Vilar (1974) Cooke *et al*. (1978) Parvinen *et al*. (1978) Farnsworth and Waller (1982)
Centella asiatica Urban syn. (*Hydrocotyle asiatica* L.) Umbelliferae	Whole plant	Saponin, according to provenance, asiaticoside or brahminoside	At a concentration of 2% the total saponins were active in vitro against rat and human sperm but pure saponins showed no spermicidal[b] or spermostatic action	N Dutta and Basu (1967) Appa Rao *et al*. (1969) Rao and Seshadri (1969) Setty *et al*. (1976)
Gossypium spp. Malvaceae	Seed oil	Gossypol (dimeric sesquiterpene)	Spermicidal[b] action studied; widely used in men in China (see text)	I Aizikov and Kurmukov (1973) Kalla and Vasudev (1980) Waller *et al*. (1980) Ridley and Blasco (1981a, b) Farnsworth and Waller (1982)
Hibiscus rosa-sinensis L. Malvaceae	Flower petals,	Ethanol extract, hibiscetin??	Ethanol and benzene extracts have antispermatogenic effect (see text)	Kholkute *et al*. (1972; 1976a, b; 1977) Kholkute and Udupa (1974, 1976, 1978a,b) Kholkute (1977) Farnsworth and Waller (1982)
Leucena glauca (L.) Benth. syn. (*Mimosa glauca*) Mimosaceae	Leaves, seeds	Mimosine (3% in leaves, 5% in seeds) = cytotoxic amino acid	Feeding of male rats prior to mating with a ration containing 15% of leafmeal caused an important decrease in pregnancies. Leafmeal produces alopoecia in sow and rabbits; addition of a mimosine-inactivating agent impairs the effect	Willet *et al*. (1942–44) Luna (1963) Farnsworth and Waller (1982)
Lupinus tassilicus Maire syn. (*L. termis* Forsk.) Fabaceae	Seeds	Debittered seeds through hot water treatment	Animals treated with a ration containing 27% of debittered seeds for nine weeks showed decreased sperm production. Possibility of limited testosterone production is mentioned	I Tannous and Nayfeh (1969)
Mimusops elengi L. (generally cultivated) (Fig. 6.2) Sapotaceae	Seeds, (bark)	Bassic acid saponin	A 6 : 100 dilution of the saponin in Sörensen's isotonic phosphate buffer solution had spermicidal[b] activity on human semen	Tripathi and Dasgupta (1974) Farnsworth and Waller (1982) Bannerji *et al*. (1978a, 1979a, b)

(*Table continued*)

Table 6.2. (Continued)

Plant Family	Part used	Active constituent(s)	Observed activity	References[a]
Momordica charantia L. (African cucumber, bitter gourd) Cucurbitaceae	Fruit	Ethanol extract of fruit, momordicine ?	Subcutaneous or oral doses of 200–400 mg for two weeks of an alcoholic extract of the fruits to gerbils and of 1.75 mg daily for 60 days to dogs reduced testicular weights and spermatogenesis without affecting the seminal vesicles or prostate glands. Additional studies required	I Dhalla *et al.* (1961) Sucrow (1966) Morton (1967) Stepka *et al.* (1974) Olaniji (1975) Dixit *et al.* (1978) Khanna *et al.* (1981) Farnsworth and Waller (1982)
Phytolacca dodecandra l'Herit. (Endod, soapberry) Phytolaccaceae	Berry	Oleanic acid saponin, lemmatoxin	Crude extracts as well as lemmatoxin and oleanoglycotoxin are more active spermicidals than nonoxynol-9, a widely used vaginal contraceptive. They avoid pregnancy in most cases or reduce embryonic counts in the horns of treated rats to low levels. Doses used are 500 and 100 μg of saponin extract or 500 μg of lemmatoxin on Days 1, 4 and 6 post coitum or 100 μg of oleanoglycotoxin on Day 4 only	I Ahmed *et al.* (1949) Shaaban and Ahmed (1959) Stolzenberg and Parkhurst (1974) Parkhurst and Stolzenberg (1975) Stolzenberg *et al.* (1975)
Samanea saman Jacq. & Merr. syn. (*Pithecellobum saman* Benth. jacq., *Albizzia flavovirens* Hoyle) Mimosaceae	Aerial parts	Glycoside of acacic acid (4 different sugars)	Spermicidal[b] to rat and human sperm (in spot and IPPF[c] tests[d] minimum activity with 1–5% in vitro). The spermicidal activity of the plants tested by Setty *et al.* (1977) was found to be associated with the β-amyrin C_{28}-carboxylic acid type of sapogenin linked to a particular sequence of sugar moieties	Garg *et al.* (1976) Setty *et al.* (1976) Setty *et al.* (1977) Varshney and Khanna (1978)
Sapindus trifoliatus L. (naturalized) Sapindaceae	Fruit	Saponin	The concentrate of a water–ethanol extract (3–4 extractions) was systematically fractionated. A 2% concentration of the *n*-butanol fraction of a water-soluble part of the concentrate had spermicidal[b] activity against human sperm in vitro. The saponin fraction used as a vaginal cream was non-toxic in rabbits and rhesus monkeys	Bodhankar *et al.* (1974) Garg *et al.* (1976) Setty *et al.* (1976) Dixit and Gupta (1983)

Trigonella foenum graecum L. (cultivated in the north) Papilionaceae	Seeds	Hederagin glycosides	A 2% concentration of the saponin fraction (isolated like that of *Sapindus*) showed in vitro spermicidal[b] action against rat and human semen. Minimum effective spermicidal doses were 0.25 and 1.5% in spot and IPPF tests, respectively	Setty *et al.* (1976, 1977)
Withania somnifera (L.) Dunal syn. (*Physalis somnifera* L.) Solanaceae	Root	Flavonoids? withaferin, withanolides??	25 mg dried root powder administered daily for 10 days to male and female mice decreased the number of pregnancies and produced small litters. Additional studies required. Locally used in South African tribes to remove retained conception product in miscarriage	N Watt and Breyer-Brandwijk (1962) Garg and Prasar (1965) Menssen and Stapel (1973) Abraham *et al.* (1975) Setty *et al.* (1976) Baer and Haensel (1982) Farnsworth and Waller (1982)

[a]Where a plant has been discussed in another chapter, reference is made to that chapter. Thus CV refers to Chapter 2 (Cardiovascular plants), N refers to Chapter 3 (The nervous system) and I refers to Chapter 4 (Anti-infection therapy).
[b]No revival of sperm motility by buffered glucose solution.
[c]International Planned Parenthood Federation.
[d]Tests based on activity against human sperm in vitro.

Abrus precatorius L. (Fig. 6.3) FABACEAE
Crabs eye, lucky bean

L (See also Chapter 2 and Sweetening agents (Chapter 7).) Under the name of *jequirity* the seeds have long been known for their topical effect in the treatment of chronic granular conjunctivitis. However, absorption of the granulations may cloud the cornea, resulting in blindness. This use has therefore been abandoned as it is too dangerous (Dalziel, 1937, p. 124).

C The seeds contain a toxalbumin, abrin, which agglutinates erythrocytes and causes haemolysis and enlargement of the lymphatic glands (Oliver, 1960). An aqueous

Fig. 6.2. *Mimusops elengi* L.

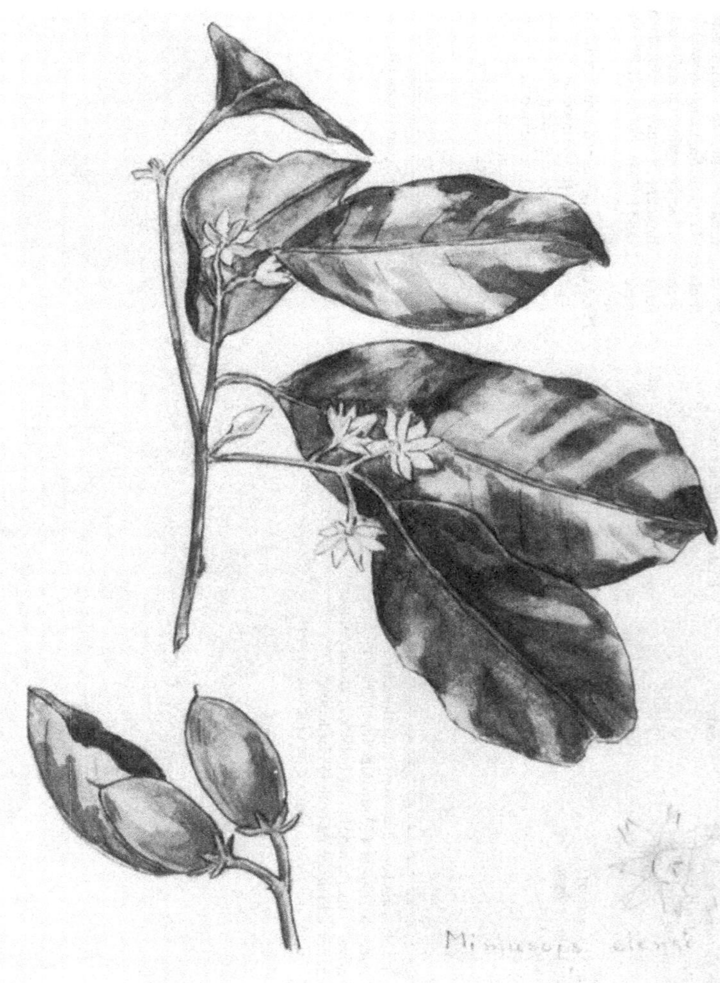

extract of the seeds was found to contain five protein fractions (Desai and Sirsi, 1966). The seeds also contain a steroidal oil, terpene, β-sitosterol, stigmasterol and three indole derivatives, *N*-methyltryptamine, *N*-methyltryptophan and hypaphorine (Desai and Rupawala, 1966; Gupta *et al.*, 1968). The roots, leaves and stems contain glycyrrhizin in higher doses than *Glycyrrhiza* root and are used for its extraction (Boiteau *et al.*, 1964).

P The seeds were found to be abortive and teratogenic (Desai and Rupawala, 1966). Detailed histopathological studies on the fetuses, placentae and uteri of rats and mice showed that oral administration of 150 mg/kg of body weight of the steroid oil of the seeds on Days 1 and 5 post-coitum produced 100% sterility; such administration on Days 1 and 2 or on Day 5 post coitum produced 80% interception. The LD_{50} of the

Fig. 6.3. *Abrus precatorius* L.

aqueous extract in mice was 2 g/kg (Desai and Sirsi, 1966; Desai and Rupawala, 1967). Petroleum ether and alcoholic extracts of the roots (100 mg/kg body weight) administered on Days 1 and 5 post coitum produced 75–80% of anti-implantation effect in rats and mice. The aqueous extracts of the seeds were found to have high cytotoxicity in vitro: 0.072 g inhibited growth by 50% as compared to the control (Dijkman *et al.*, 1966; Oliver-Bever, 1971). Oral administration of 300 mg/kg of body weight of alcoholic extracts of the roots to rats prevented implantation in 100% of the cases (Agrawal *et al.*, 1969, 1970).

As Abrus roots have (like the leaves and stems) a high glycyrrhizin content, the oestrogen antagonistic action of this substance should also be taken into consideration; it seems to point to significant inhibition of the oestrogenic response (Agrawal *et al.*, 1969, 1970). Interference with enzyme systems essential for either oestrogen absorption or oestrogenic stimulation of the target organs, which is mediated through the hypophysis, appears to Kraus and Kaminskis (1969) to be the most likely explanation of this action.

Gossypium spp. MALVACEAE

Cotton plant

L In Nigeria the root is believed to be an emmenagogue with an action like that of ergot; the active principle is in the rootbark (Dalziel, 1937). 'The use of the root to procure abortion is known amongst American negroes but not in the Orient, the knowledge of this property appears to be indigenous to Africa' (Dalziel, 1937, p. 126). Women of the Wolof tribe in Senegal use a decoction of the root to treat amenorrhoea (Kerharo and Adam, 1974, p. 521). The South American Africans used a decoction of the roots as a contraceptive (Brondegaard, 1973).

C The rootbark contains a resin, 1–2% gossypol, vitamin E and a 'not yet isolated' oxytocic and vasoconstrictive compound (Paris and Moyse, 1967, Vol. II, p. 254). Gossypol is an orange-red polyphenolic pigment which is insoluble in water and soluble in ether and chloroform. It is a dinaphthyl derivative with six phenolic OH groups and two aldehyde functions, and exists in three isomeric forms (α, β and γ). The compound is unstable and can be destroyed by heating and thus eliminated from the oil-cake on which it confers a certain toxicity. The oil-cake contains about 1% of gossypol, is rich in protides and contains a small quantity of flavonoid pigments (Paris and Moyse, 1967, Vol. II, p. 254).

P Oral administration of gossypol acetic acid (5–10 mg/kg daily for 12 weeks) induced sterility in male hamsters and rats (Hadley *et al.*, 1981; Hahn *et al.*, 1981). Dead and abnormal sperm were observed in the male tracts and there was a decrease in implantation sites and failure of pregnancy in the females mated to the gossypol-treated males. Similar treatment of male rabbits (with doses varying from 1.25–10 mg/kg for 5–14 weeks) produced poor motility of the sperm (Ridley and Blasco, 1981a, b) but had no other consequences (Chang *et al.*, 1980) and after the treatment was stopped recovery of fertility was observed (Zhou and Lei, 1981).

The emmenagogic properties of the root of *Gossypium* were confirmed by Planchon and Bretin (1946), who also reported that it acts as a haemostatic on the uterus and produces uterine contractions that are similar to the spontaneous contractions of the uterus during childbirth, but without the tetanizing effect of rye ergot. Gossypol is a nerve and cellular poison and has been tested as an anticancer agent (Paris and Moyse, 1967, Vol. II, p. 255).

Studies on gossypol for human contraception were started in China in the early 1970s but were not published in English until 1978. The status of gossypol as a male contraceptive was summarized by Zatuchni and Osborne (1981) and Prasad and Diczfalusy (1981). In 1972 semen analyses were carried out on five male subjects. After administration of gossypol for 35–42 days, at a dose of 60–70 mg daily, four of the men were azoospermic and one was necrospermic (Wu, 1972; Qian *et al.* in Turner, 1981 (from Farnsworth and Waller, 1982); Xue, 1981).

By 1980 more than 8000 men in the Republic of China had been treated with gossypol, gossypol acetic acid (more stable) and gossypol formic acid (there appears to be no significant difference in the action of the three compounds). The usual dose administered was 20 mg daily for 60–70 days followed by a maintenance dose of about 60 mg/week (Liu *et al.*, 1981; Prasad and Diczfalusy, 1981). In 99% of the cases a marked decrease in sperm count, usually to 4 million/ml or less, was noted 2–3 months after the dosing began.

Gossypol was reported to cause a degeneration of the germ cells in the seminiferous tubules of Man and animals when given orally, and to lead to absence of spermatogenesis. It might, however, find application as a vaginal contraceptive, and Waller *et al.* investigated the possibilities of appropriate chemical forms for this purpose (Waller *et al.*, 1980, 1981).

Results of chromosome studies reportedly showed that gossypol was not mutagenic. Gossypol is a lipid-soluble compound which has been shown to be eliminated from the body slowly and cumulative toxicity is a possibility (de Peyster and Wang, 1979). Clinically, the chief side-effect reported was passing weakness occurring early in therapy. The relationship, if any, between gossypol intake and the hypokalemia (Qian, 1980), mild electrocardiographic changes and transient elevations in serum glutamic pyruvic transaminase level observed in some patients remains to be determined (Bingel and Farnsworth, 1980).

Optically active gossypol has also been obtained from *Thespesia populnea* (L.) Soland. ex Corr., (Malvaceae) a coastal tree (King and de Silva, 1968).

Hibiscus rosa-sinensis L. (Fig. 6.4) MALVACEAE

L In nine different countries, including China, India, Indonesia, New Guinea, Trinidad and Peru, the flowers are used as an emmenagogue and in a number of these countries the leaves are used as an oxytocic. The leaves and flowers together are said to be labour-inducing; in Indonesia, New Caledonia and Vietnam the flowers are said to be abortifacient, whilst in India an antifertility effect has been attributed to them (Farnsworth *et al.*, 1983).

C Cyanidin diglucoside has been reported to be present in the flowers by Hayashi (1944). Agrawal and Rastogi (1971) have isolated taraxeryl acetate and sitosterol from the leafy stems. Furthermore, cyanidin, quercetin, hentriacontane, cyanidin diglucoside, calcium oxalate, thiamin, riboflavin, niacin and ascorbic acid have been detected in this plant but none of these has been found to have antifertility activity (Singh *et al.*, 1982).

P Kholkute and coworkers observed in 1972 that extracts of *H. rosa-sinensis* flowers showed antispermatogenic activity in rats. Feeding an ethanol extract of the flowers to male rats for 30 days in doses of 50, 150 and 250 mg/kg produced practically no effect at the lower doses. With the highest dose, however, histological examinations showed that after 30 days there was shrinkage of the seminiferous tubules, destruction of spermatogonial cells and complete disorganization of the testicular tissue. Leydig cells were absent and germinal cells were distinctly affected, whereas Sertoli cells were hardly affected and the seminal vesicles and prostate were unchanged (Kholkute, 1977). Later, the investigators noted in male rats a loss of secretory activity and degranulation of the gonadotrophs in the pituitary gland. The authors suggested that the antifertility results were obtained through inhibition of gonadotrophin release. They noted that the effects were reversible after discontinuation of the treatment (Kholkute *et al.*, 1976a, b, 1977, 1978a, b).

Injection of a single 7.5 mg dose of *H. rosa-sinensis* flower extract subcutaneously in a group of reproductively active *male* bats (*Rhinopoma kinneari*) followed by measurement of the six testicular lactate dehydrogenase (LDH) isozymes showed that lactate dehydrogenase (LDH_x) disappeared from the isozymograms on Days 2–4 but reappeared on Day 5 after the injection. In further investigations (Kholkute, 1977) in which a benzene extract of the flowers and seeds was screened for post-coital antifertility in *female* albino rats (Kholkute and Udupa, 1978a, b) the authors observed that only the flowers were active, the leaves, stem and stembark being

Fig. 6.4. *Hibiscus rosa-sinensis* L.

devoid of activity (Kholkute *et al.*, 1976a). Allied *Hibiscus* spp. failed to show any significant effect. Seasonal variations were important, the highest activity being seen in winter (100%) whilst in spring and summer this was reduced (to 75% and 50%, respectively) (Kholkute and Udupa, 1974; Farnsworth and Waller, 1982). Of various extracts of *H. rosa-sinensis* flowers studied in female albino rats for their antifertility effect, the benzene extract was found to be the most effective (Kholkute and Udupa, 1976).

In an attempt to isolate the active constituents of this benzene extract by fractionation, it was noticed that the ether-soluble portion of the water-insoluble fraction showed significant anti-implantation and abortifacient activities. Thus, after doses of 73 mg/kg body weight daily from the first to the tenth day of pregnancy, no implantations were found on the eleventh day (day of laparotomy). For confirmation of the abortifacient effect ten pregnant rats were given a daily dose of 73 mg/kg from the fourteenth to the twentieth day of pregnancy. In no case was a delivery recorded (Singh *et al.*, 1982).

The antifertility effect of the flower petals of *H. rosa-sinensis* had formerly been reported by Nadkarni and Nadkarni (1954) and Batta and Santhakumari (1970), who noted an antifertility activity of 80% (as shown by an 80% reduction in implantation sites on the tenth day of pregnancy) of the benzene extract in fertile female rats and mention an effect on the sex ratio of the pups born to experimental animals treated with the ethanol extract.

Rats receiving 50–250 mg/kg doses of benzene extract daily for thirty days showed follicular atresia and reduced ovarian, uterine and pituitary weights; this demonstrated anti-oestrogenic as well as oestrogenic activities of the extract (Kholkute *et al.*, 1976a). Experimental studies could thus show the anti-oestrogenic, antifertility, anti-implantation and oestrous-cycle disrupting effect of the benzene and ethanol extracts given orally to rats (Farnsworth *et al.*, 1983).

Some clinical tests confirmed the contraceptive value of *H. rosa-sinensis* (Tewari, 1974; Tewari *et al.*, 1976).

Achyranthes aspera L. AMARANTHACEAE

This plant, already described in Chapter 2, is a common weed, very variable in height (36–180 cm) and sometimes woody. (These details are mentioned because it is reported by Pakrashi *et al.* (1975b), Pakrashi and Bhattacharya (1977) and by Kamboj and Dhawan (1982) that an extract of the stembark prevents pregnancy. Perhaps the botanical identity should be controlled.)

Fifty mg of the benzene extract of the plant tested by these authors per kg body weight inhibited pregnancy by 100% when given orally on Days 1 or 6 post coitum in mice. The benzene extract was also active in rabbits on Day 8 of pregnancy. In the 1–7 days post-coitum schedule the benzene extract and a 50% ethanol extract (200 mg/kg) were inactive in rats (Pakrashi *et al.*, 1975b; 1977). Administration of 75 mg/kg in mice every twenty-first day for 6 months had no toxic effects, nor had a single dose of 1 g/kg body weight in mice (Pakrashi *et al.*, 1975b; Pakrashi and Bhattacharya, 1977).

The benzene extract was reported to be devoid of oestrogenic, anti-oestrogenic and androgenic properties in mice (Kamboj and Dhawan, 1982). The active component is believed to be oleanolic acid 3β-glucoside (Pillay *et al.*, 1977).

Sapindus trifoliatus L. i. SAPINDACEAE

This plant is also described in Chapter 2.

C The plant contains a saponin (Chopra *et al.*, 1956, p. 221).

P In rats ethanolic extracts produce 80% inhibition of implantation when administered to rats 1–7 days post coitum in 100 mg/kg doses and 100% inhibition when administered in 500 mg/kg doses. (Taking into consideration the fact that 10–20 mg/kg produce a dose-dependent decrease in blood pressure and heart rate (Singh *et al.*, 1978), this treatment seems less recommendable.) A methanol fraction of the ethanol extract had oral antifertility action at doses of 25 mg/kg when given on Days 4 and 5 of pregnancy (Bodhankar *et al.*, 1974; Garg *et al.*, 1978; Kamboj and Dhawan, 1982).

Garg *et al.* (1980) took out a patent for the use of saponin- (including oleanolic aglycones) bearing plants such as *Sapindus trifoliatus*, *Blighia sapida*, *Samaena saman* based on their spermicidal activity.

Momordica charantia L. CUCURBITACEAE

Bitter gourd

(See also Chapter 7 (Hypoglycaemic plants).)

P Antifertility activity of this plant has been studied in both males and females. Concerning the spermatocidal action, Dixit *et al.* (1978) observed that an ethanol extract of *M. charantia* fruits administered orally or subcutaneously to adult male gerbils (*Meriones hurrianae*) in doses ranging from 200 to 400 mg/kg during two weeks disrupted spermatogenesis and reduced testicular weights. The seminal vesicles and prostate did not seem to be significantly affected by this administration.

Chronic administration of an ethanol extract of *M. charantia* fruit (1.75 g daily) orally to male dogs for 60 days caused testicular lesions resulting in mass atrophy of the spermatogenic elements: 75 $\pm$ 8.9% of the seminiferous tubules completely lacked step 1–8 spermatids. Serum enzymes, serum protein, cholesterol, bilirubin, blood sugar and blood urea levels were within normal ranges. Haematological studies showed no deviations except a slight rise in total leucocyte count, which has been attributed to emotion or strenuous exercise at the time of blood sample collection (Dixit *et al.*, 1978). The authors conclude that the extract induced a state of infertility without altering general metabolic activities.

The roots have been found to have an abortifacient effect in females (Jamwall and Anand, 1964). Several constituents of the fruits, such as charantin, 5-hydroxytryptamine, diosgenin and β-sitosterol, were observed to act as stimulants of the uterus. In the rabbit charantin has been found to cause uterine haemorrhage and abortion and in rats it inhibited foetal development (Jamwall and Anand, 1964). The cucurbitacin-like triterpenes also found in the plant have been shown to possess a potent cytotoxic activity (Weniger *et al.*, 1982).

Recently, a Mexican plant, **Montanoa tomentosa** (Compositae) with a non-steroidal fertility regulating substance (zoapatanol, an oxepane diterpene) for use by women has attracted much attention. It was the subject of over thirty patents between 1971 and 1982 (Farnsworth *et al.*, 1983). Unfortunately, it is not found in tropical West Arica. In India, the alkaloid vasicine from **Adhatoda vasica** (related species in tropical West Africa) has been extensively tested for the termination of mid-trimester pregnancy. Intra-amniotic instillation of 60 mg or more produces abortion within 48 h, with minimal side-effects (Atal, 1980). The presence of vasicine in African species has not been reported so far.

Enzyme-inhibiting contraceptives and spermicides of plant origin

In order to find more selective sperm inhibitors, Farnsworth and Waller (1982) have investigated sperm-specific enzyme inhibitors. Numerous enzyme systems in viable sperm are susceptible to inhibition, for example those involving glycolysis, energy production and myosin contraction. Investigations into the biochemistry of fertilization have shown that the enzymes of major importance for fertilization are acrosin and hyaluronidase and that 'plant derived acrosin inhibitors are few, they include soybean and other trypsin inhibitors as well as a large number of monosaccharides' (Farnsworth and Waller, 1982). These authors quote recent publications by Kennedy *et al.* (1982) and van der Ven *et al.* (1982) reporting the inhibition of acrosin by gossypol and another, by Lee and Malling (1981), reporting the selective inhibition of sperm-specific lactate dehydrogenase (LDH_x) by gossypol. In addition they mention as a hyaluronidase inhibitor the essential oil of the root of *Curcuma domestica*, a species also cultivated in West Africa.

Active spermicidal compounds encompass a number of chemical structures. A majority of plant-derived spermicides are triterpene saponins and, more rarely, some are steroid saponins. Since all saponins show surface-active properties, it has been suggested that the most likely mechanism for their action is a disruption of the sperm plasma cell membrane (David and Heftman, 1981). A currently marketed spermicide, 'nonoxynol-9', has been based on this theory (Ad Elbary and Nour, 1979; Farnsworth and Waller, 1982).

In females using contraceptive steroids, changes have been reported in red blood cell enzymes requiring B vitamins for their coenzymes and in steroid-metabolizing enzymes of the uterus (Briggs and Briggs, 1981).

Plants with an action on menstruation

Plants used to treat dysmenorrhoea can act in different ways. Plants such as *Polygonum* spp., which contain haemostatic flavonoids, are used for the treatment of menorrhagia whereas other plants, such as *Alchornea cordifolia* and some members of the Solanaceae, which contain atropine, papaverine and yohimbine, are used because of the antispasmodic action of these alkaloids. Sometimes progesterone or androgens are used.

In some forms of dysmenorrhoea and in amenorrhoea, plants such as *Petroselinum* which contain essential oils (apiol) that produce contractions of the uterus and are abortive in high doses are sometimes given (Paris and Moyse, II, 1967, p. 476). Usually, however, oxytocic agents based on oestrogenic hormones are used. These plants can act as abortifacients in high doses. In folkloric medicine their mechanism of action is mostly unknown and the same lack of information as is evident for the plant oestrogens also applies in this field. A number of these plants are described as antifertility plants (*Achyranthes aspera*, *Diospyros* spp., *Grewia* spp. (Paris and Theallet, 1961), *Gossypium herbaceum* and spp. and *Withania somnifera*, etc.). Some of these plants which act as emmenagogues exert their hormonal action via the hypothalamus and the pituitary. This property was reported as early as 1950 (Gupta *et al.*, 1950) for one West African plant (*Pergularia daemia*).

Pergularia daemia (Forsk.) Chiov. syn. (*P. extensa* N.E. Br., *Daemia extensa* R. Br.)
ASCLEPIADACEAE

L The plant is believed to have emmenagogic properties. In Ghana the leaves are given to women immediately after childbirth. In Nigeria the latex or a poultice of the leaves is applied to boils and abscesses, a use also known in India. The juice of the leaves is also used to treat sore eyes (Dalziel, 1937, pp. 388–9).

C The plant contains a bitter resin, two bitter principles and a glycoside (Chopra *et al.*, 1956, p. 188).

P The glycoside has a physiological action on the uterus similar to that of the pituitary gland.

Aloe barbadensis Mill. c. syn. *vera* L., *A. vulgaris* Lam., *A. indica* Royle)
LILIACEAE

L In Nigeria the juice of the leaves of *A. schweinfurthi* Bak. is used in the treatment of guinea worm infestation and in Ghana it is used for the treatment of vitiligo (Dalziel, 1937, p. 476). In India extracts of *A. barbadensis* are considered a tonic in small doses and a purgative, emmenagogue and anthelmintic in larger doses. The dried juice is cathartic and is used to relieve constipation. The pulp is used to reduce or prevent menstruation (Chopra *et al.*, 1956, p. 13).

C The juice of the leaves contains anthraquinone derivatives (emodin, aloin, barbaloin, isobarbaloin and resins (Chopra *et al.*, 1956, p. 13; Shah and Mody, 1967). Whole leaves, the rind and the pulp contain oxidases, catalases and sugars.

P *A. barbadensis* powder has been widely used as a purgative and was found to have a beneficial effect in the healing of thermal burns and of radiation therapy (Singh *et al.*, 1973). When given in doses of 60 mg/kg intragastrically to female rabbits this powder increased not only the fertility rate but also the litter size of the animals (Sharma *et al.*, 1972).

Aloe compound was reported to be very useful in cases of disturbed menstrual function and functional sterility (Bhaduri *et al.*, 1968; Garg *et al.*, 1970). This was confirmed by Gupta (1972), who found in 250 cases of sterility that Aloe powder improved fertility in 85% of the cases and that menstrual functions improved in 44.6% of them. However, when plants were tested for interceptive action, it was found that a 50% ethanol extract of the leaves, in contrast to all other extracts tested, had a strong interceptive action when given in high doses (100–200 mg/kg) on Days 1–7 post coitum (Casey, 1960; Gupta *et al.*, 1971).

Plants with an action on lactation

Plants with latex, such as *Euphorbia hirta* and other *Euphorbia* spp., and *Alstonia scholaris*, often have the folkloric reputation of acting as galactogogues. These properties are still reported by reviewers today. Although the constituents and pharmacological properties of these plants have been studied in more detail in the last few years, a scientific confirmation of their galactogogue action has not been reported.

It has long been known that prolactin, the lactation stimulating hormone of the anterior pituitary can increase lactation. There is, however, often a certain reluctance nowadays to manipulate the hypothalamus–pituitary axis because of the other pharmacological effects produced.

Sarcostemma viminale R.Br. ASCLEPIADACEAE

L In some South African tribes this plant is used as a galactogogue for both women and cows (Watt and Breyer-Brandwijk, 1962).

C The stems contain 2% of a mixture of five pregnane glycosides. One of two recently reported genins contains a dihydroacetone group like cortisol but has different stereochemistry at C-14 and C-17 (Schaub *et al.*, 1968; Stöckel *et al.*, 1969).

P Not only does this plant seem to act via the adrenal–pituitary axis but convulsions similar to those produced by strychnine, which result in paralysis and death, have been observed in cattle (sheep) after they have eaten the plant (Steyn, 1937). Therefore this plant should not be considered for human or veterinary applications!

Vitex agnus castus L.c. VERBENACEAE

C The fruit contains an essential oil, cineol, flavonoids and the glycoside aucubin and its hydroxybenzoic ester agnuside. These melanin-forming glycosides are unstable and polymerize to form black pigments when exposed to the air (Steinegger and Hänsel, 1968, p. 497; Paris and Moyse, 1971, Vol. III, p. 255).

P *V. agnus castus* has been reported to have a luteinizing action as a result of inhibition of the gonadotrophic action of the posterior lobe of the pituitary. Extracts of the fruits have been on the market for the treatment of dysmenorrhoea and as a galactogogue (Steinegger and Hänsel, 1968, p. 497).

Other Verbenaceae like **Verbena officinalis** are also said to have galactogogue properties. The active principle is aucubin (a glycoside of a lactonic monoterpene). It has weak parasympathomimetic properties and causes contraction of the smooth muscle of the uterus (Steinegger and Hänsel, 1968, pp. 497–8). In mammals *V. officinalis* produces a strong and prolonged milk secretion. The twigs, leaves and flowers yield verbenalin (McIlroy, 1951, p. 110).

Ipomoea mauritiana Jacq. syn. (*I. digitata* of F.T.A.) CONVOLVULACEAE

L In Nigeria the whole root is used as a purgative and a galactogogue (Dalziel, 1937,

p. 438). The same uses are mentioned for the very closely related *I. digitata* L. syn. (*I. paniculata* R.Br.) in India, which is also claimed to be an aphrodisiac (Chopra *et al.*, 1956, p. 142).

C The roots of the Indian species contain 1.3% of fixed oil (glycerides of oleic, palmitic, linoleic and linolenic acids). In addition β-sitosterol and a heteroside, paniculatin, which is soluble in water and thermostable, have been isolated (Matin *et al.*, 1969).

P The heteroside of the Indian species acts as a stimulant of the smooth muscles of the myocardium, bronchi and bowel and is also oxytocic; when it is given intraperitoneally, the LD_{50} in mice is 867.4 mg/kg (Mishra and Datta, 1962; Matin *et al.*, 1969).

Kigelia africana (Lam.) Benth. syn. (*K. pinnata* (Jacq.) DC., (*K. aethiopica* Decne, *K. abyssinica* Rich., *K. elliotii* Sprague, etc.) (16 synonyms!) BIGNONIACEAE
Sausage tree

L In South Africa the fruits are used as a dressing for ulcers or to increase the flow of milk in lactating women (Watt and Breyer-Brandwijk, 1962). In Ghana the fruit and roots are boiled together with the 'tassels' of plantain flowers as a 'women's remedy' (Dalziel, 1937). In Northern Nigeria the fruit is used in some districts as a purgative, whereas in other regions it is used to treat dysentery (Dalziel, 1937, p. 443). In Cape Verde the fruit is rubbed on the breast of young girls to enhance their development. The fruit also has many superstitious uses. In Kenya a decoction of bark and leaves is drunk as an abortifacient. The fruit is commonly added to beer and claimed to be an aphrodisiac (Kokwaro, 1981). The unripe fruit is said to be very toxic.

C From the fruits and leaves El Sayyad (1981) isolated and identified the flavonoids 6-hydroxyluteolin-7-O-glucoside, luteolin-7-O-glucoside, luteolin and quercitin. From the roots Govindachari *et al.* (1971) have obtained dihydro-isocoumarins, lapachol and sterols and Alamelu and Bhuwan (1974) have reported the presence of iridoid glycosides.

Other plants growing in Nigeria that are reported to act as galactogogues are **Allophyllus africanus** Beauv. (Sapindaceae) (the leaves are also used to treat piles) and **Alternanthera repens** (L.) Link. (Amaranthaceae) (the leaves of which are also used as an abortifacient). The presence of flavonoids and the absence of steroids, terpenes, alkaloids, saponosides, tannins, quinones and cyanogenetic glucosides has been reported for the bark of the Congolese specimen of *Allophyllus africanus* (Bouquet, 1972). The chemistry of *Alternanthera* appears not to have been studied.

Trigonella foenum graecum L. (Fabaceae) is considered a galactogogue in the Sudan (Ayoub and Svendsen, 1981).

Indirect stimulation of lactation has also been obtained with dried thyroid gland or thyroxine via the pituitary gland (Robinson, 1947; Naish, 1954).

Jasminum sambac (L.) Ait. c. OLEACEAE
The flowers are used in India to stop the secretion of milk in women in the puerperal state after childbirth (in cases of threatened abscess) by applying them unmoistened to the breasts. The roots are claimed to be emmenagogic (Chopra *et al.*, 1956, p. 144).

C An iridoid glycoside as well as jasminin, quercetrin, isoquercetrin, rutin, quercetin-3-dirhamnoglycoside, kaempferol-3-rhamnoglycoside, mannitol, α-amyrin, β-sitosterol and sucrose have been reported to be present in the leaves of this plant (Ross *et al.*, 1982). The flowers contain an essential oil.

(c) Male sex hormones (androgens)

The main androgen is testosterone. It is secreted by the interstitial cells of the testis under the influence of luteinizing hormone and is required for spermatogenesis. In fact both androgen secretion and spermatogenesis are controlled by the hypothalamic–pituitary axis. In the liver a large proportion of the testosterone is degradated into androsterone and etiocholanone. These two substances and testosterone glucoronide are eliminated by the kidneys. The androgens are also secreted from the adrenal cortex and the ovaries. They cause retention of water and of nitrogen, sodium, calcium, chloride, sulphate and phosphorus ions and also increase protein metabolism (Burgen and Mitchell, 1972; Lechat *et al.*, 1978).

Testosterone can only be administered by injection as it is absorbed by the digestive mucous membranes and immediately inactivated by the liver when given perorally. Dissolving testosterone in oil and making chemical substitutions in the molecule can to a certain extent slow down the hepatic catabolism.

Exogenous oestrogen produces an anti-testosterone effect, mainly owing to its action on the hypothalamus and pituitary (Lechat *et al.*, 1978, pp. 548–51). The need to administer the hormone parenterally explains the dearth of published reports and indications in folkloric medicine of the use of plant androgens. Aphrodisiacs seem to be the main indication mentioned. The use of a leaf decoction of *Lonchocarpus cyanescens* for the treatment of venereal disease and semen insufficiency has, however, been reported (Ainslie, 1937).

Apart from being given in replacement therapy when normal secretions are reduced or absent, androgens are used to stimulate anabolic activity, encouraging protein anabolism, in the treatment of osteoporosis and sometimes to inhibit the growth of breast carcinoma. For their use as anabolic agents the difficulty lies in obtaining compounds with high anabolic and low androgenic activity, as in female patients there is the danger of masculinization (Burgen and Mitchell, 1972). A West African plant used in the treatment of osteoporosis is *Cissus quadrangularis* (see Chapter 3).

II Thyroid hormones

The thyroid gland plays an important part in the organism as it stimulates metabolic activity and also controls growth. The two hormones secreted by the gland are thyroxine and triiodotyronine, and they are present in a proportion of 50 : 1; they are both iodine-containing amino acids and are synthesized in the gland itself, which takes up iodine from the blood (Burgen and Mitchell, 1972).

Hypoactivity of the thyroid gland causes myxoedema and a 20–30% decrease in basal metabolism. Lack of iodine is often the basis of this insufficiency and treatment is usually effected by administration of thyroxine or thyrotrophin (anterior pituitary hormone) or of components rich in organic or inorganic iodine, for example algae like *Fucus vesiculosis* L., which contains 0.03–0.1% iodine (Chesney and Webster, 1928; Steinegger and Hänsel, 1968). Hyperactivity of the thyroid function, e.g. in patients with Basedow's disease, can be inhibited by drugs which inhibit the uptake of iodine by the gland. Thiocyanates inhibit this uptake and have a weak inhibitory action on iodide binding (Burgen and Mitchell, 1972, p. 148). Temporary treatment is obtained with iodine as it causes retention of thyroxine in the gland and so it is not released into the blood. (This effect is used in preparing hyperthyroid patients for thyroidectomy.) Other drugs which inhibit, by substrate competition, the peroxidase enzyme which releases iodine from circulating iodides and, in addition, inhibit the enzymes responsible for iodination of tyrosine and coupling of the iodinated derivatives are imidazoles and thiouracils (Turner and Richens, 1978, p. 168; Kagihara, 1980).

Other substances which reduce hyperactivity are extracts of **Brassica oleraceae** L. and other **Brassica** spp. as had already been observed in 1928. In 1949 an active 'goitrogenic' principle was isolated from the turnip, *B. napa*; it was called goitrine and was found to be a glucoside of 1,5-vinyl-2-thio-oxazolidone (Astwood *et al.*, 1949). Subsequently it was shown that the seeds, roots and aerial parts of many *Brassica* spp. contain progoitrine. Through hydrolysis by enzymes such as myrosinase, a common enzyme in the Cruciferae, but also under the influence of bacteria present in the digestive tracts of animals, progoitrine liberates the isothiocyanate of 2-hydroxy-3-butenyl. Cyclization of this leads to the active principle, goitrine (Virtanen, 1961). The mechanism of action of goitrine is believed to be based on its intervention in the biosynthesis of thyroxine and not on the accumulation of iodine in the thyroid gland (Murti *et al.*, 1964).

Diverse polyphenols are also capable of fixing iodine and of removing this element from the thyroid gland. The goitrogenic effect of *Glycine soya* is thought to be related to the presence of phytoagglutinins, which prevent normal intestinal absorption of thyroxine. Other plants which contain phenols of the flavonoid type such as quercetin, rutin and catechin also have an action on the thyroid gland (Steinegger and Hänsel, 1968).

An indirect inhibitory effect on thyroid secretion by *Lithospermum ruderale* and *L. officinale* (not in W. Africa) has been reported (Train *et al.*, 1941; Kemper and Loeser, 1957); they act via the thyrotrophic (and also gonadotrophic) hormone of the pituitary gland.

In West Africa it is mainly *Arachis hypogeae* which provides the goitrogenic principle although some Capparidaceae and **Carica papaya** seeds are likely to have a similar effect.

Arachis hypogeae L. FABACEAE

Groundnut, peanut

(See also Chapter 2.) A goitrogenic and an oestrogenic factor have been isolated from the seeds of this species.

P The goitrogenic principle was found to be insoluble in oil and thermostable as it is retained in the oil-cake and in the roasted seeds (Greer and Astwood, 1948; Buxton *et al.*, 1954; Busson, 1965; Adrian and Jacquot, 1968). If the normal diet of the rat is supplemented by 20% with the oil-cake, the weight of the thyroid gland is increased from 9 to 24 mg whilst its content of mineral iodine goes up from 5 to 20.5%. Addition of 1 g of water per day per animal completely re-establishes the thyroid metabolism. The goitrogenic factor is believed to be located in the chromogenic tegument of the nut. Arachoside and glycosides isolated from the nuts inhibit the formation of inorganic iodine and thyroxine and result in a major increase in the urinary secretion of iodine and phenols (Mudgal *et al.*, 1957, 1958).

The oestrogenic factor is soluble in oil and is retained after refining. Introduction of refined peanut oil to form 10% of the food ration of immature mice increases uterine weight from 9.5 to 15.9 mg (refined olive oil produces an increase to 16.7 mg) (Booth *et al.*, 1960). A substance antagonistic to aldosterone which modified sodium reabsorption and urinary secretion of sodium and potassium could also be detected in arachis oil; 0.05 ml of the oil largely antagonized the action of 10 μg aldosterone (Kumar *et al.*, 1962).

Kumar *et al.* (1962) noticed in earlier investigations that the effects of desoxycorticosterone acetate on the urinary secretion of sodium and potassium ions in adrenalectomized animals were abolished when the 'deoxycortone' was dissolved in arachis oil. They therefore undertook a systematic investigation of the aldosterone-antagonistic action of various vegetable oils. Consistent and marked antagonism was seen only with arachis oil at all doses. The potent aldosterone antagonist is believed to be contained in the unsaponifiable portion of the oil (Kumar *et al.*, 1962).

Manihot esculenta Crantz. syn. (*M. utilissima* Pohl.) EUPHORBIACEAE
Cassava. manioc

L In West Africa this species occurs mainly as the bitter cassava (dark root) but a sweet variety is cultivated in the northern parts of West Africa. Cassava is a major foodplant which requires cooking as it contains a thiocyanate which is rendered inert by heat. Many varieties exist. The meal is called *fufu*. *Gari* is a coarse powder obtained by dry methods of scraping, pounding, etc. A poultice, often used as a substitute for linseed meal, is put on burns, ulcers, etc. (Dalziel, 1937, pp. 150–4).

C Cassava is rich in starch and poor in nitrogen: there is 39% starch, 1.3% protein and 0.5% fat in Ghanaian cassava root after cooking. Linamarin is the source of the thiocyanate found in manioc meal.

P Antithyroid effects of cassava and the thiocyanate were observed in rats. In some areas of Zaire endemic goitre and cretinism was attributed to endogenous formation of thiocyanate resulting from ingestion of manioc; this aggravated iodine deficiency in these areas (Ermans, 1979). In Nigeria a degenerative ataxic neuropathy (polyneuritis of variable location) with deafness and bilateral optic atrophia in 320 patients could be traced to a chronic intoxication by hydrocyanic acid. The acid resulted from hydrolysis of cyanogenetic glycosides contained in manioc, which is a basic food in the country (Osuntokun *et al.*, 1969, 1970). Thus chronic intoxications can result from regular ingestion over long periods from boiled manioc. It had been observed that increasing thiocyanate in the blood of humans and rats aggravated the loss of iodine from the thyroid. Renal elimination of thiocyanate becomes apparent when the blood level exceeds 10–15 μg/ml, 'which was not adequate for prevention of thyroid anomalies' (Ermans, 1979; Ermans *et al.*, 1980). Ermans *et al.* (1980) recommend, in endemic goitre studies, completion of the iodine level examination by determination of thiocyanate levels in the blood and urine and checking of the intake of manioc.

7

Oral hypoglycaemic action

Insulin, a hormone secreted by the β-cells of the Islets of Langerhans in the pancreas, is of utmost importance for the correct metabolism of carbohydrates and fats. It induces the oxidative breakdown of glucose, has a stimulating effect on the synthesis of liver glycogen from glucose and inhibits the formation of liver glycogen from protein and fat.

Diabetes, a condition characterized by hyperglycaemia, is caused by insufficient secretion of insulin or to insufficiency of its peripheral efficacy. Excess of insulin leads to hypoglycaemia; this is easily counteracted if dextrose or a few lumps of sugar are taken at once.

Modern investigations into the biochemistry of diabetes show that its causes and the sites of intervention in the biochemical processes are diverse. Somatostatin, the pituitary and sex hormones, corticosteroids, prostaglandins and vascular modifications of the pancreas can all be involved, together with a straightforward inadequacy in insulin production (Randle *et al.*, 1963; Gupta *et al.*, 1966; Burkhard *et al.*, 1968).

A number of plants have constituents which have antidiabetic properties when taken orally (Oliver-Bever and Zahnd, 1979). There is great diversity in the nature and action of these constituents, but a number of them do seem to belong to certain chemical groups, such as sitosterol glycosides, alkaloids, sulphur oils and flavonoids, and in this chapter I have assembled the plants described into a few groups based on their possible active chemical constituents while at the same time taking into consideration their relation to, or identity with, hypoglycaemic plant constituents found elsewhere. Some of the plants appear to have several constituents, each of a different chemical group, which may all contribute to their activity.

In each chemical group some details are given on one or a few plants only and the others are listed in Tables 7.1–7.6. Since a major hypoglycaemic action has been found in several currently cultivated, introduced plants, which are often foodcrops, these have also been included.

In an extensive comparative evaluation of Indian antidiabetic plants, including clinical trials, Chaudhury and Vohora (1966) recommended further tests, in order of prioıity on *Momordica charantia*, *Gymnema sylvestre*, *Syzygium cumini* and *Coccinia indica*. These plants and *Syzygium guineense* varieties (Fig. 7.1) figure also amongst the plants with possible hypoglycaemic action selected on the basis of the literature for West Africa and they will be treated in more detail hereunder.

I Plants containing hypoglycaemic phytosterin glycosides

(See Table 7.1.)

The active constituent from the rootbark of hypoglycaemic *Ficus* species (*Ficus glomerata* and *F. religiosa*) was found in India to be β-sitosterol D-glycoside; this had a peroral hypoglycaemic effect in fasting and alloxan-diabetic rabbits and in

Fig. 7.1. *Syzygium guineenses*. var. *littorale* Keay.

pituitary-diabetic rats comparable to the effect of tolbutamide (Modak and Rajarama Rao, 1966; Ambike and Rajarama Rao, 1967; Vohora, 1970).

Momordica charantia L. CUCURBITACEAE
African cucumber or balsam pear

L The African cucumber, *M. charantia*, and the allied **M. foetida** are used in many countries in folk medicine as a remedy for diabetes.

C A hypoglycaemic principle could be isolated from *M. charantia* fruit and was called charantin. It appeared to be a phytosterolin (Rivera, 1941, 1942). Sucrow (1965) found that charantin is a mixture in equal parts of β-sitosterol-βD-glucoside and $\Delta^{5.25}$ stigmastadien 3β-ol. 5-Hydroxytryptamine and α-amino-butyric acid have also been reported to be present in the fruit (Dhalla *et al.*, 1961). The fruit of *M. foetida* contains foetidin, which has been isolated and characterized as a chromatographically homogeneous product consisting also of β-sitosterol-βD-glucoside and $\Delta^{5.25}$-stigmastadien 3β-ol-glucoside in equal parts and was thus shown to be identical with charantin (Olaniji, 1975; Olaniji and Marquis, 1975). Recently, a polypeptide has been isolated from the fruit, seeds and tissue of *M. charantia*. It has been called p-insulin and amino-acid analysis has revealed that it has 166 residues composed of 17 different amino acids and a molecular weight of approximately 11000. Methionine was the only amino acid not also found in bovine insulin (Khanna *et al.*, 1981). Immunoassays did not, however, indicate any cross-reaction between p-insulin and bovine insulin (Weniger *et al.*, 1982).

P Extensive investigations have shown that an extract of the dried fruits of *M. charantia* has marked hypoglycaemic properties, giving good results in clinical trials (Vad, 1960). Charantin has a more potent hypoglycaemic action than tolbutamide in equivalent doses. The action is less pronounced in depancreatized cats but seems to subsit nervetheless, indicating the existence of a slight extra-pancreatic as well as a pancreatic action. Doses of 400 mg/kg given intraperitoneally are not lethal in mice. Charantin probably does not account for all the properties of the fruit as the favourable results obtained from 50–60 ml of fresh juice consumed daily could not be attributed only to the few mg of charantin the juice contained (Lotlikar and Rajarama Rao, 1960, 1966; Kulkarni and Gaitonde, 1962). A consistent hypoglycaemic effect was produced when 0.5 g of dried powdered fruit of *M. charantia* per kg of body weight was given at 10-h intervals to normal and alloxan diabetic rabbits (Athar, 1979; Akhtar *et al.*, 1981).

Subcutaneous administration of the polypeptide p-insulin isolated from the fruit produced hypoglycaemic activity in diabetic gerbils, langurs and humans. In juvenile diabetic patients the peak effect was observed after 4–8 h as compared to 2 h for bovine insulin (Khanna *et al.*, 1981).

Foetidin from *M. foetida* has been shown to lower the blood glucose level in normal rats but, contrary to results published earlier, Marquis *et al.* (1977) noticed no significant effect in diabetic animals.

Phytosterin glycosides are found together with an alkaloid in *Syzygium cumini* (also containing tannins) and with sulphur compounds in *Phaseolus vulgaris*.

Table 7.1. *Plants containing hypoglycaemic phytosterin glycosides*

Plant Family	Part used	Possible active constituent(s)	Observed activity	References
Hygrophila auriculata (Schum.) Heine syn. (*H. spinosa* And., *Asteracantha longifolia* Nees, *Barleria auriculata* Schum., *B. longifolia* L.) Acanthaceae	Whole plant	The root contains 0.25–0.33% lupeol (= hygrosterol) and an essential oil. The seeds contain 4 steroid glucosides (phytosterols), 2 water-soluble bases, 23% semi-drying oil and enzymes (diastase, lipase and portease). The leaves contain the alkaloid vasicine (= peganine)	Oral hypoglycaemic action in cats and rabbits; through decoction of whole plant, it is mainly diuretic	McMillan (1954) Basu and Rakhit (1957) Chatterjee and Shrimani (1957) Govindachari *et al.* (1957) Martindale (1958) Dhar *et al.* (1968, p. 365)
Momordica charantia L. *M. foetida* Schum. & Thonn. (African cucumber, Bitter gourd) Cucurbitaceae	Fruit	Phytosterin glycosides: charantin, momordicin, foetidin. (Foetidin = βsitosterol, βD-glucoside plus $\Delta^{5.25}$ stigmastadiene-3β-ol-glucoside	50 mg/kg doses of charantin reduce hyperglycaemia by 42% in rabbits. Charantin possesses pancreatic and extrapancreatic action and is more active than tolbutamide. It has slight antispasmodic and anticholinergic effects and controls but does not heal diabetic patients (160 cases)	Lotlikar and Rajarama Rao (1960, 1966) Kulkarni and Gaitonde (1962) Chatterjee (1963) Gupta (1963a) Sucrow (1965) Olaniji (1975) Olaniji and Marquis (1975) Marquis *et al.* (1977) Athar (1979) Akhtar *et al.* (1981)

Phaseolus vulgaris L. Kidney or haricot bean *P. multiflorescens* Papilionaceae	Bean-husks	Hypoglycaemic principle in aqueous infusion called phaseolan, stigmasterin, querceturon (glycoside: quercetin + glucuronic acid), sulphur compound, indole acetic acid oxidase inhibitor	An aqueous infusion reduces hyperglycaemia in mild diabetics and restores carbohydrate balance	Hartleb (1932) Lyass and Vovski (1932) Sachser (1961)
Syzygium cumini (L.) Skeels syn. (*S. jambolanum* DC., *Eugenia jambolana* Lam., *E. cumini* Merr.) (Jambul, Java plum) Myrtaceae	Seeds	Hypoglycaemic principle, antimellin (glycoside). Known constituents in seeds: phytosterin, jambosine (alkaloid), jambolan, essential oil, galli and ellagi-tannins. Constituent in flowers and fruits: cyanidine-2-rhamnoglucoside	In cats and rabbits the alcoholic extract has oral hypoglycaemic action comparable to that of tolbutamide. In alloxan-diabetic rats the purified principle abolishes hyperglycaemia and glycosuria within seven days. A single dose of an aqueous extract of the seeds reduces blood sugar by 15–25% in 4–5 h	Mercier and Bonnafous (1940) Venkateswarlu (1952) Sepaha and Bose (1956) Nair and Subramanian (1962) Mukherjee *et al.* (1963) Shrotri *et al.* (1963) Sengupta and Das (1965) Sicognau-Jagodzinski *et al.* 1966) Jain and Sharma (1967) Laroche Navaron (1968)

II Plants containing hypoglycaemic alkaloids
(See Table 7.2)

Catharanthus roseus (L.) Don (Fig. 7.3) APOCYNACEAE
Madagascar periwinkle
In the folk medicine of several countries, such as the Philippines, Jamaica, South Africa, India and Australia, an infusion of the leaves is given in the treatment of diabetes.

American research workers, anxious to study the hypoglycaemic properties of this plant, lost a great many rats which had been given an extract of the leaves through

Fig. 7.2. *Securinega virosa* (Roxb. ex Willd.) Baill.

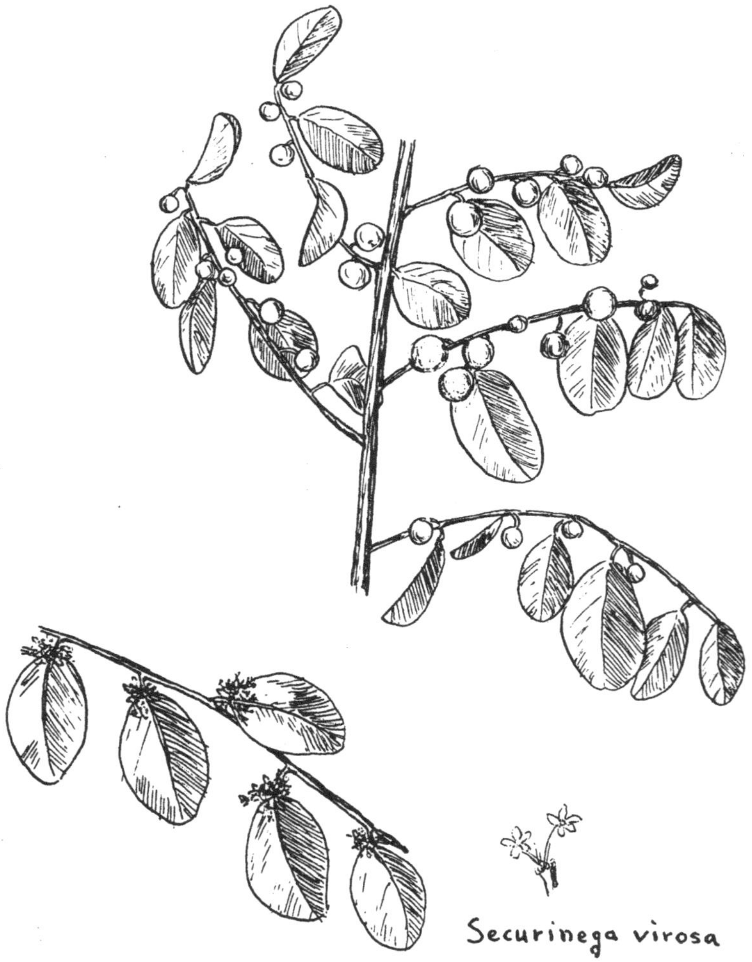

Pseudomonas infection. Investigation revealed that the rats had forfeited their resistance to a severe fall in the number of their lymphocytes. This discovery led to research into the treatment of leukemia by *Catharanthus* alkaloids (Noble *et al.*, 1958).

Research on the hypoglycaemic effect was not abandoned, however, and the different *Catharanthus* alkaloids were administered in doses of 100 mg/kg to rats with fasting hyperglycaemia. Three of the alkaloids, leurosine, vindoline and vindolinine, proved more potent than tolbutamide at equivalent doses, whilst three other alkaloids had a less pronounced effect (see Table 7.2). Leurosine alone had a slight effect on cell division. Before possible use in diabetes the hypoglycaemic alkaloids will have to be completely isolated from the others as many of these are cytotoxic (Svoboda *et al.*, 1964).

Fig. 7.3. *Catharanthus roseus* (L.) Don.

Table 7.2. *Plants containing hypoglycaemic alkaloids or amino acids*

Plant Family	Part used	Constituents	Observed activity	References
Catharanthus roseus G. Don syn. (*Lochnera rosea* (L.) Reichb., *Vinca rosea* L.) (Madagascar periwinkle) Apocynaceae	Leaves	Hypoglycaemic alkaloids: catharanthine (HCl) (1), lochnerine (2), tetra-hydroalstonine (3), leurosine sulphate (4), vindoline (HCl) (5), vindolinine (2 HCl) (6) Anthocyanins in the leaves	The action of (4), (5) and (6) is stronger than that of tolbutamide at equivalent doses. 1–3 had a less pronounced action. The hypoglycaemic effect starts slowly and is relatively long lasting	Noble *et al.* (1958) Mukherjee *et al.* (1963) Shrothri *et al.* (1963) Svoboda *et al.* (1964) Svoboda (1969)
Lupinus tassilicus Maire syn. (*L. termis* Forsk.) Fabaceae	Seeds	Sparteine and lupanine in hypoglycaemic fraction (quinolizidine alkaloids)	Alkaloids are moderately hypoglycaemic in alloxan-diabetic but not in normal rats. Short lasting effect	Shani *et al.* (1974)
Securinega virosa (Roxb. ex Willd.) Baill. syn. (*Fluggea virosa* (Roxb. ex Willd.) Baill., *F. micro-carpa* Blume, *S. micro-carpa* Pax and Hoffm., Phyllanthus virosus Roxb. ex Willd.) (Fig. 7.2) Euphorbiaceae	Seeds	Hypoglycaemic fraction: alcoholic and aqueous extracts of seeds Known constituents: numerous alkaloids in this and allied spp. (fluggeine, securinine, norsecurinine, virosine, etc.)	In doses of 0.5 mg/kg both extracts produce in cats and rabbits a decrease of blood sugar level lasting for about 4 h	Kjaer and Friis (1962) Satoda (1962) Iketobosin and Mathieson (1963) Hericz *et al.* (1964) Saito *et al.* (1964a, b) Chatterjee and Roy (1965)
Tecoma stans (L.) H.B.K. Bignoniaceae	Leaves	Alkaloids: tecomine and tecostanine	Tecomine (citrate) and tecostanine (hydro-chloride) given perorally (both in doses of 20 mg/kg, calculated on the free base) to alloxan-diabetic rabbits seems rather untoxic but do not act in total absence of active β cells of the pancreas	Garcia and Colin (1926) Guerra (1946) Hammouda and Motawi (1959) Hammouda *et al.* (1963, 1964) Jones *et al.* (1963) Hammouda and Amer (1966) Hammouda and Khallafallah (1971)

Trigonella foenum graecum L. Fabaceae	Seeds	Alkaloid: trigonelline (*N*-methylnicotinic acid). Also coumarin and nicotinic acid	Used in Israel as an oral insulin substitute. In rats trigonelline counteracts the hyperglycaemic effect of cortisone given 2 h before or simultaneously. It produces a variable effect in alloxan-diabetic rats and in diabetic patients. Nicotinic acid has a stronger hypoglycaemic action but of shorter duration whilst the effect of coumarin in diabetic rats persists for 24 h	Sulman and Menczel (1962) Menczel *et al.* (1965) Mishinsky *et al.* (1967) Varsney and Sharma (1968) Hardman and Fazli (1972) Shani *et al.* (1974)
Hordeum vulgare L. syn. (*H. sativum* Pers.) (Barley) Gramineae	Germinating seeds (radicle)	Hypoglycaemic principle has to be freed from hyperglycaemic fraction (sugars and hordenine) and vitamin B. Known constituents. alkaloids; hordenine, gramine; others: amylase, vitamins of group B, glucides, protides, lipids	Hypoglycaemic, reduces blood sugar level in fasting rabbits by 37.9% and in diabetic patients by 25%. Produces strong reduction in elimination of acetone and oxybutyric acid, which occurs in diabetic patients, and improves general condition. Is also diuretic and emollient	Donard and Labbé (1933) Labbé (1936) Dhar *et al.* (1968)
Blighia sapida Koenig (Akee apple) Sapindaceae	Fruit aril and seeds Seeds only	Hypoglycin A = αamino-2-methylene-L-cyclopropylpropionic acid Hypoglycin B = γ-glutamyl-hypoglycin A	Hypoglycins active in most animals; act also in depancreatized and adrenalectomized rats. The hypoglycins block oxidation of long-chain fatty acids leading to loss of energy production for the organism, which reacts by increasing oxidation of glucose. Hypoglycin B produces congenital malformations	Holt *et al.* (1956, 1966) Goldner (1958) Ucciani *et al.* (1964) Sherrat *et al.* (1970) Ashurst (1971) Persaud (1972)

Hypoglycaemic alkaloids are also reported in *Tecoma stans* (sometimes cultivated in West Africa). A hypoglycaemic betain was isolated from *Trigonella foenum-graecum* and called trigonelline; it is a methylbetaine of nicotinic acid and was considered to be the active constituent of fenugreek until 1974. Shani *et al.* (1974) then found that coumarin and nicotinic acid seemed to be the main hypoglycaemic constituents of all components isolated from the active fraction of the seeds. Trigonelline was found to be less effective but more persistent in its action. The seeds also contain sterols.

Many other hypoglycaemic plants contain alkaloids but it has not been shown that these are the active constituents (e.g. *Hordeum vulgare*).

Some active plant constituents contain amino nitrogen, e.g. hypoglycins in *Blighia sapida*. They are, unfortunately, too poisonous for clinical use. Their interest lies in the knowledge acquired about the ways in which they act.

Blighia sapida Koenig (Fig. 7.4) SAPINDACEAE

Akee apple

The ripe fruit is edible and tastes much like a chestnut, but is poisonous if immature or decaying (ichthyotoxic in Africa). In Jamaica there used to be annual epidemics of 'vomiting sickness', which were marked by vomiting, prostration, convulsions and finally coma. This condition was eventually traced to the ingestion of unripe akees, and in particular to the ingestion of two constituents, hypoglycin A and hypoglycin B. These substances have strong hypoglycaemic and emetic action in most animals and Man. Hypoglycin A has proved to be twice as active as hypoglycin B. However, the livers of animals treated by these substances show fatty degener-

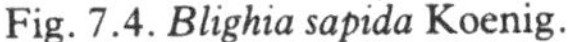

Fig. 7.4. *Blighia sapida* Koenig.

ation and reduced glycogen content. Hypoglycins appear to act through inhibition of the β-oxidase enzymes, thereby blocking the oxidation of long-chain fatty acids; this causes accumulation of unmetabolized fatty acids and makes them unavailable for energy production. The organism reacts by oxidizing glucose in large quantities, thus causing a decrease in liver glycogen and a drop in blood glucose to hypoglycaemic levels (Tanaka *et al.*, 1972).

III Plants containing hypoglycaemic organic sulphur compounds

(See Table 7.3.)

Allium cepa LILIACEAE

Onion

(See also Chapters 2 and 4). Already in 1923 Collip noticed that a totally (?) depancreatized dog could be kept alive for 66 days on three injections of crude onion extract. Later, a number of research workers confirmed that the onion and its extracts have a distinct, slowly developing hypoglycaemic action and that the effect is also obtained when the extract is given by mouth (Hermann, 1956, 1958; Jain and Vyas, 1974).

The action of a petroleum ether extract of sliced dried onion was equivalent to 62% of that of a standard dose (0.5 g) of tolbutamide and further purification led to an extract equivalent to 76.6% of this standard) (Brahmachari and Augusti, 1962). Later, two active disulphides were isolated from fresh onions by steam distillation and solvent extraction. They were allyl propyl disulphide (APDS) and allicin (diallyl disulphide oxide) (Augusti and Benaim, 1974; Augusti, 1975, 1976a, b; Augusti *et al.*, 1975).

With APDS, blood glucose and glycosuria were significantly decreased in alloxan-diabetic rabbits and glucose tolerance was also improved. In a 4-h test in fasting human subjects APDS caused a marked fall of blood sugar (hourly controls) plus an increase of serum insulin levels, whilst the free fatty acid levels remained the same. In a control trial there was no fall of blood glucose, but the serum insulin levels decreased and the free fatty acid levels increased considerably.

The authors (Augusti and Benaim, 1974; Augusti, 1976a, b) explain these results as follows. Insulin is a disulphide protein and its inactivation by compounds and albumins rich in SH groups has been established. APDS probably removes insulin-inactivating compounds by competing with insulin for the SH group(s) in these compounds, thus producing an insulin-sparing effect and preventing the increase of free fatty acids on fasting.

In clinical trials, 100 mg allicin per kg body weight produced a significant drop in fasting blood glucose levels with a concomitant rise in serum insulin levels. In 12 subjects the action lasted for about 12 h. Long-term feeding to normal rats of 100 mg of allicin per kg of body weight produced a large reduction in the lipid constituents of the blood and liver and in this respect allicin might have an advantage over tolbutamide (Augusti and Benaim, 1974). Cyanidin- and paeonidin-glycosides are also present in onion bulbs, and could well be partly responsible for the effect of the crude extract.

Table 7.3. *Plants containing hypoglycaemic organic sulphur compounds*

Plant Family	Part used	Possible active constituents	Observed activity	References
Allium cepa L. (Onion) Liliaceae	Bulbs	Allyl propyl disulphide (APDS), allicin (diallyl disulphide oxide), methylalliin in fresh juice (latter also bacteriostatic) Also in bulbs: flavon glycosides; kaempferol, quercetin- and phloroglucin-derivatives	In fasting humans, APDS in doses of 0.125 g/50 kg produced a significant fall of hyper-glycaemia, a rise in serum insulin levels and no change in free fatty acid levels. APDS is believed to remove insulin-inactivating compounds by competing with insulin for the SH group, thus having an insulin sparing effect. The activities of petroleum ether and ethyl ether extracts of sliced dried onion were equivalent to 62 and 76% of that of a tolbutamide standard respectively. APDS does not act in totally depancreatized rabbits	Bhandahari and Mukerje (1959) Brahmachari and Augusti (1961a, b, 1962a) Jain and Sharma (1967) Augusti and Benaim (1974) Jain and Vyas (1974) Augusti (1975, 1976a, 1976b) Augusti *et al.* (1975) Matthew and Augusti (1975)
Allium sativum L. (Garlic) Liliaceae	Dried flower heads	Organic sulphur compounds	The activity of an ethyl ether extract of dried *A. sativum* heads is 58% of that of a tolbutamide standard	
Brassica oleraceae L. var. *capitata* L. (Cabbage) Cruciferae	Leaves	Hypoglycaemic fraction, called vegulin, loses activity in 1 month. Thioglycosides (methyl and ethyl propyldi-sulphides and goitro-genic indole-myrosin glycoside (neo-gluco-brassicin)	Hypoglycaemic principle tied to a hyper-glycaemic fraction from which it can be separated	MacDonald and Wislicki, (1938) Lewis (1950) Johnson *et al.* (1971) Vohora *et al.* (1973)

Garlic (*Allium sativum*) (Brahmachari and Augusti, 1962b), *Brassica oleraceae* and *Phaseolus vulgaris* also contain organic sulphur compounds.

Allium cepa, *A. sativa*, *Brassica oleraceae*, *Phaseolus vulgaris* and some non-African plants contain, in addition to a hypoglycaemic factor, a hyperglycaemic factor; this has to be eliminated before the hypoglycaemic effect can be estimated. This simultaneous presence of the two antagonistic factors could explain the discrepancy in results obtained in older publications.

IV Hypoglycaemic plants containing anthocyanins, catechols or flavonoids, or their glycosides and/or tannins (See Table 7.4)

In diabetic patients the basal membrane of the small blood vessels, which has an important metabolic function, is thickened (by an accumulation of glycoproteins), thus causing a disturbance of metabolic exchange. Also, the capillary wall becomes permeable (diabetic angiopathy). It has been shown that *Vaccinium* anthocyanosides can inhibit or slow down this evolution of the capillary walls particularly at the onset of the diabetic disease process. Treatment over 6 months (a starting dose of 600 mg per day) reduces the number of affected capillaries from 34 to 24% and the average surface of glycoprotein accumulation from 14 to 8 μm (Pourrat, 1977; Pourrat *et al.*, 1977, 1978).

The active constituent of Euopean *Vaccinium myrtillus* berries has proved to be a glycoside of 7-methyl-delphinidin and the traditional use of the berries in the treatment of diabetes in Europe has been justified by pharmacological and clinical trials which have shown that the effect of a single dose can last up to two weeks and more. The active principle allowed a gradual decrease in the use of insulin in a number of patients (Allen, 1927).

In West Africa several antidiabetic plants with similar constituents have been reported. Thus leucodelphinidin and leucocyanidin are found in the flowers and also in other parts of banana plants and **Musa sapientum** pigment contains deoxyxanthin cyanidin.

In evaluating the hypoglycaemic effect of extracts or products of 56 Indian plants reputed to be antidiabetic in native medicine, by their reduction of the normal fasting blood-sugar level in rabbits, Jain and Sharma (1967) found that an extract of the flowers of a variety of *M. sapientum* (Ney Poovan) was second in order of efficacy (*Allium cepa* being first).

A solution of 10 mg/kg of the dried residue of this extract produced a hyperglycaemia of 15–24 mg, compared to one of 20–30 mg for *Allium cepa*, one of 12–23 mg for *Syzygium cumini*, one of 13–21 mg for *Ficus glomerata* and one of 10–18 mg for both *Momordica charantia* and *Gymnema sylvestre*.

Saponifiable and mainly non-saponifiable fractions of the extract of the flowers of *Musa sapientum* var. Ney Poovan have hypoglycaemic properties.

The leaves of **Morus alba** contain cyanidin and delphinidin glucosides in addition to phytosterol glycosides. Similarly, in the fruits and seeds of **Syzygium cumini** (Table 7.1), which are used as antidiabetics, cyanidin rhamnoglucosides and galli- and ellagi tannins, respectively, are found, as well as phytosterol glycosides (Venkateswarlu, 1952).

Table 7.4. *Hypoglycaemic plants containing anthocyanins, catechols or flavonoids, or their glycosides and/or tannins*

Plant Family	Part used	Possible active constituents	Observed activity	References
Anacardium occidentale L. Anarcardiaceae	Leaves	Quercetin and kaempferol glycosides	The glycoside fraction normalizes glycaemia	Dhar *et al.* (1968) Laurens and Paris (1976)
Bridelia ferruginea Benth. Euphorbiaceae	Bark, roots, leaves	Tannins, flavonoids and biflavonoids based on apogenin and kaempferol moieties	Injection of water and methanol extracts of leaves lowers hyperglycaemia in fasting rats, but only protects alloxan-diabetic rats when given 1 h before alloxan. In clinical trials 8 out of 10 patients saw their blood sugar reduced from 230 mg% to under 120 mg% when taking a decoction or maceration of the leaves perorally	Githens (1949) Iwu (1980, 1983)
Ceiba pentandra (L.) Gaertn. syn. (*Eriodendron anfractuosum* DC.) (Silk cotton tree) Bombacaceae	Juice, roots, bark	Quercetol and kaempferol glucosides, traces of gossypiol, methylglucuronoxylan. In the seeds, βsitosterol	The glycosides are reported to be antidiabetic in India, needs checking	Currie and Timell (1959)
Centaurea perrottetti DC. syn. (*C. aspera* L., *C. calcitrapa* Chev., *C. alexandrina* L.) Compositae	Inflorescence	*Centaurea* spp. contain glycosides of flavones (apigenin, baicalein, luteolin, etc.) and/or of flavonols (centaureidin, jadein, quercetin). They also contain β-sitosterin, β-amyrin, peptides, cnicin (= centaurin (sesquiterpenic lactone) and an alkaloid, stizolphine	The peptide fraction reduces glycaemia in rabbits by 28–42 mg/24 h Centaurin has antibiotic properties, mainly towards Brucella	Labo and Puig (1953) Viguera and Casabuena (1965) Ahmed *et al.* (1971) Monya and Racz (1974) Masso *et al.* (1979)
Coccinia grandis L. Voigt syn. (*C. indica* W & A., *Cephalandra indica* Naud., *C. cordifolia* (L.) Cogn.) Cucurbitaceae	Tuberous roots	There is a hypoglycaemic fraction in alcoholic and aqueous extracts of roots. Known constituents of roots: caffeic acid, quercetin, kaempferol, β-sitosterol	Hypoglycaemic effect of alcoholic extract in rabbits = 58.9% of that of tolbutamide. Effect in alloxan-diabetic rabbits is comparable to that of tolbutamide	Chopra *et al.* (1956) Currie and Timell (1959) Brahmachari and Augusti (1963) De and Mukherjee (1963) Trivedi (1963) Jain and Sharma (1967) Mukerjèe and Ghosh (1972)

Morus alba L. *M. nigra* L. (White and black mulberry) Moraceae	Leaves	Hypoglycaemic fraction = alcoholic and aqueous extract of leaves. Known constituents: cyanidin and delphinidin glucosides, rutin, moracetin (= quercetin triglycoside), βsitosterin, sitosteryl-carpate and palmitate	Leaf extracts are hypoglycaemic in tests in rats with experimental diabetes and in the fasting animal, increases glycosuria. Has also slight, antispasmodic and hypotensive action.	Leclerc (1934) Sharaf and Mansour (1964) Talyshinski (1967) Deshpande (1968) Naito (1968) Nomora and Fukai (1981)
Musa paradisiaca L. var. *sapientum* (L.) Kuntze syn. (*M. sapientum* (L.)) (Banana) Musaceae	Juice of flowers	Liquid extract of flowers, mainly 3rd fraction of non-saponifiable portion, is hypoglycaemic. In bracts, anthocyanidins; in fruits, hydroxytryptamine, glucides	The extract of the flowers is hypoglycaemic in tests on fasting rabbits (reduction of 15–24 mg in blood sugar)	Hood and Lowburry (1954) Simmonds (1954) Sinha *et al.* (1962) Jain (1968) Jain (1969)
Phyllanthus niruri L. Euphorbiaceae	Leaves	Flavonoids (phyllanthin and hypophyllanthin), lignanes, quercetoside 4 alkaloids (norsecurinine isomers). In the bark: lupeol	Antidiabetic in India, needs checking	Vila (1940) Krishnamurti and Seshadri (1946) Jain and Sharma (1967) Stanislas *et al.* (1967) Dhar *et al.* (1968) Rouffiac and Perello (1969) Nara *et al.* (1977)
Rhizophora racemosa Mey. and spp. (Mangrove) Rhizophoraceae	Bark, roots	Tannins and catechins (15–42%)	A decoction is hypoglycaemic in diabetes in India	McMillan (1954) Chopra *et al.* (1956, p. 212) Jain and Sharma (1967)
Sclerocarya birrea (Rich.) Hochst. syn. (*Spondias birrea* Rich., *Poupartia birrea* (Rich.) Aubrev.) Anacardiaceae	Leaves	Tannins and flavonoids	A decoction or maceration of the leaves is distinctly hypoglycaemic when administered by mouth or intraperitoneally. Low toxicity in rats	Busson (1965) Gueye (1973)
Scoparia dulcis L. (Sweet broom weed) Scrophulariaceae	Whole plant	Hypoglycaemic bitter principle named amellin. Known constituents: scoparol (=3′*O*-methyl-luteolin), Scoparoside (=8 glycosyl-scopanol)	Reduces hyperglycaemia and glycosuria in human diabetics. When chewed, plant tastes bitter, then sweet	Nath (1943) Nath and Bannerjee (1943) Nath and Chowdurry (1943, 1945) Nath *et al.* (1943, 1945) Whittacker (1948)

In France a hypoglycaemic drug based on *Syzgium* has been put on the market: 14 g of active substance are obtained from 100 g of dried seeds by triple extraction using 95% alcohol or boiling water. (It is used in doses of 1–2.5 g/day; the LD_{50} in mice is 4 g/kg). Hypoglycaemic action, possibly through flavonoids and tannins, has been found in the bark and roots of *Bridelia ferruginea* and *Rhizophora mucronata* and in the leaves of *Sclerocarya birrea*.

Bridelia ferruginea Benth. EUPHORBIACEAE

L The bark, leaves and roots of this tree are ingredients of the Yoruba *agbo* infusion and are used in the preparation of a popular mouthwash and as a remedy for thrush in children. The roots are used as a chewstick and in Togo externally for the treatment of skin diseases and eruptions. In Northern Nigeria it is also used as an antidote to arrow poison (Dalziel, 1937, p. 137). In the Congo the bark is used as an anthelmintic for roundworm and is given in the treatment of cystitis (Githens, 1949).

C Tannins have been reported to be present in the bark (Githens, 1949) and roots of both *B. ferruginea* and *Rhizophora mucronata*. Flavonoids and biflavonoids based on apigenin and kaempferol moieties were isolated together with their glycosides from the methanolic extract of this plant. 'It is however not clear which of these compounds is responsible for the antidiabetic properties of this plant' (Iwu, 1983). An ethyl acetate-insoluble fraction yielded coumestans (Iwu, 1983).

P One hour after intraperitoneal injection or oral administration of leaf extracts to 12-h-fasted albino rats there was a significant reduction in fasting blood sugar but the animals were not protected against alloxan-induced diabetes. There was, however, a significant reduction in the expected hyperglycaemia in alloxan-diabetic rats when administration took place 1 h before alloxan injection. (The extracts were water and methanol extracts.) Clinical evaluation at a herbal home has revealed that eight out of ten patients benefited from *Bridelia* therapy. Their blood sugar was reduced from 230 mg% to less than 120 mg% and remained at this level for 8 weeks after they had taken daily a decoction or maceration of the leaves perorally (Iwu, 1980).

Sclerocarya birrea (Rich.) Hochst. syn. (*Spondias birrea* Rich., *Poupartia birrea* (Rich.) Aubrev.) (Fig. 7.5) ANACARDIACEAE

L Locally a fermented beverage like cider is prepared from the expressed juice of the fruit. The tree is mainly found in Northern Nigeria, Ghana and Gambia, and the Hausas use a cold infusion of the bark along with native natron as a remedy for dysentery. The leafy branches are cut for fodder during periods of drought (Dalziel, 1937). In Senegal the leaves and rootbark are used together with *Securidaca longepedunculata*, sometimes mixed with other plants as well, as a plaster against venoms, e.g. those in snake bites (Kerharo and Adam, 1974).

C Analysis of the leaves revealed the presence of tannins and flavonoids. No alkaloids, steroids or triterpenoids were detected (Gueye, 1973). The fruit is rich in vitamin C, but the juice is rather toxic. The seeds are rich in oil, mostly oleic acid (64%) but also myristic and stearic acids. They also contain amino acids with a predominance of glutamic acid and arginine (Busson, 1965).

P A decoction or maceration of the leaves of the tree has a distinct hypoglycaemic action when administered by mouth or intraperitoneally. The extracts show low toxicity in rats. The leaf extract is believed to have a direct action on the glycaemia-regulating system as well as a stimulating action on the peripheral assimilation of glucose, in particular by the muscular tissues (Gueye, 1973).

Some plants contain quercetin-, kaempferol- and/or luteolin-glycosides as active constituents. Examples are *Anacardium occidentale*, *Ceiba pentandra*, *Centaurea perrottetti*, *Coccinia grandis*, *Phyllanthus niruri* and *Scoparia dulcis*. The activity of two of these plants reputed to be antidiabetic in Indian folk medicine, *Ceiba pentandra* and *Phyllanthus niruri*, needs scientific confirmation according to Jain and Sharma (1967).

Fig. 7.5. *Sclerocarya birrea* (Rich.) Hochst.

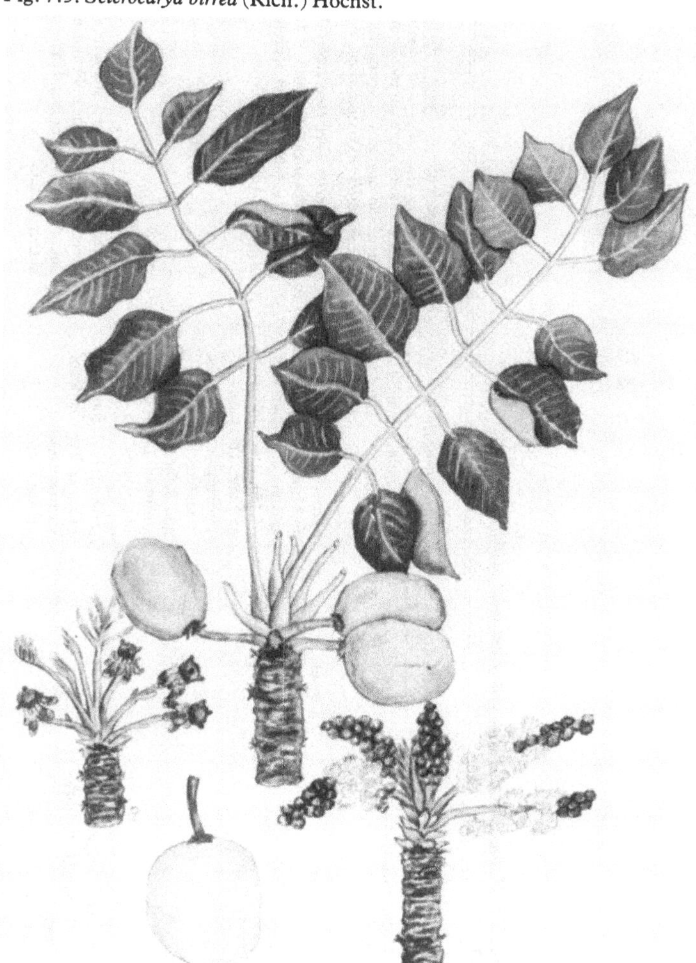

Table 7.5. *Hypoglycaemic plants containing other active constituents*

Plant Family	Part used	Known constituents	Observed activity	References
Azadirachta indica Juss. Meliaceae	Leaves	Acetylnimbin, nimbolid (lactone)	Fresh leaf extract is hypoglycaemic in dogs with adrenaline- or glucose-hyperglycaemia when given by intravenous injection	Luscombe and Taha (1974) See also Chapter 4
Corchorus olitorius L. (Jute) Tiliaceae	Leaves	A hypoglycaemic extract free from pectins, sugars and fats, contains traces of elemental sulphur and zinc. There are cardiac glycosides in the seeds (corcherosides A and B, their genin is strophanthidin olitoriside)	An aqueous extract decreases hyperglycaemia in mice, guinea-pigs and rabbits. It does not act in the absence of the pancreas; it promotes the degradation of glucose in vitro	Uzan and Dziri (1952) Frèrejacque and Durgat (1954) Peters (1957) Goldner (1958) Samilova and Lagodich (1977)
Gymnema sylvestre (Retz) Schultes syn. (*Periploca sylvestris*) Periplocaceae	Leaves	Gymnemic acid; consists of at least 9 glycosides of narrowly related constitution	Reduces glycosuria and destroys the sense of sweetness and bitterness. Found in tropical and South Africa, Australia and India. Also has antiviral action	Gupta (1961, 1963a, b) Trivedi (1963) Gupta and Vatiyar (1964) Stoecklin *et al.* (1967) Stoecklin (1968, 1969) Kuriharli (1969) Cochran and Maasab (1970) Sinsheimer *et al.* (1970) Mitra *et al.* (1975)

Lagerstroemia speciosa (L.) Pers. syn. (*L. flos regina* Retz.) Lythraceae	Old leaves and ripe fruit	Thermostable hypoglycaemic principle. Terpenes and saponins in the leaves	A decoction of 20 g of dried leaves has an activity equivalent to 6–7.7 units of insulin. Even high doses are not toxic	Garcia (1941)
Mucuna pruriens (L.) DC. var. *pruriens* syn. (*Dolichos pruriens* L.) (Cow-itch) Fabaceae	Seeds	A powder of the decoated seeds contains 27% protein, 8.3% lipids, 47.8% carbohydrates, two alkaloids (mucunine and mucunadine) and two soluble bases (prurienine and prurienivrine)	When the powder formed 96.5% of the diet (+ salt and vitamins), it produced a lowering of 39% of the blood sugar, as compared to standard diet, in fasted rats and a lowering of 61% of the cholesterol level	Majumdar and Zalani (1953) Dhar *et al.* (1968) Pant *et al.* (1968) Ghosal *et al.* (1971)
Vigna unguiculata (L.) Walp. syn. (*Dolichos unguiculatus* L., *D. biflorus* L., *D. sinensis* L., V. *sinensis* (L.) Savi & Hask) Fabaceae	Seeds	A powder of the decoated seeds contains 19% protein, 2% lipids, 53.18% carbohydrates; the 'mineral content' is 0.1–0.2% calcium, 0.37% phosphorus and traces of iron and vitamins	When the powder formed 96.5% of the diet, it produced a lowering of 42% of the blood sugar of fasted rats and a lowering of 26.7% of their cholesterol level	Pant *et al.* (1968) Kerharo and Adam (1974, pp. 480–1) Tella and Ojihomon (1980)

V Hypoglycaemic plants containing other active constituents

(See Table 7.5).

Among these we find *Azadirachta indica* and *Gymnema sylvestre*. The hypoglycaemic effect in dogs with adrenaline- or glucose-induced hyperglycaemia of the intravenous injection of a fresh-leaf extract of *A. indica* has been reported by Luscombe and Taha (1974). However, the need exists mainly for oral antidiabetics. The leaves contain acetylnimbin and a lactone, nimbolid.

The leaves of *Gymnema sylvestre*, which are chewed in India to reduce glycosuria, normalize the blood sugar in diabetic subjects in about 3–4 weeks, but when they were used in combination with insulin a prompt response was observed where insulin alone in that particular dose had failed (Gupta, 1961). The active principle has been called gymnemic acid and consists of a complex mixture of heterosides of diverse organic acids. Administration of the dried leaf powder regulates the blood sugar levels also in alloxan-diabetic rabbits and increases the activities of the enzymes commanding the utilization of glucose by insulin-dependent pathways. Thus control of phosphorylase levels, gluconeogenic enzymes and sorbitol dehydrogenase has been observed. *G. sylvestre* appears to correct the metabolic derangements in the diabetic liver, kidneys and muscles (Shanmugasundaram *et al.*, 1983).

G. sylvestre and *Scoparia dulcis* are taste-modifiers, a property which is also found in sweetening agents. The latter are a useful contribution to the diet of diabetics and are discussed briefly below. The powdered seeds of two Fabaceae, *Mucuna pruriens* and *Vigna unguiculata*, have been shown to lower the level of cholesterol as well as that of the blood sugar.

VI Mechanism of action of hypoglycaemic plants

The plant constituents which have a hypoglycaemic action act in various ways. We have seen that plants rich in organic sulphides seem to remove insulin-inactivating compounds through their SH groups by competing with insulin, which is a disulphide protein, for the SH groups in these inactivating compounds. Similarly, nicotinic acid is known to be an insulinase inhibitor (Ben David *et al.*, 1963).

We have also seen that anthocyanosides appear to act by improving the vascularization of the pancreas. Insulin is fixed on proteins at the surface of the cells, thus cellular membranes play an important part in diabetes. Vascular troubles develop gradually during the course of the disease, entailing a thickening of the basal membrane of the walls of the small blood vessels and an increase of their permeability (Osuntokun, 1975). Anthocyanosides have been found to inhibit or slow down these changes in the capillary walls, especially in the early stages of the disease. Thus the improvement in a patient's condition that is obtained thanks to some plants rich in anthocyanosides might possibly be due to recovery of the vascularization of the pancreas. Other plant flavonoids, which also appear to act on the capillaries, could have a similar effect.

In a number of cases the plants seem to intervene in oxido-reduction phenomena. We have seen that hypoglycins appear to act through inhibition of the β-oxidase enzymes. Galegine, like synthetic biguanides, acts in blocking oxidative enzymes of the Krebs cycle (succinic dehydrogenase and cytochrome oxidase), thus increasing anaerobic glycolysis and decreasing gluconeogenesis, and entailing an increased rate of transfer of glucose from the blood to tissues (see also Oliver-Bever and Zahnd, 1979).

The action of other plants seems to be related to the quantities of metal ions or plant-growth regulators they contain (see Oliver-Bever and Zahnd, 1979).

It thus seems that hyperglycaemic patients can be treated by a variety of plants which are active at different points in the glucose-metabolism cycle; the success of the treatment may well be governed by the particular abnormalities of the individual case.

The majority of the plants mentioned are moderate in their effects and can be useful in mild cases of diabetes; their use might perhaps be combined with a diet that is poor in glucides. In some instances their prolonged administration might delay the establishment of a more serious diabetic condition. In more severe cases of diabetes, these plants might constitute a form of secondary therapy, with the result that the frequency or dosage of insulin (or other orthodox drugs) could be reduced.

Several of these plants depend for their efficacy on the presence of insulin, be it exogenous or endogenous.

VII Sweetening agents

Abrus precatorius L. PAPILIONACEAE

This slender climber is rendered conspicuous by its red seeds, which are black at the base and visible among clusters of ripe, burst beans. These seeds are used locally as beads and in the manufacture of masks and shields and are so similar in size and weight that they were employed in Ashanti (Ghana) and in India as standard weights by goldsmiths.

Only the leaves (but to some extent also the roots) can be used as sweetening agents, which is what concerns us here. The dried leaves contain 6–10% of glycyrrhizin; the dried roots contain 1.5%. This saponoside, first found in the roots and rhizomes of liquorice, is a diglycuronide of glycyrrhetic acid. The substance is widely used as a sweetening agent, being sixty times sweeter than saccharose (Milhet *et al.*, 1978). It also has anti-inflammatory, expectorant, antitussive and antibiotic properties. During prolonged intake glycyrrhetic acid produces oedema through water retention with retention of sodium and chloride ions (Na^+ and Cl^-) and increased elimination of potassium ions, (K^+), an effect similar to that produced by cortisone.

In France the cultivation and propagation of *Abrus* in greenhouses as a source of extraction of glycyrrhizin was studied by Milhet and coworkers in 1978.

The seeds contain a highly toxic protein, abrin, and have been used to treat trachoma and certain forms of cancer (Dupaigne, 1974).

Dioscoreophyllum cumminsii (Stapf) Diels MENISPERMACEAE

The most potent sweetening agent known today seems to be the protein isolated from the berries of *Dioscoreophyllum*. This is a high forest climber found from Guinea to the Cameroons. The red berries of this tree have been called 'serendipity berries' as their intensely sweet taste was discovered fortuitously. The active principle, first reported by Inglett and May (1968, 1969) was isolated by van der Wel (1972) but is as yet difficult to obtain in any great quantity. It is a basic protein with a net positive charge at neutral pH; it has been called monellin and has a sweetness relative to sucrose on a weight basis of 3000 and on a molecular weight basis of about 9000.

Monellin loses its sweetening ability above pH 10 and below pH 2 and the tertiary structure is important with regard to this sweetening property, as has been shown by denaturation of the protein (van der Wel, 1972).

Preliminary trials have shown that monellin is a useful adjunct in the preparation of drinks poor in calories and in non-glucidic diets. Being required in low concentrations only, it is likely to prove completely inoffensive.

Synsepalum dulcificum (Schum. & Thonn.) Daniell. syn. (*Richardella dulcifica*) SAPOTACEAE

The small, red, single-seeded fruits of this small tree are commonly called 'miraculous berries'. Their pulp is used locally to sweeten palm wine and imparts a sweet taste to anything eaten soon afterwards. The sweetening agent is a glycoprotein which was named miraculin by Brouver *et al.* (1968). Miraculin has a molecular weight of 44 000 and the constituent sugars are arabinose and xylose. It causes sour substances and acids to taste sweet for up to several hours after a 2.3×10^{-6} M solution of it is held in the mouth for 5 min. After this time the sensitivity to acid taste is recovered. Salt and bitter tastes are not affected (Brouver *et al.*, 1968; Kurihari and Beidler, 1968). Miraculin is also said to be an appetite depressant and a patent has been issued for its use as an anorexic agent.

Miraculin is thought to modify the taste receptors of the tongue, which take a certain time, varying with the individual, to recover their acid taste sensation (Kurihari and Beidler, 1968, 1969; Morris and Juscy, 1976).

Thaumatococcus daniellii (Benn.) Benth. MARANTHACEAE

Miraculous fruit

This is a herb which grows up to ten feet tall and has papery leaves that are commonly used to wrap kola or food. In the crimson, underground fruits, an intensely sweet substance, which is used locally to sweeten palm wine or acid fruits; is formed. It is found in the jelly-like aril surrounding the hard black seed and has been called thaumatin (Inglett and May, 1968). Van der Wel and Loeve (1972) found that the sweet substance consisted of two proteins with almost identical amino acid composition; and molecular weights of 21 000 and 20 000. Thaumatin loses sweetness below pH 2.5 and upon heating, reduction of disulphide bonds or when digested by proteolytic enzymes. It is somewhat similar to liquorice in taste and is said to be 1600 times sweeter than saccharose by some and 4000 times sweeter than saccharose by others. Adesina and Harborne (1978) identified the polyphenols occurring with the protein sweetener and found flavonol and flavones, apparently absent in the fruits, in other parts of the plant.

A possible connection between electrostatic charge and taste activity might be explored as the three taste-active proteins, monellin, miraculin and thaumatin, all have isoelectric points in the alkaline region (Morris and Juscy, 1976).

NOTES

Chapter 2: The cardiovascular system

1 Arrow poisons were prepared by rubbing certain seeds between two stones until they formed a paste to which were added saliva and the juice of different toxic plants. A vesicant latex, for example from *Euphorbia*, was often used as this damaged the skin, thus facilitating penetration and absorption (Dalziel, 1937).

2 In trials by ordeal (e.g. trial of Lander in Badagry in 1827) a man suspected of evil influence or action was forced with much ceremony to swallow a dose of poison. If he survived it was the wish of the tribal gods and he was considered innocent. If he died this was evidence of guilt (Dalziel, 1937).

REFERENCES

General references

These references are referred to throughout the book. Those references marked with an asterisk (*) in the individual chapter lists give general information about plants mentioned in that chapter.

Pharmacology

Burgen, A. S. V. and Mitchell, J. F. (1972) *Gaddum's Pharmacology*, 7th edn. Oxford University Press, New York and Toronto, 251 pp.

Lechat, P., Bisseliches, F., Bournerias, F., Dechy, H., Juillet, Y., Lagier, G., Meyrignac, C., Rouveix, B., Sterin, P., Warnet, A. and Weber, S. (1978) *Abrégé de Pharmacologie Médicale*, 3rd edn. Masson et Cie, Paris, New York, Barcelona and Milan. 1 vol., 677 pp.

Turner, P. and Richens, A. (1978) *Clinical Pharmacology*, 3rd edn. Churchill Livingstone, Edinburgh and London. 1 vol., 254 pp.

Botany

Hutchinson, J., Dalziel, J. M. and Keay, R. W. J. (1958) *Flora of West Tropical Africa*, 2nd edn, Vol. 1, Parts 1 and 2. Crown Agents, London. 2 vols., 828 pp.

Hutchinson, J., Dalziel, J. M. and Hepper, F. N. (1963) *Flora of West Tropical Africa*, 2nd edn, Vol. 2. Crown Agents, London. 1 vol., 544 pp.

Hutchinson, J., Dalziel, J. M. and Hepper, F. N. (1972) *Flora of West Tropical Africa*, 2nd edn, Vol. 3, Parts 1 and 2. Crown Agents, London, 2 vols., 574 pp.

Dalziel, J. M. (1955) *The Useful Plants of West Tropical Africa*, 2nd reprint. Crown Agents, London, 612 pp.

Chemistry

Henry, T. A. (1949) *The Plant Alkaloids*, 4th edn. J. & A. Churchill, London. 1 vol., 804 pp.

Karrer, W. (1958) *Konstitution und Vorkommen der organischen Pflanzenstoffe (exclusive Alkaloide)*. Birkhäuser, Basel. 1 vol.

McIlroy, R. J. (1950) *The Plant Glycosides*. Arnold, London. 1 vol., 138 pp.

Manske, R. H. F. and Holmes, H. L. (1950–71) *The Alkaloids*. Academic Press, New York and London. 13 vols.

Microbiology (medical)

Manson-Bahr, Sir Philip H. (1952) *Synopsis of Tropical Medicine*. Cassell and Company Ltd, London. 248 pp.

Pharmaceutical uses (for local uses see Chapter 1)

Kerharo, J. and Adam, J. G. (1974) *La Pharmacopée Sénégalaise traditionnelle*. Vigot, Paris. 1011 pp.

Martindale. *The Extra Pharmacopoeia*, 24th edn, 1958; 25th edn 1969. The Pharmaceutical Press, London.

Paris, R. and Moyse, H. (1965–71) *Précis de Matière Médicale*, 3 vols. Vol. 1 (1965), 416 pp.; Vol. 2 (1967), 510 pp.; Vol. 3 (1971), 508 pp. Masson et Cie, Paris.

Chapter references

References marked with an asterisk (*) give general information about the plants mentioned in that chapter.

Chapter 1: Introduction

Ainslie, J. R. (1937). *A list of plants used in native medicine in Nigeria*. Imperial Forestry Institute, Oxford. No. 7, 109 pp.
Ayensu, E.S. (1978) *Medicinal Plants of West Africa*. Reference Publications Inc. Algona, MI. 1 vol., 330 pp.
Bouquet, A. (1969) *Féticheurs et Médicines traditionnelles du Congo-Brazzaville*. ORSTOM, Paris. 306 pp.
Chevalier, A. (1905–13) *Les végétaux Utiles de l'Afrique Tropicale Française*. Fasc. 1–8. Challamel, Paris.
Chopra, R. N., S. L. Nayar, and I. C. Chopra (1956) *Glossary of Indian Medicinal Plants*. Council of Scientific and Industrial Research. New Delhi. 1 vol., 330 pp.
Dalziel, J. M. (1937) *The Useful Plants of West Tropical Africa*. Crown Agents, London.
Githens, T. S. (1949) *Drug Plants of Africa*. The University Museum, University of Pennsylvania Press. 125 pp.
Grier, James (1937) *A History of Pharmacy*. London.
Gunther, R. T. (1934) *The Greek Herbal of Dioscorides*. Oxford University Press.
Hardie, A. D. K. (1963) Okoubaka, a rare *juju* tree. *The Nigerian Field*, **28**, 70–2.
Harley, G. W. (1941) *Native African Medicine with Special References to its Practice in the Mano Tribe of Liberia*. Harvard University Press, Cambridge Massachusetts. 294 pp.
Holland, J. H. (1908–29) *The Useful Plants of Nigeria*, pp. 3–352. Kew Bulletin, London.
Irvine, F. R. (1930) *Plants of the Gold Coast*. Oxford University Press, London. 1 vol., 521 pp.
Jayaweera, D. M. A. (1945, 1952, 1954) Indigenous and exotic drugplants growing in Ceylon. *Tropical Agriculturalist (Ceylon)*, **101**, 130–5; **108**, 109–15; **110**, 105–16.
Kerharo, J. and J. G. Adam (1974) *La Pharmacopée sénégalaise Traditionnelle*. Vigot, Paris. 1 vol., 1011 pp.
Kerharo, J. and A. Bouquet (1950) *Plantes Médicinales de la Côte d'Ivoire et Haute Volta*. Vigot, Paris. 297 pp.
Lloyd, J. U. (1921) *Origin and History of all the Pharmacopoeal Vegetable Drugs etc.*, Vol. I. Cincinnati.
Oliver, B. (1960) *Medicinal Plants in Nigeria*. Nigerian College of Arts, Science and Technology. 1 vol., 138 pp.
Oliver-Bever, B. (1983) the West African Juju man and the tools of his trade. *International Journal of Crude Drug Research*, **21**, 97–120.
Pobéguin, H. (1912) *Les Plantes Médicinales de la Guinée*. Paris (through Dalziel, 1937).
Rohde, E. S. (1922) *The Old English Herbals*. London.
Sébire, R. P. A. (1899) *Les Plantes Utiles du Sénegal*. Baillière et Fils, Paris. 1 vol., 341 pp.
Staner, P. and R. Boutique (1937) Matériaux pour l'etude des plantes médicinales du Congo belge. *Mémoires de l'Institut Royal Belge. Section Sciences Naturelles et Médicales*, **5**, 3–228.
Steenis-Kruseman, M. J. van (1953) Select Indonesian Medicinal Plants. *Bulletin of Organization of Scientific Research*, **18**, 30 pp.
Talbot, P. A. (1926) *The People of Southern Nigeria*, vols. I–IV. Oxford University Press.
Woodward, M. (1931) *Leaves from Gerarde's Herbal*. London.

Chapter 2: The cardiovascular system

Abbot, B. J., L. Hartwell, J. Leiter, S. A. Spitzman and S. A. Schepartz (1967) Screening data from the Cancer Chemotherapy National Service Center Screening Laboratories. X.I.L.: Plant Extracts. *Cancer Research*, **27** (Suppl.), 51–4, 364–536.

Abbot, B. J., J. Leiter, L. Hartwell, M. E. Caldwell, J. L. Beal, R. E. Perdue and S. A. Schepartz, Jr (1966) Screening data from the Cancer Therapy National Service Center Screening Laboratories. Plant Extracts. *Cancer Research*, **26** (Suppl.), Part 1 (2 vols.).

Abisch, E. and T. Reichstein (1960) Orientierende chemische Untersuchungen über Apocynaceen. *Helvetica Chimica Acta*, **43**, 1844–61.

Abisch, E. and T. Reichstein (1962a) Orientierende chemische Untersuchungen über Apocynaceen. Nachtrage. *Helvetica Chimica Acta*, **45**, 1375–9.

Abisch, E. and T. Reichstein (1962b) Orientierende chemische Untersuchungen einiger Asclepiadaceae und Periplocaceae. *Helvetica Chimica Acta*, **45**, 2092–116.

Adesogan, E. K. and I. A. Olatunde (1974) Pharmacology of *Adenia cissampeloides. West African Journal of Pharmacology and Drug Research*, **1**, 39.

Adrian, J. and R. Jacquot (1968) *Valeur Alimentaire de l'Arachide et de ses dérivés*. Maisonneuve and Larose, Paris, 549 refs. (cited in Kerharo and Adam, 1974).

Aebi, L. and T. Reichstein, (1950) *Cryptostegia* Glykoside. *Helvetica Chimica Acta*, **33**, 1013.

Aleshkina, J. A. and V. V. Berezhinskajà (1962) Pharmacologie des glycosides de *Thevetia peruviana. Farmakologiya i Toksikologiya SSSR*, **25**, 720–5.

Algeier, H., E. Weiss and T. Reichstein (1967) Cardenolide von *Mansonia altissima*. Die Struktur von Mansonin und Strophothevosid. *Helvetica Chimica Acta*, **50**, 456–62.

Alles, G. A. (1952) A comparative study of the pharmacology of certain crytopine alkaloids. *Journal of Pharmacology*, **104**, 1253–63.

Almquist, H. S. and A. A. Klose (1939) The antihemorrhagic activity of certain naphthoquinones. *Journal of the American Chemical Society*, **61**, 1923–4.

Ambrosia, L. and A. Mangieri (1955) Thevetin in the treatment of cardiac insufficiency. *Journal of the American Medical Association*, **157**, 394.

Ansa-Asamoa, G. E. A. (1967) Pharmacology of some oxyindole and bis-benzyl-isoquinoline alkaloids of some Ghanaian plants, Thesis Pharm., Kumasi, Ghana.

Archibald, R. G. (1933–4) *Transactions of the Royal Society of Tropical Medicine and Hygiene*, **27**, 207 (cited in Watt and Breyer-Brandwijk, 1962).

Arora, R. B., J. N. Sharma and M. L. Bhatia (1967) Pharmacological evaluation of peruvoside, a new cardiac glycoside from *Thevetia neriifolia*; with a note on its clinical trials in patients with congestive heart failure. *Indian Journal of Experimental Biology*, **5**, 31–6.

Atal, C. K. and P. D. Sethi (1962) Proteolytic activity of some Indian plants. II. Isolation, properties and kinetic studies of calotropain. *Planta Medica*, **10**, 77–90.

Attanasi, O. and C. Caglioti (1970) I costituente dell anacardio il liquido del guscio della noce, *Industrie Agrarie Italia*, **8**, 28–34.

Augusti, K. T. (1976a) Gas chromatographic analysis of onion principles and a study of their hypoglycaemic action. *Indian Journal of Experimental Biology*, **14**, 110–12.

Augusti, K. T. (1976b) Chromatographic identification of certain sulphoxides of cysteine present in onion (*Allium cepa* L.). *Current Sciences*, **45**, 863–4.

Augusti, K. T. and P. T. Mathew (1974) Lipid lowering effect of allicin (diallyl disulphide oxide) on long term feeding to normal rats. *Experientia*, **30**, 468.

Augusti, K. T. and P. T. Mathew (1975) Effect of allicin on certain enzymes of liver after a short time feeding to normal rats. *Experientia*, **31**, 148–9.

Badger, C. M., J. W. Cook and P. A. Ongley (1950) *Mitragyna* alkaloids. *Journal of the Chemical Society*, 867–73.

Bamgbose, S. O. (1974) Preliminary studies of water-soluble extracts of *Erythrophleum sauveolus* (Guill. & Perr.) Brenam bark. *West African Journal of Pharmacology and Drug Research*, **1**, 32–41.

Bapat, S. K., K. Ansari, A. C. Jauhari and V. Chandra (1970) Hypoglycaemic effect of two indigenous plants. *Indian Journal of Pharmacology*, **14**, 28–34.

Bartlett, M. F. and W. I. Taylor (1963) Burnamicine, a cryptopine-like analog of corynantheol. *Journal of the American Chemical Society*, **85**, 1203–4.

Bartlett, M. F., W. I. Taylor and R. Raymond-Hamet (1959) Sur la constitution de quatre alkaloïdes isolés des écorces de *Hunteria eburnea* Pichon; l'éburnamine, l'iso-éburnamine, l'éburnamenine et l'éburnamonine. *Comptes Rendus de l'Académie des Sciences*, **249**, 1259–60.

Bartlett, M. F., R. Sklar, A. F. Smith and W. I. Taylor (1963) The alkaloids of *Hunteria eburnea* Pichon. III. The tertiary bases. *Journal of Organic Chemistry USA*, **28**, 2197–8.

Basu, N. K., H. K. Singh and O. P. Aggarwal (1957) Chemical investigations of *Achyranthes aspera*. *Journal and Proceedings of the Institution of Chemists, India*, **29**, 55–8.

Bate-Smith, E. C. (1962) Phenolic constituents of plants. *Journal of the Linnean Society of London*, **58**, 95–173.

Beauquesne, L. (1947) Le Samagoura (*Swartzia madagascariensis*), Légumineuse africaine. *Annales Pharmaceutiques Françaises*, **5**, 470–83.

Beckett, A. H., E. J. Shellard and A. N. Tackie (1963) The *Mitragyna* species of Ghana. The alkaloids of the leaves of *M. stipulosa* (DC.) Ktze. *Journal of Pharmacy and Pharmacology GB*, **15**, 158T–165T.

Berthold, R., N. Wehrli and T. Reichstein (1965) Die Cardenolide von *Parquetina nigrescens* (Afzel.) Bull. *Helvetica Chimica Acta*, **48**, 1634–65.

Bevan, C. W. L. and A. U. Ogan (1964) Biogenesis of carpaine in *Carica papaya*. *Phytochemistry*, **3**, 591–4.

Bezanger-Beauquesne, L. (1956, 1961) Les substances polyuroniques (gommes mucilages, pectines, pseudocelluloses). *Annales pharmaceutiques françaises*, **14**, 795–812 (1956); **19**, 771–91 (1961).

Bezanger-Beauquesne, L. and M. Pinkas (1971) Propriétés antitumorales et antileucémiques des plantes. *Revue Française de la Pharmacie*, **9**, 42–52.

Bhandari, P. R. and B. Mukerjee (1959) *Garlic (Allium sativum) and its medicinal values*, Nagarjun, India, 1959, 121 (cited in Indian Council 1976).

Bhargava, U. C. and B. A. Westfall (1969) The mechanism of bloodpressure depression by ellagic acid. *Proceedings of the Society for Experimental Biology and Medicine*, **132**, 754–64.

Bhargava, U. C., B. A. Westfall and D. J. Siehr (1968) Preliminary pharmacology of ellagic acid from *Juglans nigra*. *Journal of Pharmaceutical Sciences*, **57**, 1728–32.

Bisset, N. G. (1961) Paper chromatographic study of the glycosides from *Thevetia peruviana*. *Annales Bogoriensis*, **4**, 145–52.

Bisset, N. G. (1962) Cardiac glycosides. Part VI. Moraceae. The genus *Antiaris* Lesch. *Planta Medica*, **10**, 143–51.

Bisset, N. G., J. Euw, M. Frèrejacque, S. Rangaswami, O. Schindler and T. Reichstein (1962) Die Cardenolide von *Thevetia peruviana*. *Helvetica Chimica Acta*, **45**, 938–43.

Blaise, H. (1932) Les *Crossopteryx* Africains. Etude botanique, chimique et pharmacologique. *Travaux des Laboratoires de Matière Médicale de la Faculté de Pharmacie, Paris*, **4**, 1–81.

Blanpin, O., A. Quevauviller and Cl. Pontus (1961) Apocynacées. Sur la voacangine, alcaloïde du *Voacanga africana* Stapf. *Thérapie*, **16**, 941–5.

Bloch, R., S. Rangaswami and O. Schindler (1960) Die Konstitution von Cerberosid (Thevetin B), Thevetin A und Peruvosid. *Helvetica Chimica Acta*, **43**, 652–8.

Boakiji Yiadom, K. (1979) Antimicrobial properties of West African plants. II. Antimicrobial activity of aqueous extracts of *Cryptolepis sanguinolenta* Schltr. *Quarterly Journal of Crude Drug Research*, **17**, 78–80.

Booth, A. N., E. M. Bishoff and G. O. Kohler (1960) Estrogen-like activity in vegetable oils and mill by-products. *Science (Washington)*, **131**, 1807–8.

Bose, B. C., G. C. Sepaha, R. Vijayvargiya and Q. Saifi (1961) Observations on the pharmacological actions of *Jatropha curcas*. *Archives Internationales de Pharmacodynamie*, **130**, 28–31.

Bose, B. C., R. Vijayvargiya, A. Q. Saifi and S. K. Sharma (1963) Chemical and pharmacological studies of *Argemone mexicana*. *Journal of Pharmaceutical Sciences*, **52**, 1172.

Boudreaux, H. B. and V. L. Frampton (1960) A peanut factor for haemostasis in haemophilia. *Nature (London)*, **185**, 469–70.

Boudreaux, H. B., R. M. Boudreaux, M. Brandon, V. L. Frampton and L. S. Lee (1960) Bioassay of a haemostatic factor from peanuts. *Archives of Biochemistry and Biophysics*, **89**, 276–80.

Bouquet, A. and Fournet, A. (1975) Recherches récentes sur les plantes médicinales congolaises. *Fitoterapia*, **46**, 243–6

Bowden, K. (1962) Isolation from *Paullinia pinnata* L. of material with action on the frog's isolated heart. *British Journal of Pharmacology and Chemotherapy*, **18**, 173–4.

Bowden, K. and K. J. Ross (1963) The local anaesthetic in *Fagara xanthoxyloides*. *Journal of the Chemical Society*, 3503–5.

Broadbent, J. L. (1962) Cardiotonic action of two tannins. *British Journal of Pharmacology and Chemotherapy*, **18**, 167–72.

Büchi, G., Manning, R. E. and Monti, S. A. (1953) Voacamine. *Journal of the American Chemical Society*, **85**, 1893–4.

Büchi, G., Manning, R. E. and Monti, S. A. (1964) Voacamine and voacorine. *Journal of the American Chemical Society*, **86**, 4631–41.

Buxton, J., M. Grundy, D. C. Wilson and D. G. Jamison (1954) *British Journal of Nutrition*, **8**, 170 (cited in Kerharo and Adam, 1974).

Caiment-Leblond, J. (1957) Contribution à l'étude des plantes médicinales de l'A.O.F. et de l'A.E.F. *Thèse Doc. Pharm.*, Paris, 1957.

Calderwood, J. M. and F. Fish (1966) Screening for tertiary and quaternary alkaloids in some African *Fagara* species. *Journal of Pharmacy and Pharmacology*, **18**, 119S–125S.

Cepelak, V. and Z. Horacova (1963) Protease inhibitors from groundnut skins. *Nature (London)*, **198**, 295.

Chakrabasti, J. K. and N. K. Senn (1954) Corchorogenin, a new cardiac-active aglycone from *Corchorus olitorius* L. *Journal of the American Pharmaceutical Association*, **76**, 2390.

Chakravarti, R. N., P. C. Marti and T. K. Saha (1954) A preliminary note on fractionating of *Argemone* alkaloids. *Bulletin of the Calcutta School of Tropical Medicine*, 2.

Chandrasekhar, N. V., C. S. Vaidyanathan and M. Sirsi (1961) Some pharmacological properties of the blood anticoagulant obtained from the latex of *Carica papaya*. *Journal of Scientific and Industrial Research*, **20**, 213–15.

Chatterjee, A. and M. N. Mitra (1951) Isolation of spirochine from the roots of *Moringa oleifera*. *Science and Culture*, **17**, 43.

Chen, K. K. and F. G. Henderson (1962) Cardiac activity of Apocynaceous glycosides and aglycones. *Archives Internationales de Pharmacodynamie et de Thérapie*, **140**, 8–19.

Chen, K. K., C. I. Bliss and E. Brown Robbins (1942) The digitalis-like principles of *Calotropis* compared with other cardiac substances. *Journal of Pharmacology and Experimental Therapy*, **74**, 223–34.

Chevalier, A. (1947) Les Mostuea africains et leurs propriétes stimulantes. *Revue de Botanique Appliquée et d'Agriculture Tropicale*, **27**, 104–9.

Chopra, R. N., S. L. Nayar and I. C. Chopra (1956) *Glossary of Indian Medicinal Plants*, Council of Scientific and Industrial Research, New Delhi, 330 pp.

Chopra, R. N., I. C. Chopra, K. Handa and L. D. Kapur (1938) *Chopra's Indigenous Drugs of India*, Dhur, Calcutta, 366 pp.

Cliffton, E. E., D. Agostino and A. Girolami (1965) Prevention of traumatic bleeding by ellagic acid in rats. *Proceedings of the Society for Experimental Biology and Medicine*, **120**, 179–80.

Cotten, M. de V., L. J. Goldberg and R. P. Walton (1952) Cassaine and cassaidine quantitative measurements of heart contractile force *in situ*. *Journal of Pharmacology and Experimental Therapeutics*, **106**, 94–102.

Crout, D. H. G., C. H. Hassal and T. L. Jones (1964) Cardenolides IV. Uscharidin, calotropin and calotoxin. *Journal of the Chemical Society*, 2187–94.

Crout, D. H. G., H. F. Curtis, C. H. Hassal and T. L. Jones (1963) The cardiac glycosides of *Calotropis procera*. *Tetrahedron Letters, GB*, **2**, 63–6.
Dalma, G. (1939) Zur Kenntnis der *Erythrophleum* Alkaloide. I. Cassain, ein krystallisiertes Alkaloid aus der Rinde von *E. guineensis* Don. *Helvetica Chimica Acta*, **22**, 1497–1512.
Dalziel, J. M. (1937) *The Useful Plants of West Tropical Africa*, Crown Agents, London.
Dang Van Ho (1955) Traitement et prevention de l'hypertension et de ses complications cérébrales par l'extrait total de *Morinda citrifolia*. *Presse Médicale*, 1878.
Dar, R. N., L. C. Garg and R. D. Pathak (1965) Anthelmintic activity of *Carica papaya* seeds. *Indian Journal of Pharmacy*, **27**, 335–6.
Darwish Sayed, M. (1980) Traditional medicine in health care. *Journal of Ethnopharmacology*, **2**, 19–22.
Das, B. R., P. A. Kurup and P. L. Narasimha Rao (1957a) Antibiotic principle from *Moringa pterygosperma*. Part VII. Antibacterial activity and chemical structure of compounds related to pterygospermine. *Indian Journal of Medical Research*, **45**, 191–6.
Das, B. R., P. A. Kurup, P. L. Narasimha Rao and A. S. Ramaswany (1957b) Antibiotic principle from *Moringa pterygosperma*. Part VIII. Some pharmacological properties and *in vivo* action of pterygospermine and related compounds. *Indian Journal of Medical Research*, **45**, 197–206.
Datta, S. K. and P. C. Datta (1977) Pharmacognosy of *Thevetia peruviana* bark. *Quarterly Journal of Crude Drug Research*, **15**, 109–24.
Delaveau, P. (1966) *Rauwolfias*. *Pharmaceutisch Weekblad*, **101**, 73–88 (227 refs.).
Derasari, H. R. and G. F. Shah (1965) Preliminary pharmacological investigation of the roots of *Calotropis procera* R. Br. *Indian Journal of Pharmacy*, **27**, 278–80.
Dhar, M. L., M. M. Dhar, B. N. Dhawan, B. N. Mehrotra and C. Ray (1968) Screening Indian plants for biological activity. Part I. *Indian Journal of Experimental Biology*, **6**, 232–47.
Doskotch, R. M., M. Y. Malik, Ch. D. Hufford, J. E. Trent and W. Kubelka (1972) Antitumor agents. V. Cytotoxic cardenolides from *Cryptostegia grandiflora* (Roxb.) R. Br. *Journal of Pharmaceutical Sciences*, **61**, 570–3.
Dossaji, S. F., M. G. Kairu, A. T. Gondwe and J. H. Ouma (1977) On the molluscicidal properties of *Polygonum senegalense* forma *senegalense*. *Lloydia*, **40**, 290–3.
Duret, S. and R. R. Paris (1972) Apocynacées: Sur les flavonoïdes et les acides-phenols de quelques Echitoïdes appartenant aux genres *Baissea, Echitea, Nerium* et *Strophanthus*. *Plantes Médicinales et Phytothérapie*, **6**, 210–15.
Duret, S. and R. R. Paris (1977) Sur les flavonoïdes de divers Bauhinias: *B. Vahlii* Wright, *B. variegata* Lindl. et *B. malabarica* Roxb. *Plantes Médicinales et Phytothérapie*, **11**, 213–21.
Dutta, A. and Gosh, S. (1947) Chemical examination of *Daemia extensa*. *Journal of the American Pharmaceutical Association*, **36**, 250–2.
Dykman, M. J., M. L. Boss, W. Lichter, M. M. Sigel, J. E. O'Connor and R. Search (1966) Cytotoxic substances from tropical plants. *Cancer Research*, **26** *Suppl.* 2, 1121–30.
Ekong, D. E. U. and O. G. Idemudia (1967) Constituents of some West-African members of the genus *Terminalia*. *Journal of the Chemical Society*, 863–4.
Euw, J. V. and T. Reichstein (1950a) Cardenolide von *Strophanthus hispidus*. *Helvetica Chimica Acta*, **33**, 1546–50.
Euw, J. V. and T. Reichstein (1950b) Die Glykoside der Samen von *Strophanthus sarmentosus* D.C. *Helvetica Chimica Acta*, **33**, 2153–7.
Farnsworth, N. R. and G. A. Cordell (1976) A review of some biologically active compounds isolated from plants as reported in the 1974–75 literature. *Lloydia*, **39**, 420–55.
Fattorusso, V. and O. Ritter (1967) *Dictionnaire de Pharmacologie Clinique*, Masson, Paris (cited in Kerharo and Adam, 1974).
Fauconnet, L. and P. L. Pouly (1962) *Les Cardénolides du laurier-rose. Volume Jubilaire A. Mirimanoff*, Société Suisse de Pharmacie, Geneva, pp. 41–8.
Fechtig, B., J. V. Euw, O. Schindler and T. Reichstein (1960) Die Struktur der Sarmentoside. Glykoside von *Strophanthus sarmentosus* D.C. *Helvetica Chimica Acta*, **43**, 1570–84.

Feng, P. C., L. J. Haynes, K. E. Magnus and J. R. Plimmer (1964) Further pharmacological screening of some West Indian medicinal plants. *Journal of Pharmacy and Pharmacology*, **16**, 115–17.

Fieser, L. F., H. Tischler, W. L. Sampson and S. Woodford (1941) Vitamin K activity and structure. *Journal of Biological Chemistry*, **137**, 659–92.

Fish, F. and P. G. Waterman (1971) Alkaloids from two Nigerian species of *Fagara*. *Journal of Pharmacy and Pharmacology*, **23**, 123S–125S.

Fish, F. and P. G. Waterman (1972) Methanol soluble quaternary alkaloids from *Fagara* species. *Phytochemistry*, **11**, 3007–14.

Foussard-Blanpin, O., A. Quevauviller and J. Bretaudeau (1969) Etude pharmacodynamique des effets cardiaques de la triacanthine. *Annales pharmaceutiques françaises*, **27**, 1257–9.

Frèrejacque, M. (1947) Thevetine, neriifoline et monoacetylneriifoline. *Comptes Rendus de l'Académie des Sciences*, **225**, 695–6.

Frèrejacque, M. and M. Durgeat (1954) Poisons digitaliques des graines de jute. *Comptes Rendus de l'Académie des Sciences*, **238**, 507–9.

Frèrejacque, M. and M. Durgeat (1971) Structure de la thévéfoline. *Comptes Rendus de l'Académie des Sciences, Série D*, **272**, 2620–1.

Fuhrer, H., R. F. Zürcher and T. Reichstein (1969) Sarverogenin, vermutliche Struktur. *Helvetica Chimica Acta*, **52**, 616–21.

Garg, L. C., P. K. Roy and A. Dutta (1963) Anthelmintic activity of calotropain and bromelin. *Indian Journal of Pharmacy*, **25**, 422–3.

Geiger, U. P., E. Weiss and T. Reichstein (1967) Die Cardenolide der Samen von *Strophanthus gratus*. *Helvetica Chimica Acta*, **50**, 179–206.

Gellert, E. and H. Schwartz (1951) Die Isolierung von Sempervirin aus *Mostuea buchholzii*. *Helvetica Chimica Acta*, **34**, 779–80.

Gellert, E., R. Raymond-Hamet and E. Schlitter (1951) Die Konstitution des Alkaloids Cryptolepin. *Helvetica Chimica Acta*, **34**, 642–51.

Giono, P., A. Laurens, P. Dreyfus and H. Giono (1971) Recherches sur l'action antihypertensive d'un extrait d'*Anacardium occidentale*. Communication, *Journées Médicales Dakar* (cited in Kerharo and Adam, 1974).

Girolami, A., D. Agostino and E. E. Cliffton (1966) The effect of ellagic acid on coagulation *in vivo*. *Blood*, **27**, 93–102.

Githens, T. S. (1949) *Drug Plants of Africa*. University of Pennsylvania Press, The University Museum, Philadelphia, 125 pp.

Gopalachari, R. and M. L. Dhar (1958) Studies in the constitution of the saponins from the seeds of *Achyranthes aspera*. Identification of the sapogenin. *Journal of Scientific and Industrial Research, India*, **17B**, 276–8.

Goutarel, R. (1964) *Les Alcaloïdes Stéroïdiques des Apocynacées*, Hermann, Paris.

Guedel, J. (1955) Contribution à l'étude des ictères en AOF. Le diekouadio (Côte d'Ivoire). *Notes Africaines IFAN, Dakar*, **66**, 50–3.

Gupta, S. S., A. W. Bhagwat and A. K. Ram (1972a) Cardiac stimulant activity of the saponin of *Achyranthes aspera* L. *Indian Journal of Medical Research*, **60**, 462–71.

Gupta, J. C., P. K. Roy, G. K. Ray and A. Dutta (1950) Pharmacological action of an active constituent isolated from *Daemia extensa* L. *Indian Journal of Medical Research*, **38**, 75–82.

Gupta, S. S., S. C. L. Verma, A. K. Ram and R. M. Tripathi (1972b) Diuretic effect of the saponin of *Achyranthes aspera*. *Indian Journal of Pharmacology*, **4**, 208.

Hakim, S. A. E. (1954) Argemone oil, sanguinarine, and epidemic dropsy glaucoma. *British Journal of Ophthalmology*, **272**, 193–216.

Harborne, J. B., T. J. Mabry and H. Mabry (1974) *The Flavonoids*. Chapman and Hall, London.

Hardman, R. and E. A. Sofowora (1972) A reinvestigation of *Balanitis aegyptiaca* as a source of steroidal sapogenins. *Economic Botany*, **26**, 169–73.

Hariharan, V. and S. Ranjaswani (1970) Structure of saponins A and B from the seeds of *Achyranthes aspera* L. *Phytochemistry*, **9**, 409.

Harley, G. W. (1941) *Native African Medicine with Special Reference to its Practice in the Mano Tribe of Liberia*. Harvard University Press, Cambridge, MA.

Hauth, H. (1971) Cardioactive synthetic cassaine analogs. *Lloydia*, **33**, 490.

Heal, R. F. and E. F. Rogers (1950) A survey of plants for insecticidal activity. *Lloydia*, **13**, 89–162.

Heckel, E. and F. Schlagdenhaufen (1888) Sur le Batjentjor (*Vernonia nigritiana* S. et H.) de l'Afrique occidentale et sur son principe actif, la vernonine, nouveau poison du coeur. *Comptes Rendus de* l'*Académie des Sciences*, **106**, 1448–9.

Hegnauer, R. (1962–8) *Chemotaxonomie der Pflanzen*, 5 vols., Birkhäuser, Basel and Stuttgart.

Henry, T. A. (1949) *The Plant Alkaloids*. Churchill, London, 4th edn, 803 pp.

Hess, J. C. and A. Hunger (1953) Identifizierung von Honghelosid G mit Somalin. *Helvetica Chimica Acta*, **36**, 85–7.

Hesse, C. and G. Ludwig (1960) Calotropogenin. *Annalen der Chemie*, **632**, 158.

Hocking, G. M. (1976) *Asclepias curassavica* Herba et Radix. *Quarterly Journal of Crude Drug Research*, **14**, 61–3.

Holland, J. H. (1908–29) *The Useful Plants of Nigeria*, Kew Bulletin, London, pp. 3–352.

Hood, A. M. and E. J. L. Lowburry (1954) Anthocyanins in bananas. *Nature (London)*, **173**, 402–3.

Hörhammer, L. and H. Wagner (1962) Citrusbioflavonoids, *Deutsche Apotheker-Zeitung*, **102**, 759–65.

Hufford, Ch. D. and B. O. Oguntimein (1978) Nonpolar constituents of *Jatropha curcas*. *Lloydia*, **41**, 161–5.

Hunger, A. and T. Reichstein (1950) Glykoside aus *Adenium honghel* D.C. *Helvetica Chimica Acta*, **33**, 76–99.

Idemudia, O. G. and D. E. U. Ekong (1968) Constituents of some West African members of the genus *Terminalia*. *First Inter-African Symposium on Traditional Pharmacopoeias and African Medicinal Plants*. ◡AU/STRC, Dakar, Publ. No. 104.

Indian Council of Medical Rese[illegible]h (1976) *Medicinal Plants of India*, Vol. I, New Delhi, p. 12.

International Commission for the Standardization of Pharmaceutical Enzymes (1965) *World Journal of Pharmacy*, 5–32.

Irvine, F. R. (1930) *Plants of the Gold Coast*, Oxford University Press, London, 521 pp.

Isaac-Sodeye, W. A., E. A. Sofowora, R. O. Williams, V. O. Marquis, A. A. Adekunle and C. O. Anderson (1975) Extracts of *Fagara zanthoxyloides* root in sickle cell anaemia. *Acta Haematologica*, **53**, 58.

Jacquemain, H. (1970, 1971) Recherches sur les anthocyanes foliaires de trois arbres tropicaux: *Mangifera indica* L., *Theobroma cacao*, L., *Lophira alata* Banks ex Gaertn., *Plantes Médicinales et Phytothérapie*, **4** (1970), 230–59, 306–41; **5** (1971), 45–94.

Jain, R. C. and C. R. Vyas (1974) Hypoglycaemic action of onion on rabbits. *British Medical Journal*, **2**, 730.

Jain, S. R. (1968) *Musa sapientum*. *Planta Medica*, **16**, 43–7.

Janot, M. M., A. Cave and R. Goutarel (1959) Togholamine, holamine, holaphyllamine et holaphylline. Trois nouveaux alcaloïdes retirés des feuilles d'*Holarrhena floribunda*. *Bulletin de la Société Chimique de France*, 896–900.

Janot, M. M., A. Cave and R. Goutarel (1960) Alcaloïdes stéroïdes. Holaphyllamine et holamine, alcaloides de l'*Holarrhena floribunda* (G. Don) Dur. et Schinz. *Comptes Rendus de l'Académie des Sciences*, **251**, 559–61.

*Jayaweera, D. M. A. (1945–54) Indigenous and exotic drug plants growing in Ceylon. *Tropical Agriculturist (Ceylon)*, **101** (1945), 130–5; **108** (1952), 109–15; **110** (1954), 105–16.

Kapoor, V. K. and H. Singh (1966) Isolation of betaine from *Achyranthes aspera*. *Indian Journal of Chemistr*[illegible], **4**, 461.

Keller, L. and C. Tamm (1959) Die Glykoside der Samen von *Strophanthus hispidus* DC. *Helvetica Chimica Acta*, **42**, 2467–8.

Kerharo, J. (1968) Revue des plantes médicinales et toxiques du Sénégal, *Plantes Médicinales et Phytothérapie*, **2**, 108–46.

Kerharo, J., and J. G. Adam (1974) *La Pharmacopée Sénégalaise Traditionnelle*, Vigot, Paris, 1011 pp.

Kerharo, J. and A. Bouquet (1950) *Plantes Médicinales de la Côte d'Ivoire et Haute Volta*, Vigot, Paris, 297 pp.

Kiteava, R. I. (1966) *Farmakologiya i Toksikologiya SSSR*, **29**, 438–41.

Knipel, M., B. Brusoni and A. Cadel (1971) Modificazioni emodinamiche dopo sommistrazione endovenosa di ajmalina in pazienti cardiopatici. *Bollettino della Societa Italiana di Cardiologia*, **16**, 47–56.

Kohli, J. D. and M. M. Vohra (1960) Pharmacological studies on peruvoside, a new cardiac glycoside from *Thevetia neriifolia* Juss. *Archives Internationales de Pharmacodynamie*, **126**, 412–25.

Kon, G. A. R. and W. T. Weller (1939) Sapogenins IV. The sapogenin of *Balanites aegyptiaca* Wall. *Journal of the Chemical Society*, (1939) 800–801.

Kondagbo, B. and P. Delaveau (1974) Chimiotaxonomie des Capparidaceae. *Plantes Médicinales et Phytothérapie*, **8**, 96–103.

Krewson, C. F. and J. Naghski (1953) Occurrence of rutin in plants. *American Journal of Pharmacy*, **125**, 190–200.

Kupchan, S. M., J. R. Knox, J. E. Kelsey and J. A. Saenz-Renauld (1964) Calotropin, a cytotoxic principle from *Asclepias curassavica*. *Science*, **146**, 1685–6.

Kupchan, S. M., M. Mokotoff, R. S. Sandhu and L. E. Hokin (1967) *Journal of Medicinal Chemistry*, **10**, 1025 (Cited in Doskotch *et al.*, 1972).

Kuritzkes, A. M., C. Tamm, C. Jäger and T. Reichstein (1963) Die Glykoside von *Xysmalobium undulatum* R. Br. *Helvetica Chimica Acta*, **46**, 8–23.

Kurup, P. A. and P. L. Narasimha Rao (1954) Antibiotic principle from *Moringa pterygosperma* II. Chemical nature of pterygospermine. *Indian Journal of Medical Research*, **42**, 85–95, 115–23.

Kurup, P. A., P. L. Narasimha Rao and R. S. Ramaswany (1957) Antibiotic principle from *Moringa pterygosperma*. Part VIII. Some pharmacological properties and *in vivo* action of pterygospermine and related compounds. *Indian Journal of Medical Research*, **45**, 197–206.

La Barre, J. and L. Gillo (1955) A propos des propriétés cardiotoniques de la voacangine et de la voacanginine. *Bulletin de l'Académie Royale de Belgique*, **20**, 194.

La Barre, J. and L. Gillo (1958) Sur les caractères chimiques de la réserpiline et sur l'absence de propriétés ulcérigènes gastriques. *Comptes Rendus des Séances de la Société de Biologie Paris*, **152**, 530–1.

La Barre, J. and J. J. Demarez (1958) A propos de l'action neuroleptique de la raumitorine. *Comptes Rendus des Séances de la Société de Biologie Paris*, **152**, 1272–3.

La Barre, J. and M. J. Hans (1958) Sur les caractères chimiques et les propriétés hypotensives de la raumitorine. *Comptes Rendus des Séances de la Société de Biologie Paris*, **152**, 1269–70.

La Barre, J. and C. Wirtheimer (1962) Etude comparative des effets hypotenseurs des extraits et dérivés du *Rauvolfia vomitoria* et du *Morinda lucida* chez le rat éveillé. *Archives Internationales de Pharmacodynamie et de Thérapie*, **139**, 596–603.

La Barre, J., C. S. Liber and J. Castiau (1958) A propos des effets de la raumitorine sur la sécrétion et la mobilité gastriques. *Comptes Rendus des Séances de la Société de Biologie Paris*, **152**, 1270–1.

Lambin, S. and J. Bernard (1953) Action de quelques alcaloïdes sur le *Mycobacterium tuberculosis*. *Comptes Rendus des Séances de la Société de Biologie Paris*, **147**, 638–41, 760–70.

Lampertico, M. (1971) Valutazione della terapia ajmalina in 187 pazienti. *Minerva Medica*, **62**, 1797–809.

Latif, A. (1959) Isolation of a vitamin K-active compound from the leaves of *Lawsonia*. Chemical composition of the air-dried leaves. *Indian Journal of Agricultural Science*, **29**, 147–50.

Latour, R. (1957) Contribution à l'étude de quelques quinones d'origine végétale. Thèse Doc. Pharm., Paris.

Laurens, A. and R. R. Paris (1976) Sur les polyphenols d'Anacardiacées africaines et malgaches, *Poupartia* species and *Anacardium occidentale*. *Plantes Médicinales et Phytothérapie*, **11**, 16–24.

Le Double, G., L. Olivier, M. Quirin, J. Lévy, J. Le Men and M. M. Janot (1964) Alcaloïdes du *Picralima nitida* Stapf. VIII. Etude des feuilles et des racines: isolation de 2 alcaloïdes nouveaux, picraphylline, picracine. *Annales pharmaceutiques françaises*, **22**, 35–9, 463–8.

Le Men, J. and L. Olivier (1978) Alcaloïdes de deux espèces du genre *Hunteria: H. elliotii* Stapf Pichon et *H. congolana* Pichon. *Plantes Médicinales et Phytothérapie*, **12**, 173–85.

Lindwall, O., F. Sandberg and R. Thorsen (1965) Erythrophleguin, a new alkaloid from the bark of *E. guineense* G. Don. *Tetrahedron Letters*, **47**, 4203–8.

McIlroy, R. J. (1950) *The Plant Glycosides*. 1 vol. Arnold, London. 138 pp.

Malcolm, S. A. and E. A. Sofowora (1969) Antimicrobial activities of selected Nigerian folk remedies and their constituent plants. Antimicrobial properties of *Balanites*. *Lloydia*, **32**, 512–17.

Mameesh, M. S. (1963) Contraceptive principles in *Jatropha* seeds and fruits. *Planta Medica*, **11**, 98.

Manske, R. H. F. and H. L. Holmes (1950–71) *The Alkaloids*, 13 vols., Academic Press, New York and London.

Marker, R. E. (1947) New sources of sapogenin. *Journal of the American Chemical Society*, **69**, 2242.

Martell, M. J., T. O. Soine and L. B. Kier (1963) The structure of argemonine, identification as (−)-N-methylpavine. *Journal of the American Chemical Society*, **85**, 1022–3.

Martindale (1958, 1969) *The Extra Pharmacopoeia*, Pharmaceutical Press, London, 24th edn 1958, 25th edn 1969.

Martinez, M. (1959) *Las Plantas Medicinales de Mexico*, Andres Botas, Mexico D. F., 657 pp.

Massion, L. (1934) Contribution à l'étude de la mitraphylline. *Archives Internationales de Pharmacodynamie et de Thérapie*, **48**, 217–26.

Mauli, R. and Ch. Tamm (1957) Die Glykoside von *Periploca nigrescens* Afzel. *Helvetica Chimica Acta*, **40**, 299–305.

Merck Index of Chemicals and Drugs (1960) Merck, Rahway, NY, 7th edn.

Messmer, W. M., M. Tin Wa, H. S. Fong, C. Bevelle and N. R. Farnsworth (1972) Fagaronine a new tumor inhibitor isolated from *Fagara zanthoxyloides*. *Journal of Pharmaceutical Sciences*, **61**, 1858–9.

Mittal, O. P., Ch. Tamm and T. Reichstein (1962) The glycosides of *Pergularia extensa* Jacq. Glycosides and aglycones. *Helvetica Chimica Acta*, **45**, 907–24.

Morfaux, A. M., R. Olivier, J. Lévy and J. Le Men (1969) Alcaloïdes des feuilles d'*Hunteria eburnea* Pichon. *Annales Pharmaceutiques Françaises*, **27**, 679–86.

Morfaux, A. M., J. Vercauterin, J. Kerharo, L. Le Men Olivier and J. Le Men (1978) Alcaloïdes des feuilles d'*Hunteria elliotii*. *Phytochemistry*, **17**, 167–9.

Mourgue, M., R. Baret, R. Kassab and J. Reynaud (1961a) Etude des proteines de la graine de *Jatropha curcas* L. *Bulletin de la Société de Chimie Biologique*, **43**, 505–16.

Mourgue, M., J. Delphaut, R. Baret and R. Kassab (1961b) Etude de la toxalbumine (curcine) des graines de *Jatropha curcas*. *Bulletin de la Société de Chimie Biologique*, **43**, 517–31.

Muelenaere, H. J. H. de (1965) Toxicity and haemagglutinating activity of legumes. *Nature (London)*, **206**, 827–8.

Mühlrad, P., E. Weiss and T. Reichstein (1965) Die Cardenolide der Samen von *Antiaris toxicaria* Lesch. II. Strukturermittlungen Glykoside and Aglykone. *Liebig's Annalen der Chemie*, **685**, 253–61.

Müller, J. M., E. Schlitter and H. J. Bein (1952) Reserpin, der sedative Wirkstoff aus *Rauwolfia serpentina* Benth. *Experientia*, **8**, 338.

Murayama and Makyo (1972) Sickle cell hemoglobin (HbS). *West African Journal of Pharmacology and Drug Research*, **1**, 6P–14P.

Narayanan, K. N., G. S. Bains and D. S. Bhatia (1963) Lipoxidase inhibitor in groundnut testa. *Chemistry and Industry (London)*, 1558–9.

Neogi, N. C., R. D. Garg and R. S. Rathor (1970) Preliminary pharmacological studies on achyranthine. *Indian Journal of Pharmacy*, **32**, 43–6.

Noble, I. G. (1946–7) Fruta bomba (*Carica papaya*) in hypertension. *Annales de la Academia de Ciencias Medicias, Fisicas y Naturales de la Habana*, **85**, 198–203.

Ojha, D., S. N. Tripathi and G. Singh (1966) Role of an indigenous drug (*Achyranthes aspera*) in the management of reactions of leprosy. Preliminary observations. *Leprosy Review*, Indian Council of Medical Research Publication, 1966, pp. 37 and 115.

Oliver, B. (1960) *Medicinal Plants in Nigeria*. Nigerian College of Arts, Science and Technology, 138 pp.

Oliver-Bever, B. (1967) Quelques Apocynacées et Asclépiadacées cardiotoniques. *Quarterly Journal of Crude Drug Research*, **7**, 982–91.

Oliver-Bever, B. (1968) Selecting local drug plants in Nigeria. Botanical and chemical relationship in three families. *Quarterly Journal of Crude Drug Research*, **8**, 1194–211.

Oliver-Bever, B. (1971) Vegetable drugs for cancer therapy. *Quarterly Journal of Crude Drug Research*, **11**, 1665–83.

Oliver-Bever, B. and G. R. Zahnd (1979) Plants with oral hypoglycaemic action. *Quarterly Journal of Crude Drug Research*, **17**, 139–96.

Olivier, L., J. Lévy, J. Le Men and M. Janot (1965) Structure et configuration de picraline pseudo-akuammigine, akuammine et akuammiline (dérivés indoliniques). *Bulletin de la Société Chimique de France*, 868–76.

Olson, M. O. J. and I. E. Liener (1967) Some physical and chemical properties of concanavalin A, the phytohemagglutinin of Jackbean. *Biochemistry*, **6**, 105–12.

Ongley, P. A. (1953) Les alcaloïdes des *Mitragyna*. *Annales pharmaceutiques françaises*, **11**, 594–602.

Pandey, Y. N. (1975) A study on an important drug plant: *Tephrosia purpurea*. *Quarterly Journal of Crude Drug Research*, **13**, 65–8.

Paris, R., Les flavonoïdes (1971) *Pharmaceutisch Weekblad*, **106**, 214–23.

Paris, R. (1977) Plantes à flavonoïdes. Introduct. au colloque du 23.4.1977 sur médicaments d'origine naturelle et maladies vasculaires. *Plantes Médicinales et Phytothérapie*, **11** (Suppl.), 129–132.

Paris, R. and P. Delaveau (1966) Possibilités et limites de la Chimiotaxonomie. *Travaux des Laboratoires de Matière Médicale de la Faculté de Pharmacie, Paris*, **51**, 43–149.

Paris, R. and P. Delaveau (1977) Métabolisme et pharmacocinétique des flavonoïdes. *Plantes Médicinales et Phytothérapie*, **11** (Suppl.), 198–204.

Paris, R. and S. Duret (1973) Sur les polyphenols de diverses espèces d'*Holarrhena*. *Plantes Médicinales et Phytothérapie*, **7**, 145–50.

Paris, R. and S. Etchepare (1966) Sur le noircissement des feuilles et des fruits de *Thevetia peruviana* Pers. Isolement d'un chromogène identifié à l'aucuboside. *Comptes Rendus de l'Académie des Sciences, Serie D*, **262**, 1239–41.

Paris, R. and S. Etchepare (1967) Sur les flavonoïdes des feuilles de *Rauvolfia vomitoria* Afzel. *Annales pharmaceutiques françaises*, **25**, 783–96.

Paris, R. and S. Etchepare (1968) Presence de C-flavonosides chez une Rutacée africaine, *Teclea sudanica* Chev. *Annales pharmaceutiques françaises*, **26**, 51–3.

Paris, R. and R. Letouzey (1960) Répartition des alcaloïdes chez le *Johimbe*. *Journal d'Agriculture Tropicale et de Botanique Appliquée*, **7**, 256.

Paris, R. and J. Moury (1964) Action sur la perméabilité capillaire de divers types de flavonoïdes. *Annales pharmaceutiques françaises*, **22**, 489.

Paris, R. and H. Moyse-Mignon (1947) Etude préliminaire du *Fagara xanthoxyloides* Lam. *Annales pharmaceutiques françaises*, **5**, 410–20.

Paris, R. and H. Moyse-Mignon (1949a) Pouvoir antimicrobien et présence de plumbagol chez deux *Diospyros* africaines. *Comptes Rendus de l'Académie des Sciences*, **228**, 2063–4.

Paris, R. and H. Moyse-Mignon (1949b) Etude chimique préliminaire d'une Loganiacée du Gabon, *Mostua stimulans* Chev. *Comptes Rendus de l'Académie des Sciences*, **229**, 86–8.

Paris, R. and H. Moyse-Mignon (1951) A propos des feuilles de baobab, *Adansonia digitata* L. Composition chimique et action physiologique. *Annales pharmaceutiques françaises*, **9**, 472–9.

Paris, R. and H. Moyse (1967) *Précis de Matière Médicale*, Vol. 2, Masson, Paris.

Paris, R. and H. Moyse (1971) *Précis de Matière Médicale*, Vol. 3, Masson, Paris.

Paris, R. and M. Rigal (1940) Recherches préliminaires sur l'écorce d'*Erythrophleum guineense*. *Bulletin des Sciences Pharmacologiques*, **47**, 79–87.

Paris, R. and M. Rigal (1941) Recherches sur les *Erythrophleum*. II. Les *Erythrophleum* de l'Afrique occidentale. *Bulletin des Sciences Pharmacologiques*, **48**, 362–72.

Paris, R., J. P. Grammond and R. Rousselet (1972) Sur l'analyse des citroflavonoïdes. *Plantes Médicinales et Phytothérapie*, **6**, 292–8.

Patel, M. B. and J. M. Rowson (1964) Investigations of certain Nigerian medicinal plants. Part I. Preliminary pharmacological and phytochemical screenings for cardiac activity. *Planta Medica*, **12**, 34–42.

Patel, M. B., J. Poisson, J. L. Pousset and J. M. Rowson (1964) Alkaloids of the leaves of *Rauwolfia vomitoria* Afz. *Journal of Pharmacy and Pharmacology*, **16**, 163T–165T.

Patel, P. N., A. Tye, J. W. Nelson and J. L. Beal (1963) A study of the alkaloids of *Thalictrum*. Pharmacology of an extract of *T. revolutum*. *Lloydia*, **26**, 299–35.

Patnaik, G. K. and B. N. Dhawan (1971) Asclepine, a new cardioactive glycoside from *Asclepias curassavica*. *Indian Journal of Pharmacy*, **3**, 18.

Perrot, S. and M. Leprince (1909) Sur l'*Adenium honghel*, poison d'épreuve du Soudan français. *Comptes Rendus l'Académie des Sciences*, **149**, 1393.

Plouvier, V. (1948) Sur la recherche des itols et du saccharose chez quelques Sapindales. *Comptes Rendus de l'Académie des Sciences*, **227**, 85–7.

Pobéguin, H. (1912) *Les Plantes Médicinales de la Guinée*, Paris.

Poisson, J. (1964) Recherches récentes sur les alcaloïdes du *Pseudocinchona* et du *Yohimbe*. *Annales de Chimie (Paris)*, **9**, 99–121.

Pourrat, H. (1977) Drogues à anthocyanes et maladies vasculaires. *Plantes Médicinales et Phytothérapie*, **11** (Suppl.), 143–51.

Pousset, J. L., M. Debray and R. R. Paris (1970) Sur le *Baissea leonensis* Benth (Apocynacées). Presence d'un nouvel hétéroside couminique, le baisseoside. *Comptes Rendus de l'Académie des Sciences, Serie D*, **271**, 2320.

Pousset, J. L., J. Poisson, L. Olivier, J. Le Men and M. Janot (1965) Sur la structure de la desacetyl- et desformo-akuammiline et de l'akuammiline (alcaloïdes indoliques). *Comptes Rendus de l'Académie des Sciences*, **261**, 5538–41.

Puech, P., H. Latour, J. Herbault and R. Grolleau (1964) L'ajmaline injectable dans les tachycardies paroxistiques et le syndrome de W.P.W., comparaison avec la procainamide. *Archives des Maladies du Coeur et des Vaisseaux*, **52**, 897–918.

Puiseux, F., M. P. Patel, J. M. Rowson and J. Poisson (1965) Alcaloïdes des *Voacanga*. *Voacanga bracteata* Stapf. *Annales pharmaceutiques françaises*, **23**, 33–9.

Quevauviller, A. and O. Blanpin (1957a) Etude pharmacodynamique de la voacamine, alcaloïde du *Voacanga africana*. Apocynacées. *Thérapie*, **12**, 635–47.

Quevauviller, A. and O. Blanpin (1957b) Etude pharmacodynamique comparée de la voacamine et de la voacorine, alcaloïdes du *Voacanga africana* Stapf (Apocynacées). *Annales pharmaceutiques françaises*, **15**, 617–30.

Quevauviller, A. and O. Blanpin (1961) Action de la triacanthine, alcaloïde dérivé naturel de l'adénine, particulièrement sur le système nerveux central. *Thérapie*, **16**, 782–90.

Quevauviller, A., O. Blanpin and Y. Takenaka (1963) Sur la pharmacodynamie de la rauvanine, nouvel alcaloïde extrait du *Rauvolfia vomitoria* Afzel. *Annales pharmaceutiques françaises*, **21**, 399–404.

Quevauviller, A., O. Foussard-Blanpin and J. Pottier (1965) Sur la vobtusine, alcaloïde du *Voacanga africana* Stapf. Apocynacées. *Comptes Rendus des Séances de la Société de Biologie Paris*, **159**, 821–5.

Quevauviller, A., G. Sarrazin and Y. Takenaka (1971, 1972) Action sur le système cardiovasculaire de la rauvanine, alcaloïde du *Rauvolfia vomitoria*. *Annales pharmaceutiques françaises*, **29**, 507 (1971); **30**, 81–2 (1972).

Quisumbing, E. (1951) *Technical Bulletin of Philippine Department of Agriculture* and *Natural Resources*, **16** (through Watt, 1967).

Rabaté, J. (1940) Etude du *Schwenkia americana* L. *Journal de Pharmacie et de Chimie*, **1**, 234–40.

Ram, A. K., W. Bhagwata and S. S. Gupta (1971) Phosphorylase activity of the saponin of *Achyranthes aspera* (Apamaya). *Indian Journal of Physiology and Pharmacology*, **15**, 6.

Ramachandran, C., K. V. Peter and P. K. Gopalakrishnan (1980) Drumstick, (*Moringa oleifera*) a multi-purpose Indian vegetable. *Economic Botany*, **34**, 276–82.

Ravina, A. (1964) Corps à action vitaminique P et flavonoïdes. *Presse Médicale*, **72**, 2855–7.

Ravina, A. and H. Wenger (1957) Le traitement local des ulcères de jambe par l'association de papaine et de penicilline. *Presse Médicale*, **65**.

Raymond-Hamet, R. and L. Millat (1934) Sur quelques effets physiologiques de l'échitamine, *Comptes Rendus des Séances de la Société de Biologie Paris*, **116**, 1022–5.

Raymond-Hamet, R. (1937) Sur quelques propriétés physiologiques des alcaloïdes du *Cryptolepis sanguinolenta* Schltr. *Comptes Rendus des Séances de la Société de Biologie Paris*, **126**, 768–70.

Raymond-Hamet, R. (1938) Sur les effets hypotenseurs et vasodilatateurs de la cryptolepine. *Comptes Rendus de l'Académie des Sciences*, **207**, 1016.

Raymond-Hamet, R. (1941) Action reno-dilatatrice de l'échitamine, *Comptes Rendus des Séances de la Société de Biologie Paris*, **135**, 1565–7.

Raymond-Hamet, R. (1944) *Picralima nitida*. Thèse Doc. Pharm., Paris, 1944.

Raymond-Hamet, R. (1951) Sur une drogue remarquable de l'Afrique tropicale, le *Picralima nitida* (Stapf) Th. & R. Dur. *Revue de Botanique Appliquée*, **31**, 465.

Raymond-Hamet, R. (1955) Sur quelques propriétés physiologiques d'une Apocynacée africaine. *Hunteria eburnea* Pichon. *Comptes Rendus de l'Académie des Sciences*, **240**, 1470–2.

Raymond-Hamet, R. and R. Goutarel (1965) L'*Alchornea floribunda* Müller Arg. doit-il à la yohimbine ses effets excitants chez l'homme? *Comptes Rendus de l'Académie des Sciences*, **216**, 3223–4.

Raymond-Hamet, R. and L. Millat (1934) Sur un nouvel alcaloïde des *Mitragyna*. *Comptes Rendus de l'Académie des Sciences*, **199**, 587–9.

Reichstein, T. (1963) Chemische Rassen von *Strophanthus sarmentosus* D.C. *Planta Medica*, **11**, 292–9 (61 refs.).

Renner, U. (1963) Hunteriamin, ein neues Alkaloid mit hypotensiver Wirkung aus *Hunteria eburnea* Pichon. *Hoppe-Seyler's Zeitschrift für Physiologische Chemie*, **331**, 105–8.

Rigal, M. (1941) Recherches botaniques, chimiques et pharmacologiques sur *l'Erythrophleum* de l'Afrique occidentale. Thèse Doc. Pharm., Paris.

Rigaud, A., J. Brisou and R. Babin (1956) Thérapeutiques locales à base de papaïne. *Presse Médicale*, **64**, 722.

Roberts, E. A. H. and M. Myers (1959) The phenolic substances of manufactured tea. Their origin as enzymatic products of fermentation. *Journal of the Science of Food and Agriculture*, **9**, 216–23.

Roberts, E. A. H. and D. M. Williams (1959) The phenolic substances of manufactured tea. Ultra violet and visible absorption spectra. *Journal of the Science of Food and Agriculture*, **9**, 212–16.

Robinson, B. and G. Spitteler (1964) Structure of eseramine. *Chemistry and Industry (London)*, 459.

Rowson, J. M. (1965) Recherches sur quelques plantes médicinales nigériennes. *Annales pharmaceutiques françaises*, **23**, 125–35.

Ruben, J. M., J. Shapiro, P. Muehlbauer and M. Grolnick (1965) Shock reaction following ingestion of mango. *Journal of the American Medical Association*, **193**, 397–8.

Rumen, N. M. (1975) Inhibition of sickling in erythrocytes by amino acids. *Blood*, **45**, 45.

Saluja, M. P., R. S. Kapil and S. P. Popli (1978) Studies in medicinal plants. Part IV. Chemical constituents of *Moringa oleifera* Lamk. *Indian Journal of Chemistry Sect. B*, **16B**, 1044–5.

Santi, R. (1939) Die pharmakologische Wirkung neuer Alkaloide die aus dem *Erythrophleum guineense* isoliert wurden. *Archiv. der experimentellen Pathologie und Pharmakologie*, **193**, 152.

Santi, R. and B. Zweifel (1936) Ricerche farmacologiche sui nuovi alcaloïdi isolati dall *Eritrofleo guineense* e da quallo del Madagascar. *Bulletin Societico Italiano Biologia Sperimentale*, **11**, 758.

Schaub, F., H. Kaufmann, W. Stöcklin and T. Reichstein (1968) Die Pregnanglykoside der oberirdischen Teile von *Sarcostemma viminale* (L.) R.Br. *Helvetica Chimica Acta*, **51**, 738–67.

Schmersal, P. (1969) Uber das Vorkommen von Helveticosid in den Samen von *Corchorus capsularis* L. und *C. olitorius* L. *Tetrahedron Letters*, **10**, 789–90.

Schulz, O. E. and H. L. Mohrman (1965) Beitrag zur Analyse der Inhaltstoffe von Knoblauch. *Pharmazie*, **20**, 441.

Sébire, R. P. A. (1899) *Les Plantes Utiles du Sénégal*, Paris, 341 pp.

Sen Gupta, K. P., N. C. Ganguli and N. R. S. Bhattacharjee (1956) Bacteriological and pharmacological studies of a vibriocidal drug derived from an indigenous source. *Antiseptic*, **53**, 287–92 (cited in Kerharo and Adam, 1974).

Senn, N. K., J. K. Chakrabarti, W. Kreis, Ch. Tamm and T. Reichstein (1957) Die Glykoside der Jutesamen, *Corchorus capsularis* L. and *Corchorus olitorius* L. Identifizierung von Corchorin, Corchorogenin and Corchsularin mit Strophantidin. *Helvetica Chimica Acta*, **40**, 588–92.

Seshadri, T. R. and S. S. Subramanian (1950) Chemical examination of Indian squill. *Journal of Scientific and Industrial Research, Section B*, **9**, 114–18.

Shellard, E. J. and M. Z. Alam (1968) The quantitative determination of some *Mitragyna* oxyindole alkaloids by U.V. spectrophotometry. *Planta Medica*, **16**, 127–36.

Shellard, E. J. and K. Sarpong (1969) The alkaloids of the leaves of *Mitragyna inermis* (Willd.) Ktze. *Journal of Pharmacy and Pharmacology*, **21**, 113S–117S.

Shellard, E. J. and K. Sarpong (1970) The alkaloid pattern in the leaves, stembark and rootbark of *Mitragyna* species from Ghana. *Journal of Pharmacy and Pharmacology*, **22**, 34S–39S.

Shellard, E. J. and K. Sarpong (1971) The isolation of speciogynine from the leaves of *Mitragyna inermis* Willd. *Journal of Pharmacy and Pharmacology*, **23**, 559–60.

Shellard, E. J., P. J. Houghton, K. Sarpong and P. K. Sarpong (1976) The alkaloids of *Mitragyna stipulosa* (DC.) Kuntze. *Journal of Pharmacy and Pharmacology*, **28**, 664.

Singh, B. and R. P. Rastogi (1969) Chemical investigation of *Asclepias curassavica* L. *Indian Journal of Chemistry*, **7**, 1105.

Singh, N., S. P. Singh, J. N. Sinha and R. P. Kohli (1978) An analysis of hypotensive response to *Sapindus trifoliatus*. *Quarterly Journal of Crude Drug Research*, **16**, 96–102.

Sinha, J. N., *et al.* (1962) 5-Hydroxytryptamine in bananas. *Biological Abstracts*, **39**, 16587.

Smith, E. R. J. (1963) Reserpine. Les rapports avec l'adrenaline et la noradrenaline. *Journal de Pharmacologie Expérimentale et Thérapie*, **139**, 321.

Sofowora, E. A. and W. A. Isaac-Sodeye (1971) Reversal of sickling and crenation in erythrocytes by the root-extract of *Fagara zanthoxyloides*. *Lloydia*, **34**, 383–5.

Sofowora, E. A., W. A. Isaac-Sodeye and L. O. Ogunkoya (1975) Isolation and characterization of an antisickling agent from *Fagara zanthoxyloides* root. *Lloydia*, **38**, 169–71.

Sravasta, O. P., A. Khare and M. P. Khare (1982) Structure of calocin (pregnan-glycoside) from *Periploca calophylla (Parquetina nigrescens)*. *Journal of Natural Products*, **45**, 211–15.

Steinmetz, E. F. (1976) *Pseudocinchonae africanae* Arboris cortex, Rubiaceae. *Quarterly Journal of Crude Drug Research*, **14**, 68.

Stöckel, K., K. Hürzeler and T. Reichstein (1969) Viminolon, Strukturbeweis. *Helvetica Chimica Acta*, **52**, 1086–91.

Swallow, W. (1951) Preservation of eserine solutions. *Pharmaceutical Journal*, **166**, 11.

Takagi, K., E. H. Park and M. Kato (1980) Antiinflammatory activities of hederagenin and crude saponins isolated from *Sapindus mukurossi* Gaertn. *Chemical and Pharmaceutical Bulletin*, **28**, 1183–8.

Tayeau, F. and J. Masquelier (1949) Les pigments de la graine d'arachide. Le chromogene constitution chimique, propriétés physiologiques. *Bulletin de la Société de Chimie Biologique*, **31**, 72–5.

Thoms, H. and F. Thumen (1911) Ueber das Fagaramid einen neuen Stickstoff haltigen Stoff aus der Wurzelrinde von *Fagara xanthoxyloides* Lam. *Berichte der Chemischen Gesellschaft*, **44**, 3717–30.

Terrioux, J. (1952) Thèse Doc. Pharm., Paris.

Torto, G., P. Sefcovic, D. A. Dadson and J. A. Mensah (1969) Alkaloids from *Fagara* species. *Ghana Journal of Science*, **9**, 3–8.

Toubiana, R. (1969) Structure de l'hydroxyvernolide, nouvel ester sesquiterpenique isolé du *Vernonia colorata* Drake. Composées. *Comptes Rendus de l'Académie des Sciences, Serie C*, **268**, 82–5.

Toubiana, R. (1975) Isolement du vernolepin à partir de *Vernonia guineenis*. *Phytochemistry*, **14**, 775.

Trabucchi, E. (1937) Sull'azione anestetica di alcuni dell'*Erythrophleum*. *Archive di Farmacologia Sperimentali e Science Affini*, **64**, 97–129.

Tschesche, R., D. Forstman and V. K. Mohan Roa (1958) Aglykone und Kardenolide von *Asclepias curassavica* L. *Chemische Berichte*, **91**, 1204–11.

Turova, A. D. (1962) *Farmakologiya i Toksikologiya SSSR*, **24**, 548–54.

Tyler, V. L. Jr. (1966) The psychological properties and chemical constituents of some habit-forming plants. *Lloydia*, **29**, 275–93.

Umarova, R. U., V. A. Maslemnicova and N. K. Abubakirov (1968) *Khimiya Prirodnykh Soedinenii*, **5**, 325–6.

Unesco (1960) *Les Plantes Médicinales des régions arides*, p. 48 (cited by Jawalekar in Caiment-Leblond, 1957).

Vaquette, J., J. L. Pousset and A. Cavé (1974) Alcaloïdes des feuilles de *Teclea unifolia* Baill. (Rutacées). *Plantes Médicinales et Phytothérapie*, **8**, 72–5.

Visnawadham, N., B. S. Sasky and E. V. Rao (1970) Reexamination of the flower petal of *Bauhinia tomentosa* L. as a commercial source of rutin. 22nd Indian Pharmaceutical Congress, Calcutta, 1970 (cited in *Medicinal Plants of India*, Indian Council of Medical Research, 1976).

Vogel, G. and H. Ströecker (1966) Die Wirkung von Flavonoiden und Escin auf den Lymphfluss und die Permeabilität der intakten Plasma-Lymphschränke von Ratten. *Arzneimittel-Forschung*, **16**, 1630–4.

Vogel, G. and H. Uebel (1961) Zur Pharmakologie der Alkaloide von *Voacanga*. *Arzneimittel-Forschung*, **11**, 787–93, 941–81.

Wall, M. E., J. W. Garvin, J. J. Williman, Q. Jones and B. G. Schuberti (1961) Steroidal sapogenins LX. Survey of plants for sapogenin and other constituents. *Journal of Pharmaceutical Sciences*, **50**, 1001–34.

Wan, Tra Liao, J. L. Beal, Wu Nan Wu and R. W. Doskotch (1978) Alkaloids of *Thalictrum*. XXVII. New hypotensive benzylisoquinoline-derived dimeric alkaloids from *Thalictrum minus*. *Lloydia*, **41**, 271–6.

Watt, J. M. (1967) African plants potentially used in mental health. *Lloydia*, **30**, 1–22.

Watt, J. M. and M. G. Breyer-Brandwijk (1962) *The Medicinal and Poisonous Plants of Southern and Eastern Africa*, Livingstone, Edinburgh, 2nd edn (7000 refs.), 1457 pp.

Wehrli, M., O. Schindler and T. Reichstein (1962) Die Glykoside des Milchsaftes von *Antiaris toxicaria* Lesch. aus Malaya sowie von *A. africana* aus Kenya. Isolierung von Glykoside und Aglycone. *Helvetica Chimica Acta*, **45**, 1183–205.

Woodson, R. E., H. W. Youngken, E. Schlitter and J. A. Schneider (1957) *Rauwolfia, Botany, Pharmacognosy, Chemistry and Pharmacology*, Little Brown, Boston.

Zetler, G. (1964) Einige pharmakologische Eigenschaften von Indol-Alkaloiden aus tropischen Apocynaceen des Sub-tribus *Tabernaemontaninae*. *Arzneimittel-Forschung*, **14**, 1277–86.

Chapter 3: The nervous system

Abbot, B. J., J. Leiter, M. E. Caldwell, J. Beal, R. E. Perdu and S. A. Schepartz, Jr (1966) Screening data from the Cancer Chemotherapy National Service Center Screening Laboratories. Plant extracts. *Cancer Research*, **26** (Suppl.) Part 1, 364–536.

Addae-Mensah, I., F. G. Torto, C. D. Dimonyeka, I. Baxter and J. K. M. Sanders (1977a) Novel amide alkaloids from the roots of *Piper guineense*. *Phytochemistry*, **16**, 757–9.

Addae-Mensah, I., F. G. Torto, I. V. Oppong, I. Baxter and J. K. M. Sanders (1977b) *N*-isobutyl-trans-2-trans-eicosadienamide and other alkaloids of fruits of *Piper guineense*. *Phytochemistry*, **16**, 483–4.

Adesina, S. K. (1979) Anticonvulsant properties of the rootbark of *Boerhaavia diffusa* L. *Quarterly Journal of Crude Drug Research*, **17**, 84–6.

Adesina, S. K. (1983) Chemical examination of *Khaya ivorensis* and *Khaya senegalensis*. *Fitoterapia* **54**, 141–3.

Adesina, S. K. and E. I. Ette (1982) The isolation and identification of anticonvulsant agents from *Clausena anisata* and *Afraegle paniculata*. *Fitoterapia*, **53**, 63–6.

Adesina, S. K. and Sofowora, E. A. (1979) The isolation of an anticonvulsant glycoside from the fruit of *Tetrapleura tetraptera*. *Planta Medica*, **36**, 270–1.

Adesogan, E. K. (1968) The chemical constituents of *Khaya senegalensis*. Communication au Symposium interafricain: *Pharmacopées traditionnelles et Plantes médicinales africaines*, Dakar, mars.

Adesogan, E. K. and J. I. Durodola (1976) Antitumor and antibiotic principles of *Annona senegalensis*. *Phytochemistry*, **15**, 1311–12.

Adesogan, E. K. and I. A. Olatunde (1974) Extractives of *Adenia cissampeloides* and *Anthocleista procera*. *West African Journal of Pharmacology and Drug Research*, **1**, 38a–39a.

Adesogan, E. K. and D. A. H. Taylor (1968) Extractives from *Khaya senegalensis* (Desr.) A. Juss. *Journal of the Chemical Society*, 1974–83.

Adesogan, E. K. and D. A. H. Taylor (1970) Limonoid extractives from *Khaya ivorensis*, *Journal of the Chemical Society, C*, 1710–14.

Adesogan, E. K., J. W. Powel and D. A. H. Taylor (1967) Extractive from the seed of *Khaya senegalensis*. *Journal of the Chemical Society*, 554–6.

Ahmed, Z. F., A. M. El Moghazi Shoaib, G. H. Wassel and S. M. El Sayyad (1972) Phytochemical study of *Lantana camara* (Terpenes and lactones). *Planta Medica*, **21**, 282–8; **22**, 34–7.

Ainslie, J. R. (1937) *A list of Plants used in Native Medicine in Nigeria*, Imperial Forestry Institute, Oxford, No. 7, 109 pp.

Almeida Silva, L., L. Nogeira Prista and A. Correia Alves (1963) Primeiros ensaios quimicos executados com a raiz de *Sarcocephalus esculentus* Afz. *Garcia de Orta*, **11**, 88–95.

Angenot, L. (1978) Nouveaux alcaloides oxindoliques du *Strychnos usambariensis* Gilg. *Plantes Médicinales et Phytotherapie*, **12**, 123–9.

Angenot, L., N. G. Bisset and A. Denoel (1973) Alcaloides de Loganiacées. Sur l'isolement de l'harmane à partir du *Strychnos usambariensis* Gilg. *Plantes Médicinales et Phytothérapie*, **7**, 33–6.

Angenot, L., A. Denoël and M. Goffart (1970) Activité curarisante d'un Strychnos africain, le *Strychnos usambariensis* Gilg. du Rwanda. *Journal de Pharmacie Belge*, **25**, 73–7.

Angenot, L., C. Coune and M. Tits (1978a) New alkaloids from *Strychnos usambariensis* leaves. *Journal de Pharmacie Belge*, **33**, 11–23.

Angenot, L., C. A. Coune, M. J. G. Tits and K. Yamada (1978b) Alkaloids of *Strychnos usambariensis*. Revised structure of usambarine. *Phytochemistry*, **12**, 1687–9.

Angenot, L., M. Dubois, Ch. Ginion, W. Dorsier and A. van Dresse (1975) Chemical structure and pharmacological (curarizing) properties of various indole alkaloids extracted from an African Strychnos. *Archives Internationales de Pharmacodynamie*, **215**, 246–58.

Ansa-Asamoa and E. A. Gyang (1967) Pharmacology of some oxindole and bis-benzylisoquinoline alkaloids of some Ghanaian plants. Thesis, Department of Pharmacy, University of Kumasi, Ghana.

Appa Rao, M. V. R., K. Srinivasan and K. T. Rao (1973) The effect of *Centella asiatica* on the general mental ability of mentally retarded children. *Journal of Research on Indian Medicine*, **8**, 9–12.

Appa Rao, M. V. R., S. A. Rajagopalan, V. R. Srinavasan and R. Sarangan (1969) Study of Mandookaparni *(Centella asiatica)* for the anabolic effect on normal healthy adults. *Nagarjun*, **12**, 33.

Arora, R. B., L. Gupta, R. C. Sharma and G. Tayal (1976) Pharmacodynamic and toxicological aspects of different Cannabis preparations. Seminar on long-term effects of Cannabis use in India. In: *Medical Plants in India*, Indian Council of Medical Research, New Delhi, p. 174.

Aspinal, G. O. and A. K. Bhattacharjee (1970) Plant gums of the genus *Khaya*. Pt IV. *Journal of the Chemical Society, C*, 361–5; Pt V, *Journal of the Chemical Society, C*, 365–9.

Aspinal, G. O. and T. B. Christensen (1961) *Anogeissus schimperi* gum. *Journal of the Chemical Society*, 3461–8.

Ayim, J. S. R., D. Dwuma Badu, N. Y. Fiagbe, A. M. Ateya, D. J. Slatkin, J. E. Knapp and P. L. Schiff, Jr (1977) Constituents of West African Plants XXI. Tiliafunimine a new imino bisbenzylisoquinoline alkaloid from *Tiliacora funifera*. *Lloydia*, **40**, 561–5.

Bailleul, F., A. Delaveau and M. Koch (1980) Iridoides du *Feretia apodanthera*. *Planta Medica*, **39**, 267–8.

Bailleul, F., A. Rabason, M. Koch and P. Delaveau (1979) Nouveaux iridoides du *Feretia apodanthera*. *Planta Medica*, **37**, 316–24.

Balbaa, S. I., A. M. Kenawy, A. A. Dessouky and N. Farrag (1979) Item 8/76 p. 283. In: *Symposium of African Unity in Cairo* 1975, ed. The Organisation of African Unity, Lagos.

Barakat, I., A. H. Jackson and M. I. Abdulla (1977) Further studies on Erythrina alkaloids. *Lloydia*, **40**, 471–5.

Barros, G. S. G., F. J. A. Matos, J. E. V. Vieira, M. P. Sousa and M. C. Medeiros (1970) Pharmacological screening of some Brazilian plants. *Journal of Pharmacy and Pharmacology*, **22**, 116–22.

Barton, D. H. R. and J. Elad (1956) Columbo root bitter principles. The functional group of columbin II. The constitution of columbin. *Journal of the Chemical Society*, 2085–95.

Barton, D. H. R. and J. Elad (1962) Diterpenoid bitter principles. *Journal of the Chemical Society*, 4809–15.

Basu, D. K. (1970) Studies on curariform activity of hayatinin methochloride, an alkaloid of *Cissampelos pareira*. *Japanese Journal of Pharmacology*, **20**, 246–52.

Beauquesne, L. (1938) Recherches sur quelques Menispermacées médicinales des genres

Tinospora et *Cocculus*. Thèse de Pharmacie, Université de Paris 1937 et *Bulletin des Sciences Pharmacologiques*, **45**, 7–14.

Beckeley, V. A. (1936) Essential oils II. Oils from indigenous plants. *East African Agriculture*, 1936, 468–70 (through *Chemical Abstracts*, **30**, 6510).

Bell, E. A. and D. H. Jansen (1971) Medical and ecological considerations of L-Dopa and 5HTP in seeds. *Nature (London)*, **229**, 136–7.

Bell, E. A., J. R. Nulu and C. Cone (1971) *l*-DOPA and *l*-3-carboxy-6,7-dihydroxy-1,2,3,4-tetrahydroisoquinoline, a new imino acid from seeds of *Mucuna mutisiana*. *Phytochemistry*, **10**, 2191–4.

Benjamin, T. V. (1979) Investigation of *Borreria verticillata*, an anti-eczematic plant of Nigeria. *Quarterly Journal of Crude Drug Research*, **17**, 135–6.

Bennet, H. (1950) Leaves and bark of *Alchornea cordifolia*. *Colonial Plant and Animal Products*, 132–4.

*Berghe, A. D., M. van den Ieven, F. Mertens and A. J. Vlietinck (1978) Screening of higher plants for biological activities and antiviral activity. *Lloydia*, **41**, 463–71.

Bevalot, F., M. Leboeuf, A. Cavé and A. Bouquet (1976) *Pachypodanthium confine*. Isolement du triméthoxy-2-4-5-styrène du *P. confine*. Engl. Diels. *Plantes Médicinales et Phytothérapie*, **10**, 179–81.

Bevalot, F., M. Leboeuf and A. Cavé (1977) Alcaloides des Annonacées XXL. Alcaloides du *Pachypodanthium staudtii* Engl. Diels. *Plantes Médicinales et Phytothérapie*, **11**, 315–22.

Bevan, C. W. L. and J. Hirst (1958) A convulsant alkaloid of *Dioscorea dumetorum*. *Chemistry and Industry*, 103.

Bevan, C. W. L. and A. U. Ogun (1964) Studies on West African plants III. Constituents of the genus *Afrormosia* Harms. *Journal of the West African Science Association*, **2**, 1–12.

Bevan, C. W., J. L. Broadbent and J. Hirst (1956) A convulsant alkaloid of *Dioscorea dumetorum*. *Nature (London)*, **177**, 925–6.

Bevan, C. W. L., D. E. U. Ekong and D. A. H. Taylor (1965) Extractives from West African members of the family of Meliaceae. *Nature (London)*, **206**, 1323–5.

Bevan, C. W. L., J. W. Powell and D. A. H. Taylor (1963) West African Timbers, Part IV, Petroleum extracts of the genera *Khaya*, *Guara*, *Carapa* and *Cedrela*. *Journal of the Chemical Society*, 980–2.

Bevan, C. W. L., M. B. Patel, A. H. Rees and D. A. H. Taylor (1964) An alkaloid from *Phyllanthus discoides*. *Chemistry and Industry*, 838–9.

Bhakuni, D. S. and P. P. Joshi (1975) Alkaloids of *Cocculus pendulus* (Forst.) Diels. *Tetrahedron Letters*, **31**, 2572–9.

Bhakuni, D. S., N. C. Gupta and M. M. Dhar (1970) Cocsulin, a new bisbenzylisoquinoline alkaloid from *Cocculus pendulus* Diels. *Experientia*, **26**, 241–2.

Bhakuni, D. W., S. Tewari and M. M. Dhar (1972) Aporphine alkaloids of *Annona squamosa*. *Phytochemistry*, **11**, 1819.

*Bhakuni, D. S., M. L. Dhar, M. M. Dhar, B. L. Dhawan and B. N. Mehrotra (1969a) Screening of Indian plants for biological activity. Part II. *Indian Journal of Experimental Biology*, **7**, 250–62.

Bhakuni, D. S., M. L. Dhar, M. M. Dhar, B. L. Dhawan, B. Gupta and R. C. Shrimal (1969b) Screening of Indian plants for biological activity, Part III. *Indian Journal of Experimental Biology*, **9**, 91–102.

Bhandari, P. R. and J. C. Bose (1954) Chemical examination of Varuna (*Crateva religiosa* Forst.) I. Crystalline constituents of the stembark. *Journal of Scientific and Industrial Research (India)*, **13B**, 773–75.

Bhatnagar, A. K. and S. P. Popli (1967) Chemical examination ofthe roots of *Cissampelos pareira* L. Part V. Structure and stereochemistry of hayatidin. *Experientia*, **23**, 242–3.

Bhatnagar, A. K., S. Bhattacharya, A. C. Roy, S. P. Popli and M. L. Dhar (1967) Chemical examination of the roots of *Cissampelos pareira* L. Structure and stereochemistry of hayatin. *Journal of Organic Chemistry*, **32**, 819.

Bhatnagar, A. K., S. Bhattacharya, A. C. Roy, S. P. Popli and M. L. Dhar (1971) Chemical examination of the roots of *Cissampelos pareira* L., Part III. Biological activity of the roots of *Cissampelos pareira* L. *Indian Journal of Experimental Biology*, **9**, 91–102.

Bhattacharya, S. K. and B. Lythgoe (1949) Derivatives of *Centalla asiatica* used against leprosy. Triterpenic acids. *Nature*, **163**, 258–9.

Bhattacharya, S. K. and A. K. Sanyal (1976) Curariform activity of some Indian medicinal plants, *Nagarjun* **13**, 19 (through Indian Council of Medical Research, (1976) and Refs. therein).

Bhattacharya, S. K., A. B. Ray and S. C. Dutta (1975a) Psychopharmacological investigation of the 4-methoxy indole alkaloids of *Alstonia venenata*. *Planta Medica*, **27**, 164–70.

Bhattacharya, S. K., P. K. Debnath, V. P. Pandy and A. K. Sanyal (1975b) Pharmacological investigations of *Elaeocarpus ganitrus*. *Planta Medica*, **28**, 174–5.

Bhattacharya, S. K., P. K. Debnath, A. K. Sanyal and S. Ghosal (1971) Pharmacological studies of *Erythrina variegata*. *Journal of Research on Indian Medicine*, **6**, 235.

Biberfeld, J. (1910) Ueber die Wirkung der Colombo Alkaloide. *Zeitschrift für experimentelle Pathologie und Therapie*, **7**, 569–76.

Binet, L., P. Binet, M. Miocque, H. Morin, C. Pechery and M. Moux (1972) Le farnésol, substance psychosedative et spasmolytique. *Thérapie*, **27**, 893–905.

Bissent, N. G. and J. D. Phillipson (1970) The African species of Strychnos. Part I. The Ethnobotany. *Lloydia*, **33**, 201–43.

Bisset, N. G. and J. D. Phillipson (1973) The African species of Strychnos Part II. The alkaloids. *Lloydia*, **36**, 1–60.

Blanc, P., P. Bertrand, G. de Saqui Sanner and M. Ane (1972) Identification par chromatographie et étude spectrale de quelques acides phénols, acides ellagique, gallique, chlorogénique, caféique dans une Euphorbiacée exotique; *Euphorbia hirta* L. *Annales de Pharmacie françaises*, **30**, 720–1.

Blanpin, O., M. Pais and A. Quevauviller (1963) Etude pharmacodynamique de l'adouétine z, alcaloide de *Waltheria americana* (Sterculiacées). *Annales de Pharmacie françaises*, **21**, 127–50.

Boissier, J. R., G. Combes, R. Perret and C. Dumont (1965) Menispermaceous alkaloids of Madagascar: *Cissampelos pareira, Cyclea madagascariensis. Anisocyclea grandieri* and *Spirospermus penduliferum*. *Lloydia*, **28**, 191–8.

Boissier, J. R., A. Bouquet, G. Combes, C. Dumont and M. M. Debray (1963) Phaeanthine dans les Ménispermacées africaines. *Triclisia patens–quelques dérivés* d'ammonium quaternaires. *Annales de Pharmacie françaises*, **21**, 767–72.

Boiteau, P. and A. R. Ratsimamanga (1956) Asiaticoside isolé de *Centella asiatica* et ses emplois thérapeutiques. *Thérapie*, **11**, 125–50.

Boiteau, P., M. Dureuil and A. R. Ratsimamanga (1949) Contribution à l'étude des propriétés antituberculeuses de l'oxyasiaticoside (dérivé hydrosoluble de l'asiaticoside extrait de *Centalla asiatica*). *Comptes Rendus de l'Académie des Sciences, Paris, Série D*, **228**, 1165–7.

Bontemps (1942) Gazette médicale. *Madagascariensis*, **5**, 29 (in Oliver, 1960).

Bouquet, A. (1970) Plantes médicinales du Congo Brazzaville. Uvariopsis, Pauridiantha, Diospyros. Thèse Pharmacie Université, Paris.

Bouquet, A. and A. Cavé (1971) Note sur l'*Epinetrum villosum*. *Plantes Médicinales et Phytothérapie*, **5**, 131–3.

Bouquet, A. and A. Fournet (1975a) Recherches chimiques préliminaires sur les plantes médicinales du Congo Brazzaville, Cairo Symposium 1975. *Fitoterapia*, **46**, 175–7.

Bouquet, A. and A. Fournet (1975b) ORSTOM. Recent research on Congolese medicinal plants, Item 4/11, Cairo Symposium, 1975. *Fitoterapia*, **46**, 243–6.

Bowden, K. (1962) Isolation from *Paullinia pinnata* L. of material with action on the frog's isolated heart. *British Journal of Pharmacology and Chemotherapy*, **18**, 173–4.

Bowden, K., B. G. Brown and J. E. Batty (1954) 5-Hydroxytryptamine, its occurrence in cowhage. *Nature (London)*, **174**, 925–6.

Broadbent, J. L. (1953) Observations on itching produced by cowhage and on the part played by histamine as a mediator of the itch sensation. *British Journal of Pharmacology*, **8**, 263.

Broadbent, J. L. and J. Schnieden (1958) A comparison of some pharmacological properties of dioscorine and dioscine. *British Journal of Pharmacology and Chemotherapy*, **13**, 213–15.
Büchi, G. (1945) Review (*Stephania*). *Schweizer Apotheker Zeitung*, **83**, 198.
Bukhari, A. Q. S. and I. Khan (1963) Effects of chaksine chloride on cholinergic and tryptamine receptors in the isolated guinea pig ileum. *Pakistan Journal of Industrial Research*, **6**, 285–9.
Burger, A. and S. Nara (1965) *In vitro* inhibition studies with homogeneous monoamino oxidases. *Journal of Medicinal Chemistry*, **8**, 859–62.
Busson, F. (1965) *Plantes Alimentaire de l'Ouest African*, 1 Vol., 568 pp. Leconte, Marseille.
Caiment-Leblond, J. (1957) Contribution à l'étude des plantes médicinales de l'A.O.F. et de l'A.E.F. Thèse de Doctorat en Pharmacie, Université de Paris, 144 pp.
Canonica, L. (1962) Trasformazione del genziopicrosido in genzianina. *Gazzetta Chimica Italiana*, **92**, 293–300.
Cavé, A., A. Bouquet and R. R. Paris (1973) pachypodol, genine flavonique isolé de *Pachypodanthium staudtii*. *Comptes Rendus de l'Académie des Sciences. Série D., Paris*, **276**, 1899–901.
Cavé, A., N. Kunesch, M. Leboeuf, F. Bevalot, A. Chiaroni and C. Riche (1980) Alcaloides des Annonacées XXV. Staudine nouvel alcaloide isoquinolinique du *Pachypodanthium staudtii*. Engl. Diels. *Journal of Natural Products (Lloydia)*, **43**, 103–95.
Chakravarti, R. N., D. Chakravarti and R. Banerjee (1959) Triterpenes from *Crataeva religiosa*. *Bulletin of the Calcutta School of Tropical Medicine*, **7**, 105.
Chang, H. Y. (1974) Toxicity of securinine and comparison with strychnine. *Chung Hua I Hsueh Tsa Chih*, **54**, 234 (*Chemical Abstracts*, **82**, 68178n).
Chapelle, J. P. (1974) Constituants chimiques des feuilles d'*Anthocleista vogelii*. *Planta Medica*, **26**, 301–4.
Chatterjee, M. S. and S. R. Bhattacharya (1964) Securinega alkaloids. Allosecurinine in Indian *Securinega* species. *Journal of the Indian Chemical Society*,. **41**, 163.
Chatterjee, M. S. and D. Roy (1965) Pharmacological studies of the seeds of *Securinega securidaca*. Effect on normal blood sugar of cat and rabbit. *Bulletin of the Calcutta School of Tropical Medicine*, **13**, 12–14.
Chaudhury, R. R., M. Haq and M. Gupta (1980) Review of plants screened for antifertility I–IV. *Bulletin Medical and Ethnobotanical Research*, **1**, 408–545.
Cheema, M. A. and O. D. Priddle (1965) Pharmacological investigation of isochaksine, an alkaloid isolated from the seeds of *Cassia absus* L. *Archives Internationaux de Pharmacodynamie et de Thérapie, Belgique*, **158**, 307–13.
Chen-Yu Sung, Hsiu Chuan Chi and Keng Tao Liu (1958) *Scheng Li Hsueh Pao*, **22**, 201 (*Chemical Abstracts*, **53** (1959) 13415).
Chevalier, A. (1947) Les *Mostuea* africains et leurs propriétés stimulantes. *Revue de Botanique appliquéé*, **27**, (Nos. 291–2) 104–9.
Chopra, R. N. and De S. Ghosh (1923) The Pharmacology and Therapeutics of *Boerhaavia diffusa* (Punarnava). *Indian Medical Gazette*, **58**, 203.
Chopra, R. N., S. L. Nayar and I. C. Chopra (1956) *Glossary of Indian Medicinal Plants*, 1 Vol., Council of Scientific and Industrial Research, New Delhi, 330 pp.
Chopra, R. N., I. C. Chopra, K. Handa and L. D. Kapur (1938) *Chopra's Indigenous Drugs of India*, Vol. 1, Dhur and Sons, Calcutta, 366 pp.
Cocker, W., T. B. H. McMurray and P. A. Staniland (1965) A synthesis of dimethylpterocarpin. *Journal of the Chemical Society*, 1034–7.
Cocker, W., T. Dahl, C. Dempsey and T. B. H. McMurray (1962) Extractives of wood from *Andira inermis* Wright (H. B. K.). *Journal of the Chemical Society*, 4906–9.
Combier, H., M. Becchi and M. Cavé (1977) Alcaloides de *Guiera senegalensis* L. *Plantes Médicinales et Phytothérapie*, **11**, 251–3.
Cooper and Gun (1931) Harmalol in the treatment of Parkinsonism. *Lancet*, **221**, 901.
Correia Alves, and L. Noguera Prista (1962) Gentistic acid precursor in the biosynthesis of

yohimbine. *Anais Faculdada da Farmacia do Porto*. **22**, 27–33.

Correia da Silva, A. C. and M. Quiteria Paiva (1964) Nota sobre a acçao curarisante dos alcaloides do *Cissampelos mucronata* A. Rich. *Revista portuguesa da farmacia*, **14**, 413–16.

Correia da Silva, A. C. and M. Quiteria Paiva (1970) Ensalos sobre acçao farmacologica dos extractos da casca de *Pentaclethra macrophylla* Benth. *Anais Faculdada da Farmacia do Porto*, **30**, 5–18.

Correia da Silva, A. C., A. Correia Alves and L. Noguera Prista (1960) Nota previa acerca da acçoa da casca de *Pentaclethra macrophylla* Benth sobre o utero da cobaia isolado. *Anais da Faculdada da Farmacia do Porto*, 21–5.

Correia da Silva, A. C., M. Quiteria Paiva and A. Costa (1962) Algunos aspetos da acçao fisiologica dos alcaloides da *Dioscorea dumetorum*. *Garcia de Orta, Lisboa*, **10**, 667–71.

Correia da Silva, A. C., A. Costa and M. Quiteria Paiva (1964) Nota previa sobre alguna aspectos da actividade farmacodinamica do alcaloido do *Sarcocephalus esculentus* Afz. *Garcia de Orta*, **12**, 309–15.

Correia da Silva, A. C., A. Costa and M. Quitero Paiva (1966) Algunos aspettos da actividade farmacodinamica dos alcaloides de *Newbouldia laevis* Seem. *Garcia de Orta*, **14**, 91–6.

Dalziel, J. M. (1937) *The Useful Plants of West Tropical Africa*. Crown Agents, London, 1 vol., 612 pp.

Das, P. K. and A. K. Sanyal (1964) Studies on *Cissus quadrangularis* L. Acetylcholine-like action of the total extract. *Indian Journal of Medical Research*, **52**, 63–6.

Das, P. K., V. Nath, K. D. Gode and A. K. Sanyal (1964) Preliminary phytochemical and pharmacological studies of *Cocculus hirsutus*. *Indian Journal of Medical Research*, **52**, 300.

Dass, B., E. Fellion and M. Plat (1967) Alkaloids in *Conopharyngia durissima* seeds. *Comptes Rendus de l'Académie des Sciences, Serie C*, **264**, 1765–7.

Dass, B. (1966) Chemical examination of *Cardiospermum halicacabum* L. *Bulletin of the Botanical Society of India*, **8**, 357.

Debray, M. (1966) Contribution à l'étude du genre *Epinetrum* (Ménispermacées) *E. cordifolium* Mangenot et Miège et *Epinetrum mangenotii* Guil. and Debray de Côte d'Ivoire. *Mémoires ORSTOM*, **18**, 74 pp.

Debray, M., M. Plat and J. Le Men (1966) Alcaloides des Ménispermacées africaines *Epinetrum cordiofolium* et *E. mangenotii*–Isolation de cycléanine, norcycléanine et isochondrodendrine. *Annales de pharmacie françaises*, **24**, 551–8.

Debray, M., M. Plat and J. Le Men (1967) Alcaloides des Ménispermacées africaines II *Stephania dinklagei* (Engl.) Diels; isolement de la (+)corydine, de la (+)isocorydine et de la (−)roemerine. *Annales pharmaceutiques françaises*, **25**, 237–42.

Delphaut, J. and J. Balansard (1941) Sur les propriétés du nénuphar (*Nymphaea alba* L.) *Comptes Rendus de la Société de Biologie Paris*, **135**, 1665–70.

Delphaut, J. and J. Balansard (1943) Recherches pharmacologiques sur le nénuphar blanc (*N. alba* L.). *Revue de Phytothérapie*, **7**, 83–5.

Denoël, A. (1958) *Matière Médicale Végetale*, 2 Vols., Presses Universitaires, Liège (Belgique).

*Deshpande, J. (1973) A decade of progress in Indian medicine. *Bulletin of Indian Medicine*, 72.

Dhar, M. L., M. M. Dhar, B. N. Dhawan, B. N. Mehrotra and C. Ray (1968) Screening Indian plants for biological activity, Part I. *Indian Journal of Experimental Biology*, **6**, 232–47.

Dhillon, K. S. and B. S. Paul (1971) Clinical studies of *Lantana camara* L. poisoning in buffalo calves with special reference to its effects on rumen motility. *Indian Journal of Animal Sciences*, **4**, 1034–6.

Dickel, D. F., C. L. Holden, R. C. Maxfield, L. E. Paszek and W. I. Taylor (1958) The alkaloids of *Tabernaemontana iboga*, Part III, Isolation studies. *Journal of the American Chemical Society*, **80**, 123–5.

Dimitrienko, G. I., D. G. Murray and S. McLean (1974) Nauclefine and naucletine, two new alkaloids of indolo-quinolizidine type isolated from *N. latifolia*. *Tetrahedron Letters*, **23**, 1961–4.

Dimitrov, S. (1965) Effet pharmacologique des alcaloides du *Nuphar luteum*. *Veterinae Medica Nauki Bulgaria*, **2**, 752–62 (in *Bulletin Signaletique* CNRS (1966) **27** (no.) 13–7066).

Duah, F. K., P. D. Owusi, J. E. Knapp, D. J. Slatkin and P. L. Schiff, Jr (1981) Constituents of West African medicinal plants XXIX, Quaternary alkaloids of *Heptacyclum zenkeri*, *Planta Medica*, **42**, 275–8.

Dublin, L. (1965) Le colatier (*Cola nitida*) en République Centrafricaine. *Café, Cacao et Thé*, **2**, 97–112.

Durand, E., E. V. Ellington, P. C. Feng, J. L. Haynes, K. E. Majorus and B. Philips (1962) Simple hypotensive and hypertensive principles from some West Indian medicinal plants. *Journal of Pharmacy and Pharmacology*, **14**, 562–6.

Durodola, J. F. (1975) Antitumour effects against sarcoma 180 ascites of fractions of *Annona senegalensis*. *Planta Medica*, **28**, 32–6.

Dutta, S. C., S. K. Bhattacharya and A. B. Ray (1976) Flower alkaloids of *Alstonia scholaris*, *Planta Medica*, **30**, 86–92.

Dutta, T. and D. K. Basu (1967) Terpenoids IV. Isolation and identification of asiatic acid from *Centella asiatica* L. *Indian Journal of Chemistry*, **5**, 586.

Dwivedi, S. K., G. A. Schivnani and H. C. Joshi (1971) Clinical and biochemical studies in *Lantana* poisoning in ruminants. *Indian Journal of Animal Sciences*, **4**, 948–53.

Dwuma Badu, D., J. S. K. Ayim, C. A. Mingle, A. N. Tackie, D. J. Slatkin, P. L. Schiff, Jr and J. E. Knapp (1975a) Constituents of West African medicinal plants X. Alkaloids of *Cissampelos pareira*. *Phytochemistry*, **14**, 2520–3.

Dwuma Badu, D., J. S. K. Ayim, A. N. Tackie, M. A. Elsohly, J. E. Knapp, D. J. Slatkin and P. L. Schiff, Jr (1975b) Constituents of West African plants, XII. Trigilletimine, a new bisbenzyl isoquinoline alkaloid from *Triclisia* species. *Experientia*, **31**, 1251–2.

Dwuma Badu, D., J. S. Ayim, A. N. Tackie, J. E. Knapp, K. J. Slatkin and P. L. Schiff, Jr (1975c) Additional alkaloids of *Triclisia patens* and *Triclisia subcordata*. *Phytochemistry*, **14**, 2524–5.

Dwuma Badu, D., J. S. Ayim, T. T. Dabra, H. N. Elsohly, J. E. Knapp, D. J. Slatkin and P. L. Schiff, Jr (1975d) Constituents of West African medicinal plants IX. Dihydrocubebin a new lignan from *P. guineense*. *Lloydia*, **38**, 343–5.

Dwuma Badu, D., J. S. Ayim, T. T. Dabra, H. N. Elsohly and J. E. Knapp (1976a) Constituents of West African medicinal plants XIV. Constituents of *Piper guineense* Schum. & Thonn. (Piperaceae). *Lloydia*, **39**, 60–5.

Dwuma Badu, D., J. S. K. Ayim, N. Y. Fiagbe, A. N. Tackie, J. E. Knapp, D. J. Slatkin and P. L. Schiff, Jr (1976b) Constituents of West African medicinal plants VII. Alkaloids of *Tiliacora dinklagei*. *Lloydia*, **39**, 213–17.

Dwuma Badu, D., T. U. Okarter, A. N. Tackie, J. A. Lopez, D. J. Slatkin, J. E. Knapp and P. L. Schiff, Jr (1977) Constituents XIX. Funiferine *N*-oxide a new alkaloid from *Tiliacora funifera*. *Planta Medica*, **16**, 158–65; *Journal of Pharmaceutical Sciences*, **66**, 1242–44.

Dwuma Badu, D., J. S. K. Ayim, N. I. Y. Fiagbe, J. E. Knapp, P. L. Schiff, Jr and D. Slatkin (1978) Constituents of West African medicinal plants XX. Quindoline from *Cryptolepis sanguinolenta*. *Journal of Pharmaceutical Sciences*, **67**, 433–4.

Dwuma Badu, D., S. F. Withers, S. A. Ampofo, M. M. El Azizi, D. J. Slatkin, J. E. Knapp and P. L. Schiff, Jr (1979) Constituents XXIII. The position of the phenolic function in dinklacorine – a confirmation of structure. *Journal of Natural Products (Lloydia)*, **42**, 92–5.

Dwuma Badu, D., J. S. K. Ayim, S. F. Withers, N. O. Agyemarg, A. M. Atoya, M. M. El Azizi, J. E. Knapp, D. J. Slatkin and P. L. Schiff, Jr (1980) Constituents of West African medical plants XXVII. Alkaloids of *Rhigiocarya racimifera* and *Stephania dinklagei*. *Journal of Natural Products (Lloydia)*, **43**, 123–9.

Efron, D. H., B. Holmstedt and N. S. Kline (1967) *Ethnopharmacological Research for Psychoactive Drugs*. Proceedings of a Symposium held in San Francisco, California, 1967. US Department of Health, Education and Welfare, Washington, DC, p. 468 (new edn 1979, Raven Press, New York).

Ekong, D. E. U., E. O. Olagbeni and A. E. Spiff (1968) Cycloeucalenol and 2-4-methyl-encycloartanol in wood oil of the family Meliaceae. *Chemistry and Industry, London*, 1808.

Ekong, D. E. U., C. O. Fakunle, A. K. Fasina and J. I. Okogun (1969) Meliacins (Limonoids) Nimbolin A and B, two meliacin cinnamates from *Azadirachta indica* L. and *Melia azedarach* L. *Journal of the Chemical Society*, **20D**, 1166–7.

El Nagger, L., J. L. Beal, L. M. Parks, K. N. Salman, P. Patel and A. Schwarting (1978) A note on the isolation and identification of two pharmacologically active constituents of *Euphorbia pilulifera* L. *Lloydia*, **41**, 73–5.

El Olemy, A. A., B. C. Ali and A. El Mottaleb (1978) Erythrina alkaloids. The alkaloids and seeds of *E. variegata*. *Lloydia*, **41**, 342–7.

El Olemy, A. A. and A. E. Schwarting (1965) Simulated biosynthesis of anhygrine. *Experientia*, **21**, 249.

El Said, F., A. Sofowora and A. A. Olaniyi (1968) *Study of certain Nigerian plants used in fever*. Communication at Symposium on Interafrican Pharmacopaeia Traditional and Medicinal Plants, Dakar.

Emboden, W. A. (1981) Transcultural use of narcotic waterlilies in ancient Egyptian and Maya drug ritual. *Journal of Ethnopharmacology*, **3**, 39–83.

Escalante, M. G., R. Nico and V. Oleachea (1963) Sustitucion de ipecacuanha por raiz de *Borreria verticillata* (L.) Mey. *Revista farmaceutica Buenos Aires*, **105**, 27–43.

Esdorn, I. (1961) *Alstonia* Arten als Arzneipflanzen. *Pharmazeutica Acta Helvetica*, **36**, 6–9.

Farnsworth, N. R. (1969) Hallucinogenic plants. *Science*, **162**, 1086–92.

*Farnsworth, N. R., J. P. Bederka, Jr and M. Moses (1974) Modern approaches for selecting biologically active plants. Central nervous system depressants. *Journal of Pharmaceutical Science*, **63**, 457–9.

*Farnsworth, N. R., L. K. Henry, G. H. Svoboda, R. N. Blomster, M. J. Yates and K. L. Euler (1966) Biological and phytochemical evaluation of plants. Biological test procedures and results from two hundred accessions. *Lloydia*, **29**, 101–22.

*Farnsworth, N. R., L. K. Henry, G. H. Svoboda, N. H. Blomster, N. S. Fong, M. W. Quimby and M. J. Yates (1968) Biological and phytochemical evaluation from an additional two hundred accessions. *Lloydia*, **31**, 237–48.

Fattorusso, V. and O. Ritter (1967) *Dictionnaire de Pharmacologie Clinique*, 1 vol., 829 pp., Masson, Paris.

Ferreira, M. A., A. C. Alves and L. N. Prista (1963a) Estudo quimico da *Newbouldia laevis* Seem. I. Isolamento de bases indolicas. *Garcia de Orta*, **11**, 477–86.

Ferreira, M. A., A. C. Alves and L. M. Prista (1963b) Ensaios sobre as raizes de *Alchornea cordifolia* (Schum) Muell. Arg. *Garcia de Orta, Lisboa*, **11**, 265–74.

Ferreira, M. A., L. Nogueira Prista, A. C. Alves and A. Spinola Rogue (1965) Estudo quimico de *Cissampelos mucronata* A. Rich. Isolamento de *D. isochondrodendrina*. *Garcia de Orta, Lisboa*, **13**, 395–405.

Folkers, K. and F. Koniuszy (1940) Isolation and characterization of erysodine, erysovine, erysopine and erysocine. *Journal of the American Chemical Society*, **62**, 1677–83.

Folkers, K. and K. Unna (1939) Chazuta curare, its botanical components and other plants of curare interest. *Archives Internationales de Pharmacodynamie*, **61**, 370–9.

Folkers, K., J. Shavel and F. Koniuszy Jr (1941) Erythrina alkaloids X. Isolation and characterization of erysonine and other liberated alkaloids. *Journal of the American Chemical Society*, **63**, 1544–9.

Fong, H. H. S., M. Trojankova, J. Trojanek and N. R. Farnsworth (1972) Alkaloid screening II. *Lloydia*, **35**, 117–49.

*Fong, H. H., N. R. Farnsworth, L. K. Henry, G. H. Svoboda and N. J. Yates (1972) Biological and phytochemical evaluation of plants X. Central nervous system depressant principles. Test results from a third 200 accessions. *Lloydia*, **35**, 35–48.

Fontaine, R. and A. Erdös (1976) Zur zentralen Wirkung verschiedener *Withania*-Extrakte nach oraler Applikation am Tier. *Planta Medica*, **30**, 242–50.
Forgaes, P., J. F. Desconolois, D. Mansard, J. Provost, R. Tiberghien, J. Tocker and A. Touché (1981) Dopamine et alcaloides tetrahydroisoquinoliques d'*Annona reticulata* L. (Annonacées). *Plantes Médicinales et Phytothérapie*, **15**, 10–15.
Forrest, J. E. and R. A. Heacock (1972) Nutmeg and mace. The psychotropic species from *Myristica fragrans* Houtt. *Lloydia*, **35**, 440–9.
Fournier, G., M. R. Paris, M. C. Fourniat and M. Quero (1978) Bacteriostatic activity of *Cannabis sativa* essential oil. *Annales pharmaceutiques françaises*, **36**, 603–6.
Foussard-Blanpin, O., A. Quevauviller and P. Bourrinet (1967) Sur la phyllochrysine, alcaloide du *Phyllanthus discoides*, Euphorbiacées. *Thérapie*, **22**, 303–7.
Fraga de Azevedo, J. and L. de Medeiros (1963) L'action molluscicide d'une plante de l'Angola, la *Securidaca longepedunculata* Fres. *Bulletin de Société Pathologique exotique*, **56**, 68–76.
Games, D. E., A. H. Jackson, N. A. Khan and D. S. Millington (1974) Alkaloids of some African, Asian, Polynesian and Australian species of *Erythrina*. *Lloydia*, **37**, 581–8.
Gaoni, Y. and R. Mechoulam (1964) The structure and synthesis of cannabigenol, a new hashish constituent. *Proceedings of the Chemical Society*, 82–3.
Garg, S. C. and H. L. Kasera (1984) Neuropharmacological studies of the essential oil of *Anacardium occidentale*. *Fitoterapia*, **55**, 131–6.
Garnier, J., M. Koch and M. Plat (1974) Loganiacées de la Côte d'Ivoire X. Alcaloides de *Strychnos camptoneura* Gilg et Busse. *Plantes Médicinales et Phytothérapie*, **8**, 281–6.
Gaudin, O. and R. Vacherat (1938) Recherches sur le rotenone et le pouvoir ichthyotoxique de quelques plantes du Soudan. *Bulletin de la Société de Pharmacologie*, **40**, 385–94.
Geran, R. I., N. H. Greenberg, M. H. MacDonald, A. M. Schumacher and B. J. Abbott (1972) Protocols for screening chemical agents and natural products against animal tumours and other biological systems. *Cancer Chemotherapy Report*, **3**, 25.
Ghosal, S. and P.K. Banerjee (1968) Indole alkaloids of *Desmodium gangeticum* DC. *Indian Journal of Pharmacy*, **30**, 280.
Ghosal, S. and S. K. Bhattacharya (1972) Desmodium alkaloids Part II. Chemical and pharmacological evaluation of *D. gangeticum*. *Planta Medica*, **24**, 434–440.
Ghosal, S., S. Banerjee and A. W. Frahm (1979a) Prostalidins A, B, C and retrochinensin–a new antidepressant, 4-aryl 2,3-naphthalidine lignane, from *Justizia prostata*. *Chemistry and Industry*, 854–5.
Ghosal, S., S. Banerjee and D. K. Srivastava (1979b) Simplexolin. A new lignan from *Justicia simplex* Don. *Phytochemistry*, **18**, 503–5.
Ghosal, S., S. K. Bhattacharya and R. Mehta (1972) Naturally occurring and synthetic β-carbolines as cholinesterase inhibitors. *Journal of Pharmaceutical Sciences*, **61**, 808.
Ghosal, S., K. Dutta and S. K. Bhattacharya (1972) Erythrina chemical and pharmacological evaluation. *Journal of Pharmaceutical Science*, **61**, 1274–7.
Ghosal, S., S. Singh and S. K. Bhattacharya (1971) Alkaloids of *Mucuna pruriens*. Chemistry and Pharmacology. *Planta Medica*, **19**, 279–84.
Ghosal, S., P. S. Rama Ballav and R. Mehta Schauhan (1975) *Sida cordifolia*. *Phytochemistry*, **14**, 830–2.
Ghosal, S., P. V. Sharma, R. K. Chaudhuri and S. K. Bhattacharya (1973) Chemical constituents of the Gentianaceae V. Tetraoxygenated xanthones of *Swertia chirata* Buch. *Journal of Pharmaceutical Science*, **62**, 926–30.
Ghosal, S., A. K. Srivastava, R. S. Srivastava, S. Chattopadhyey and M. Maitra (1981) Justice saponin, a new triterpenoid saponin from *Justicia simplex*. *Planta Medica*, **42**, 279–83.
Gilmore, C. J., R. F. Bryan and S. M. Kupchan (1976) Conformation and reactivity of the macrocyclic tumor-inhibiting alkaloid tetrandine. *Journal of the American Chemical Society*, **97**, 1947–51.

Goodman, L. S. and A. Gilman (1980) *The Pharmacological Basis of Therapeutics*. Macmillan, New York; Baillière and Tindall, London, 6th edn, 1843 pages (5th edn 1976).

Goodson, J. A. and T. A. Henry (1925) Echitamine. *Journal of the Chemical Society*, **127**, 1640.

Gorman, M., N. Neuss, N. J. Cone and J. A. Deyrup (1960) Alkaloids from Apocynaceae III. Alkaloids of *Tabernaemontana and Ervatamia*. The structure of coronaridine, a new alkaloid related to ibogamine. *Journal of the American Chemical Society*, **82**, 1142–5.

Gottlieb, O. R. (1979) Chemical studies on medicinal Myristicaceae from Amazonia. *Journal of Ethnopharmacology*, **1**, 309–23.

Goutarel, R. (1964) *Les Alcaloides Stéroidiques des Apocynacées*, 1 vol., 289 pp., Hermann, Paris.

Goutarel, R. French patent 2 087 982 of 5.5.1970.

Guedel, J. (1955) Contribution à l'étude des ictères en Afrique Orientale Française. Le *diekouadio* (Côte d'Ivoire). *Notes Africaines*, IFAN, Dakar, **66**, 50–3.

Guha, K. P., B. Mukherjee and R. Mukherjee (1979) Bisbenzylisoquinoline alkaloids. A Review. *Lloydia*, **42**, 1–84.

Guinaudeau, H., M. Leboeuf and A. Cavé (1975) Aporphine alkaloids I. *Lloydia*, **38**, 275–338 (400 references); Aporphine alkaloids II. *Lloydia*, **38**, 339–60 (178 references).

Gunatilaka, A. A. L., S. Sotheeswaran, S. Balasubramanian, I. Chandrasekara and H. T. Badra Sryani (1980) Studies on medicinal plants of Sri Lanka III. Pharmacologically important alkaloids of some *Sida* species. *Planta Medica* **39**, 66–72.

Gupta, K. L. and I. C. Chopra (1953) A short note on antibacterial properties of chaksine, an alkaloid from *Cassia absus*. *Indian Journal of Medical Research*, **41**, 459–60.

Gupta, D. R. and S. K. Garg (1966) A chemical examination of *Euphorbia hirta* L. *Bulletin of the Chemical Society of Japan*, **39**, 2532–4.

Gupta, N. C., D. S. Bhakuni and M. M. Dhar (1970) Penduline, a new biscoclaurine alkaloid from *Cocculus pendulus* Diels. *Experientia*, **26**, 12–13.

Gyang, E. A. and H. W. Kosteilitz (1966) Agonist and antagonist actions of morphine-like drugs on the guinea pig's isolated ileum. *British Journal of Pharmacology and Chemotherapy*, **27**, 514–27.

Gyang, E. A., H. W. Kosteilitz and G. H. Leeds (1964) The inhibition of autonomic neuro-effector transmission by morphine-like drugs and its use as a screening test for narcotic and analgesic drugs. *Archiv fuer Experimentelle Pathologie und Pharmakologie*, **248**, 231–46.

Hager's Handbuch der Pharmazeutischen Praxis (1967–80) 4th edn, P. L. List and L. Hörhammer (Eds.), 12 vols., Springer, Berlin and New York.

Haller, A. and A. Heckel (1901) Sur l'ibogaine, principe actif d'une plante du genre *Tabernaemontana* originaire du Congo. *Comptes Rendus de l'Académie des Sciences*, **133**, 850–3.

Hallet, F. P. and L. M. Parkes (1953) Observations on the antispasmodic principle of *Euphorbia pilulifera*. *Journal of the American Pharmaceutical Association*, **42**, 607–9.

Hannonière, M., M. Leboef and A. Cavé (1974) Oliveridine et oliverine, nouveaux alcaloides aporphiniques du *Polyalthia oliveri* Engl. (Annonaceae). *Comptes Rendus de l'Académie des Sciences, Serie C*, **278**, 921–4.

Hannonière, M., M. Leboef, A. Cavé and R. R. Paris (1975) Alcaloides des Annonacées. Alcaloides d'*Enantia chlorantha* Oliv. *Plantes Médicinales et Phytothérapie*, **9**, 296–303.

Hänzel, R., Ch. Leuckert and G. Schulz (1966) *Zeitschrift fuer Naturforschung*, **21B**, 530 (cited in Kerharo and Adam, 1974).

Harada, M. and Y. Ozaki (1976) Effect of indole alkaloids from *Gardneria* genus and *Uncaria* genus on neuromuscular transmission in the rat limb in situ. *Chemical and Pharmaceutical Bulletin*, **24**, 211–14.

Harada, M. and Y. Ozaki (1978, 1979) *Uncaria rhynchophylla*. Dose dependent muscular relaxation or stimulation. *Chemical and Pharmaceutical Bulletin*, **26**, 48; **27**, 345–9.

Harada, M., Y. Ozaki and H. Ohno (1979) Effect of indole alkaloids from *Gardneria mutans* Siel & Zucc. and *Uncaria rhynchophylla* Miq. on guinea pig urinary bladder preparation in situ. *Chemical and Pharmaceutical Bulletin*, **27**, 1069–74.

Hargreaves, R. T., R. D. Johnson, D. S. Millington, M. H. Mondal, W. Beavers, L. Becker, C. Young and K. L. Rinehart, Jr (1974) Alkaloids of American species of Erythrina. *Lloydia*, **37**, 569–80.

Harley, G. W. (1941) *Native African Medicine with Special References to its Practice in the Mano Tribe of Liberia*. Harvard University Press, Cambridge, Massachusetts, 294 pp.

Heal, R. F. and E. F. Rogers (1950) A survey of plants for insecticidal activity. *Lloydia*, **13**, 89–162.

Heckel, T., B. C. Knight, C. Remington, H. D. Ritchie and I. J. Williams (1960) Studies on biliary secretion in the rabbit. The effect of icterogenin and rehmannic acid on bile flow and the excretion of bilirubin, phylloerythrin, coproporphyrin and sulphalein. *Proceedings of the Royal Society*, **53R**, 47–9.

Hegnauer, R. (1962–68) *Chemotaxonomie der Pflanzen*, 5 vols., Birkhäuser Verlag, Basel, Stuttgart.

Hellerman, R. C. and L. W. Hasleton (1950) The antispasmodic action of *Euphorbia pilulifera*. *Journal of the American Pharmaceutical Association*, **39**, 142–6.

Henry, T. A. (1949) *The Plant Alkaloids*, 1 vol., 804 pp., 4th edn, Churchill, London.

Hill, T. R. and C. Worster-Drought (1929) Harmine in the treatment of chronic epidemic encephalitis. *Lancet*, 217, 647.

Hockemiller, R., P. Cabalion, A. Bouquet and A. Cavé (1977) Isopiline, a new aporphine alkaloid from *Isolona pilosa* Diels. Annonaceae. *Comptes Rendus de l'Académie des Sciences, Serie C*, **285**, 447–50.

Hockemiller, R., P. Cabalion, J. Bruneton and A. Cavé (1978) Alcaloides des Annonacées XXIII. Alcaloides des écorces d'*Isolona campanulata* Engl. Diels. *Plantes Médicinales et Phytothérapie*, **12**, 230–4.

Hofer, A., H. Osmond and J. Smythens (1950) Schizophrenia, a new approach. *Journal of Mental Science*, **100**, 29–45.

Holmstedt, B. (1967) *Historical Survey* (cited in Efron *et al.*, 1967).

Holmstedt, B., J. E. Lindgren, L. Rivier and T. Plowman (1979) Cocaine in blood of coca chewers. *Journal of Ethnopharmacology*, **1**, 69–78.

*Hooper, P. A. and B. E. Leonard (1965) Pharmacological properties of some West Indian medicinal plants. *Journal of Pharmacy and Pharmacology*, **17**, 98–107.

Hotellier, F., P. Delaveau and J. L. Pousset (1975) Nauclèfine et nauclétine, deux nouveaux alcaloides de type indoloquinolizidine isolés du *Nauclea latifolia*. Sm. *Phytochemistry*, **14**, 1407–9.

Hotellier, F., P. Delaveau and J. L. Pousset (1977) Isolement de l'isovincoside lactame (strictosamide) des écorces de racine de *Nauclea latifolia* Sm. *Plantes Médicinales et Phytothérapie*, **11**, 106–8.

Hotellier, F., P. Delaveau and J. L. Pousset (1979) Alcaloides et glucoalcaloides des feuilles de *Nauclea latifolia* Sm. *Planta Medica*, **35**, 242–6.

Hsiu-Chuan Chi, Keng Toa Liu and Chen-Yu Sung (1959) *Scheng Li Hsueh Pao*, **23**, 203 (in *Chemical Abstracts*, **57** (1962) 15764).

Hutchinson, J. and J. M. Dalziel (1954–72) *Flora of West Tropical Africa*. Crown Agents, Milbank, London, 2nd edn, 3 vols., published in 5 parts.

Iketubosin, G. D. and D. W. Mathieson (1963) The isolation of hordenine and norsecurinine from *Securinega virosa* Baill. *Journal of Pharmacy and Pharmacology*, **15**, 810–15.

Indian Council of Medical Research (1976) *Medicinal Plants of India*, Vol. 1, New Delhi, 487 pp.

Irvine, F. R. (1930) *Plants of the Gold Coast*, Oxford University Press, London, 521 pp.

Irvine, F. R. (1961) *Woody Plants of Ghana*. 2nd edn, Oxford University Press, London, 868 pp.

Jain, S. R. and M. R. Jain (1972) Therapeutic utility of *Ocimum basilicum* var. album. *Planta Medica*, **22**, 136–9.

Janot, M. M., J. Mainil and R. Goutarel (1958) La Phyllocrisine, nouvel alcaloide isolé du *Phyllanthus discoideus*. *Annales pharmaceutiques françaises*, **16**, 148.

Jentzch, K. (1953) Production of alkaloids in Solanaceae. *Scientia Pharmaceutica*, **21**, 285–9.

Joshi, P. P., D. S. Bhakuni and M. M. Dhar (1974) Structure and stereochemistry of cocsuline. *Indian Journal of Chemistry*, **12**, 517–18.

Ju-Ichi, M., Y. Ando, J. Yoshida, J. Kunitomo and T. Shingu (1978) Alkaloids of *Cocculus trilobus* DC. Isolation and structure of erythrinan alkaloids. *Yakugaku Zasshi*, **98**, 886–90.

Kapoor, V. K. and H. Singh (1967) Investigation of *Achyranthes aspera* L. *Indian Journal of Pharmacy*, **29**, 285–8.

Kariyone, T. (1971) *Atlas of Medicinal Plants*, Chemical Industries Press, Osaka, Japan, p. 58.

Karnick, C. R. and M. D. Saxena (1970a) On the variability of alkaloid production in *Datura* species. *Planta Medica*, **18**, 266–9.

Karnick, C. R. and M. D. Saxena (1970b) *Datura*, the famous narcotic from the East. *Quarterly Journal of Crude Drug Research*, **10**, 1493–1516.

Kerharo, J. (1968) Revue des plantes médicinales et toxiques du Sénégal. *Plantes Médicinales et Phytothérapie*, **2**, 108–46.

Kerharo, J. (1970) Une drogue des Pharmacopées africaines réputée anti-venimeuse, la *Securidaca longepedunculata* Fres. *Afrique Medicale*, **9**, 401–3.

Kerharo, J. and J. G. Adam (1974) *La Pharmacopée Sénégalaise Traditionelle*, Vigot, Paris, 1 vol., 1011 pp.

Kerharo, J. and A. Bouquet (1950) *Plantes Médicinales de la Côte d'Ivoire et Haute Volta*, 1 vol., Vigot, Paris, 297 pp.

Kerharo, J., A. Bouquet and R. Heintz (1948) Le wilinga des Mossis (*Guiera senegalensis* Lam) ses usages therapeutiques indigènes et son application au traitement des diarrhées cholériformes. *Acta Tropica*, **5**, 345–8.

Kerharo, J., F. Guichard and A. Bouquet (1960–62) Les végétaux ichthytoxiques. *Bulletin et Mémoires de la Faculté de Médicine et Pharmacie, Dakar*, **8**, 313–19; **9**, 355–86; **10**, 223–42.

Kettenes-van den Bosch, J. J., C. A. Salemink, J. van Noordwyk and I. Khan (1980) Biological activity of the Tetrahydrocannabinols. *Journal of Ethnopharmacology*, **2**, 197–231.

Khan, I., A. Q. S. Bukhari and M. A. Khan (1963) Some pharmacological actions of chaksine chloride and isochaksine. *Pakistan Journal of Industrial Research*, **6**, 97–102.

Khanna, K. L., A. E. Schwarting, A. Rother and J. H. Bobbit (1963) Occurrence of tropine and pseudotropine in *Withania somnifera*. *Lloydia*, **26**, 258–78.

Khanna, N. K., B. R. Madan, O. P. Mhatma and S. C. Surana (1972) Some psychopharmacological actions of *Stephania glabra* (Roxb) Miers. An Indian indigenous herb. *Indian Journal of Medical Research*, **60**, 472–80.

Kjaer, A. and P. Friis (1962) Isothiocyanates from *Putranjeva roxburghii* Wall including 2-methylbutyl isothiocyanate, a new mustard oil of natural derivation. *Acta Chemica Scandinavica*, **16**, 936–46.

Kjaer, A. and J. Thomsen (1963) Isothiocyanate producing glucosides in species of Capparidaceae. *Phytochemistry*, **2**, 29–32.

Koch, M. (1965) Gentianine et swertiamarine de l'*Anthocleista procera* Leprieur ex Bureau (Loganiacées). Thèse de Doctorat de Pharmacie, Université de Paris.

Koch, M., E. Fellion and M. Plat (1973) Loganiacées de la Côte d'Ivoire XI. Alcaloides du *Strychnos usambariensis* Gilg. *Annales pharmaceutiques françaises*, **31**, 45–8.

Koch, M., J. Garnier and M. Plat (1972) Loganiacées de la Côte d'Ivoire IX. Alcaloides du *Strychnos camptoneura* Glig. *Annales pharmaceutiques françaises*, **30**, 299–306.

Koch, M., M. Plat, J. Le Men and M. M. Janot (1964) Heterosides de l'*Anthocleista procera*. II. Monoterpenoides. Contribution à la détermination de la structure de la swertiamarine. *Bulletin de la Société Chimique de France*, **2**, 403–6.

Kondagbo, B. and P. Delaveau (1974) Chimiotaxonomie des Capparidacées. *Plantes Médicinales et Phytothérapie*, **8**, 96–103.
Koumaré, M., J. Cros and G. Pitet (1968) Recherches sur les constituants chimiques du *Guiera senegalensis* (Combretacée). *Plantes Médicinales et Phytothérapie*, **2**, 204–9.
Krishna Rao, R. V., T. Satyanaryana and B. V. Kameswara Rao (1984) Phytochemical investigations on the roots of *Sida acuta* growing in Waltair. *Fitoterapia*, **55**, 249–50.
Krishna Rao, R. V., J. V. L. N. Seshagiri Rao and U. Vimladevi (1979) Phytochemical investigation of *Cassia absus*. *Journal of Natural Products (Lloydia)*, **42**, 299–300.
Kronlund, A., K. Kristianson and F. Sandberg (1970) Occurrence of phaenthine and *N,N'*-dimethylphaenthine in *Triclisia dictyophylla* and *T. patens*. New simple method for estimation of muscle relaxant effect. *Acta Pharmaceutica Suecica*, **7**, 279–84.
Krukoff, B. A. (1977) Notes on the species of *Erythrina*. *Lloydia*, **40**, 407–11.
Kučera, M., V. O. Marquis and H. Kucerova (1972) Contribution to the knowledge of Nigerian medicinal plants I. TLC separation and quantitative evaluation of *Alstonia boonei* alkaloids. *Planta Medica*, **21**, 343–6.
Kučera, M., V. O. Marquis and A. O. Okuyemi (1973) Contribution to the knowledge of Nigerian medicinal plants II. Pharmacology of the alkaloids of *Alstonia boonei*. *African Journal of Pharmacy and Pharmaceutical Sciences*, **3**, 228.
Kupchan, S. M., A. C. Patel and E. Fujita (1965) Tumor inhibitors VI. Cissampareine, a new cytotoxic alkaloid from *Cissampelos pareira*. Cytotoxicity of bisbenzylisoquinoline alkaloids. *Journal of Pharmaceutical Sciences*, **54**, 580–3.
Lang, E. and H. Horster (1977) Production and accumulation of essential oils in *Ocimum basilicum* callus and suspension cultures. *Planta Medica*, **31**, 112–18.
Lavie, D. and R. Taylor Smith (1963) Isolation of gentianin from *Anthocleista procera* (Loganiaceae). *Chemistry and Industry*, **63**, 781–2.
Lavie, D., E. Glotter and Y. Shvo (1965) Constituents of *Withania somnifera* Dun. Part IV. Structure of withaferine A. *Journal of the Chemical Society*, 7517–31.
Leboeuf, M. and A. Cavé (1972a) Alcaloides des Annonacées. Alcaloides des écorces de tronc de l'*Uvariopsis guineensis* Keay. *Phytochemistry*, **11**, 2833–40.
Leboeuf, M. and A. Cavé (1972b) Alcaloides des Annonacées. Sur l'isolement de la *l*(+)isocorydine à partir des feuilles de l'*Enantia polycarpa* Engl. & Diels. *Annales pharmaceutiques françaises*, **30**, 211–22.
Leboeuf, M. and A. Cavé (1980) Alcaloides des Annonacées XXVIII. Alcaloides de l'*Uvaria chamae* Beauv. *Plantes Medicinales et Phytothérapie*, **14**, 143–7.
Leboeuf, M., A. Cavé, A. Touche, J. Provost and P. Forgaes (1981) Isolement de l'higémanine à partir de l'*Annona squamosa;* intérêt des résines adsorbantes macromoléculaires en chimie végetale extractive. *Lloydia*, **44**, 53–60.
Leboeuf, M., C. Legueut, A. Cavé, J. F. Deconclois and P. Forgaes (1980) Anomurine et anomuricine, deux nouveaux alcaloides isoquinoleiques de l'*Annona muricata*. *Planta Medica*, **39**, 204–5.
Lechat, P., F. Bisseliches, F. Bournerias, H. Dechy, Y. Juillet, G. Lagier, C. Merignac, B. Rouvex, P. Sterin and S. Weber (1978) *Abrégé de Pharmacologie Médicale*, 3rd edn, 1 vol., Masson, Paris, 677 pp.
Le Men, J. and L. Olivier (1978) Alkaloides de deux espèces du genre *Hunteria: H. elliotii* Stapf Pichon and *H. congolana* Pichon. *Plantes Médicinales et Phytothérapie*, **12**, 175–85.
Levèque, J., J. L. Pousset and A. Cavé (1975) Le lyaloside, nouveau glucoalcaloide isolé de *Pauridiantha lyalli* Brem, Rubiacées. *Comptes Rendus de l'Académie des Sciences, Serie C*, **280**, 593.
* Lockwood, H. (1975) Preliminary pharmacological screening of plant extracts. *Symposium of the Organisation of African Unity, Cairo*, item 6/33, 164–71.
* Lockwood, H. (1976) Tests of plant extracts for potential anticonvulsant activity. *International Congress on Research on Medicinal Plants*, Munich.

Lofgren, F. V. and D. L. Kinsley (1942) A study of a *Strychnos* species. *Journal of the American Pharmaceutical Association*, **31**, 295–8.

Louw, P. G. J. (1948) Isolation of lantaden B and the oxygen functions of lantaden A and lantaden B. *Onderstepoort Journal of Veterinary Science and Animal Industry*, **23**, 233–8.

Louw, P. G. J. (1949) Lantaden A, the active principle of *Lantana camara* II. *Onderstepoort Journal of Veterinary Science and Animal Industry*, **24**, 321–9.

Lutomsky, J. and T. Wrocinski (1960) Pharmacodynamic properties of *Passiflora incarnata* preparations, effect of flavonoid and alkaloid components on pharmacodynamic properties of the raw material. *Biuletyn Instytutu Roslin Leczniczych* **55**, 176–84 (*Chemical Abstracts* (1961) 6785e).

*McIlroy, R. J. (1950) *The Plant Glycosides*, Vol. 1, Arnold, London.

McIsaac, N. M., P. A. Khairallah and I. M. Page (1961) 10-Methoxyharmalan, a potent serotonin antagonist which affects conditioned behaviour. *Science*, 674–5.

McLean, S. and D. G. Murray (1970) Isolation of indole (β-carboline), pyridine and indole pyridine alkaloids from *Nauclea diderichii*. *Canadian Journal of Chemistry*, **48**, 867–8.

Majumdar, D. N. and C. D. Zalani (1953) Alkaloid constituents III. Isolation of water-soluble alkaloids and a study of their chemical and physiological characterization. *Mucuna pruriens* DC. *Indian Journal of Pharmacy*, **15**, 62–5.

Malcolm, S. A. and E. A. Sofowora (1969) Antimicrobial activity of selected Nigerian folk remedies and their constituent plants. *Lloydia*, **32**, 512–17.

Malhotra, C. L., P. K. Das and N. S. Dhalla (1960a) Studies on *Withania ashwaganda* (somnifera) I. Effect of total extract on CNS and smooth muscles. *Indian Journal of Physiology and Pharmacology*, **4**, 35–48.

Malhotra, C. L., P. K. Das and N. S. Dhalla (1960b) Studies on *Withania ashwaganda* (somnifera) II. Effect of total extract on cardiovascular system, respiration and skeletal muscles. *Indian Journal of Physiology and Pharmacology*, **4**, 49–64.

Manske, R. H. F. and H. L. Holmes (1950–71) *The Alkaloids*, 13 vols., Academic Press, New York, London.

Marderosian, A. der (1967) Hallucinogenic indole compounds from higher plants. *Lloydia*, **30**, 23–38.

Marini Bettolo, G. B., E. Miranda delle Monache, C. Galeffi, M. A. Ciasca Rendina and A. Villar Del Fresno (1970) On the alkaloids of Strychnos. *Strychnos panamensis* Seem (akagerine). *Phytochemistry*, **11**, 372–81.

Marini Bettolo, G. B., M. Nicoletti, I. Messana, M. Patamia, C. Galeffi, J. U. Oguakwa, G. Portalone and A. Vacagio (1983) Research on African Medicinal Plants IV. Boonein, a new C-9 terpenoid lactone from *Alstonia boonei:* A possible precursor in the indole alkaloid biogenesis. *Tetrahedron*, **39**, 323–9.

Marquis, V. O. (1975) Organization of Research in Africa. Pharmacological screening techniques (*Alstonia boonei*). Organization of African Unity Symposium, Cairo, item 6/32, 159–63, ed. OAU, Lagos, 1979.

Marquis, V. O. and J. A. O. Ojewole (1976) Neuromuscular actions of the alkaloid echitamine contained in *Alstonia boonei*. *Nigerian Journal of Science*, in OAU Symposium, Cairo, ed. OAU, Lagos, 1979.

Martin, M., J. Biot, J. Ridet, L. Porte, A. Chartol and A. Bezon (1964) Action thérapeutique de l'extrait d'*Euphorbia hirta* dans l'amibiase intestinale (à propos de 150 observations). *Médicine Tropicale, France*, **24**, No. 3, pp. 250–61.

Martindale (1958) *The Extra Pharmacopoeia*, 24th edn, (25th edn 1969), Pharmaceutical Press, London.

Massion, L. (1934) Contribution à l'étude de la mitraphylline. *Archives Internationales de Pharmacodynamie et Thérapie*, **16**, 782–90.

Mathis, C. and P. Duquenois (1963) Contribution à l'identification par chromatographie sur papier des alcaloides de quelques Strychnos d'Afrique et d'Asie. *Annales pharmaceutiques françaises*, **21**, 17–26.

Mehrotra, P. K. and V. P. Kamboj (1978) Hormonal profile of coronaridine hydrochloride an antifertility agent of plant origin. *Planta Medica*, **33**, 345–9.

Merlini, L., G. Nasini and R. E. Haddock (1972) Indole alkaloids from *Uncaria gambir*. *Phytochemistry*, **11**, 1525–6.

Merck (1976) *The Merck's Index*, 7th edn, Merck, Rahway, NY.

Meyer, Th. M. (1941) The alkaloids of *Annona muricata*. *Ingenieur, Nederlandsch-Indie (Indonesia)*, **8**, 64–6 (in Chemical Abstracts, **35** (1941) 8206).

Michel, S., F. Tillequin, M. Koch and L. Ake Assi (1980) L'ellipticine, alcaloide majeur des écorces de *Strychnos dinklagei*. *Journal of Natural Products (Lloydia)*, **43**, 294–5.

Michl, H. and F. Haberler (1954) Uber die Bestimmung von Purinen in kaffein-haltigen Drogen. *Monatshefte*, **85**, 779–96.

Mirand, C., C. Delaude, J. Levy and L. Le Men Olivier (1979) Alcaloides de *Strychnos aculeata*. *Plantes Médicinales et Phytothérapie*, **13**, 84–6.

*Mitscher, L. A., R. P. Leu, M. S. Bathala, W. N. Wu and J. L. Beal (1972) Antimicrobial agents from higher plants. I. Introduction, Rationale and Methodology. *Lloydia*, **35**, 157–66.

Monseur, X. and M. L. van Bever (1955) Recherches chimiques sur l'écorce d'*Alstonia congensis*. *Journal de Pharmacie de Belgique*, **10**, 93–102.

Moreira, E. A. (1963) Pesquiza e identificaçao da emetina un algunas especies do genero *Borreria* (Rubiaceae). *Boletim da Universidade Parana Pharmacognosia*, **2**, 1–60.

Moti, N. T. and B. S. Deshmanhar (1972) Some preliminary pharmacological investigations on *Cardiospermum halicacabum* seeds. *Indian Journal of Pharmacy*, **34**, 76.

Moyse-Mignon, H. (1942) Recherches sur quelques Méliacées africaines et sur leurs principes amers. Thèse de Doctorat en Pharmacie, Université de Paris.

Mudgal, V. (1975) Studies on medicinal properties of *Convolvulus pluricaules* and *Boerhaavia diffusa*. *Planta Medica*, **28**, 62–8.

Muraveva, V. I. and A. I. Bankovski (1956) Chemical studies of *Securinega suffruticosa* (Pall.) Rehd. *Doklady Akademii Nauk SSSR*, **110**, 998–1000 (*Chemical Abstracts* (1957), 8121a).

Nadkarni, A. K. (1954) *Indian Materia Medica*, 3rd edn, Popular Books Depot India, 1 vol. (through Indian Council of Medical Research, 1976, p. 164).

Namjoshi, A. N. (1955) Studies on the pharmacognosy of *Tinospora cordifolia* Miers. *Bulletin of the National Institute of Sciences of India*, **4**, Proceedings 113–17.

Nara, T. K., J. Cleye, L. Lavergne de Cerval and E. Stanislas (1977) Flavonoides de *Phyllanthus niruri* L. *P. urinaria* L. et *P. orbiculatus* L, Rich. *Plantes Médicinales et Phytothérapie*, **11**, 82–6.

Naranjo, C. (1967) Psychotropic properties of the Harmala alkaloids. In Efron *et al.* (1967) *Ethnopharmacological Research for Psychoactive Drugs*, p. 385.

Ndir, O. and J. L. Pousset (1981) Plantes africaines VII. Essais *in vitro* d'*Euphorbia hirta* sur *Entamoebia histolytica*. *Plantes Médicinales et Phytothérapie*, **15**, 113–25.

Neu, R. (1954, 1956) Inhaltstoffe von *Passiflora incarnata*, *Arzneimittel-Forschung*, **4**, 292–4; **4**, 601–6; **6**, 94–8.

Nicholas, H. J. (1961) Determination of sterol and triterpene contents of *Ocimum basilicum* and *Salvia officinalis* at various stages of growth. *Journal of Pharmaceutical Sciences*, **50**, 645–7.

Nickel, L. G. (1959) Antimicrobial activity of vascular plants. *Economic Botany*, **13**, 281–318.

Nicto, M., A. Cavé and M. Leboeuf (1976) Alcaloides des Annonacées. Composition des écorces de tronc et de racines d'*Enantia pilosa*. *Lloydia*, **39**, 350–6.

Noamesi, B. K. and E. A. Gyang (1980) Effects of methylflavinantine on the response of coaxial stimulation of guinea pig ileum. *Planta Medica*, **38**, 138–43.

Odebiji, O. O. (1978) Antibacterial property of tetramethylpyrazine from the stem of *Jatropha podagrica*. *Planta Medica*, **38**, 144–6.

Oguakwa, J. U., C. Galeffi, M. Nicoletti, I. Messana, M. Patamia and G. B. Marini-Bettolo (1980) On the alkaloids of Strychnos XXXIV. The alkaloids of *Strychnos spinosa* Lam. *Gazzetta Chimica Italiana*, **110**, 97–100.

Ojewole, J. A. O. (1980) Studies on the pharmacology of tetramethylpyrazine from the stem of *Jatropha podagrica*. *Planta Medica*, **39**, 238.

Ojewole, J. A. O. (1981) Effects of tetramethylpyrazine on isolated atria of the guinea pig. *Planta Medica*, **42**, 223–8.

Ojewole, J. A. O. (1983a) Autonomic pharmacology of echitamine, an alkaloid from *Alstonia boonei* de Wild. *Fitoterapia*, **54**, 99–113.

Ojewole, J. A. O. (1983b) Antibronchoconstrictor and antiarrhythmic effects of chemical compounds from Nigerian medicinal plants. *Fitoterapia*, **54**, 158–61.

Ojewole, J. A. O. (1983c) Blockade of autonomic transmission by scopoletin. *Fitoterapia*, **54**, 203–11.

Ojewole, J. A. O. and S. K. Adesina (1983) Effect of hypoxanthine-9-L-arabinofuranoside, a nucleoside from the roots of *Boerhavia diffusa* L. (Nyctaginaceae), on isolated coronary artery of the goat. *Fitoterapia*, **54**, 163–9.

Ojewole, J. A. O. and O. O. Odebiji (1980) Neuromuscular and cardiovascular action of tetramethylpyrazine from the stem of *Jatropha podagrica*. *Planta Medica*, **39**, 238.

Ojewole, J. A. O. and O. Odebiji (1984) Some studies on the pharmacology of tetramethylpyrazine an alkaloid from the stembark of *Jatropha podagrica*. *Fitoterapia*, **55**, 213–25.

Olaniji, A. A. and W. N. A. Rolfsen (1980) Two new tertiary indole alkaloids of *Strychnos decussata*. *Journal of Natural Products (Lloydia)*, **43**, 595–7.

Olaniji, A. A., E. A. Sofowora and B. O. Oguntimein (1975) Phytochemical investigation of some Nigerian plants against fevers II. *Cymbopogon citartus*. *Planta Medica*, **28**, 186–9.

Oliver, B. (1958) Nigeria's useful plants I. Plants yielding fibres, fats and oils. *The Nigerian Field*, **23**, 147–71.

Oliver, B. (1960) *Medicinal Plants in Nigeria*, ed. Nigerian College of Arts, Science and Technology. 1 vol., 138 pp.

Oliver-Bever, B. (1961) Pharmacognosie de l'Epoque pharaonique. *Quarterly Journal of Crude Drug Research*, **6**, 853–67.

Oliver-Bever, B. (1968) Selecting local drug plants in Nigeria, botanical and chemical relationship in three families. *Quarterly Journal of Crude Drug Research*, **8**, 1194–211.

Oliver-Bever, B. (1972) Drug plants in ancient and modern Mexico. *Quarterly Journal of Crude Drug Research*, **11**, 1957–72.

Oliver-Bever, B. and G. R. Zahnd (1979) Plants with oral hypoglycaemic action. *Quarterly Journal of Crude Drug Research*, **17**, 139–96.

Pais, M., F. Mainil and R. Goutarel (1963) Les adouétines x, y et z. Alcaloides de *Waltheria americana* L. (Sterculiacées). *Annales pharmaceutiques françaises*, **21**, 139–46.

Panichpol, K., R. D. Waigh and P. G. Watermann (1977) Chemical studies on the Annonaceae Part 3. Chondrofoline from *Uvaria ovata*. *Phytochemistry*, **16**, 621–2.

Pant, M. C., I. Uddin, U. R. Bardwaj and R. D. Tewari (1968) Blood sugar and total cholesterol lowering effect of *Glycine soja*, *Mucuna pruriens* DC. and *Dolichos biflorus* L. seed diets in normal fasting albino rats. *Indian Journal of Medical Research*, **56**, 1808–12.

Parello, J. (1966) Les alcaloides du *Phyllanthus discoideus* Mull. Arg. (Euphorbiacées). Isolement et détermination des structures. Thèse de Doctorat Sc. Phys., Paris.

Parello, J., A. Melera and L. Goutarel (1963) Phyllochrysine et securinine, alcaloides du *Phyllanthus discoides* Muell. Arg. *Bulletin de la Société Chimique de France*, 197–8.

Paris, M. and R. Paris (1972) Sur les polyphenols de *Virectaria multiflora* Brenack Rubiacée d'Afrique occidentale. *Plantes Médicinales Phytothérapie*, **6**, 225–9.

Paris, R. R. (1956) Sur une Tiliacée africaine ocytocique le *Grewia elyseoi*. *Annales pharmaceutiques françaises*, **14**, 348.

Paris, R. (1963) Sur l'action tranquillisante de quelques plantes médicinales. *Annales pharmaceutiques françaises*, **21**, 389–97.

Paris, R. R. and L. Beauquesne (1938) *Tinospora cordifolia*. *Bulletin de la Société de Pharmacie Française*, **45**, 7; **46**, 736.

Paris, R. R. and L. Cosson (1965) Sur la biogenèse des alcaloides du *Datura*, étude des variations ontogéniques à l'aide d'une méthode chromatographique chez le *Datura metel* Sims. *Comptes Rendus de l'Académie des Sciences. Serie D*, **260**, 3148–51.

Paris, R. and R. Goutarel (1958) Les Alchorneas africains. Présence de yohimbine chez *Alchornea floribunda* (Euphorbiacées) *Annales pharmaceutiques françaises*, **16**, 15–20.

Paris, R. R. and J. Le Men (1955) Sur un Stephania d'A.O.F., le *Stephania dinklagei* Diels (Menispermacées). *Annales pharmaceutiques françaises*, **13**, 200–4.

Paris, R. R. and H. Moyse (1963) *Abrégé de Matière Médicale*, Vigot, Paris. 1 vol., 192pp.

Paris, R. R. and H. Moyse (1965–71) *Précis de Matière Médicale*, Vol. I, 1965; Vol. II, 1967; Vol. III, 1971. Masson et Cie, Paris, 1436 pp.

Paris, R. and H. Moyse-Mignon (1939) Sur quelques Meliacées réputées fébrifuges. *Bulletin des Sciences Pharmacologiques*, **46**, 104–8.

Paris, R. and H. Moyse-Mignon (1956) Characterisation de la choline chez quelques plantes médicinales. *Annales pharmaceutiques françaises*, **14**, 464–9.

Paris, R. and J.-P. Théallet (1961) Nouvelles recherches sur la composition chimique et l'activité ocytocique de divers *Grewias* (Tiliacées) d'origine africaine. *Annales de Pharmacie françaises*, **19**, 20–3.

Paris, R. R., E. Henri and M. Paris (1976) Sur les C-flavonoides du *Cannabis sativa*. *Plantes Médicinales et Phytothérapie*, **10**, 144–54.

Paris, R. R., H. Moyse and J. Le Men (1955) Sur une Euphorbiacée à alcaloides, le *Fluggea virosa* Baill. *Annales pharmaceutiques françaises*, **13**, 245–9.

Patel, H. B. and J. M. Rowson (1964) Investigations of certain Nigerian medicinal plants Part I. Preliminary pharmacological and phytochemical screenings for cardiac activity. *Planta Medica*, **12**, 34–42.

Patel, M. B., C. Miet and J. Poison (1967) Alkaloides de *Tabernaemontana pachysiphon*. *Annales pharmaceutiques françaises*, **25**, 379–84.

Paton, W. D. M. (1975) Pharmacology of Marijuanha. *Annual Review of Pharmacology*, **15**, 191–220.

Persinos, G. L. and M. W. Quimby (1967a) Nigerian plants III. Phytochemical screening of *Leptactina densiflora* Hook. (Rubiaceae). *Bulletin de la Société Chimique de France*, 750.

*Persinos, G. L. and M. W. Quimby (1967b) Nigerian plants III. Phytochemical screening for alkaloids, saponins and tannins. *Journal of Pharmaceutical Sciences (USA)*, **56**, 1512–15.

Persinos, G. L., M. W. Quimby and J. W. Schermerhorn (1964) A preliminary pharmagnostical study of ten Nigerian plants. *Economic Botany*, **18**, 329–41.

Phillipson, J. D. and S. R. Hemingway (1973) *Uncaria* species as sources of the alkaloids gambirine and the roxburghines. *Journal of Pharmacy and Pharmacology*, **25**, 143.

Phillipson, J. D. and S. R. Hemingway (1975) Alkaloids from *Uncaria* species. *Phytochemistry*, **14**, 1855.

Phillipson, J. D., S. R. Hemingway, C. A. Ridsdale and B. A. Krukoff (1973) Alkaloids of *Uncaria* P.V. Their occurrence and chemotaxonomy. *Lloydia*, **41**, 503–67.

Pichi Sermoli, R. E. (1955) *Tropical East Africa*. Cited in Recherches sur la zone aride XIII. Les plantes médicinales des zones arides, pp. 302–60. UNESCO publication (private), 1960.

Plat, M., M. Koch, A. Bouquet, J. Le Men and M. M. Janot (1963) Présence d'un hétéroside générateur de gentianine dans l'*Anthocleista procera*. Monoterpenoids I. *Bulletin de la Société Chimique de France*, **6**, 1302–5.

Pletscher, A., H. Besendorf, H. P. Bächtold and K. F. Guy (1959) Uber pharmakologische Beeinflussung des Central nervensystems durch kurzwirkende MAO hemmer aus der Gruppe der Harmala Alkaloide. *Helvetica Physiologica Acta*, **17**, 202–14.

Pobéguin, H. (1912) *Plantes Médicinales de la Guinée*, Paris.

Poisson, J., R. Cahen, J. Nadaud, N. Lambert, M. Sautai, F. Bernard, D. Bonnet, C. Boudet, A. Pessonier and N. Duval (1972) Etude de quelques plantes médicinales malgaches. *Annales pharmaceutiques françaises*, **30**, 241–54.

Pope, H. C., Jr (1969) *Tabernanthe iboga* an African narcotic plant of social importance. *Economic Botany*, **23**, 174–84.

Popp, F. D., J. M. Wefer, D. P. Chakraborty, G. Rosen and A. C. Casey (1968) Investigations of African plants for alkaloids, antimalarial agents and antineoplastic agents. *Planta Medica*, **16**, 343–7.

Pousset, J. L., A. Bouquet, A. Cavé and R. Paris (1971) Structure de deux nouveaux, alcaloides isolés du *Pauridanthia callicarpoida* Brem. *Comptes Rendus de l'Académie des Sciences, Serie C*, **272**, 665–7.

Pousset, J. L., J. Kerharo, G. Maynard, X. Monseur, A. Cavé and R. Goutarel (1973) La borrerine, nouvel alcaloide isolé du *Borreria verticillata*. *Phytochemistry*, **12**, 2310–12.

Pousset, J. L., L. Levesque, A. Cavé, F. Picot, P. Potier and R. R. Paris (1974) Etude chimique du *Pauridanthia lyallii* (Rubiacée). *Plantes Médicinales et Phytothérapie*, **8**, 51–6.

Pradhan, S. N. and N. N. De (1959) Comparative pharmacological activities of some derivatives of hayatin. *Archives Internationales de Pharmacodynamie et de Thérapie*, **120**, 136.

Pradhan, S. N., K. S. Varadan, C. Ray and N. N. De (1953) Pharmacological investigation of chaksine, an alkaloid from *Cassia absus*. *Journal of Scientific and Industrial Research*, **12B**, 258–63.

Prasad, G. and K. N. Udupa (1963) Effect of *Cissus quadrangularis* on the healing of cortisone-treated fractures. *Indian Journal of Medical Research*, **51**, 667–76.

Prasad, G. C. and K. N. Udupa (1972) Pathways and site of action of phytogenic steroid from *Cissus quadrangularis*. *Journal of Research in Indian Medicine*, **7**, 29.

Prasad, G. C., S. C. Chatterjee and K. N. Udupa (1970) Effect of the phytogenic steroid of *C. quadrangularis* on endocrine glands after fracture. *Journal of Research in Indian Medicine*, **4**, 182.

Prasad, S. and C. L. Malhotra (1968) *Withania ashwaganda* Kaul. Effect of the alkaloidal fractions (acetone, alcohol, water-soluble) on the central nervous system. *Indian Journal of Physiology and Pharmacology*, **12**, 175–81.

Prema, P. (1968) Pharmacological studies of *Desmodium gangeticum* DC. Thesis University of Kerala Trivandrum, India.

Prista, I. N. and A. Correia Alves (1958) Estudo farmacnosico, botanico quimico e farmacodynamico da *Securidaca longepedunculata* Fresen. *Garcia de Orta Lisbon*, **6**, 131–47.

Quevauviller, A. and O. Blanpin (1959) Etude pharmacodynamique de la phyllochrysine, alcaloide de *Phyllanthus discoides*. Euphorbiacées. *Thérapie*, **15**, 619–24.

Quevauviller, A. and O. Blanpin (1960) Propriêtés pharmacodynamiques comparées de l'holaphylline et de l'holaphyllamine, alcaloides stéroidiques de l'*Holarrhena floribunda*. Thérapie, **15**, 1212–21.

Quevauviller, A. and O. Blanpin (1961) Action de la triacanthine, alcaloide dérivé naturel, particulièrement sur le système nerveux central. *Thérapie*, **16**, 782–90.

Quevauviller, A. and M. Hannonière (1977) Action des principaux alcaloides de *Polyalthia oliveri* Engl. sur le système nerveux central et le système cardiovasculaire. *Comptes Rendus de l'Académie des Sciences, Serie D*, **284**, 93–6.

Quevauviller, A. and G. Sarasin (1967) Sur un alcaloide papaverinique extrait du *Stephania dinklagei* Diels, Ménispermaceae. *Annales pharmaceutiques françaises*, **25**, 371–7.

Quevauviller, A., P. Bourrinet and M. Bezançon (1977) Observations pharmacodynamiques avec les alcaloides totaux de l'*Uvariopsis guineensis* Keay, Annonacées. *Thérapie*, **32**, 215–18.

Quevauviller, A., O. Foussard-Blanpin and D. Coignard (1965) Un alcaloide *de Phyllanthus discoides* (Euphorbiacées) la phyllalbine, sympathomimétique central et péripherique. *Thérapie*, **20**, 1033–41.

Quevauviller, A., O. Foussard-Blanpin and P. Bourrinet (1967) Pharmacodynamie de la securinine, alcaloide présent dans le *Phyllanthus discoides*. *Thérapie*, **22**, 297–302.

Quisumbing, E. (1951) Technical Bulletin of the Philippines Department of Agriculture and Natural Resources No. 16.

Qureshi, A. W., A. M. Ahsam and G. Hahn (1964) Constituents of *Cassia absus* L. Isolation and characterisation of a glycoside from the seeds. *Pakistan Journal of Scientific and Industrial Research*, **7**, 219–20.

Raina, K. L., K. L. Dhar and O. K. Atal (1976) Occurrence of *N*-isobutyl-eicosa-*trans*-2-*trans*- and *trans*-4-dienamide in *Piper nigrum*. *Planta Medica*, **30**, 198–200.

Ramjelal, L., R. S. Rathor, R. Chakravarty and P. K. Das (1972) Preliminary studies on the antiinflammatory and antiarthritic activity of *Crataeva nurvala* (religiosa). *Indian Journal of Pharmacology*, **4**, 122.

Rao, P. S. and T. R. Seshadri (1969) Variations in the chemical composition of Indian samples of *Centella asiatica*. *Current Science*, **38**, 77–9.

Raven, P. H. (1974) Erythrina (Fabaceae) Introduction to Erythrina symposium II. Achievements and opportunities. *Lloydia*, **37**, 321–31.

Ray, A. B., R. M. Chattopahijay, R. P. Tripathi, S. S. Gambir and P. K. Das (1979a) *Stephania* alkaloids. *Planta Medica*, **35**, 167–73.

Ray, A. B., R. M. Chattopahijay, R. P. Tripathi, S. S. Gambir and P. K. Das (1979b) Isolation and pharmacological action of epistephanine, an alkaloid of *Stephania hernandifolia*. *Planta Medica*, **35**, 488–91.

Raymond-Hamet, R. (1933) Sur les manifestations initiales de l'action sympatholytique de la yohimbine. *Comptes Rendus de l'Académie des Sciences Paris*, **198**, 977–98.

Raymond-Hamet, R. (1934) Sur quelques effets physiologiques de l'echitamine. *Comptes Rendus des Scéances de la Société de Biologie Paris*, **116**, 1022–5.

Raymond-Hamet, R. (1936) Bulbocapnine, typical of a new group of medicinals. *Comptes Rendus de l'Academie des Sciences*, **202**, 357–9.

Raymond-Hamet, R. (1937) Sur quelques propriétés physiologiques du *Sarcocephalus esculentus* Afz. *Comptes Rendus des Scéances de la Société de Biologie Paris*, **126**, 488–91.

Raymond-Hamet, R. (1941) Action réno-dilatatrice de l'échitamine *Comptes Rendus des Scéances de la Société de Biologie Paris*, **135**, 1565–7.

Raymond-Hamet, R. (1954) Sur les effets tensio-vasculaires d'une Euphorbiacée africaine: l'*Alchornea floribunda*. *Comptes Rendus des Scéances de la Société de Biologie Paris*, **148**, 655–8.

Raymond-Hamet, R. (1964) Sur l'alcaloide extrait d'une Rubiacée Nauclée africaine l'*Ourouparia africana* (Don) Baillon. *Comptes Rendus des Scéances de la Société de Biologie Paris*, **259**, 3872–4.

Raymond-Hamet, R. and D. Vincent (1960) Sur quelques effets pharmacologiques de trois alcaloides du *Tabernanthe iboga*. *Comptes Rendus des Scéances de la Société de Biologie Paris*, **154**, 2223–7.

Ridet, J. and A. Chartol (1964) Les propriétés antidysentériques de l'*Euphorbia hirta*. *Médecine Tropicale*, **24**, No. 2, 119–43.

Rolfsen, W. and L. Bohlin (1978) New indole alkaloids of *Strychnos Dale* and *Strychnos eleaocarpa*. *Lloydia*, **41**, 656.

Rolfsen, W. N. A., A. A. Olaniyi and P. J. Hylands (1980b) Isolation of five tertiary indole alkaloids from the stembark of *Strychnos decussata*. *Journal of Natural Products (Lloydia)*, **43**, 97–102.

Rolfsen, W. N. A., A. A. Olaniyi, F. Sandberg and A. N. Koick (1980a) Muscle relaxant activity of decussine, a new indole alkaloid of *Strychnos decussata*. *Acta Pharmaceutica Suecica*, **17**, 105–11.

Rommelspacher, H. (1981) The β-carbolines (harmanes), a new class of endogenous compounds, their relatives for the pathogenesis and treatment in psychiatry and neurotropic diseases. *Pharmacopsychiatry*, **18**, 117–25.

Roychoudhury, A. (1972) Chemical investigations on *Cissampelos pareira*. *Science and Culture*, **38**, 358–9.

Russel, T. A. (1955) The Kola of Nigeria and the Cameroons. *Tropical Agriculture*, **32**, 210–40.

Saito, S., T. Iwamoto, T. Tanaka, C. Matsumura, N. Sugimoto, Z. Horii and Y. Tamatura

(1964a) Two new alkaloids, viroallosecurinine and virosine isolated from *Securinega virosa* Pax Hoffm. *Chemistry and Industry*, **28**, 1263–4.

Saito, S. *et al.* (1964b) Structure and stereochemistry of norsecurinine and dihydrosecurinine. *Chemical and Pharmaceutical Bulletin*, **18**, 786–96 (in *International Pharmaceutical Abstracts*, 1966).

Sandberg, F. (1980) Medicinal and toxic plants from equatorial Africa, a pharmacological approach. *Journal of Ethnopharmacology*, **2**, 105–8.

Sandberg, F. and A. Cronlund (1982) An ethnographical inventory of medicinal and toxic plants from Equatorial Africa. *Journal of Ethnopharmacology*, **5**, 187–204.

Sandberg, F. and K. Kristianson (1970) A comparative study of the convulsant effects of Strychnos alkaloids. *Acta Pharmaceutica Suecica*, **7**, 329–36.

Sandberg, F., E. Lunell and K. J. Ryrberg (1969b) Pharmacological and phytochemical investigations of African Strychnos species. *Acta Pharmaceutica Suecica*, **6**, 79–102.

Sandberg, F., R. Verpoorte and A. Cronlund (1971) Screening of African Strychnos species for convulsant and muscle relaxant effects. *Acta Pharmaceutica Suecica*, **8**, 341–50.

Sandberg, F., K. Roos, K. J. Ryrberg and K. Kristianson (1969a) The pharmacologically active alkaloids of *Strychnos icaja* Baill. Strychnine and a new alkaloid, 4-hydroxy-strychnine. *Acta Pharmaceutica Suecica*, **6**, 103–8.

Sanity, O., F. Bailleul, P. Delaveau and H. Jacquemin (1981) Iridoides du *Borreria verticillata*. *Planta Medica*, **42**, 260–4.

Satoda, I., M. Murrayama, D. Tsuji and E. Yoshii (1972) Studies on securinine and allosecurinine *Tetrahedron Letters*, **25**, 1199–202.

Schimmel *et al.* (1914) Ethereal oils. Semi-annual report April 1914 (through *Chemical Abstracts*, **8**, 2598–9).

Schlag, J., E. Philippot, M. J. Dallemagne and G. Troupin (1959) Propriétés pharmacologiques d'un stimulant central, l'extrait de *Dioscorea dumetorum*. *Journal de Physiologie*, **51**, 553–64.

Schneider, J. A. and E. B. Sigg (1957) Neuropharmacological studies of ibogaine, an indole alkaloid with central stimulant properties *Annals of the New York Academy of Science*, **66**, 765–76.

Schröter, H. B., D. Neumann, A. R. Katritzky and F. J. Swinborne (1966) Withasomnine, a pyrazole alkaloid from *Withania somnifera* Dun. *Tetrahedron (London)*, **22**, 2895–7.

Schulz, O. A. (1964) Der gegenwärtiger Stand der Cannabis Forschung. *Planta Medica*, **12**, 371–82.

Schwarting, A. E., J. M. Bobbit, A. Rother, C. K. Atal, K. L. Khanna, J. D. Leary and W. G. Walter (1963) The alkaloids of *Withania somnifera*. *Lloydia*, **26**, 258–73.

Seawright, A. A. (1963) Studies on experimental intoxication of sheep with *Lantana camara*. *Australian Veterinary Journal*, **9**, 340–4.

Seawright, A. A. and J. G. Allen (1972) Pathology of the liver and kidneys in *Lantana* poisoning of cattle. *Australian Veterinary Journal*, **48**, 323–33.

Sébire, R. P. (1899) *Les Plantes Utiles du Senegal*, 1 vol., Baillière, Paris, 341 pp.

Sen, R. N. and S. N. Pradhan (1963) Cissampelos alkaloids. Action of hayatine derivatives on the central nervous system of cats and dogs. *Archives Internationales de Pharmacodynamie*, **152**, 106.

Sen, S. P. (1966) Studies on the active constituents of *Cissus quadrangularis II*. *Current Science*, **35**, 317.

Shah, C. S. and P. N. Khanna (1963, 1964) Chemical investigation of *Datura metel* and *D. metel* var. *fastuosa*. *Indian Journal of Pharmacy*, **25**, 370–2; **26**, 140.

Shah, S. and P. N. Khanna (1965a) Alkaloid estimation of roots of *Datura metel* and *D. metel* var. *fastuosa*. *Lloydia*, **28**, 71–2.

Shah, C. S. and P. N. Khanna (1965b) A note on the alkaloidal content of *Datura innoxia* Miller. *Journal of Pharmacy and Pharmacology*, **17**, 115.

Shah, C. S. and A. N. Saoje (1967) Alkaloidal estimation of roots of *Datura arborea* L. *Indian Journal of Pharmacy*, **29**, 199.

Shohat, B. (1967) Withaferin. *Cancer Chemotherapy Report*, **51**, 271–7.

Shulgin, A. T. (1966) Possible implication of myristicin as a psychotropic substance. *Nature*, **210**, 380–4.

Shulgin, A. T., T. Sargent and C. Naranjo (1967) The chemistry and psychopharmacology of nutmeg and related phenylisopropylamines (in Efron *et al.* (1967) 202–14).

Siddiqui, S. and Z. Ahmad (1935) Alkaloids from the seeds of *Cassia absus*. *Proceedings of the Indian Academy of Science*, **2A**, 421–5.

Singh, R. H. and K. N. Udupa (1972a) Studies on the Indian indigenous drug Punarnava (*Boerhavia diffusa* L.) Part II. Preliminary phytochemical studies. *Journal of Research on Indian Medicine*, **7**, 13.

Singh, R. H. and K. N. Udupa (1972b) Studies on the Indian indigenous drug Punarnava (*Boerhavia diffusa* L.) Part III. Experimental and pharmacological studies. *Journal of Research on Indian Medicine*, **7**, 17.

Singh, R. H. and K. N. Udupa (1972c) Studies on the Indian indigenous drug Punarnava. Part IV Preliminary controlled clinical trial in nephrotic syndrome. *Journal of Research on Indian Medicine*, **7**, 28.

Slotkin, T. A., V. Distefano and W. Y. Au (1970) Bloodlevels and urinary excretion of harmine and its metabolites in man and rats. *Journal of Pharmacology and Experimental Therapeutics*, **173**, 26–30.

Smith, E. J. R. (1963) Reserpine. Les rapports avec l'adrenaline et la noradrenaline. *Journal de Pharmacologie expérimentale et de Thérapie*, **139**, 321.

Smolenski, S. J., H. Silinis and R. N. Farnsworth (1972) Alkaloid screening. *Lloydia*, **35**, 3–4.

Sofowora, A. (1980) The present status of the plants used in traditional medicine in Western Africa. A medical approach and a chemical evaluation. *Journal of Ethnopharmacology*, **2**, 109–18.

Sofowora, E. A., W. A. Isaac-Sodeye and L. O. Ogunkoya (1975) Isolation and characterisation of an antisickling agent from *Fagara zanthoxyloides* root. *Lloydia*, **38**, 169–71.

Sollmann, T. (1957) *A Manual of Pharmacology*, 8th edn, 1 vol., Saunders & Cie, Philadelphia, 1535 pp.

Sorer, H. and O. Pylko (1965) Effect of γ-amino-butyric acid upon brucine convulsions. *Journal of Pharmacy and Pharmacology*, **17**, 249–50.

Spiff, A. I., V. Zabel, W. H. Watson, M. A. Zemaitis, A. M. Ayeya, D. J. Slatkin, J. E. Knapp and P. L. Schiff, Jr. (1981) Constituents of West African medicinal plants XXX. Tridictyophylline, a new morphinan alkaloid from *Triclista dictyophylla*. *Journal of Natural Products (Lloydia)*, **44**, 160–5.

Srivastava, R. M. and M. P. Khare (1964) Uber wasserlösliche Alkaloide aus der wurzelrinde von *Cissampelos pareira* L. *Chemische Berichte*, **97**, 2732–41.

Staner, P. and R. Boutique (1937) *Matériaux pour l'Etude des Plantes Médicinales du Congo belge*, Belgian Government, Bruxelles.

Staunton, J. (1979) Biosynthesis of isoquinoline alkaloids (Erythrina). *Planta Medica*, **36**, 1–20.

Steinegger, E. and Th. Weibel (1951) Biologische Untersuchung des Gentianins. *Pharmaceutica Acta Helvetica*, **26**, 333–42.

Su, K. L., Y. Abul Haji and L. Staba (1975) Antimicrobial effect of aquatic plants from Minnesota. *Lloydia*, **38**, 175–338.

Subbu, V. S. V. (1970) Pharmacological and toxicological evaluations of an active principle obtained from the plant *Vitis quadrangularis*. *Indian Journal of Pharmacology*, **2**, 91.

Subbu, V. S. V. (1971) Mechanism of action of *Vitis* glucoside on myocardial tissue. *Indian Journal of Medical Sciences*, **25**, 400–2.

Suzuki, O., Y. Katsumata, M. Oya, V. M. Chari, B. Vermes, H. Wagner and K. Hostettmann (1981) Inhibition of type A and type B monoamine oxidases by naturally occurring xanthones. *Planta Medica*, **42**, 17–21.

Tackie, A. N. and A. Thomas (1968) Alkaloids of *Tiliacora funifera*. *Planta Medica*, **16**, 158–65.

Tackie, A. N., D. Dwuma Badu, T. Okarter, J. E. Knapp, D. J. Slatkin and P. L. Schiff, Jr (1973a) Constituents of West African medical Plants I. Trigilletine and Tricordatine. Two new bisbenzyl isoquinoline alkaloids from *Triclisia* species. *Phytochemistry*, **12**, 2509–11.

Tackie, A. N., D. Dwuma Badu, J. E. Knapp and P. L. Schiff, Jr (1973b) The structure of funiferine, a biphenyl alkaloid from *Tiliacora funifera*. *Planta Medica*, **16**, 158–65; and *Lloydia*, **36** (1972) 66–71.

Tackie, A. N., D. Dwuma Badu, J. E. Knapp and P. J. Schiff, Jr (1973c) Nortiliacorinine A and nortiliacorine A from *Tiliacora funifera*. *Phytochemistry*, **12**, 203–5.

Tackie, A. N., D. Dwuma Badu, T. Okarter, J. E. Knapp, D. J. Slatkin and P. L. Schiff, Jr (1974a) Constituents of West African plants II. The isolation of alkaloids from selected *Triclisia* species. *Lloydia*, **37**, 1–5.

Tackie, A. N., D. Dwuma Badu, L. Lartey, P. L. Schiff, Jr, J. E. Knapp and D. J. Slatkin (1974b) Constituents of West African medicinal Plants III. Alkaloids of *Stephania dinklagei* (Menispermaceae), *Lloydia*, **37**, 6–9.

Tackie, A. N., D. Dwuma Badu, D. J. Slatkin, P. L. Schiff, Jr (1974c) Constituents of West African medicinal plants IV. *O*-Methylflavinanthine from *Rhigiocarya racemifera*. *Phytochemistry*, **13**, 2884–5.

Tackie, A. N., D. Dwuma Badu, T. T. Daha, J. E. Knapp, D. J. Slatkin and P. L. Schiff, Jr (1974d) Constituents of West African medicinal plants V. Tiliageine, a new bisbenzylisoquinoline alkaloid from *Tiliacora dinklagei*. *Experientia*, **30**, 847–8.

Tackie, A. N., D. Dwuma Badu, J. S. K. Ayim, H. N. El Sohly, J. E. Knapp, D. J. Slatkin and P. L. Schiff, Jr (1975a) *N*-Isobutyloctadeca-*trans*-2-*trans*-4-dienamide, a new constituent of *Piper guineense*. *Phytochemistry*, **14**, 1888–9.

Tackie, A. N., D. Dwuma Badu, J. S. K. Ayim, T. T. Dabra, J. E. Knapp, D. J. Slatkin and P. L. Schiff, Jr (1975b) Constituents of West African medicinal plants XV. Dinklacorine, a new biphenyldibenzo-dioxin alkaloid from *Tiliacora dinklagei*. *Lloydia*, **38**, 210–17.

Tackie, A. N., D. Dwuma Badu and J. S. K. Ayim (1979) The potential Menispermaceous plants of West Africa as tumor inhibitors, muscle relaxants, antitussives and analgesics. In: *Second International Symposium*, item 6/21, Cairo 1975, ed. Organisation of African Unity, Lagos, 1979.

Taylor, W. I. (1957) Iboga alkaloids II. The structure of ibogaine, ibogamine, and tabernanthine. *Journal of the American Chemical Society*, **79**, 3298–9.

Taylor-Smith, R. E. (1965) Investigation on plants of West Africa II. Isolation of anthocleistin from *Anthocleista procera*. *Tetrahedron Letters*, **21**, 3721–5.

Tourova, A. D. (1957) L'utilisation curative de la securinine (en russe). *Klinischeskala Medicina*, **2** (in Kerharo and Adam, 1974).

Truitt, E. B. Jr (1967) The pharmacology of myristicin and nutmeg (in Efron *et al.* (1967) 215–22).

Turner, C. E., M. A. Elsohly and E. G. Boeren (1980) Constituents of *Cannabis sativa* L. XVII. A review of the natural constituents. *Journal of Natural Products (Lloydia)*, **43**, 169–234.

Turner, P. and A. Richens (1978) *Clinical Pharmacology*, 3rd edn, 1 vol., 154 pp. Churchill Livingstone, Edinburgh, London, New York.

Tyler, V. I. Jr (1966) The psychological properties and chemical constituents of some habit-forming plants. *Lloydia*, **29**, 275–93.

Udupa, K. N. and G. C. Prasad (1964) Biochemical and ^{45}Ca studies on the effect of *Cissus quadrangularis* on fracture repairs. *Indian Journal of Medical Research*, **52**, 480–7.

Udupa, K. N. and R. H. Singh (1964) A study on intense therapy of calcium and ascorbic acid in healing of fractures. *Indian Journal of Medical Research*, **52**, 519–27.
Udupa, R. N., G. C. Prasad and S. P. Sen (1965) The effect of phytogenic anabolic steroid in the acceleration of fracture repair. *Life Sciences*, **4**, 317.
Usdin, E. and D. H. Efron (1976) *Psychotropic Drugs and Related Compounds*, Vol. 1, 2nd edn, Government Printing Office, Washington, 791 pp.
Verpoorte, R. (1980) Alkaloids from *Strychnos floribunda* Gilg. *Planta Medica*, **39**, 236.
Verpoorte, R. (1981) Further alkaloids from *Strychnos floribunda* Gilg. *Planta Medica*, **42**, 32–6.
Verpoorte, R. and L. Bohlin (1976) Screening of African Strychnos species for convulsant and muscle relaxant effects. *Acta Pharmaceutica Suecica*, **13**, 245–50.
Verpoorte, R. and F. Sandberg (1971) Alkaloids of *Strychnos camptoneura*. *Acta Pharmaceutica Suecica*, **8**, 119–22.
Verpoorte, R. A. and B. A. Svendsen (1976) *Strychnos dolichothyrsa* Gilg (ex Onochie and Hepper) alkaloids. *Lloydia*, **39**, 357–62.
Wada, K., S. Marumo and K. Munakata (1967) An insecticidal alkaloid, cocculidine from *Cocculus trilobus* I. The structure of cocculidine. *Agricultural and Biological Chemistry*, **31**, 452–60.
*Walker, A. R. (1953) Usages pharmaceutiques des plantes spontanées du Gabon. *Institut d'Etudes Centrafricaines*, **5**, 19–40; **6**, 275–329.
Watanabe, H., M. Ikeda, K. Watanabe and T. Kikuchi (1981) Effects on central dopaminergic systems of *d*-coclaurine and *d*-reticuline extracted from *Magnolia salicylifolia*. *Planta Medica*, **42**, 213–22.
Watt, J. M. (1967) African plants potentially useful in mental health. *Lloydia*, **30**, 1–22.
Watt, J. M. and M. G. Breyer-Brandwijk (1962) *The Medicinal and Poisonous Plants of Southern and Eastern Africa*, 2nd edn, 1 vol., Livingstone, London, 1457 pp. (7000 Refs.).
Weil, A. T. (1965) Nutmeg as a narcotic. *Economic Botany*, **19**, 194–217.
Weil, A. T. (1967) Nutmeg as a psychoactive drug (in Efron *et al.* (1967) 188–201).
Weiss, G. (1960) Hallucinogenic and narcotic-like effects of powdered myristicin (nutmeg). *Psychiatric Quarterly*, **34**, 346–56.
Wenzel, D. G. and G. Y. Koff (1956) The effect of triterpenes on the excretion of sodium and potassium by rats. *Journal of the American Pharmaceutical Association, Scientific Edition*, **45**, 372–3.
Wolfson, S. L. and T. M. G. Solomons (1964) Poisoning by fruit of *Lantana camara*. An acute syndrome observed in children following ingestion of the green fruit. *American Journal of Diseases of Children*, **107**, 173–6.
Wooley, H. E. D. (1962) *The Biochemical Basis of Psychoses or the Serotonin Hypothesis about Mental Diseases*, John Wiley and Sons, New York, 331 pp.
Xiao, P. G. (1983) Recent developments on Medicinal Plants in China. *Journal of Ethnopharmacology*, **7**, 95–109.

Chapter 4: Anti-infective activity of higher plants

Abbot, B. H., L. Hartwell, J. Leiter, S. A. Spitzman and S. A. Schepartz (1966) Screening data from the Cancer Chemotherapy National Service Center Screening Laboratories. XIL. Plant extracts 51–54. *Cancer Research*, **26**, 364–536.
Abd El-Gawad, M. M. and S. A. El Zait (1981) Flavonoids of *Polygonum senegalense* growing in Egypt. *Fitoterapia*, **52**, 239–40.
Abd El Malik, L. Y., M. A. El-Leithy, F. A. Reda and M. Khalil (1973) *Lawsonia inermis*, antimicrobial activity of the leaves. *Zentralblatt für Bakteriologie*, Abteilung II, **128**, 61.
Abdullah, W. A., H. Kadry, S. G. Mahran, E. H. El-Raziky and S. Nakib (1978) *Egyptian Journal of Bilharzia*, **4**, 19 (in Seida *et al.*, 1981).

Acharya, T. K. and I. B. Chatterjee (1974) Isolation of chrysophanic acid-9-anthrone, a fungicidal compound from *Cassia tora*. *Science Culture*, **40**, 316.

Acharya, T. K. and I. B. Chatterjee (1975) Isolation of chrysophanic acid-9-anthrone, the major antifungal principle of *Cassia tora*. *Lloydia*, **38**, 218.

Adawadkar, P. D. and M. A. El-Sohly (1981) Isolation, purification and antimicrobial activity of anacardic acids from *Ginkgo biloba* fruits. *Fitoterapia*, **52**, 129–35.

Addea Mensah, I., F. G. Torto, C. D. Dimoneyeka, I. Baxter and J. K. M. Sanders (1977) Novel amide-alkaloids from the roots of *Piper guineense*. *Phytochemistry*, **16**, 757–9.

Adesina, S. K., C. O. Adewunmi and V. O. Marquis (1980a) Phytochemical investigations of the molluscicidal properties of *Tetrapleura tetraptera* Taub. *Journal of African Medicinal Plants*, no. **3** (International Research Congress on Natural Products and Medicinal Agents, Strasbourg 1980).

Adesina, S. K., B. J. Oguntimein and D. D. Akinwusi (1980b) Phytochemical and biological examination of the leaves of *Acalypha wilkesiana* (red acalypha). *Quarterly Journal of Crude Drug Research*, **18**, 45–8.

Adesogan, E. K. (1973) Anthraquinones and anthraquinols from *Morinda lucida*. The biogenetic significance of oruwal and oruwalol. *Tetrahedron Letters*, **29**, 4099–102.

Adesogan, E. K. and A. L. Okunade (1979) A new flavone from *Ageratum conyzoides*. *Phytochemistry*, **18**, 1863–4.

Adewunmi, C. O. and V. O. Marquis (1980) Molluscicidal evaluation of *Jatropha* species grown in Nigeria. *Quarterly Journal of Crude Drug Research*, **18**, 141–5.

Adewunmi, C. O. and E. A. Sofowora (1980) Preliminary screening of some plant extracts for molluscicidal activity. *Planta Medica*, **39**, 57–65.

Adgina, V. V. (1972) Antifungal activity of sanguiritin (sanguinarine). *Chemical Abstracts*, **82**, 133506j.

Agrawal, G. D., S. A. I. Rizoi, P. C. Gupta and J. D. Tewari (1972) Structure of fistulic acid, a new colouring matter from the pods of *Cassia fistula*. *Planta Medica*, **21**, 150–5.

Ahmed, Z. T., C. J. Zufall and G. L. Jenkins (1949) Chemistry and toxicology of the root of *Phytolacca americana*. *Journal of the American Pharmaceutical Association*, **38**, 443–8.

Ahmed, Z. F., A. M. Rizk, F. M. Hammouda and M. M. Seif el Nasr (1972) Naturally occurring glucosinolates with special reference to those of the family of Capparidaceae. *Planta Medica*, **21**, 35–60.

Aizikov, M. I. and A. G. Kurmukov (1973) Toxicology of gossypol. *Lokl. Akad. Nauk Uzb. SSR*, **30**, 28 (through *Chemical Abstracts*, **83**, 1923d).

Alain, M., E. Massal, R. Touzin and L. Porte (1949) Le traitement de l'amibiase par la conessine. Résultats d'une expérimentation faite à l'hôpital Michel Lévy, Hôpital d'instruction de l'Ecole d'Application du Service de Santé des troupes coloniales. *Médicine Tropicale*, **9**, 5–38.

Albert *et al.* (1972) *Ageratum conyzoides*, activités anthelmintiques *in vitro*. *Annales de l'Université et de l'ARERS, Reims*, **10**, 101–3, cited in Kerharo and Adam, 1974, p. 800).

Al-Shamma, A. and L. A. Mitscher (1979) Comprehensive survey of indigenous Iraqi plants for potential economic value. Screening results of 327 species for alkaloids and antimicrobial agents. *Lloydia*, **42**, 633–42.

Al-Shamma, A., S. Drake, D. L. Flynn, L. A. Mitscher, Y. H. Park, G. S. R. Roa, A. Simpson, J. K. Swayze, T. Veysoglu and S. T. S. Wu (1981) Antimicrobial agents from higher plants. Antimicrobial action of *Peganum harmala*. *Journal of Natural Products (Lloydia)*, **44**, 745–7.

Amin, M. A., A. A. Daffala and O. A. El Monein (1972) Preliminary report of the molluscicidal properties of Habbat El-Mollock (*Jatropha* spp.). *Transactions of the Royal Society of Tropical Medicine and Hygiene*, **66**, 805–6.

Amoros, M., B. Fauconnier and L. Girre (1977) Propriétés antivirales de quelques extraits de plantes médicinales. *Annales pharmaceutiques françaises*, **35**, 371–6.

Amoros, M., B. Fauconnier and L. Girre (1979) Proprietes antivirales du mouron rouge *Anagallis arvensis*. *Plantes Médicinales et Phytothérapie*, **13**, 122.
Anchel, M. (1949) Identification of the antibiotic substance from *Cassia reticulata* as 4,5-dihydroxyanthraquinone-2-carboxylic acid. *Journal of Biological Chemistry*, **177**, 169–77.
Anisimov, M. H., V. V. Shchlegov, S. N. Dzizenko, L. I. Strigina and N. I. Uvarova (1974) Effect of some sterols on the antimicrobial activity of triterpenoid glycosides of plant and animal origin. *Chemical Abstracts*, **82**, 39064s.
Anton, R. and P. L. Duquenois (1968) L'emploi des *Cassias* dans les pays tropicaux et subtropicaux examinés d'après quelques uns des constituants chimiques de ces plantes médicinales. *Plantes Médicinales et Phytothérapie*, **2**, 255–68.
Arndt, U. (1968) Determination of the biological activity of small quantities of wood extractives against subterranean termite, *Reticuli termis*. *Holzforschung*, **22**, 104–9.
Arora, R. B., N. Ghatak and S. P. Gupta (1971) Antifertility activity of *Embelia ribes*. *Journal of Research in Indian Medicine*, **6**, 107.
Arthur, H. R. and W. H. Hui (1954) Triterpene acids from the leaves of *Psidium guaijava* L. *Journal of the Chemical Society*, 1403–6.
Aschenbach, H., B. Raffelsberger and G. U. Brillinger (1980) 19-Hydroxycoronaridine und 19-hydroxy-ibogamine, zwei antibiotisch wirksame Alkaloide vom ibogamin Typ. *Phytochemistry*, **19**, 2185–8.
Attia, I., S. Ahmad, S. H. H. Zaidi and Z. Ahmad (1972) Ethylgallate and gallic acid, the major antimicrobial principles in *Acacia* species. *Pakistan Journal of Scientific and Industrial Research*, **15**, 199.
Ayim, J. S. R., D. Dwuma Badu, N. Y. Fiagbe, A. M. Ateya, D. J. Slatkin, J. E. Knapp and P. L. Schiff, Jr (1977) Constituents of West African plants XXI. Tiliafunimine, a new imino bisbenzylisoquinoline alkaloid from *Tiliacora funifera*. *Lloydia* **40**, 561–5.
Baas, W. (1900) Triterpenes in latex of *Euphorbia pulcherrima* Willd. latex. *Planta Medica*, 31–5.
Bailenger, J. and F. Sequin (1966) Étude expérimentale de l'activité ténifuge des semences de courge. *Bulletin de la Société Pharmaceutique Bordeaux*, **105**, 189–200.
Balansard, J. (1936) Sur quelques Labiées. *Bulletin des Scéances Pharmacologiques*, **43**, 148–52.
Balansard, G., D. Zamble, G. Dumenil and A. Cremieux (1980) Mise en evidence des proprietes antimicrobiennes du latex obtenu par incision du tronc d'*Alafia multiflora* Stapf. Identification de l'acide vanillique. *Plantes Médicinales et Phytotherapie*, **14**, 99–104.
Balbar, O. D., B. L. Chowdhury, M. P. Singh, S. Khan and S. Baypai (1970) Nature of antiviral activity detected in some plant extracts screened in cell cultures infected with Vaccinia and Ranikhet disease (Newcastle) virusus. *Indian Journal of Experimental Biology*, **8**, 304–12.
Balmain, A., K. Bjåmer, J. D. Conolly and G. Ferguson (1967) The constitution and stereochemistry of caesalpin ϵ. *Tetrahedron Letters*, **49**, 5027–31.
Bamgbose, S. C. A. and B. K. Noamesi (1981) Studies on cryptolepine II. Inhibition of carageenan induced oedema by cryptolepine. *Planta Medica*, **41**, 392–6.
Barnes, D. K. and R. H. Freyre (1966a) Recovery of natural insecticides from *Tephrosia vogelii* I. Efficiency and rotenoid extraction from fresh and oven-dried leaves. *Economic Botany*, **20**, 278–84.
Barnes, D. K. and R. H. Freyre (1966b) Recovery of natural insecticides from *Tephrosia vogelii* II. Toxicological properties of rotenoids extracted from fresh and oven-dried leaves. *Economic Botany*, **20**, 368–71.
Barnes, D. K. and R. H. Freyre (1967) Recovery of natural insecticides from *Tephrosia vogelii* III. An improved procedure for sampling and assaying rotenoid content in leaves. *Economic Botany*, **21**, 93–8.
Basu, A. P. (1971) Antibacterial activity of *Curcuma longa*. *Indian Journal of Pharmacy*, **33**, 131.

Beauquesne, L. (1947) Le Samagoura (*Swartzia madagascariensis*), Légumineuse africaine. *Annales pharmaceutiques françaises*, **5**, 470–83.

Bendz, G. (1956) Gallic acid in *Lawsonia inermis* leaves. *Physiologia Plantarum (Copenhagen)*, **9**, 243.

Benjamin, T. V. (1979) Investigation of *Borreria verticillata*, an antieczematic plant in Nigeria. *Quarterly Journal of Crude Drug Research*, **17**, 135–6.

Benjamin, T. V. and A. Lamiranka (1981) Investigation of *Cassia alata*, a plant used in Nigeria in the treatment of skin diseases. *Quarterly Journal of Crude Drug Research*, **19**, 93–6.

Berghe, D. A., van den and A. Boeije (1973) In situ fragmentation of RNA in poliovirus. *Archiv der Gesellschaft für Virusforschung*, **40**, 215.

Berghe, D. A., van den, M. Ieven, F. Mertens, A. J. Vlietinck and E. Lammens (1978) Screening of higher plants for biological activities II. Antiviral activity. *Lloydia*, **41**, 463–71.

Bertho, A. (1944) Pharmakologische Prüfung der Extrakte und Alkaloiden aus *Holarrhena antidysenterica*. *Archiven der Experimentellen Pathologie und Pharmakologie*, **203**, 41–6.

Bevan, C. W. L., D. E. U. Ekong and D. A. H. Taylor (1965) Extractives from West African members of the family of Meliaceae. *Nature (London)*, **206**, 1323–5.

Bevan, C. W. L., J. W. Powel and D. A. H. Taylor (1963) West African timbers, Part IV. Petroleum extracts of the genera *Khaya, Guara, Carapa* and *Cedrela*. *Journal of the Chemical Society*, 980–2.

Beveridge, W. I. B. (1977) *Influenza: the Last Great Plague*. Heinemann, London, 1 vol., 124 pp.

Beveridge, W. I. B. (1981) *Animal Health in Australia. Vol. 1. Viral Diseases of Farm Livestock*. Australian Government Publishing Service, 197 pp.

Bézanger-Beauquesne, L. and J. Vanlerenberghe (1955) Contribution à l'étude du *Droséra*. *Annales de Pharmacie françaises*, **13**, 204–7.

Bézanger-Beauquesne, L., M. Pinkas, M. Torck and F. Trottin (1980) *Les Plantes Médicinales des Regions Temperées*, Maloine, Paris, 439 pp.

*Bézanger-Beauquesne, L., M. Torck and F. Trottin (1981) Conquêtes possibles en Phytothérapie. *Plantes Médicinales et Phytothérapie*, **15**, 25–74.

Bhaduri, A. P., R. P. Rastogi and N. M. Khanna (1968) Biologically active carissone derivatives. *Indian Journal of Chemistry*, **6**, 405.

Bhakuni, D. S., M. L. Dhar, M. M. Dhar, B. N. Dhawan and B. N. Mehrotra (1969) Screening of Indian plants for biological activities Part II. *Indian Journal of Experimental Biology*, **7**, 250–62.

Bhakuni, D. S., S. Tewari and M. M. Dhar (1972) Aporphine alkaloids of *Annona squamosa*. *Phytochemistry*, **11**, 1819–22.

Bhatia, B. B. and S. Lal (1933) The pharmacological action of *Plumbago zeylanica* and its active principle (plumbagin). *Indian Journal of Medical Research*, **20**, 777–88.

Bhattacharya, S. K. and B. Lythgoe (1949) Derivatives of *Centelle asiatica* used against leprosy. Triterpenic acids. *Nature (London)*, **163**, 258–9.

Bhide, M. B., S. T. Nikam and S. R. Chavan (1976) Effects of seeds of *Caesalpinia bonducella* on some aspects of the reproductive system. 16th Annual Conference on Pharmacology, Physiology and Therapy of Medicinal Plants, India. Indian Council of Medical Research (1976, p. 28).

Bizyulyavichyus, S. (1969) Deworming of the patients against tapeworms. *Acta Parasitica Lithuanica*, **9**, 11–18.

Blaise, H. (1932) Les Crossopteryx africains. Thèse de Doctorat en Pharmacie, Université de Paris.

Blanchon, E., L. de Saint Rat and P. Bonnet-Maury (1948) Quelques essais thérapeutiques par l'action du plumbagol. *Bulletin Académie de Médicine*, **112**, 125–8.

Boakaiji Yiadom, K. (1977) Antimicrobial properties of some West African medicinal plants. *Quarterly Journal of Crude Drug Research*, **15**, 201–3.

Boakaiji Yiadom, K. (1979) Antimicrobial properties of West African plants II. Antimicrobial activity of aqueous extracts of *Cryptolepis sanguinolenta* Schltr. *Quarterly Journal of Crude Drug Research*, **17**, 78–80.

Boakaiji Yiadom, K. and S. M. Heman Ackah (1979) Cryptolepine hydrochloride effect on *Staphylococcus aureus*. *Journal of Pharmaceutical Sciences*, **68**, 1510–14.

Boakaiji Yiadom, K. and G. H. Konnig (1975) Incidence of antibacterial activity in the Connaraceae. *Planta Medica*, **28**, 397–400.

Boakaiji Yiadom, K., N. I. Fiagbe and J. S. K. Ayim (1977) Antimicrobial activity of xylopic acid and other constituents of the fruits of *Xylopia aethiopica* (Annonaceae). *Lloydia*, **40**, 543–5.

Boiteau, P., M. Dureil and A. Rakto Ratsimamanga (1949) Contribution à l'étude des propriétés antituberculeuses de l'oxyasiaticoside (dérivé hydrosoluble de l'asiaticoside extrait de *Centella asiatica*). *Comptes Rendus de l'Académie des Sciences*, **228**, 1165–7.

Boukef, K., G. Balansard, M. Lallemand and P. Bernard (1976) Étude des acides phénols isolés des feuilles d'*Eucalyptus globulus* Labill. *Plantes Médicinales et Phytothérapie*, **10**, 24–9.

Bouquet, A. (1972) Plantes médicinales du Congo Brazzaville Doctorat. Publ. No. 13 ORSTOM, 112 pp. and in Travaux Labo. *Matière Medicale*, 1972.

*Bouquet, A. and M. Debray (1974) *Plantes Médicinales de la Côte d'Ivoire*. Document ORSTOM no. 32, 232 pp.

*Bouquet, A., A. Cavé and R. Paris (1971) Medicinal plants of Congo Brazzaville III. *Plantes Médicinales et Phytothérapie*, 5, 154–8.

Bowden, K. (1962) Isolation from *Paullinia pinnata* L. of material with action on the frog's isolated heart. *British Journal of Pharmacology and Chemotherapy*, **18**, 173–4.

Brannon, D. R. and R. W. Fuller (1973) Microbiological production of pharmacologically active compounds other than antibiotica. *Lloydia*, **37** (Proc.), 134–46.

Broadbent, J. L. (1962) Cardiotonic action of two tannins. *British Journal of Pharmacology and Chemotherapy*, **18**, 167–72.

Burdick, E. M. (1971) Carpaine, an alkaloid of *Carica papaya*. Its chemistry and pharmacology. *Economic Botany*, **25**, 363–5.

Burger, A. (1960) *Medicinal Chemistry*, 2nd edn, Interscience, New York, 1243 pp.

Caiment-Leblond, J. (1957) Contribution à l'étude des plantes médicinales de l'A.O.F. Thèse de Doctorat en Pharmacie, Université de Paris.

Camp, B. J. and M. J. Norrel (1966) The phenyl-ethylamine alkaloids of native range plants. *Economic Botany*, **20**, 274–8.

Canonica, L., G. Jommi, P. Manito and U. M. Pagnoni (1966) Struttura delle Caesalpine I, II, III. *Gazetta Chimica Italiana*, **96**, 662–720.

Canonica, L., G. Jommi, P. Manito and F. Pellizoni (1963; 1964) Bitter principles of *Caesalpinia bonducella*. *Tetrahedron Letters*, **29**, 2079–86; **30**, 692.

Carrara, G. and L. Lorenzini (1946) The antibacterial activity of sulfonamide derivatives of penicillin, thiouracil and naphthoquinone derivatives. *Chemica e Industria Milan*, **28**, 15 (in Caiment-Leblond, 1957, p. 47).

Castagne, E. (1938) Contribution à l'étude des Légumineuses insecticides du Congo Belge. *Mémoires de l'Institut Royal Belge*, **6**, fasc. 3.

Chak, I. M. and G. K. Patnaik (1972) Local anaesthetic activity of *Plumeria rubra*. *Indian Journal of Pharmacy*, **34**, 10–11.

Chatterjee, Mitra and Siddiqui (1948) *Indian Journal of Scientific and Industrial Research*, **69B** (cited in Hegnauer, 1968, Vol. 5, 6).

Chattopadhyaya, M. K. and R. L. Khare (1970) Antimicrobial activity of anacardic acid and its metallic complexes. *Indian Journal of Pharmacy*, **32**, 46–8.

Chaudhury, R. K. and S. Ghosal (1971) Xanthones of *Canscora decussata* (Roxb.) Roem. & Schult. *Phytochemistry*, **10**, 2425.

Chauhan, J. S., M. Sultan and S. K. Srivastasa (1977) Two new glycoflavones from the roots of *Phyllanthus niruri*. *Planta Medica*, **32**, 217–22.

Chaumont, J. P. and P. Bourgeois (1978) Screening de plantes vis à vis de champignons pathogènes. *Lloydia*, **41**, 234–6.

Chaumont, J. P. and J. M. Senet (1978) Propriétés antagonistes des plantes supérieures vis-à-vis de champignons parasites de l'homme ou contaminant des aliments. Screening portant sur 200 Phanérogames. *Plantes Médicinales et Phytothérapie*, **12**, 186–96.

Chaumont, J. P., H. De Scheemaeker and J. Rousseau (1978) Alkaloides responsables de l'action fongistatique des Amaryllidacées. *Plantes Médicinales et Phytothérapie*, **12**, 157–61.

Chaurasia, S. C. and A. Kher (1978) Activity of essential oils of three medicinal plants against various pathogenic and non pathogenic fungi. *Indian Journal of Hospital Pharmacy*, **15**, 139–41.

Chauvin, R. (1946) Sur la substance qui dans les feuilles de *Melia azadarach* repousse les criquets. *Comptes Rendus de l'Académie des Sciences*, **222**, 412–14.

Cheema, M. A. and D. O. Priddle (1965) Pharmacological investigation of isochaksine, an alkaloid isolated from the seeds of *Cassia absus*. *Archives internationaux de Pharmacodynamie et de Thérapeutique, Belgique*, **158**, 307–13.

Chen, C. R., J. L. Beal, R. W. Doskotch, L. E. Mitscher and G. H. Swoboda (1974) A phytochemical study of *Doryphora sassafras*. II. Isolation of 11 crystalline alkaloids from the bark. *Lloydia*, **37** (Proc.), 493–500.

Chevalier, A. (1928) Notes sur l'exploration botanique de l'Afrique occidentale française (gorli). *Revue de Botanique Appliquée*, **1928**, 643–58.

Chopra, R. N., R. L. Badhwar and S. L. Nayar (1941) Insecticidal and piscicidal plants of India. *Journal of the Bombay Natural History Society*, **42**, 854.

Chopra, R. N., I. C. Chopra, K. L. Handa and C. D. Kapur (1938) *Chopra's Indigenous Drugs of India*, Dhur & Sons, Calcutta, 342 pp.

Chopra, R. N., B. B. Dikshit and J. S. Showhan (1932) The pharmacological action of berberine. *Indian Journal of Medical Research*, **19**, 1193–203.

Chopra, R. N., S. L. Nayar and I. C. Chopra (1956) *Glossary of Indian Medicinal Plants*, Council of Scientific and Industrial Research, New Delhi, 330 pp.

Cochran, K. W. and E. H. Lucas (1958–59) Chemoprophylaxis of poliomyelitis in mice through the administration of plant extracts. *Antibiotic Annals*, 104–8.

Cochran, K. W., T. Nishikawa and E. S. Bencke (1966) Botanical sources of influenza inhibitors. *Antimicrobiology and Chemotherapy*, 515.

Cohen, R. A., L. S. Kučera and E. C. Herrman, Jr (1964) Antiviral activity of *Melissa officinalis* (lemon balm) extract. *Proceedings of the Society of Experimental Biology and Medicine*, **117**, 431–4.

*Collier, W. A. and L. van der Pijl (1950) Investigation on the antibiotic activity of the leaves of plants in Java. *Chronica Naturae*, **106**, 73–80.

Crosnier, R., F. Merle, G. Bernier, Molinier and Tabusse (1948) Traitement de la dysenterie amibienne par les extraits d'*Holarrhena floribunda*. *Bulletin de l'Academie de Medicine*, **132**, 336–8.

Crowder, J. L. and R. P. Sexton (1964) Kerato conjunctivitis resulting from the sap of candelabra cactus and the pencil tree. *Archives of Ophthalmology USA*, **72**, 476–84.

Cutting, W., E. Furusawa, S. Furusawa and Y. K. Woo (1965) Antiviral activity of herbs on Columbia S.K. in mice and L SM vaccinia and Adeno type viruses *in vitro*. *Proceedings of the Society of Experimental Biology and Medicine*, **120**, 330–3.

Dale, T. and W. E. Court (1981) Amino acids of *Cassia* seeds. *Quarterly Journal of Crude Drug Research*, **19**, 25–9.

Dalziel, J. M. (1937) *The Useful Plants of West Tropical Africa*, Crown Agents, London, 612 pp.

Dar, R. N., J. L. Garg and R. D. Pathak (1965) Anthelmintic activity of *Carica papaya* seeds. *Indian Journal of Pharmacy*, **27**, 335–6.

Das, B. R., P. A. Kurup and P. L. Narasimha-Rao (1957a) Antibiotic principle from *Moringa pterygosperma* VII. Antibacterial activity and chemical structure of compounds related to pterygospermine. *Indian Journal of Medical Research*, **45**, 191–6.

Das, B. R., P. A. Kurup, P. L. Narasimha-Rao and A. S. Ramaswany (1957b) VIII. Some pharmacological properties and *in vivo* action of pterygospermine and related compounds. *Indian Journal of Medical Research*, **45**, 197–206.

Das, B. R., P. A. Kurup and P. L. Narasimha-Rao (1958) Antibiotic principle from *Moringa pterygosperma* IX. Inhibition of transaminase by isothiocyanates. *Indian Journal of Medical Research*, **46**, 75–7.

Das Gupta, B. M. and R. Dikshit (1929) *Indian Medical Gazette*, **64**, 67 (cited in Henry, 1949, p. 346).

*Dawson, J. L. (1968) Prophylactic drug control of parasitic infections. *Journal of Tropical Hygiene*, **66**, 87–98.

Debaille, G. and P. Petard (1953) Notes preliminaires sur les plantes antidysentériques du Soudan et de la Haute Volta. *Bulletin Médical de l'A.O.F.*, **10**, 11–14.

Degener, O. (1975) *Plants of Hawaii National Parks*, Braun-Brunfield Inc., pp. 299–300 (Wat *et al.*, 1980).

Delaveau, P., A. Desvignes, E. Adoux and A. M. Tessier (1979) Baguettes frotte-dents d'Afrique occidentale. Examen chimique et microbiologique. *Annales pharmaceutiques françaises*, **37**, 185–90.

Della Monache, F., L. E. Cuca Suarez and G. B. Marini Bettolo (1978) Flavonoids from the seeds of six *Lonchocarpus* species. *Phytochemistry*, **17**, 1812–13.

Denoël, A. (1958) *Matière Médicale Végétale*, 2 vols., Presses Universitaires, Liège.

Dhar, D. N. and R. C. Munjal (1976) Chemical examination of the seeds of *Bombax malabaricum*. *Planta Medica*, **29**, 148–50.

Dhar, M. L., M. M. Dhar, B. N. Dhawan, B. N. Mehotra and C. Ray (1968) Screening Indian plants for biological activity. Part I. *Indian Journal of Experimental Biology*, **6**, 232–47.

Dorsett, P. H., E. E. Kerstine and L. J. Powers (1975) Antiviral activity of gossypol and apogossypol. *Journal of Pharmaceutical Sciences*, **64**, 1073–5.

Dossaji, S. and I. Kubo (1980) Quercetin 3-(2″)-galloyl-glycoside, a molluscicidal flavonoid from *Polygonum senegalense*. *Phytochemistry*, **19**, 482.

Dossaji, S. F., M. G. Kairu, A. T. Gondwe and J. H. Ouma (1977) On the evaluation of the molluscicidal properties of *Polygonum senegalense* forma *senegalense*. *Lloydia*, **40**, 290–3.

Durodola, J. I. (1977) Antibacterial property of crude extract from a herbal wound healing remedy–*Ageratum conyzoides* L. *Planta Medica*, **32**, 388–90.

Dutta, S. K., B. N. Sharma and P. V. Sharma (1978) Buchanine, a novel pyridine alkaloid from *Cryptolepis buchanini*. *Phytochemistry*, **17**, 2047–8.

Dwuma Badu, D., J. S. K. Ayim, T. T. Dabra, H. N. El Sohly, J. E. Knapp, D. J. Slatkin and P. L. Schiff (1976) Constituents of West African medicinal plants. XLV. Constituents of *Piper guineense* Schum. & Thonn. (Piperaceae). *Lloydia*, **39**, 60–4.

Dwuma Badu, D., J. S. K. Ayim, N. I. Y. Fiagbe, J. E. Knapp, P. L. Schiff and D. J. Slatkin (1978) Constituents of West African medicinal plants XX. Quindoline from *Cryptolepis sanguinolenta*. *Journal of Pharmaceutical Sciences*, **67**, 433–4.

Eichbaum, D., Koch Weser and A. T. Leano (1950) Activity of cashew (*Anacardium occidentale*) nutshell oil in human ankylostomiasis. *American Journal of Digestive Diseases*, **17**, 370.

Eilert, U., B. Wolters and A. Nahrstedt (1980) Antibiotic principles of seeds of *Moringa oleifera* and *M. stenopetala* Lam. *Planta Medica*, **39**, 235.

Eilert, U., B. Wolters and A. Nahrstedt (1981) Antibiotic principle of seeds of *Moringa oleifera* and *M. Stenopetala* Lam. *Planta Medica*, **42**, 55–6.

Ekong, D. E. U. and A. U. Ogan (1968) Chemistry of the constituents of *Xylopia aethiopica*. The structure of xylopic acid, a new diterpene acid. *Journal of the Chemical Society, C*, 311–12.

Ekong, D. E. U. and E. O. Olagbeni (1967) Limonoids from the timber of *Pseudocedrela kotschyi*. *Tetrahedron Letters*, 3325–7.

Ekong, D. E. U., E. O. Olagbeni and F. A. Odutola (1969a) Further diterpenes from *Xylopia aethiopica* (Annonaceae). *Phytochemistry*, **8**, 1053.

Ekong, D. E. U., E. O. Olagbeni and A. E. Spiff (1968) Cycloeucalenol and 2-4-methylene cycloartanol in wood oil of the family of Meliaceae. *Chemistry and Industry* (London), 1808.

Ekong, D. E. U., C. O. Fakunle, A. K. Fasina and J. I. Okogun (1969b) The Meliacins (Liminoids) Nimbolin A and B, two meliacin cinnamates from *Azadirachta indica* L. and *Melia azedarach* L. *Journal of the Chemical Society Chemical Communications*, 20 D, 1166–7.

El Kheir, Y. M. and A. M. Salih (1979) Investigation of the nature of the molluscicidal factor of *Croton macrostachys*. 2nd OAU/STRC *Inter-African Symposium on Traditional Pharmacopoeia and African Medicinal Plants*, Publication no. 115, OAU/STRC Lagos.

El Said, F., E. A. Sofowora and T. Olaniyi (1968) Study of certain Nigerian plants used in fever. Communication at the Inter-African Symposium on Traditional Pharmacopoeia and Medicinal Plants, Dakar.

El Said, F., S. O. Fadulu, J. O. Kuye and E. A. Sofowora (1971) Native cures in Nigeria II. The antimicrobial properties of the buffered extracts of the chewing sticks. *Lloydia*, **34**, 172–4.

El-Sohly, H. N., W. L. Lasswell, Jr and C. D. Hufford (1979) Two new C-benzylated flavanones from *Uvaria chamae* and ^{13}C NMR analyses of flavonone methyl ethers. *Journal of Natural Products (Lloydia)*, **42**, 264–70.

El Tayeb, O., M. Kučera, V. O. Marquis and H. Kučerova (1974) Contribution to the knowledge of Nigerian medicinal plants. III Study on *Carica papaya* seeds as a source of reliable antibiotic, the benzylisothiocyanate. *Planta Medica*, **26**, 79–89.

Emeruwa, A. C. (1982) Antibacterial substance from *Carica papaya* fruit extract. *Journal of Natural Products*, **45**, 123–7.

Etkin, N. L. (1981) A Hausa herbal pharmacopoeia: Biomedical evaluation of commonly used plant medicines. *Journal of Ethnopharmacology*, **4**, 75–98.

Euw, J. V. and Reichstein (1968) Aristocholic acid in the swallowtail butterfly. *Israel Journal of Chemistry*, **6**, 659–70.

Ezmirly, S. T., J. C. Cheng and S. R. Wilson (1979) Saudi Arabian medicinal plants: *Salvadora persica*. *Planta Medica*, **35**, 191–2.

Fadulu, S. A. (1975) The antibacterial properties of the buffer extracts of chewing sticks used in Nigeria. *Planta Medica*, **27**, 122–6.

Farley, D. L. (1944) Canavalia, a new enzymatic bactericidal agent. *Surgery, Gynecology, Obstetrics*, **79**, 83–8.

Farnsworth, N. R. and G. A. Cordell (1976) A review of some biologically active compounds isolated from plants as reported in the 1974–1975 literature. *Lloydia*, **39**, 420–55.

Farnsworth, N. R., G. H. Svoboda and N. R. Blomster (1968) Antiviral action of perivine, an α-acylindolic alkaloid of *Catharanthus roseus*. *Journal of Pharmaceutical Sciences*, **57**, 2174–5.

Farooqi, M. I. H. and J. G. Srivastava (1968) *Salvadora persica*. *Quarterly Journal of Crude Drug Research*, **8**, 1297–9.

*Feng, P. C., L. J. Haynes, K. E. Magnus and J. R. Plimmer (1964) Further pharmacological screening of some West Indian medicinal plants. *Journal of Pharmacy and Pharmacology*, **16**, 115–17.

*Feng, P. C., L. J. Haynes, K. E. Magnus and J. R. Plimmer and H. S. A. Sherrat (1962) Pharmacological screening of some West Indian medicinal plants. *Journal of Pharmacy and Pharmacology*, **14**, 556–61.

Ferreira, M. A., L. Nogueira Prista, and A. Correia Alves (1963a) Ensaios sobre as raizes de *Alchornea cordifolia* (Schum.) Müll Arg. *Garcia de Orta, Lisboa*, **11**, 265–74.

Ferreira, M. A., A. Correia Alves and L. Nogueira Prista (1963b) Estudo quimico da *Newbouldia laevis* Seem. I. Isolamento de bases indolicas. *Garcia de Orta*, **2**, 477–86.

Finch, N. and W. D. Ollis (1960) Mundoserone. *Proceedings of the Chemical Society*, 176.

*Fong, H. H. S., N. R. Farnsworth, L. K. Henry, G. H. Svoboda and M. J. Yates (1972)

Biological and phytochemical evaluation of plants X. Test results from a third two hundred accessions. *Lloydia*, **35**, 35–48.

Fournier, G., M. R. Paris and M. C. Fourniat (1978) Bacteriostatic activity of *Cannabis sativa* essential oil. *Annales Pharmaceutiques françaises*, **36**, 603–6.

Fraenkel, G. S. (1959) Raison d'être of secondary plant products. *Science*, **129**, 1466.

Fraga de Azevedo, J. and L. de Medeiros (1963) L'action mollusicide d'une plante de l'Angola, la *Securidaca longepedunculata* Fres. *Bulletin de la Société Pathologique Exotique*, **56**, 68–76.

Freise, F. W. (1935) Pharmacologically utilizable constituents of *Bixa orellana*. *Pharmazeutische Zentralhalle*, **76**, 4–5.

Gaind, K. N. and R. L. Gupta (1969) Investigations on the leaves of *Bryophyllum pinnatum* Pers. *Indian Journal of Pharmacy*, **31**, 167.

Gaind, K. N. and T. R. Juneja (1969) Investigations on *Capparis decidua* Edgew. *Planta Medica*, **17**, 95–8.

Gaind, K. N. and A. K. Singla (1966) Antimicrobial activity of shell fibres of *Cocos nucifera* L. *Indian Drugs*, **4**, 178.

Gaind, K. N., R. D. Budhiraja and R. N. Kaul (1966) Antibiotic activity of *Cassia occidentalis* L. *Indian Journal of Pharmacy*, **28**, 248–50.

Gaind, K. N., T. R. Juneja and P. V. Bhandarkar (1972) Volatile principles of *Capparis decidua* (kinetics of in vitro activity against *Vibrio cholerae ogava*, *inaba* and *el tor*). *Indian Journal of Pharmacy*, **34**, 86–8.

Gaind, K. N., T. R. Juneja and P. C. Jain (1969a) Anthelmintic and purgative activity of *Capparis decidua* Edgew. *Indian Journal of Hospital Pharmacy*, **2**, 153.

Gaind, K. N., T. R. Juneja and P. C. Jain (1969b) Investigations on *Capparis decidua* Edgew. II. Antibacterial and antifungal studies. *Indian Journal of Pharmacy*, **31**, 24–5.

Gal, I. E. (1964) Capsicidin, eine neue Verbindung mit antibiotischer Wirksamkeit aus Gewürz paprika. *Zeitschrift für Lebensmittel Untersuchung und Forschung*, **124**, 533–6.

Gal, I. E. (1967) Nachweis und Bestimmung des Steroidsaponins capsicidin mit der Agardiffusionsmethode. *Die Pharmazie*, **22**, 120–3.

Gangadharan, P. R. J. and Sirsi, M. (1955) Studies on the antitubercular properties of *Curcurbita pepo*. *Indian Journal of Pharmacy*, **17**, 133.

Gaudin, O. and Vacherat, R. (1938) Recherches sur la rotenone et le pouvoir ichthyotoxique de quelques plantes du Soudan francais. *Bulletin des Sciences Pharmacologiques*, **40**, 385–94.

Gellerman, J. L. and H. Schlenk (1968) Anacardic acids in leaves and nuts of *Anacardium occidentale*. *Journal of Analytical Chemistry*, **40**, 741.

Gellerman, J. L., N. J. Walsh, N. K. Werner and H. Schlenk (1969) Antimicrobial effects of anacardic acids. *Canadian Journal of Microbiology*, **15**, 1219–225.

Gellert, E., R. Raymond-Hamet and E. Schlitter (1951) Die Konstitution des Alkaloids Cryptolepin. *Helvetica Chimica Acta*, **34**, 642–51.

*Georges, M. and K. M. Pandelai (1949) Investigations on plant antibiotics IV. Further search for antibiotic substances in Indian medicinal plants. *Indian Journal of Medical Research*, **37**, 169–81.

Ghosal, S. and P. K. Banerjee (1968) Indole alkaloids of *Desmodium gangeticum* DC. *Indian Journal of Pharmacy*, **30**, 280.

Ghosal, S. and R. K. Chaudhury (1975) Chemical constituents of Gentianaceae XVI. Antitubercular activity of xanthones of *Canscora decussata* Schult. *Journal of Pharmaceutical Sciences*, **64**, 888.

Ghosal, S., R. K. Chaudhury and A. N. Chaudhury (1971) Chemical constituents of *Canscora decussata* II. *Journal of the Indian Chemical Society*, **48**, 589.

Ginde, B. S., B. D. Hosangadi, N. A. Kudav, K. V. Nayak and A. B. Kulkarni (1970) Chemical investigations on *Cassia occidentalis* L. Isolation and structure of cassiolin, a new xanthone. *Journal of the Chemical Society*, **9**, 1285–9.

Githens, T. S. (1949) *Drugplants of Africa*. University of Pennsylvania Press, 125 pp.

Glotter, E., I. Kirson, D. Lavie and A. Abraham (1978) The withanolides – a group of natural steroids. *Bioorganic Chemistry*, **2**, 57–95.

Godet, R. (1950) La vaginite à *Trichomonas vaginalis*. Son traitement par la conessine. Thèse Doctorat en Médecine, Paris.

Gonzalez, A. E., O. R. Bravos, M. H. Garcia, M. R. de la Santo and M. Del Tomas (1974) Pharmacological (anthelmintic) study of *Curcurbita maxima* seeds and their active principle curcurbitin. *Anais Real Acad., Farm*, **40**, 475 (through *Chemical Abstracts*, **82**, 149446a).

Gopalchari, R. and M. L. Dhar (1958) Constitution of saponin from seeds of *Achyranthes aspera*. I Identification of the sapogenin. *Journal of Scientific and Industrial Research, India*, **17B**, 276–8.

Goutam, P. and R. M. Purohit (1973) Antibacterial activity of leaves and seeds of *Diospyros montana*. *Indian Journal of Pharmacy*, **35**, 93–4.

Govindachari, T. R., B. R. Pai and K. Nagarajan (1956) Wedelolactone from *Eclipta alba*. *Journal of Scientific and Industrial Research*, **15B**, 664–5.

Greathouse, G. A. (1939) Alkaloids from *Sanguinaria canadensis* and their influence on the growth of *Phymatotrichum omnivorum*. *Plant Physiology*, **14**, 377.

Grégoire, J. (1953) Contribution a l'etude du kinkeliba (Combretum micranthum Don.) Thèse de Doctorat en Pharmacie, Université de Marseille.

Guerrero, L. E., A. B. M. Sison, A. Makalintal, P. Villaseñor, I. Rosal and A. Ocampo (1924) *Quisqualis indica* as an anthelmintic. *Journal of the Philippine's Medical Association*, **4**, 83–7.

Gulati, A. S., U. S. Krishnamachar and B. C. Subba Rao (1964) Quaternary nitrogen germicides derived from the monophenolic components of cashewnut shell liquid. *Indian Journal of Chemistry*, **2**, 114–17.

Gunatilaka, A. A. L., S. Sotheeswaran, S. Balasuramaniam, A. I. Chandrasekara and H. T. B. Sriyani (1980) Studies on medicinal plants of Sri-Lanka III. Pharmacologically important alkaloids of some *Sida* spp. *Planta Medica*, **39**, 66–72.

Gupta, D. R. and S. K. Garg (1966) A chemical examination of *Euphorbia hirta* L. *Bulletin of the Chemical Society of Japan*, **39**, 2532.

Gupta, R. K. and V. Mahadevan (1967) Chemical examination of the stems of *Euphorbia tirucalli*. *Indian Journal of Pharmacy*, **27**, 152–9; *Indian Journal of Experimental Biology*, **9**, 91.

*Gupta, B. and R. C. Shrimal (1965) Screening of Indian plants for biological activity III. *Indian Journal of Experimental Biology*, **3**, 66.

Gupta, O. P., K. K. Anand, M. Ali, B. J. R. Ghatak and C. K. Atal (1976) Anthelmintic activity of disalts of Embelin. *Indian Journal of Experimental Biology*, **14**, 356–7.

Guru, L. V. and D. N. Mishra (1966) Effect of the alcoholic and aqueous extractives of *Embelia ribes* Burm. on patients infested by ascarides – certain clinical studies. *Journal of Research in Indian Medicine*, **1**, 47.

Haerdi, F. von (1964) Die Heilpflanzen der Eingeborenen des Ulanga Distriktes in Tanganyika (Ost Afrika). In *Afrikanische Heilpflanzen*, 1 vol. *Acta Tropica* Supplement 8, Basel.

Hager's (1967–80) *Handbuch der Pharmazeutischen Praxis*, 4th edn, 12 vols., by P. L. List and L. Hörhammer, Springer, Berlin and New York.

Haines, D. W. and F. L. Warren (1949) The Euphorbia resins II. The isolation of taraxasterol and a new triterpene tirucallol from *E. tirucalli*. *Journal of the Chemical Society*, 2554–6.

Haines, D. W. and F. L. Warren (1950) The Euphorbia resins IV. A comparative study of euphol and tirucallol. *Journal of the Chemical Society*, 1562–3.

Hanriot, M. (1907) Sur les substances actives du *Tephrosia vogelii*. *Comptes Rendus de l'Académie des Sciences*, **144**, 150–2; Sur la toxicité des principes définis du *Tephrosia vogelii*, **144**, 498–500; Sur la mode d'action de la téphrosine, **144**, 651–3.

Harborne, J. B., T. J. Mabry and H. Mabry (1975) *The Flavonoids*, Part 2, Chapman and Hall, 1 vol., 1204 pp.

Hardman, R. and E. A. Sofowora (1972) A reinvestigation of *Balanites aegyptiaca* as a source of steroidal sapogenins. *Economic Botany*, **26**, 169–73.
Harper, S. H., A. D. Kemp, W. G. E. Underwood and R. V. M. Campbell (1969) Pterocarpanoid constituents of the heartwood of *Pericopsis angolensis* and *Swartzia madagascariensis*. *Journal of the Chemical Society C*, **8**, 1109–16.
Harrison, J., A. Silva Santisteban and B. Peyes Rojas (1973) Pharmacognosis and pharmacology of *Plumeria alba*. *Boletin de la Sociedad Quimica del Peru*, **39**, 89–92.
Hart, K. N., S. R. Johns, J. A. Lamberton and R. L. Willing (1970) Alkaloids of *Alchornea javanensis*, the isolation of hexhydroimidazol (pyrimidines and guanidines). *Australian Journal of Chemistry*, **23**, 1679–93.
Heal, R. F. and E. F. Rogers (1950) A survey of plants for insecticidal activity. *Lloydia*, **13**, 89–162.
Heftman, E. (1975) Steroid hormones in higher plants. Insect molting hormones. *Lloydia*, **38**, 195–209.
Hegnauer, R. (1962–68) *Chemotaxonomie des Pflanzen*, 5 vols., Birkhauser, Basel.
Henry, T. A. (1949) *The Plant Alkaloids*, 4th edn, Churchill, London, 803 pp.
Hoizey, M. J., L. Le Men-Oliver, J. Le Men, M. Leboeuf, J. F. Fauquet, J. de Jong, E. Morel and Ch. Warolin (1978) Alcaloides à structure α-acyl indolique étude de leurs activités antivirales et anésthesiques locales de surface. *Annales pharmaceutiques françaises*, **36**, 519–28.
Hopp, K. H., L. V. Cunningham, M. C. Bromel, L. J. Schermeister and S. K. Wahba Khalil (1976) In vitro antitrypanosomal activity of certain alkaloids against *Trypanosoma lewisi*. *Lloydia*, **39**, 375–7.
Hostettmann, K., M. Hostettmann-Kaldas and K. Nakanishi (1978) Molluscicidal saponins from *Cornus florida* L. *Helvetica Chimica Acta*, **61**, 1990.
Hostettmann, K., H. Kizu and T. Tomimori (1982) Molluscicidal properties of various saponins. *Planta Medica*, **44**, 34–5.
Hufford, D. and H. N. El-Sohly (1978) Two new benzylated flavanones isolated from the root bark of *Uvaria chamae* Beauv. *Lloydia*, **41**, 652C.
Hufford, D. and W. L. Lasswell, Jr (1978) Antimicrobial activities of constituents of *Uvaria chamae*. *Lloydia*, **39**, 156–60.
Hufford, Ch. D. and Oguntimein (1978) Non polar constituents of *Jatropha curcas*. *Lloydia*, **41**, 161–5.
Hussain, N., M. H. Modan, S. Ghulam, S. Shabbir and S. A. H. Zaidi (1979) Antimicrobial principles in *Mimosa hamata*. *Lloydia*, **42**, 525–7.
Hussein Ayoub, S. M. (1983) Molluscicidal properties of *Acacia nilotica* subspecies *tomentosa* and *astringens* I. *Fitoterapia*, **54**, 183–7, 189–92.
Hussein Ayoub, S. M. (1984) Effect of the galloyl group on the molluscicidal activity of tannins. *Fitoterapia*, **55**, 343–5.
Ieven, M., J. Totte, D. van den Berghe and A. J. Vlietinck (1978) Antiviral activity of some Amaryllidaceae alkaloids. *Planta Medica*, **33**, 284.
Ieven, M., D. van den Berghe, F. Mertens, A. Vlietinck and E. Lammens (1979) Screening of higher plants for biological activities. Antimicrobial activity. *Planta Medica*, **36**, 311–21.
Ikram, M. and Inamul-Haq (1980) Screening of medicinal plants for antimicrobial activity. Parts I and II. *Fitoterapia*, **51**, 231–5; 281–4.
Inayat Khan, A. Q., S. Bukhari and Mujahid Hussain Khan (1963) Some pharmacological actions of chaksine HCl and isochaksine. *Pakistan Journal of Industrial Research*, **6**, 97–102.
Indian Council of Medical Research (1976) *Medicinal plants of India*, 1 vol., New Delhi, 487 pp.
Irvine, F. R. (1930) *Plants of the Gold Coast*. Oxford University Press, London, 521 pp.
Irvine, F. R. (1961) *Woody Plants of Ghana*, 2nd edn, Oxford University Press, London, 868 pp.
Isaac-Sodeye, W. A. (1971) Antisickling substance from the roots of *Fagara zanthoxyloides*. *Lloydia*, **34**, 383.

Jacobson, M., R. E. Redfern and G. D. Mills Jr (1975) Naturally occurring insect growth regulators II. Screening of insect and plant extracts as insect juvenile hormone mimics. *Lloydia*, **38**, 455–76.

Jacquemain, D. (1959) La noix d'Anacarde. *Oleagineux*, **14**, 527–37.

Jain, S. R. and M. R. Jain (1972) Therapeutic utility of *Ocimum basilicum* var. *album*. *Planta Medica*, **22**, 136–9.

Jain, S. R. and A. Kar (1971) The antibacterial activity of some essential oils and their combinations. *Planta Medica*, **20**, 118–23.

Jain, S. R., P. R. Jain and M. R. Jain (1974) Antibacterial evaluation of some indigenous volatile oils. *Planta Medica*, **26**, 196–9.

Jenkin, H. M. (1973) Drug for treatment of virus infections. Ger. Offen. Patent 2.305.553 (through *Chemical Abstracts*, **83**, 108631j).

Jentzch, K., P. Spiegel and L. Fuchs (1962) Untersuchungen über die Inhaltstoffe der Blätter von *Combretum micranthum* G. Don. *Planta Medica*, **10**, 1–8.

Juneja, T. R., K. N. Gaind and C. L. Dhawan (1970a) Investigations on *Capparis sepiaria*. *Research Bulletin Punjab University*, **21**, 23.

Juneja, T. R., K. N. Gaind and A. S. Panesar (1970b) Investigations on *Capparis decidua* Edgew. Study of isothiocyanate glucoside. *Research Bulletin Punjab University*, **21**, 519–21.

Junod, C. (1964) Traitement du *Taenia saginata* à l'aide d'un extrait de graines de courge. *Presse Médicale*, **20**, 1243.

Kabelik, J. (1970) Antimikrobielle Eigenschaften des Knoblauchs. *Pharmazie*, **25**, 266–70.

Kaji, N. N. (1968) Studies on *Cassia* species. *Indian Journal of Pharmacy*, **30**, 282.

Kaji, N. N., M. L. Khorana and M. M. Sanghavi (1968) Studies on *Cassia fistula* L. *Indian Journal of Pharmacy*, **30**, 8.

Kaleyra, R. R. (1975) Screening of indigenous plants for anthelmintic action against human *Ascaris lumbricoides*. *Indian Journal of Physiology and Pharmacology*, **19**, 47–9.

Kapoor, L. D., A. Singh, S. L. Kapoor and S. N. Srivastava (1975) Survey of Indian plants for saponins, alkaloids and flavonoids IV. *Lloydia*, **38**, 221–4.

Karrer, W. (1958) *Konstitution und Vorkommen der organischen Pflanzenstoffe (exclusive Alkaloide)*, 1 vol., Birkhauser, Basel and Stuttgart, 1207 pp.

Kasturi, T. R. and Manithomas (1967) Essential oil of *Ageratum conyzoides*. Isolation and structure of two new constituents. *Tetrahedron Letters*, **27**, 2573–5.

Kaustiva, B. S. (1958) Amoebicidal activity of some derivatives of *Cinchona* alkaloids related to emetine. *Journal of Scientific and Industrial Research*, **17C**, 137.

Kerharo, J. (1968) Revue des plantes medicinales et toxiques du Senegal. *Plantes Médicinales et Phytothérapie*, **11**, 108–46.

Kerharo, J. (1969) Un remède populaire sénégalais; le nébéday (*Moringa oleifera* Lam.). Thérapeutique en milieu africain. Chimie et Pharmacologie. *Plantes Médicinales et Phytothérapie*, **3**, 214–19.

Kerharo, J. and J. G. Adam (1974) *La Pharmacopee Sénégalaise Traditionnelle*, Vigot, Paris, 1 vol., 1011 pp.

Kerharo, J. and A. Bouquet (1950) *Plantes Médicinales de la Côte d'Ivoire et de la Haute Volta*, Vigot, Paris, 297 pp.

Kerharo, J., A. Bouquet and R. Heintz (1948) Le wilinwiga des Mossis (*Guiera senegalensis* Lam.) ses usages thérapeutiques indigènes et son application au traitement des diarrhées cholériformes. *Acta Tropica*, **5**, 345–8.

Kerharo, J., F. Guichard and A. Bouquet (1960) Les végétaux ichthiotoxiques (poisons de pêche). *Bulletins et Memoires de la Faculté de Médecine et de Pharmacie de Dakar*, **8**, 313–19.

Kerharo, J., F. Guichard and A. Bouquet (1961) Les végétaux ichthytoxiques. *Bulletin et Mémoires de la Faculté de Médecine et de Pharmacie, Dakar*, **9**, 355–86.

Kerharo, J., F. Guichard and A. Bouquet (1962) Les végétaux ichthiotoxiques (poisons de pêche). *Bulletins et Mémoires de la Faculté de Médecine et de Pharmacie de Dakar*, **10**, 223–42.

Khadem, H. El and Y. S. Mohammed (1958) Constituents of the leaves of *Psidium guijava* L. II. Quercetin, avicularin and guaijaverin. *Journal of the Chemical Society*, 3320–3.

Khan, M. R., M. H. H. Nkunya and H. Wevers (1980a) Triterpenoids from leaves of *Diospyros* species. *Planta Medica*, **38**, 380–1.

Khan, M. R., G. Ndaalio, M. H. H. Nkunya, H. Wevers and A. N. Sawhney (1980b) Studies on African medicinal plants I. Preliminary screening of medicinal plants for antibacterial activity. *Planta Medica*, Suppl., 91–7.

Khorrami, J. S. (1979) Dosage du lawsone dans le henné par la méthode colorimétrique. *Quarterly Journal of Crude Drug Research*, **17**, 131–4.

Kitagawa, M. and T. Tomiyana (1930) A new amino compound in the Jack bean and a corresponding new ferment. *Journal of Biochemistry, Tokyo*, **11**, 265–71.

Kjaer, A. and J. Thomson (1963) Isothiocyanate producing glucosides in species of Capparidaceae. *Phytochemistry*, **2**, 29.

Kjaer, A., O. Malver, B. El-Menshawi and J. Reisch (1979) Isothiocyanates in myrosinase treated seed extracts of *Moringa peregrina*. *Phytochemistry*, **18**, 1485–7.

Kokwaro, O. (1976) *Medicinal Plants of East Africa*, East African Literature Bureau, Nairobo, Dar es Salaam.

Kraus, W. and M. Bokel (1981) New tetranortriterpenoids from *Melia azadarach* L. (Meliaceae). *Chemische Berichte*, **114**, 267–75.

Kraus, W., W. Grimminger and G. Sawitzki (1978) New insect antifeedant from Meliaceae. *International Research Congress on Natural Products as Medicinal Agents*, Strasbourg, **2**, 115–16 (through *Chemical Abstracts* (1978) 18787v).

Krishnakumari, M. K. and S. K. Majumdar (1960) Studies on the anthelmintic activities of seeds of *Carica papaya* L. *Annals of Biochemical and Experimental Medicine, India*, **20**, Suppl., 551–6.

Krishnamurti, G. V. and T. R. Seshadri (1946) Bitter principle of *Phyllanthus niruri* L. *Proceedings of the Indian Academy of Sciences*, 357 (through *Chemical Abstracts* (1947) 2712).

Kubo, I., M. Taniguchi, A. Chapya and K. Tsujimoto (1980) An insect antifeedant and antimicrobial agent from *Plumbago capensis* (plumbagin). *Planta Medica*, Suppl., 185–7.

Kubo, I., S. P. Tanis, Y. Lee, I. Muira, K. Nakanishni and A. Chapya (1976) The structure of harrisonine from *Harrisonia abyssinica*. *Heterocycles*, **5**, 485.

Kubo, I., I. Muira, M. J. Pettei, Y. W. Lee, F. Pilkiewiez and N. Nakanishi (1977) Muzigadial and warburganal, potent antifungal, antiyeast and African army worm antifeedant agents. *Tetrahedron Letters*, **52**, 4553–6.

Kučera, L. S., R. A. Cohen and E. C. Herrman Jr (1965) Antiviral activities of extracts of lemon balm plant. *Annals of the New York Academy of Sciences*, **130**, 474–82.

Kurup, P. A. and P. L. Narasimha-Rao (1954) Antibiotic principle from *Moringa pterygosperma* II. Chemical nature of pterygospermine. *Indian Journal of Medical Research*, **42**, 85–95 and 115–23.

Kurup, P. A. and P. L. Narasimha-Rao (1954) Antibiotic principle from *Moringa pterygosperma*. Part V. Effect of pterygospermine on the assimilation of glutamic acid by *Micrococcus pyogene* var. *aureus*. *Indian Journal of Medical Research*, **42**, 109–13.

Kurup, P. A., P. L. Narasimha-Rao and R. S. Ramaswany (1957) Antibiotic principle from *Moringa pterygosperma* VIII. Some pharmacological properties and *in vivo* action of pterygospermine and related compounds. *Indian Journal of Medical Research*, **45**, 197–206.

Lahon, L. C., H. N. Khanikor, N. Ahmed and A. R. Gogoi (1978) Preliminary pharmacological and anticestodal screening of *Curcurbita maxima*. *Indian Journal of Pharmacology*, **10**, 315–17.

Lal, S. and I. Gupta (1970) Control of sarcoptic mange with chotidudhi (*Euphorbia prostrata* and *E. thymifolia*). *Indian Journal of Pharmacology*, **2**, 28.

Lambin, S. and J. Bernard (1953) Action de quelques alcaloides sur le *Mycobacterium tuberculosis*. *Comptes Rendus de la Société de Biologie, Paris*, **147**, 638–41; 760–70.

Lasswell, W. L., Jr (1977) Isolation and structure elucidation of biologically active constituents of *Uvaria chaemae*. *Chemical Abstracts*, **87**, 189345b. Thesis, 104 pp.

Laurens, P. and R. R. Paris (1977) The polyphenols of African and Madagascan Anacardiaceae. *Poupartia birrea*, *P. caffra* and *Anacardium occidentale*. *Plantes Médicinales et Phytothérapie*, **11**, 15–24.

Lavie, D. and E. C. Levy (1969) *Tetrahedron Letters*, 3525.

Lavie, D., M. K. Jain and I. Kirson (1967) Terpenoids VI. The complete structure of melianone. *Journal of the Chemical Society, C*, 1347–51.

Leboeuf, M. and A. Cavé (1980) Alcaloides des Annonacées XXVIII. Alcaloides de l'*Uvaria chamae* P. Beauv. *Plantes Médicinales et Phytothérapie*, **14**, 143–7.

Lechat, P., F. Bisseliches, F. Bournerias, H. Dechy, Y. Juillet, G. Lagier, C. Meryrignac, B. Rouveix, P. Sterin, A. Warnet and S. Weber (1978) *Pharmacologie Médicale*, 3rd edn, Masson, Paris, 677 pp.

*Lee, K. H., T. Ibuka and R. Y. Wu (1974) Beta unsubstituted cyclopentenone, a structural requirement for antimicrobial and cytotoxic activities. *Chemical and Pharmaceutical Bulletin*, **22**, 2206.

Lemma, A. (1970) Laboratory and field evaluation of the molluscicidal properties of *Phytolacca dodecandra*. *Bulletin of the World Health Organization*, **42**, 597–612.

*Lespagnol, A., A. Sevin and H. Beerens (1949) Recherches relatives à diverses substances susceptibles de présenter des proprietés antibiotiques vis à vis du bacille tuberculeux. *Comptes Rendus de l'Académie des Sciences Paris*, **229**, 483–4.

Levêque, J., J. L. Pousset and A. Cavé (1975) Le lyaloside, nouveau gluco-alcaloide isolé du *Pauridiantha lyallii* Brem. (Rubiacées). *Comptes Rendus de l'Académie des Sciences Paris, Serie, C*, **280**, 593.

Lewis, W. H. and M. P. F. Elvin-Lewis (1977) *Medical Botany*, John Wiley and Sons, New York.

Liao, W. T., J. L. Beal, W. N. Wu and R. W. Doskotch (1978) Alkaloids of *Thalictrum* XXVI and XXVII. New hypotensive and other alkaloids from *T. minus* race B. *Lloydia*, **41**, 257–76.

Lillykutty, L. and Santhakumari (1969) Antimicrobial activities of *Cassia fistula* L. *Journal of Research in Indian Medicine*, **4**, 25.

Little, J. E., M. W. Foote, W. I. Rogers and D. B. Johnstone (1953) Ethylgallate, mycobacteria-specific antibiotic isolated from *Haematoxylon campechianum*. Isolation and chemical studies. *Antibiotics and Chemotherapy*, **3**, 183–91.

*Lockwood, G. B. (1975) *Simple and quick method for the estimation of steroidal sapogenins in the leaves of plants*, Cairo Symposium.

Low, D., B. D. Rawal and W. J. Griffin (1974) Antibacterial activity of the essential oils from some Australian Myrtaceae with special references to the activity of chromatographic fractions of oil of *Eucalyptus citriodora*. *Planta Medica*, **26**, 184–9.

Lugt, Ch. B. (1980) Development of molluscicidal potency in long and short staminate racemes of *Phytolacca dodecandra*. *Planta Medica*, **38**, 68–72.

Luscombe, D. K. and S. A. Taha (1974) Pharmacological studies on the leaves of *Azadirachta indica*. *Journal of Pharmacy and Pharmacology*, **26**, 110–11.

McDonald, A. D., F. L. Warren and J. M. Williams (1949) The Euphorbia resins I. Euphol. *Journal of the Chemical Society*, Suppl., 474–83.

McIlroy, R. J. (1950) *The Plant Glycosides*, Vol. 1, Arnold and Cie, London, 138 pp.

McKee (1955) Host parasite relationships in the dry-rot disease of potatoes. *Annals of Applied Biology*, **43**, 147–8.

Mackie, A. and N. Ghatge (1958) Chemical investigation of the leaves of *Annona senegalensis* L. *Journal of the Science of Food and Agriculture*, **9**, 88.

Mackie, A. and A. L. Misra (1956) Chemical investigation of the leaves of *Annona senegalensis* L. Constituents of leaf wax. *Journal of the Science of Food and Agriculture*, **7**, 203–9.

Madran, B. R. (1960) Spermicidal action of *Canscora decussata*, an Indian indigenous drug. *Archives Internationales de Pharmacodynamie et de Thérapie*, **124**, 358.

Maichuk, Ju F., T. N. Abazov and R. I. Abramishvii (1972) Antiviral activity of gossypol in experimental herpetic keratitis. Through *Chemical Abstracts*, **82**, 670 g.

Makboul, A. M. and A. M. Abdel-Baki (1981) Flavonoids from the leaves of *Duranta plumieri*. *Fitoterapia*, **52**, 219–20.

Malathi, V., G. Ramakrishnan and M. Sirsi (1959) The immunological effect of coconut factors in experimental tuberculosis in mice. *Journal of the Indian Institute of Sciences*, **41**, 52.

Malcolm, S. A. and E. A. Sofowora (1969) Antimicrobial activities of selected Nigerian folk remedies and their constituent plants. Antimicrobial properties of Balanites. *Lloydia*, **32**, 512–17.

Malvan, G. H., S. M. Abdel Wahab and M. Salah Ahmed (1974) Isolation and quantitative estimation of plumeried from the different organs of *Plumeria rubra* and *Plumeria rubra* var. *alba*. *Planta Medica*, **25**, 226–30.

Manson, D. J. (1939) Toxicity of the berries of *Duranta repens* to anopheline and culcicine larvae. *Journal of the Malaria Institute, India*, **2**, 85.

Manson-Bahr, P. H. (1952) *Synopsis of Tropical Medicine*, Cassel et Cie, London, 248 pp.

Maradufu, A. and J. H. Ouma (1978) A new chalcone as a natural molluscicide from *Polygonum senegalense*. *Phytochemistry*, **17**, 823–4.

Martin, M., J. Ridet, A. Chartol, J. Biot, L. Porte and A. Bezon (1964) Action thérapeutique de l'extrait d'*Euphorbia hirta* dans l'amibiase intestinale. A propos de 150 observations. *Médicine Tropicale*, **24**, 250–61.

Marwick, M. G. (1963) A note on ordeal poison in East Africa. *Man (London)*, **63**, 75–6.

Maynart, G., J. L. Pousset, S. Mboup and F. Denis (1980) Action antibactérienne de la borréverine, alcaloide isolé du *Borreria verticillata*. *Comptes Rendus de la Société de Biologie*, **174**, 925–8.

Mehta, S. and J. V. Bhat (1952) Studies on Indian medicinal plants II. Bryophyllin, a new antibacterial substance from the leaves of *Bryophyllum calycinum* Salisb. *Journal of the University of Bombay*, **21**, 21.

Mela, C. (1950) Presence of substances having antibiotic action in the higher plants. *Fitoterapia*, **21**, 98–9.

Merck (1976) *The Merck Index*, An Encyclopaedia of Chemicals and Drugs, 9th edn, Merck, Rahway, NY.

Misra, R. N. and Sikhibhushan Dutt (1937) Chemical examination of the seeds of *Cleome pentaphylla* L. Isolation of cleomin. *Proceedings of the National Institute of Sciences, India*, **3**, 45–9.

Mital, H. C. and F. R. Dove (1971) The study of shea butter. *Planta Medica*, **20**, 283–8.

Mitscher, L. A., R. Leu, M. S. Bathala, W. Wu and J. L. Beal (1972a) Antimicrobial agents from higher plants: Introduction, rationale and methodology. *Lloydia*, **35**, 157–66.

Mitscher, L. A., H. D. H. Showalter, M. T. Shipchandler, R. D. Leu and J. L. Beal (1972b) Antimicrobial agents from higher plants. IV *Zanthoxylum elephantiasis*, isolation and identification of cantin-6-one. *Lloydia*, **35**, 177–88.

Mitscher, L. A., W. N. Wu, R. W. Doskotch and J. L. Beal (1972c) Antibiotics from higher plants II. Alkaloids from *Thalictrum rugosum*. *Lloydia*, **35**, 167–76.

Mitscher, L. A., Y. H. Park, D. Clark, G. W. Clark, P. P. Hammersfaler, W. N. Wu and J. L. Beal (1978) Antimicrobial agents from higher plants. An investigation of *Hunnemannia fumerariaefolia*. Pseudoalcolates of sanguinarine and chelerythrine. *Lloydia*, **41**, 145–50.

*Mitscher, L. A., Y. H. Park, D. Clark and J. L. Beal (1980) Antimicrobial agents from higher plants – Isoflavonoids and related substances from *Glycyrrhiza glabra* L. var. *typica*. *Lloydia*, **43**, 256–69.

Mitsui, S., S. Kubayashi, H. Nagahori and A. Ogiso (1976) *Alpinia galanga* seeds. *Chemical and Pharmaceutical Bulletin*, **24**, 2377–82.

Modi, N. T. and B. S. Deshmankar (1972) Some preliminary pharmacological investigations on *Cardiospermum halicacabum* seeds. *Indian Journal of Pharmacy*, **34**, 76.

Morton, J. F. (1962) Spanish needles (*Bidens pilosa* L.) as a wild food resource. *Economic Botany*, **16**, 173–9.

Mourgue, M., J. Delphaut, R. Baret and R. Kascab (1961) Etude de la toxalbumine (curcine) des graines de *Jatropha curcas* L. *Bulletin de la Société de Chimie et Biologie*, **43**, 517–31.

Moyse-Mignon, H. (1942) Recherches sur quelques Meliacées africaines et sur leurs principes amers. Thèse de Doctorat en Pharmacie, Université de Paris.

Mozley, A. (1939). The fresh water molluscs of Tanganyika Territory and Zanzibar protectorate, and their relation to human schistosomiasis. *Transactions of the Royal Society of Edinburgh*, **59**, 687–744.

Mozley, A. (1952) *Molluscicides*, H. K. Lewis & Co., London.

Muelenaere, H. J. M. de (1965) Toxicity and hemagglutinating activity of legumes. *Nature (London)*, **206**, 827–8.

Mukerjee, T. D. and R. Govind (1958) Studies on indigenous insecticidal plants II. *Annona squamosa*. *Indian Journal of Scientific and Industrial Research*, **17C**, 9–15.

Murthy, S. P. and M. Sirsi (1957) Pharmacological studies on *Melia azadirachta* I. Antibacterial, antifungal and antitubercular activity of Neem oil and its fractions. *Symposium: Utilization of Indian Medicinal Plants*, Lucknow.

Murthy, S. P. and M. Sirsi (1958) Pharmacological studies on *Melia azadirachta* II. The effect of nimbidin, nimbidol and neem oil on experimental avian malaria. *Journal of Mysore Medical Association*, **23**, 1.

Murti, V. V. S. and T. R. Seshadri (1964) Toxic amino acids in plants. A review (87 references). *Current Science India*, **33**, 323–9.

Nadkarni, A. K. (1954) *Indian Materia Medica*, Vol. I, 3rd edn, Popular Book Depot, Bombay (from Indian Council of Medical Research, 1976, p. 164).

Nandy, A. and B. K. Gupta (1968) Effects of the seasonal variations on the rotenone content of *Derris uliginosa* Benth. *Indian Journal of Pharmacy*, **30**, 284.

Nara, T. K., J. Cleye, L. Lavergne de Cerval and E. Stanislas (1977) Flavonoides de *Phyllanthus niruri* L., *P. urinaria* L. and *P. orbiculatus* L. Rich. *Plantes Médicinales et Phytothérapie*, **11**, 82–6.

Narayanan, C. R. and K. N. Iyer (1967) Isolation and characterization of desactyl nimbin. *Indian Journal of Chemistry*, **5**, 460.

Narayanan, C. R. and T. R. Seshadri (1972) Proanthocyanidins of *Cassia fistula*. *Indian Journal of Chemistry*, **10**, 379.

Nath, B. (1954) Chemical examination of the heartwood of *Melia azedarach*. *Journal of Scientific and Industrial Research (India)*, **13B**, 740–1.

Ndir, O. and J. L. Pousset (1981) Plantes médicinales africaines VII. Essais in vitro d'*Euphorbia hirta* sur *Entamoebia histolytica*. *Plantes Médicinales et Phytothérapie*, **15**, 113–25.

Neogi, N. C. *et al.* (1958) *Indian Journal of Pharmacy*, **20**, 95 (through Watt and Breyer-Brandwijk, 1962).

Ngo, Van Thu, Nguyen Van Kim, Truong Van Nhu, Dao-Boi Huan, Nguyen Thuy Ma and Dinh Thi Thahn (1979) Effectiveness of *Brucea sumatrana* plant extracts against malaria. *Duo Hoc*, 15–17 (through *Chemical Abstracts*, **92** (1980) 191396r).

Nguyen Ba Tuoc (1953) Recherches botaniques, chimiques et pharmacodynamiques sur quelques drogues vermifuges d'Indochine. Thèse de Pharmacie, Paris, 120 pp.

Nickel, L. G. (1959) Antimicrobial activity of vascular plants. *Economic Botany*, **13**, 281–318.

Odebiji, O. O. (1978) Preliminary phytochemical and antimicrobial examination of leaves of *Securidaca longepedunculata*. *Nigerian Journal of Pharmacy*, **9**, 29 (through *Chemical Abstracts*, **89**, 176331k).

Odebiji, O. O. (1980) Antibacterial property of tetramethylpyrazine from the stem of *Jatropha podagrica*. *Planta Medica*, **38**, 144–6.

*Odebiji, O. and E. A. Sofowora (1978) Phytochemical screening of Nigerian medicinal plants II. *Lloydia*, **41**, 234–6.

Odebiji, O. O. and E. A. Sofowora (1979) Antimicrobial alkaloids from a Nigerian chewing stick (*Fagara zanthoxyloides*). *Planta Medica*, **36**, 204–7.

Ogan, A. U. (1971) West African medicinal plants V. Isolation of cuminal from *Xylopia aethiopica*. *Phytochemistry*, **10**, 2823–4.

Ogan, A. U. (1972) The alkaloids in the leaves of *Combretum micranthum*. *Planta Medica*, **21**, 210–16.

Oguakwa, T. U., M. Patamia, C. Galeffi, I. Messana and M. Nicoletti (1981) Isolation of Cleomin from roots of *Ritchiea longipedicillata*. *Planta Medica*, **41**, 410–12.

*Ogunlana, E. O. and E. Ramstad (1975) Investigation into the antibacterial activities of local plants. *Planta Medica*, **27**, 354–60.

Ojewole, J. A. O. (1981) Effects of tetramethylpyrazine on isolated atria of the guinea pig. *Planta Medica*, **42**, 223–8.

Ojewole, J. A. O. and O. Odebiji (1980) Neuromuscular and cardiovascular actions of tetramethylpyrazine from the stem of *Jatropha podagrica*. *Planta Medica*, **39**, 238.

Ojha, D., S. N. Tripathi and G. Sing (1966) Role of an indigenous drug (*Achyranthes aspera*) in the management of reactions of leprosy. Preliminary observations. *Leprosy Review*, **37**, 115.

Okogun, J. I., G. O. Fakunle, D. E. U. Ekong and J. D. Connoly (1975) Chemistry of the meliacins (limonoids). The structure of meliacin A, a new protomeliacin from *Melia azedarach*. *Journal of the Chemical Society, Perkin, Transactions I*, 1352–6.

Okpanyi, S. N. and G. C. Ezeukwu (1981) Anti-inflammatory and antipyretic activities of *Azadirachta indica*. *Planta Medica*, **41**, 34–39.

Oladele Arigbabu, S. and S. G. Don Pedro (1971) Studies on some pharmaceutical properties of *Azadirachta indica* or 'Baba yaro'. *African Journal of Pharmacy and Pharmaceutical Sciences*, **1**, 181–4.

Oliver, B. (1960) *Medicinal Plants in Nigeria*, Private edn, Nigerian College of Arts, Science and Technology, 138 pp.

Oliver-Bever, B. (1968) Selecting local drug plants in Nigeria. Botanical and chemical relationship in three families. *Quarterly Journal of Crude Drug Research*, **8**, 1194–211.

Oliver-Bever, B. (1970) Why do plants produce drugs? What is their function in the plants? *Quarterly Journal of Crude Drug Research*, **10**, 1541–9.

Oliver-Bever, B. (1983) Medicinal plants in tropical West Africa III. Anti-infection therapy with higher plants. *Journal of Ethnopharmacology*, **9**,1–83.

Olsen, R. A. (1975) Triterpene glycosides as inhibitors of fungal growth and metabolism. Induced leakage of nucleotide materials. *Physiology Plantarum* (Copenhagen), **33** (through *Biological Abstracts*, **59**, 69.207).

Olson, M. O. J. and I. E. Liener (1967) Some physical and chemical properties of concanavalin A, the phytohemagglutinin of Jack Bean. *Biochemistry*, **6**, 105–12.

Olugbade, T. A., J. A. Oluwaduja and W. A. Yisak (1982) Chemical constituents of *Cnestis ferruginea* DC. *Journal of Ethnopharmacology*, **6**, 365–72.

Onuaguluchi, G. (1964) Further studies on the pharmacology of the antiascaris fraction of the ethanolic bark extract of the erin tree. (*Polyadoa umbellata* Dalz.). *West African Medical Journal*, **13**, 162.

Onuaguluchi, G. (1966) Further studies on the pharmacology of the antiascaris fraction of the ethanolic bark extract of the erin tree (*Polyadoa umbellata* Dalz.). *West African Medical Journal*, **15**, 22.

Orazi, O. O. (1946) Estudio fitoquimico de la *Borreria verticillata*. *Revista da Faculdade Ciencias y Quimica*, **19**, 17–28.

*Osborne, E. M. (1973) On the occurrence of antibacterial substances in green plants. *British Journal of Experimental Pathology*, **24**, 227–31.

Osborne, E. M. and J. L. Harper (1957) Antibacterial activity of *Cassia tora* and *C. obovata*. *Indian Journal of Pharmacy*, **19**, 70.

Pai, B. R., P. S. Subramanian and U. Rao Ramdas (1970) Isolation of plumericin and isoplumericin from *Allemanda cathartica*. *Indian Journal of Chemistry*, **8**, 851.
Pais, M., J. Marchand, X. Monseur, F. X. Jarreau and R. Goutarel (1976) Alcaloides peptidiques: Structure de l'hyménocardine, alcaloïde de l'*Hymenocardia acida* Tul. *Comptes Rendus de l'Académie des Sciences, Serie D*, **264**, 1409–11.
Pal, S. and I. Gupta (1971) In vitro studies of the antifungal activity of chotidudhi plant (*Euphorbia prostata* and *E. thymifolia*). *Indian Journal of Pharmacology*, **3**, 27.
Pal, S. N. and M. Narasimhan (1943) *Eclipta prostata*. *Journal of the Indian Chemical Society*, **181** (through Chemical Abstracts (1944) 1609).
Paris, R. (1942) Sur une Combrétacée africaine, le kinkeliba (*Combretum micranthum* Don.). *Bulletin des Sciences Pharmacologiques*, **49**, 181–6.
Paris, R. and Bézanger-Beauquesne (1956) Sur la constitution du swartziol. Identité avec le kaempferol. *Comptes Rendus de l'Académie des Sciences, Serie D*, **242**, 1761–2.
Paris, R. and S. Etchepare (1966) Sur le noircissement des feuilles et des fruits de *Thevetia peruviana* Pers. Isolement d'un chromogène identifié à l'aucoboside. *Comptes Rendus de l'Académie des Sciences, Serie D*, **262**, 1239–41.
Paris, R. and H. Moyse-Mignon (1939) Sur quelques Méliacées réputées fébrifuges. *Bulletin des Sciences Pharmacologiques*, **46**, 104–8.
Paris, R. and H. Moyse-Mignon (1949) Pouvoir antimicrobien et présence de plumbaginol chez deux *Diospyros* africains (*D. xanthochlamys* Gürke et *D. mespiliformis* Hochst.). *Comptes Rendus de l'Académie des Sciences*, **228**, 2063–4.
Paris, R. and H. Moyse-Mignon (1956) Caractérisation de la choline chez quelques plantes médicinales. *Annales de Pharmacie Françaises*, **14**, 464–9.
Paris, R. R. and H. Moyse (1963) *Abrégé de Matière Médicale. Matières Premières d'Origine Végetale*, 1 vol., Vigot, Paris, 192 pp.
Paris, R. R. and H. Moyse (1965) *Précis de Matière Médicale*, 1 vol., Masson et Cie, Paris.
Paris, R. R. and H. Moyse (1967) *Précis de Matière Médicale*, Vol. II, Masson et Cie, Paris, 510 pp.
Paris, R. R. and H. Moyse (1971) *Précis de Matière Médicale*, Vol. III, Masson et Cie, Paris, 508 pp.
*Paris, R. R. and P. Nothis (1969) Sur quelques plantes de la Nouvelle Calédonie. *Plantes Médicinales et Phytothérapie*, **3**, 274–86.
Patel, M. B., C. Miet and J. Poisson (1967) Alkaloides de *Tabernaemontana pachysiphon*. *Annales pharmaceutiques françaises*, **25**, 379–84.
Paulose, M. M., S. Venkeob Rao and K. T. Achaya (1964) Nature of the sterol in castor oil. *Indian Journal of Chemistry*, **2**, 381–2.
Pelt, J. M. (1959) Contribution à l'étude des huiles de Chaulmoogra africaines. Thèse de Doctorat en Pharmacie, Université de Nancy.
Pelter, A. and P. I. Amenechi (1969) Isoflavonoid and pterocarpanoid extractives of *Lonchocarpus laxiflorus*. *Journal of the Chemical Society*, 887–96.
Pernet, R. (1972) Phytochimie des Burseracées. *Lloydia*, **35**, 280–7.
Perrot, E., R. Raymond-Hamet and L. Millat (1936) Sur les propriétés hypothermisantes de la mitrinermine. *Bulletin de la Société Pharmacologique*, **43**, 694–6.
Persinos, G. J. and M. W. Quimby (1967) Nigerian plants III. Phytochemical screening for alkaloids, saponins and tannins. *Journal of Pharmaceutical Sciences*, USA, **56**, 1512–15.
Persinos, G. J., M. W. Quimby and J. W. Schermerhorn (1964) A preliminary pharmagnostical study of ten Nigerian plants. *Economic Botany*, **18**, 329–41.
Phillipson, J. O. and F. A. Darwish (1981) Bruceolides from Fijian *Brucea javanica*. *Planta Medica*, **41**, 209–20.
Pillai, N. C., G. S. S. Rao and M. Sirsi (1957) Plant anticoagulants. *Journal of Scientific and Industrial Research*, **160**, 106.
Planche, O. (1949) Sur le *Trichilia heudelotii* Planch. *Annales pharmaceutiques françaises*, **7**, 460–5.

Popp, F. D., J. M. Wefer, D. P. Chakraborty, G. Rosen and A. C. Casey (1968) Investigation of African plants for alkaloids, antimalarial agents and antineoplastic agents. *Planta Medica*, **16**, 343–9.

Pousset, J. L. (1981) The antimicrobial properties of *Euphorbia hirta*. Proceedings of the fourth DRPU symposium. University of Ife, Nigeria, July 1981.

Pousset, J. L., J. Kerharo, G. Maynart, X. Monseur, A. Cavé and R. Goutarel (1973) La Borrerine, nouvel alkaloide isolé du *Borreria verticillata*. *Phytochemistry*, **12**, 2310–12.

Pousset, J. L., L. Levesque, A. Cavé, F. Picot, P. Potier and R. R. Paris (1974) Etude chimique du *Pauridiantha lyallii* (Rubiaceae). *Plantes Médicinales et Phytothérapie*, **8**, 51–6.

Prakash, A., R. K. Varma and S. Ghosal (1981) Alkaloid constituents of *Sida acuta, Sida humilis, Sida rhomboifolia* and *S. spinosa*. *Planta Medica*, **43**, 384.

Prema, P. (1968) Pharmacological studies of *Desmodium gangeticum* DC. Thesis in Medicine, University of Kerala Trivandrum, India.

Preston, N. W. (1977) Cajanol, an antifungal isoflavone from *Cajanus cajan*. *Phytochemistry*, **16**, 143–4.

Prista, I. N. and A. Correia Alves (1958) Estudo farmacosico, botanico, quimico e farmacodynamico de *Securidaca longepedunculata* Fres. *Garcia di Orta Lisbon*, **6**, 131–47.

Quadry, J. S. and R. Zafar (1978) Tissue culture of some *Cassia* species. *Planta Medica*, **33**, 299.

Qudrat-I-Khuda, M. and M. Efran Ali (1964) *Caesalpinia bonducella*. Structures of α-, β- and hydrolysed γ caesalpins. *Scientific Researches (Pakistan)*, **1**, 177–83 (through *Chemical Abstracts* (1964) 61 10718).

Rahman, W., K. Ishratula, H. Wagner, O. Seligmann, V. Mohan Chari and B. G. Osterdahl (1978) Prunin-6″-O-*p*-cumarate, a new acylated flavanone glycoside from *Anacardium occidentale*. *Phytochemistry*, **17**, 1064–5.

Rai, A. and M. S. Sethi (1973) Antiviral properties of *Azadirachta indica*. *Indian Journal of Animal Science*, **42**, 1066–70.

Raina, K. L., K. L. Dhar and C. K. Atal (1976) Occurrence of *N*-isobutyl-eicosa-trans-2-trans-4-dienamide in *Piper nigrum*. *Planta Medica*, **30**, 198–200.

Ramakrishnan, P. N. (1969) A study of hypoglycaemic action of *Phyllanthus niruri* L. *Indian Journal of Pharmacy*, **31**, 175.

Ramaswany, A. and M. Sirsi (1957) (+)-Sesamine extracted from *Fagara chalybea*. *Naturwissenschaften*, **44**, 380.

Ramprasad, C. and A. Sirsi (1956) *Curcuma longa* – In vitro antibacterial activity of curcumin and of the essential oil. *Journal of Scientific and Industrial Research*, **156**, 239–41.

Rangaswani, S., E. V. Rao and M. Suryanarayana (1961) Chemical examination of *Plumeria acutifolia*. *Indian Journal of Pharmacy*, **23**, 122–4.

Rao, E. V. and T. S. R. Anjaneyulu (1967) Chemical components of the bark of *Plumeria rubra* L. *Indian Journal of Pharmacy*, **29**, 273–4.

Rao, V. R. and I. Gupta (1970) In vitro studies on the antifungal activity of some indigenous drugs against *Trichophyton mentagrophytes*. *Indian Journal of Pharmacology*, **2**, 29.

Rao, A. R., S. Sukumar, T. B. Paramasivan, S. Kamalaksi, A. R. Parastrumaran and M. Shantha (1967) Study of antiviral activity of tender leaves of Margosa tree (*Melia azadarach*) on Vaccinia and Variola virus. A preliminary report. *Indian Journal of Medical Research*, **57**, 495.

Ray, P. G. and S. K. Majumdar (1976) Antifungal flavonoid from *Alpinia officinarum* Hanc. *Indian Journal of Experimental Biology*, **14**, 712–714.

Ray, A. B., L. Chand and S. C. Dutta (1975) Salvadourea – New urea derivatives from *Salvadora persica*. *Chemistry and Industry*, **12**, 517–18.

Raymond-Hamet, R. (1940) Sur la crossoptine. *Bulletin des Sciences Pharmacologiques*, **47**, 194–202.

Ridet, J. and A. Chartol (1964) Les propriétés antidysenteriques de l'*Euphorbia hirta*. *Médecine Tropicale*, **24**, 119–143.

Ross, S. A., S. E. Megalla, D. W. Bishay and A. H. Awad (1980) Studies for determining antibiotic substances in some Egyptian plants. Part II Antimicrobial alkaloids of *Peganum harmala* L. *Fitoterapia*, **51**, 309–12.

Rouffiac, C. R. and J. Parello (1969) Etude chimique des alcaloides de *Phyllanthus niruri* L. Présence de l'antipode optique de la norsecurinine. *Plantes Médicinales et Phytothérapie*, **3**, 220–3.

Rudloff, E. von (1969) Chemical composition of the leaf oil of *Ageratum conyzoides* L. *Perfumery and Essential Oil Record*, 303–4.

Saint Rat, L. de and Ph. Luteraan (1947) Action antibiotique à l'égard de champignons pathogènes pour l'homme. *Comptes Rendus de l'Académie des Sciences*, **224**, 1587–9.

Saint Rat, L. de, H. R. Olivier and J. Chouteau (1946) Propriétes antibiotiques de la plumbagine. *Bulletin de l'Académie de Médicine*, **130**, 57–60.

Saint Rat, L. de, H. R. Olivier and J. Chouteau (1948) Propriétes antibiotiques de la plumbagine. *Bulletin de l'Académie de Médicine*, **132**, 125–8.

Sainty, D., F. Bailleul, P. Delaveau and H. Jacquemin (1981) Iridoides du *Borreria verticillata*. *Planta Medica*, **42**, 260–4.

Sakkawala, O. P. Shukla and C. R. Krishnamurti (1962) Bacteriolytic activity of plant latexes. *Journal of Scientific and Industrial Research*, **21**, 279–349.

Sankara Subramanian, S., S. Najarajan and N. Sulochana (1971) Flavonoids of some Euphorbiaceous plants. *Phytochemistry*, **10**, 2548–9.

Sannie, C., H. Lapin and I. P. Varsney (1963) Sur les sapogenines de l'*Albizia lebbeck* Benth. *Bulletin de la Société Chimique de France*, 1440–4.

Sato, N. and T. Muro (1974) Antiviral activity of scillarenin, a plant bufadienolide. *Japanese Journal of Microbiology*, **18**, 441.

Satyanarayan Murty, K., D. Narayana Rao and D. Krishna Rao (1978) A preliminary study on hypoglycaemic and antihyperglycaemic effects of *Azadirachta indica*. *Indian Journal of Pharmacology*, **10**, 247–50.

Schabort, J. C. (1978) Cucurbitacin-19-hydroxylase in *Cucurbita maxima*. *Phytochemistry*, **17**, 1062–4.

Schulte, K. E., G. Rücker and H. U. Matern (1979) Ueber einige Inhaltstoffe der Früchte und Wurzel von *Melia azadarach* L. *Planta Medica*, **35**, 76–83.

Schultes, R. E. (1979) Medicinal uses of *Swartzia madagascariensis* in the North-West Amazon. *Journal of Ethnopharmacology*, **1**, 79–87.

Seida, A. A., A. D. Kinghorn, G. A. Cordell and N. R. Farnsworth (1981) Isolation of Bergapten and Marmesin from *Balanites aegyptiaca*. *Planta Medica*, **43**, 92–103.

Sen, A. B. and Y. N. Shukla (1968) Chemical examination of *Cassia fistula*. *Journal of the Indian Chemical Society*, **45**, 744.

Sen, H. G., B. S. Joshi, P. C. Parthasarathy and V. N. Kamat (1974) Anthelmintic efficacy of diospyrol and its derivatives. *Arzneimittel forschung*, **24**, 2000–3.

Sen Gupta, K. P., N. C. Ganguli and N. R. S. Bhattacharya (1956) Bacteriological and pharmacological studies of a vibriocidal drug derived from an indigenous source. *Antiseptic*, **53**, 287–92.

Sethi, P. D., P. C. Ravindran, K. B. Sharma and S. S. Subramanian (1974) Antibacterial activity of some C_{28} steroidal lactones. *Indian Journal of Pharmacy*, **36**, 122.

Shah, C. S. and M. V. Shinde (1969) Phytochemical studies of seeds of *Cassia tora* L. and *C. occidentalis* L. *Indian Journal of Pharmacy*, **32**, 70.

Shah, C. S., S. M. J. S. Quadry and M. P. Tripathi (1968) Indian *Cassia* spp. II, Pharmacognostical and phytochemical studies of leaves of *Cassia tora* L. and *C. occidentalis* L. *Indian Journal of Pharmacy*, **30**, 282.

Shankaranarayan, D., C. Gopalkrishnan and C. Kameswaran (1979) Pharmacology of mangiferin. *Indian Journal of Pharmaceutical Sciences*, **41**, 78–9 and *Archives Internationales de Pharmacologie et Thérapie* **239**, 257–9.

Shankaranarayana, K. H., K. S. Ayar and G. S. K. Krishna Rao (1980) Insect growth inhibitor from the bark of *Santalum album*, *Phytochemistry*, **19**, 1239–40.

Shihata, M., A. Y. El-Gendi and M. M. Abd El Malik (1977) Pharmaco-chemical studies on saponin fraction of *Opilia celtidifolia*. *Planta Medica*, **31**, 60–7.

Shukla, B. and C. R. Krishnamurti (1961a) Bacteriolytic activity of plant latexes. *Journal of Scientific and Industrial Research India*, **20C**, 109–12.

Shukla, O. P. and C. R. Krishnamurti (1961b) Properties and partial purification of bacteriolytic enzyme from the latex of *Calotropis procera* (Mudar.). *Journal of Scientific and Industrial Research*, **20C**, 225–6.

Shukla, S. D., N. T. Modi and B. S. Deshmankar (1973) Pharmacological action of an alkaloidal fraction from *Cardiospermum halicacabum*. *Indian Journal of Pharmacy*, **35**, 40–1.

Siguier, F., M. Piette and J. Crosnier (1949) Etude experimentale comparative de la conessine et de l'emetine. *Médicine Tropicale*, **9**, 99–109.

*Simeray, J., J. P. Chaumont and G. Maréchal (1981) Propriétés antagonistes de cent extraits de plantes supérieures vis à vis de vingt champignons parasites de l'homme et des végétaux. *Plantes Médicinales et Phytothérapie*, **15**, 183–8.

Singh, L. B. and J. Bose (1961) Ethyl gallate in panicles of *Mangifera indica*. *Journal of Scientific and Industrial Research India*, **20B**, 296.

Skinner, F. A. (1955) The antibiotics. In *Modern Methods of Plant Analysis*, ed. K. Peach and H. V. Tracey, Vol. 3, Springer, New York, pp. 626–725.

Small, L. D., J. H. Bailey and C. J. Cavalitto (1949) Alkylthiosulphinates. *Journal of the American Chemical Society*, **69**, 1710.

*Sofowora, A. (1980) The present status of knowledge of the plants used in traditional medicine in Western Africa. A medical approach and a chemical evaluation. *Journal of Ethnopharmacology*, **2**, 109–18.

Sofowora, E. A. and A. A. Olaniji (1975) Phytochemical examination of *Dracaena manni* stembark. *Planta Medica*, **27**, 65–7.

Sofowora, E. A., W. A. Isaac-Sodeye and L. O. Ogunkoya (1975) Isolation and characterization of an antisickling agent from *Fagara zanthoxyloides* root. *Lloydia*, **38**, 169–71.

Soulimov, A. A., A. Bouquet, B. A. Glouchko, F. P. Filatov and A. Fournet (1975) Recherches preliminaires sur l'action viricide et antivirus de quelques plantes médicinales du Congo Brazzaville. *Plantes Médicinales et Phytothérapie*, **9**, 171–81.

Spencer, C. F., F. R. Koniuszi, E. F. Rogers, J. Shavel, N. R. Easton, A. Kaczka, F. A. Kuehl, R. F. Phillips, A. Walti, K. Folkers, C. Malanga and O. Seeler (1947a) Survey of plants for antimalarial activity. *Lloydia*, **10**, 145–60.

Spencer, C. F., F. R. Koniuszi, E. F. Rogers, J. Shavel, N. R. Easton, A. Kaczka, F. A. Kuehl, R. F. Phillips, A. Walti, K. Folkers, C. Malunga and O. Seeler (1947b) Survey of plants for antimalarial activity. *Lloydia*, **10**, 160–74.

Srivasta, M. C. and S. W. Singh (1967) Anthelmintic activity of *Curcurbita maxima* (Kaddu) seeds. *Indian Journal of Medical Research*, **55**, 629.

Stanislas, E., R. Rouffiac and J. J. Fayard (1967) Constituents of *Phyllanthus niruri* L. (Euphorbiaceae) *Plantes Médicinales et Phytothérapie*, **1**, 136–41.

Steenis-Kruseman, van, M. J. (1953) Selected Indonesian medicinal plants. *Bulletin of the Organization of Scientific Research Indonesia*, No. **18**, 1–90.

Steffen, K. and H. Peschel (1975) Chemische Konstitution und antifungale Wirkung von 1–4 Naphtochinonen und chemisch verwandter Stoffe. *Planta Medica*, **27**, 201–16.

Sticher, O. and B. Meier (1978) Quantitative Analyse und Isolierung von Aucubin und Catapol aus *Folium plantaginis* mit HPLC. *Planta Medica*, **33**, 295–6.

Stipanovic, R. D., A. A. Bell, M. A. Mace and C. R. Howell (1975) Antimicrobial terpenoids of *Gossypium:* 6-methoxygossypol and 6,6′-dimethoxygossypol. *Phytochemistry*, **14**, 1077.

Su, K. L., Y. Abul Hajj and E. J. Staba (1973) Antimicrobial principle from *Nymphea tuberosa*, *Lloydia*, **36**, 80–7.

Su, K. L., Y. Abul Hajj and E. J. Staba (1975) Antimicrobial effect of aquatic plants from Minnesota. *Lloydia*, **38**, 175–338.

Sullivan, J. T., C. S. Richards, H. A. Lloyd and G. Krishna (1982) Anacardic acid, molluscicide in cashew nutshell liquid. *Planta Medica*, **44**, 175–7.

Takechi, M. and Y. Tanaka (1981) Purification and characterization of antiviral substances from the bud of *Syzygium aromatica*. *Planta Medica*, **42**, 69–74.

Tanguy, F., Ch. Robin and A. Raoult (1948) Nouveaux essais de traitement de la dysenterie amibienne par le chlorhydrate de conessine. *Médecine Tropicale*, **8**, 12–31.

Taylor, A. H. (1979) A limonoid, pseudrelone B from *Pseudocedrela kotschyi*. *Phytochemistry*, **18**, 1574–6.

Tewari, D. and J. Rajbehari (1972) Chemical examination of the roots of *Cassia tora*. *Planta Medica*, **21**, 393–7.

Tewari, J. P. and M. C. Srivasta (1968) Fractionating of fatty acids of *Curcurbita maxima* seedoil with urea. *Journal of Pharmaceutical Sciences, USA*, **57**, 328–9.

Tewari, R. D., B. B. Dixit and S. S. Mishra (1965) Fractionating of fatty acids of *Cassia tora* seed oil with urea. *Journal of Pharmaceutical Sciences, USA*, **54**, 923.

Thompson, I. B. and C. C. Anderson (1978) Cardiovascular effects of *Azadirachta indica* extract. *Journal of Pharmaceutical Sciences*, **67**, 1476–8.

Thomson, F. J., C. J. Morris and J. E. Hunt (1964) The identification of L-α-δ-amino hydroxyvaleric acid and L-homoserine in Jack bean seed (*Canavalia ensiformis*). *Journal of Biological Chemistry*, **239**, 1122–6.

Tomassini, T. C. B. and M. E. O. Mathos (1979) On the natural occurrence of 15-α-tiglinoyloxy-kaur-16-en-19-oic acid. *Phytochemistry*, **18**, 663–4.

Towers, G. H. N. and C. K. Wat (1979) Phenylpropanoid metabolism. Review. *Planta*, **37**, 97–114.

Tripathi, W. J. and B. Dasgupta (1974) Neutral constituents of *Albizza lebbeck*. *Current Science*, **43**, 46.

Tubery, P. R. (1968) Heteroside from *Lasiosiphon kraussianus* roots as a cure for lepra. French Patent No. 6366 of 12.11.1968 and 7333 of 24.11.1969.

Tynecka, Z. and Z. Gos (1973) The inhibitory action of garlic (*Allium sativum*) on growth and respiration of some micro organisms. *Acta Microbiologica Polonica*, Series B, **5**, 51–62.

Uiso, F. W. (1979) Traditional Medicine Research Unit, Annual report. Muhimbili Medical centre, University of Dar es Salaam, Tanzania.

Valnet, J., Ch. Duraffoud, P. Duraffoud and J.-Cl. Lapraz (1978) L'aromatogramme: Nouveaux resultats et essai d'interprètation sur 268 cas cliniques. *Plantes Médicinales et Phytothérapie*, **12**, 43–52.

Veliky, I. A. and K. Genest (1972) Growth and metabolites of *Cannabis sativa* cell suspension cultures (antimicrobial effects). *Lloydia*, **35**, 450–6.

Venkataramann, S. and N. Radnakrishnan (1972) Antifungal activity of *Asteracantha longifolia*. *Indian Journal of Pharmacology*, **4**, 148.

Verma, V. S. (1974) Chemical compounds from *Azadirachta indica* as inhibitors of potato virus X. *Acta Microbiologica Polonica Applicata (B)*, **6**, 9–13.

Verpoorte, R., E. W. Kode, H. van Doorne and A. B. Svendsen (1978) Antimicrobial effect of the alkaloids from *Strychnos afzelii* Gilg. *Planta Medica*, **33**, 237–42.

Verpoorte, R., T. A. van Beek, P. H. A. M. Thomassen, J. Aandewiel and A. Baerheim-Svendsen (1983) Screening of antimicrobial activity of some plants belonging to the Apocynaceae and Loganiaceae. *Journal of Ethnopharmacology*, **8**, 287–302.

Vichkanova, S. A. and L. V. Gorunyunova (1972) Antiviral properties of gossypol with respect to influenza virus of the A2 Frunze strain. Through *Chemical Abstracts*, **82**, 149162e.

Vichkanova, S. A. and L. D. Shipulina (1972) Antiherpes properties of gossypol in experimental herpetic keratitis. Through *Chemical Abstracts*, **82** (1975) 133505h.

Vichkanova, S. A., V. Adgina, L. V. Gorunyunova and M. A. Rubenchik (1973a)

Chemotherapeutic properties of plumbagin isolated from *Plumbago europeae*. *Chemical Abstracts*, 66906s.

Vichkanova, S. A., L. V. Makarova and N. I. Gordeikina (1973b) Tuberculostatic activities of preparations from plants. Fitontsidy Mater. Soveshch. 6th 1969. *Chemical Abstracts*, 78, 66905i.

Vigneron, J. P. (1978) Antiappetant substances of natural origin. *Annales Zoologie et Ecologie Animale*, **10**, 663–94.

Vijver, L.-M. van der and A. R. Lötter (1971) The constituents in the roots of *Plumbago auriculata* Lam. and *P. zeylanica* L. responsible for antibacterial activity of local plants. *Planta Medica*, **20**, 8–13.

Vleggar, H. C., J. Kruger, T. M. Smalberger and A. J. van den Berg (1978) Flavonoids from *Tephrosia* XI. The structure of glabratepherin. *Tetrahedron Letters*, **34**, 1405.

Vohora, S. B., M. Rizwan and J. A. Khan (1973) Medicinal uses of common Indian vegetables. *Planta Medica*, **23**, 381–93.

Wacker, A. and H. G. Eilmes (1975) Zur Virushemmung mit hesperidin. *Naturwissenschaften*, **62**, 301.

Wat, C. K., T. Johns and G. H. N. Towers (1980) Phototoxic and antibiotic activities of plants of the Asteraceae used in folk medicine. *Journal of Ethnopharmacology*, **2**, 279–90.

Wat, C. K., R. K. Biswas, E. A. Graham, L. Bohm and G. H. N. Towers (1978) U.V. mediated antibiotic activity of phenylheptatriene in *Bidens pilosa*. *Planta Medica*, **33**, 309–10.

Wat, C. K., R. K. Biswas, E. A. Graham, L. Bohm, G. H. N. Towers and E. R. Waygood (1979) Ultraviolet mediated cytotoxic activity of phenylheptatriene from *Bidens pilosa* L. *Journal of Natural Products (Lloydia)*, **42**, 103–11.

Watt, J. M. and M. G. Breyer-Brandwijk (1962) *The Medicinal and Poisonous Plants of Southern and Eastern Africa*. Livingstone, Edinburgh and London, 1 vol., 1057 pp.

Webb, L. J. (1948a) *Guide to the Medicinal and Poisonous Plants of Queensland*, 1 vol., Council for Scientific and Industrial Research, no. 232, Melbourne.

Webb, L. J. (1948b) *Carica papaya*. *Council of Scientific and Industrial Research Organization (Australia) Bulletin*, 1948, p. 232.

Willis, E. D. (1956) Enzyme inhibition by allicin, the active principle of garlic. *Biochemical Journal*, **63**, 514.

Worsley, R. R. le G. (1936) *Mundulea sericea*. *Annals of Applied Botany*, **23**, 311.

Worsley, R. R. le G. (1937) *Mundulea sericea*. *Annals of Applied Botany*, **24**, 651.

Wu, W. N., J. L. Beal, G. W. Clark and L. A. Mitscher (1976) Additional alkaloids and antimicrobial agents from *Thalictrum rugosum*. *Lloydia*, **39**, 65–75.

Wu, W. N., J. L. Beal, A. Mitscher, K. N. Salman and P. Patel (1976) Isolation and identification of the hypotensive alkaloids of *Thalictrum lucidum*. *Lloydia*, **39**, 204–12.

*Xaio, P. (1983) Recent developments of medicinal plants in China. *Journal of Ethnopharmacology*, **7**, 95–109.

Yamada, Y., K. Hagiwara, K. Iguchi, Y. Takahasi and H. Y. Hsu (1978) Cucurbitacins from *Anagallis arvensis*. Triterpenoids, cucurbitacines B, D, E, I, L and R bitter principles. *Phytochemistry*, **17**, 1798.

Yousef, F. (1973) Separation and characterization of a new alkaloid from the fruit of *Duranta repens*. *Planta Medica*, **23**, 73–5.

Zenan, I. and J. Podkorny (1963) Gas chromatography of cyclopentenyl fatty acids. *Journal of Chromatography*, **10**, 15.

Zyl, J. J. van, G. J. H. Rall and D. G. Roux (1979) The structure, absolute configuration, synthesis and ^{13}C NMR spectra of phenylated pyranoflavonoids from *Mundulea sericea*. *Journal of Chemical Research (South Africa)*, (5) 97.

Chapter 5: Hormone secretion in Man

Abraham, A., I. Kirson, D. Lavie and E. Glotter (1975) Constituents of *Withania somnifera* XIV. Withanolides of *W. somnifera*, chemotypes I and II. *Phytochemistry*, **14**, 189–94.

Abu-Mustafa, E. A., M. B. E. Fayez, A. M. Gad and F. Osman (1960) Isolation of βcholesterol from Chufa (*Cyperus esculentus*) tubers. *Journal of Organic Chemistry*, **25**, 1269.

Adesina, S. K. and E. Etté (1982) The isolation and identification of anticonvulsant agents from *Clausena anisata* and *Afraegle paniculata*. *Fitoterapia*, **53**, 63–6.

Adesina, S. K. and E. A. Sofowora (1979) The isolation of an anticonvulsant glycoside from the fruit of *Tetrapleura tetraptera*. *Planta Medica*, **36**, 270–1.

Adinarayana, D. and T. R. Seshadri (1965) Chemical compounds of Indian seeds of *Calophyllum inophyllum*. The structure of new 4-phenyl-coumarin ponnalide. *Bulletin of the National Institute of Science, India*, **31**, 91.

Adjangba, M. S., W. A. Asomaning and W. R. Phillips (1974) Further contribution to the study of the extractives of *Afraegle paniculata* (Schum. & Thonn.) Part III. *Ghana Journal of Sciences*, **14**, 137–40.

Adjangba, M. S., W. A. Asomanning, A. Barranco, R. T. Bone and W. R. Phillips (1975) Pharmacological activities of coumarins isolated from *Afraegle paniculata*. Part II. *West African Journal of Pharmacology and Drug Research*, **2**, 83–6.

Adjanohoun, E. J. (1980) *Médicine Traditionelle et Pharmacopée. Contribution aux Etudes Ethnobotaniques et Floristiques au Niger*, ed. Agence de Coopération culturelle et technique, Paris, 1 vol., 250 pp.

Ainslie, J. R. (1937) *List of Plants used in Native Medicine in Nigeria*, Oxford University Press, 109 pp.

Arman, C. C. van, A. J. Begamy, L. M. Miller and H. H. Pless (1965) Some details of the inflammations caused by yeast and carrageenan. *Journal of Pharmacology and Experimental Therapy*, **150**, 328–34.

Arora, R. B., C. N. Mathur and S. D. S. Seth (1962) Anticoagulant and cardiovascular actions of calophyllolide, an indigenous complex coumarin. *Archives Internationales de Pharmacodynamie*, **139**, 75.

Arora, R. B., N. Basu, V. Kapoor and A. P. Jain (1971) Anti-inflammatory studies on *Curcuma longa* (turmeric). *Indian Journal of Medical Research*, **59**, 1289–95.

Arora, R. B., V. Kapoor, S. K. Gupta and R. E. Shanma (1972) Isolation of a crystalline steroidal compound from *Commiphora mukul* and its anti-inflammatory activity. *Indian Journal of Experimental Biology*, **9**, 403–4.

Awouters, F., C. J. E. Niemeegers, F. M. Lenaerts and P. A. J. Janseen (1978) Delay of Castor oil diarrhoea in rats: a new way to evaluate inhibitors of prostaglandin biosynthesis. *Journal of Pharmacy and Pharmacology*, **30**, 41–5.

Ayoub, S. M. H. and A. B. Svendsen (1981) Medicinal and aromatic plants in the Soudan. usage and exploration. *Fitoterapia*, **52**, 245.

Bach, G. L. (1978) Activity of βsitosterol against prostata adenoma and rheumatoid problems. 'Euromed', No. 1.

Bär, V. and R. Hänsel (1982) Immunomodulating properties of 5,20α(R)dihydroxy 6α,7α-epoxy-1-oxo-(5α)witha-2,24-dienolide and solasodine. *Planta Medica*, **44**, 32–3.

Bamgbose, S. O. A. and B. K. Noamesi (1981) Studies on cryptolepine II. Inhibition of carrageenan-induced oedema by cryptolepine. *Planta Medica*, **42**, 392–6.

Beisler, J. A. and Y. Sato (1971) Chemistry of the solanidane ring system. *Journal of Organic Chemistry*, *C*, 149–52.

Benitz, K. F. and L. M. Hall (1963) The Carrageenin-induced abscess as a new test for anti-inflammatory activity of steroids and non steroids. *Archives Internationales de Pharmacodynamie*, **144**, 185–95.

Benoit, P. S., H. H. S. Fong, G. H. Swoboda and N. R. Farnsworth (1976) Biological and phytochemical evaluation of plants. XIV. Anti-inflammatory evaluation of 163 spp. of plants. *Lloydia*, **39**, 160–171.

Bhalla, T. N., M. B. Gupta and K. P. Bhargava (1971) Anti-inflammatory and biochemical study of *Boerhavia diffusa*. *Journal of Research on Indian Medicine*, **6**, 11.

Bhalla, T. N., M. B. Gupta, P. K. Sheth and P. K. Bhargava (1968) Anti-inflammatory activity of *Boerhavia diffusa*. *Indian Journal of Physiology and Pharmacology*, **12**, 31.

Bhalla, T. N., R. C. Saxena, S. K. Nigam, G. Misra and K. P. Bhargava (1980) A new non-steroidal inflammatory agent, calophyllolide. *Indian Journal of Medical Research*, **72**, 762–5.

Bhattacharya, T. K., M. N. Ghosh and S. S. Subramanian (1980) A note on anti-inflammatory activity of carpestrol. *Fitoterapia*, **51**, 265–7.

Bhattacharya, S. K., A. K. Parikh, P. K. Debnath, V. B. Pandey and N. C. Neogy (1973) Pharmacological studies with the alkaloids of *Costus speciosus*. *Journal of Research on Indian Medicine*, **8**, 10–19.

Bhusan, B., S. Rhangaswani and T. R. Seshadri (1975) Calaustrin, a new 4-phenyl coumarin from the seed oil of *Calophyllum inophyllum* L. *Indian Journal of Chemistry*, **13**, 746–7.

Bianchi, M. (1962) Interaction entre la reserpine et l'axe hypophyso medulo surrénalien. *Comptes rendus de la Société de Biologie*, **156**, 179.

Biswas, M., A. B. Ray and B. Das Gupta (1975a) Changes in rat adrenal medulla following Δ^9 tetrahydro cannabinol treatment. *Acta Endocrinologica (Copenhagen)*, **80**, 329–38.

Biswas, M., A. B. Ray and B. Das Gupta (1975b) Chemical investigation of *Crateva nurvala*. Search for the anti-inflammatory principle. *Current Science*, **44**, 227–8.

Blanpin, O. and A. Quevauviller (1960a) Les effets hormonaux de quelques alcaloides steroïdiques de *Funtumia* et de *Holarrhena*. *Semaines des Hôpitaux*, **36**, 899–908.

Blanpin, O. and A. Quevauviller (1960b) Les effets hormonaux de quelques alcaloïdes steroïdiques de *Funtumia* et de *Holarrhena*. *Semaines des Hôpitaux*, **36**, 909–12.

Boakaiji Yiadom, K., N. I. Fiagbe and J. S. K. Ayim (1977) Antimicrobial activity of xylopic acid and other constituents of the fruits of *Xylopia aethiopica* (Annonaceae). *Lloydia*, **40**, 543–5.

Boiteau, P., A. R. Ratsimanga and B. Pasich (1964) Les triterpenoides en physiologie végétale et animale. Gauthier-Villars, Paris, 1 vol., 1370 pp.

Bombardelli, E., A. Bonati, B. Gabetta and G. Mustich (1974) Plants of Mozambique VII. Triterpenoids of *Terminalia sericea*. *Phytochemistry*, **13**, 2559–62.

Bombardelli, E., B. Gabetta, E. M. Martinelli and G. Mustich (1979) Quantitative evaluation of glycyrrhetic acid and gas chromatographic and mass spectrometric investigation of liquorice triterpenoids. *Fitoterapia*, **50**, 11–24.

Bouquet, A. (1969) *Féticheurs et Médicines Traditionnelles du Congo Brazzaville*, ed. ORSTOM, Paris, 1 vol., 282 pp.

Brekhman, I. I. and I. V. Dardymon (1969) Pharmacological investigation of glycosides from *Ginseng* and *Eleutherococcus*. *Lloydia*, **32**, 46–51.

Brodgen, R. N., T. M. Speight and G. S. Avery (1974) Deglycyrrhinised liquorice, a report of its pharmacological properties and therapeutic effect in peptic ulcer. *Drugs*, **8**, 330–9.

Busson, F. (1965) *Plantes Alimentaires de l'Ouest Africain*. Lecomte, Marseille, 1 vol., 568 pp.

Chakravarti, R. N., D. Chakravarti and R. Banerjee (1959) Triterpenes from *Crateva religiosa*. *Bulletin of Calcutta School of Tropical Medicine*, **7**, 105.

Chandra, D. and S. S. Gupta (1972) Anti-inflammatory and anti-arthritic activity of volatile oil of *Curcuma longa* (Haldi). *Indian Journal of Medical Research*, **60**, 138–42.

*Chaturvedi, A. K. and R. Singh (1965) Experimental studies of the anti-arthritic effect of certain indigenous drugs. *Indian Journal of Medical Research*, **53**, 71–80.

Chaturvedi, A. K., S. S. Parmar, S. C. Bhatnagar, G. Misra and S. K. Nigam (1974) Anti-convulsant and anti-inflammatory activity of natural plant coumarins and triterpenoids. *Research Communications in Chemical Pathology and Pharmacology*, **7**, 757; **9**, 11.

Chaturvedi, A. K., S. S. Parmar, S. K. Nigam, S. C. Bhatnagar, C. Misra and B. V. R. Sastry (1976) Anti-inflammatory and anti-convulsant properties of some natural plant triterpenoids. *Pharmacological Research Communications*, **8**, 199–210.

Chen-Yu, S., C. Hsiu-Chuan and L. Keng-Tao (1958) Pharmacology of gentianine I. Anti-inflammatory effect and action on pituitary adrenal function in rat. *Scheng Li Hsuch Pao*, **22**, 201–5 (though *Chemical Abstracts*, **53** (1959) 13415 g).

Chopra, R. N., S. L. Nayar and I. C. Chopra (1956) *Glossary of Indian Medicinal Plants*. Council of Scientific and Industrial Research, New Delhi.

Chul Kim, Chung, Chin Kim, Myung Suk Kim, Chang Yong Hu and Yong Su Rhe (1970) Influence of *Ginseng* on the stress mechanism. *Lloydia*, **33**, 43–8.

Cwalina, G. E. and G. L. Jenkins (1938) A phytochemical study of *Ipomea pes-caprae*. *Journal of the American Pharmaceutical Association*, **27**, 585–95.

Dalziel, J. M. (1937) *The Useful Plants of West Tropical Africa*. Crown Agents, London, 1 vol., 612 pp.

Dawalkar, M. P. and M. G. Dawalkar (1960) Studies in organic acid metabolism in *Ipomea pes-caprae*. *Journal of Biological Sciences, Bombay*, **3**, 86–91.

Dhar, M. L., M. M. Dhar, B. N. Dhawan, B. N. Mehrothra and C. Ray (1968) Screening of Indian plants for biological activity Part I. *Indian Journal of Experimental Biology*, **6**, 232.

Di Rosa, M., J. P. Giroud and D. A. Willoughby (1971) Study of the mediators of the acute inflammatory response induced in rats in different sites by carrageenan and turpentine. *Journal of Pathology*, **104**, 15–29.

Ekong, D. E. U. and O. G. Idemudia (1967) Constituents of some of the West-African members of the genus *Terminalia*. *Journal of the Chemical Society, C*, 863–4.

Ekong, D. E. U. and A. U. Ogan (1968) Chemistry of the constituents of *Xylopia aethiopica*. The structure of xylopic acid, a new diterpene acid. *Journal of the Chemical Society, C*, 311–12.

Ekong, D. E. U., E. O. Olagbeni and F. A. Odutola (1969) Further diterpenes from *Xylopia aethiopica* (Annonaceae). *Phytochemistry*, **8**, 1053–5.

Esdorn, I. (1961) *Alstonia* Arten als Arzneipflanzen. *Pharmaceutica Acta Helvetica*, **36**, 6–9.

Ezmirly, S. T., J. C. Cheng and S. R. Wilson (1979) Saudi Arabian medicinal plants. *Salvadora persica*. *Planta Medica*, **35**, 191–2.

Farnsworth, N. R. and G. A. Cordell (1976) A review of some biologically active compounds isolated from plants as reported in the 1974–75 literature. *Lloydia*, **39**, 420–55.

Farooqi, M. I. H. and J. G. Srivastava (1968) Toothbrush tree (*Salvadora persica*). *Quarterly Journal of Crude Drug Research*, **8**, 1297–9.

Ferreira, S. H. and J. R. Vane (1974) New aspects of the mode of action of non-steroid anti-inflammatory drugs. *American Review of Pharmacology*, **14**, 57–73.

Fuegner, A. (1973) Hemmung immunologisch bedingter Entzündungen durch das Pflanzensteroid Withaferin A. *Arzneimittel forschung*, **23**, 932–5.

Gaind, K. N. and T. R. Junega (1970) Investigations on *Capparis decidua* Edgew. Phytochemical study of flowers and fruits. *Research Bulletin of Punjab University, India*, **21**, 67–71.

Gal, I. E. (1964) Capsicidin, eine neue Verbindung mit antibiotischer Wirksamkeit aus Gewürzpaprika. *Lebensmittel Untersuchung und Forschung*, **124**, 533–6.

Gal, I. E. (1967) Nachweis und Bestimmung des Steroidsaponins capsicidin mit der Agardiffusions Methode. *Pharmazie*, **22**, 120–3.

Garg, S. K. (1971) Effect of *Curcuma longa* L. on fertility in female albino rats. *Bulletin P. G. I. (Chandigahr)*, **5**, 178.

Ghosal, S. B. and P. K. Bannerjee (1971) Indole alkaloids of *Desmodium gangeticum* DC. *Indian Journal of Pharmacy*, **30**, 280.

Ghosal, S. and S. K. Bhattacharya (1972) Desmodium alkaloids Part II. Chemical and Pharmacological evaluation of *Desmodium gangeticum*. *Planta Medica*, **24**, 434–40.

Ghosal, S. and K. Biswas (1979) Two new 1,3,5,6,7-pentoxygenated xanthones from *Canscora decussata*. *Phytochemistry*, **18**, 1029–31.

Ghosal, S., R. K. Chaudhuri and R. R. Chaudhuri (1971) Chemical constituents of the roots of *Canscora decussata* II. *Journal of the Indian Chemical Society*, **48**, 589.

Gibson, M. R. (1978) Glycyrrhiza in old and new perspectives. *Lloydia*, **41**, 348–53.

Goodman, L. S. and A. Gilman (1976) Hypothalamic control of the anterior pituitary. In *The Pharmacological Basis of Therapeutics*, Macmillan, New York, 5th edn, p. 1391.

Gopalakrishnan, C., D. Shankaranarayanan, S. K. Nazimudeen, S. Viswanathan and L. Kameswaran (1980) Anti-inflammatory and CNS depressant activities from *Calophyllum inophyllum* and *Mesua ferrea*. *Indian Journal of Pharmacology*, **12**, 181–91.

Govindachari, T. R., B. R. Pai, N. Mutukumaraswany, R. U. Rao and N. N. Rao (1968) Chemical components of the heartwood of *Calophyllum inophyllum* L. Part I Isolation of mesuaxanthone B and a new xanthone, calophyllin B. *Indian Journal of Chemistry*, **6**, 57.

Govindachari, T. R., N. Viswanathan, B. R. Pai, U. R. Rao and Srinivasan (1967) Triterpenes of *Calophyllum inophyllum* L. *Tetrahedron*, **23**, 1901–10.

Gujral, M. L., K. Sareen, K. K. Tangri, M. K. P. Amma and A. K. Roy (1960) Anti-arthritic and anti-inflammatory activity of gum guggul. *Indian Journal of Physiology and Pharmacology*, **4**, 267.

Gunde, B. G. and T. P. Hilditch (1939) The seed fats of *Salvadora oleoides* and *Salvadora persica*. *Journal of the Chemical Society*, 1015–16.

Gupta, S. S., D. Chandra and N. Mishra (1972) Antiinflammatory and anti-hyaluronidase activity of volatile oil of *Curcuma longa* (Haldi). *Indian Journal of Physiology and Pharmacology*, **16**, 264.

Gupta, M. B., R. N. Lal and Y. N. Shukla (1981) 5α-stigmast-9(11) en-3β-ol, a sterol from *Costus speciosus* roots. *Phytochemistry*, **20**, 2557–9.

Gupta, M. B., N. Singh, T. K. Palit and L. Bhargava (1970) Anti-inflammatory activity of an active constituent of *Cyperus rotundus*. *Indian Journal of Pharmacology*, **2**, 23.

Gupta, M. B., R. Nath, N. Srivasta, K. Shanker, K. Kishor and K. P. Bhatgava (1980) Anti-inflammatory and antipyretic activities of β-sitosterol. *Planta Medica*, **39**, 157.

Hager's Handbuch der Pharmazeutischen Praxis (1967–80) 4th edn, ed. P. L. List and L. Hörhammer: 12 vols, Springer, Berlin and New York.

*Harborne, J. B., T. J. Mabry and H. Mabry (1975) *The Flavonoids*, Chapman and Hall, London, 1 vol., 1204 pp.

Hardman, R. and E. A. Sofowora (1971) A reinvestigation of *Balanites aegyptiaca* as a source of steroidal sapogenins. *Economic Botany*, **26**, 169–73.

Hegnauer, R. (1962–68) *Chemotaxonomie der Pflanzen*, 5 vols., Birkhäuser Verlag, Basel, Stuttgart.

Hikino, H., T. Taguchi, H. Fugimura and Y. Hiramatsu (1977) Anti-inflammatory principles of *Caesalpinia sappan* wood and of *Haematoxylum campechianum* wood. *Planta Medica*, **31**, 214–20.

Hye, H. K. M. A. and M. A. Gafur (1975) Anti-inflammatory and anti-arthritic activity of a substance isolated from *Dalbergia volubilis*. *Indian Journal of Medical Research*, **63**, 93–100.

Idemudia, O. G. and D. E. U. Ekong (1970) Constituents of some West African members of the genus *Terminalia*. *Phytochemistry*, **9**, 2401.

Indian Council of Medical Research (1976) *Medicinal Plants of India*, Vol I, 487 pp. New Delhi.

Indira, M., P. S. N. Murthy and M. Sirsi (1956) Estrogenic activity, antibacterial action and some pharmacodynamic properties of *Cyperus rotundus* L. *Journal of the Mysore Medical Association*, **21**, 1.

Irvine, F. R. (1961) *Woody plants of Ghana*, 2nd edn. Oxford University Press, London, 868 pp.

Iwu, M. M. (1982) Steroidal constituents of *Costus afer*. *Planta Medica*, **44**, 413–15.

Iwu, M. M. and B. N. Anyanwu (1982a) Anti-inflammatory and anti-arthritic properties of *Terminalia ivorensis*. *Fitoterapia*, **52**, 25–34.

Iwu, M. M. and B. N. Anyanwu (1982b) Phytotherapeutic profile of Nigerian herbs I. Anti-inflammatory and anti-arthritic agents. *Journal of Ethnopharmacology*, **6**, 263–74.

Iwu, M. M. and F. C. Ohiri (1980) Antiarthritic triterpenoids of *Lonchocarpus cyanescens*. *Canadian Journal of Pharmaceutical Sciences*, **15**, 39–42.

Iyer, A., V. Bhasin and B. C. Joshi (1974) Chemical investigation of *Leptadenia spartium*. *Herba Polonica*, **20**, 321–4. (Syn. of *L. pyrotechnicum*).

Jacquemain, H. (1970, 1971) Recherches sur les anthocyanes foliaires de trois arbres tropicaux (*Mangifera indica* L., *Theobroma cacao* L., *Lophira alata* Banks ex Gaertn. Thèse Paris. *Plantes Médicinales et Phytothérapie*, (1970) **4**, 230–59, 306–41; (1971) **5**, 45–94.

Jentzch, K., P. Spiegel and L. Fuchs (1962) Untersuchungen über die Inhaltstoffe der Blätter von *Combretum micranthum* Don. *Planta Medica*, **10**, 1–8.

Kalsi, P. S., M. L. Chandi and I. S. Bhatia (1969) Chemical studies on the essential oil from *Cyperus rotundus* L. *Journal of Research of Punjab Agricultural University*, **6**, 383.

Kapadia, V. H., N. G. Naik, M. S. Wadia and S. Dev (1967) Sesquiterpenoids of *Cyperus rotundus*. *Tetrahedron*, **47**, 4661–7.

Keng-Tao Liu, C. Hsiu-Chuan and S. Chen-Yu (1959) *Scheng Li hsueh Pao*, **23**, 203 (through *Chemical Abstracts*, **57**, 15764).

Kerharo, J. (1968) Revue des plantes médicinales et toxiques du Sénégal. *Plantes Médicinales et Phytothérapie*, **2**, 108–46.

Kerharo, J. and J. G. Adam (1974) *La Pharmacopée Sénégalaise Traditionnelle*, Vigot Freres, Paris, 1011 pp.

Kishore, P. and S. N. Tripathi (1966) *Dalbergia lanceolaria* (Gaurakha) in the management of rheumatoid arthritis. A clinical and experimental evidence. *Journal of Research on Indian Medicine*, **1**, 29.

Kraus, S. D. (1949) Impairment of the pituitary adrenal response to acute stress in alloxan diabetes. *Acta Endocrinologica*, **54**, 328–34.

Kraus, S. D. (1960) Glycyrrhetinic acid, a triterpene with anti-oestrogenic and anti-inflammatory activity. *Journal of Pharmacy and Pharmacology*, **12**, 300–6.

Kumigai, A. (1969) Effect of glycyrrhizin on cortisone action. *Chemical Abstracts*, **70**, 95365e.

Kumigai, A., A. S. Nanaboshi, T. Yagura, K. Nishino and Y. Yamamura (1967a) Effect of glycyrrhizin on thymolytic and immunodepressive action of cortisone. *Endocrinology, Japan*, **14**, 39–42 (through *Chemical Abstracts*, **67**, 107072k).

Kumigai, A., K. Nishino, M. Yamamoto, M. Manaboshi and Y. Yamamura (1967b) An inhibiting effect of glycyrrhizin on metabolic actions of cortisone. Through *Chemical Abstracts*, **66**, 102232g.

Kumigai, A., Y. Asanuma, S. Yano, K. Tekeuchi, Y. Morimoto, T. Uemura and Y. Yamamura (1967c) Effect of glycyrrhizin on the suppressive action of cortisone on the pituitary adrenal axis. *Endocrinology, Japan*, **13**, 234–44 (through *Chemical Abstracts*, **66**, 27559q).

Kupchan, S. M., A. C. Patel and A. Fujita (1965) Tumor inhibitors VI. Cytotoxicity of bisbenzylisoquinoline alkaloids. *Journal of Pharmaceutical Sciences*, **54**, 580–3.

Lavie, D. and R. Taylor Smith (1963) Isolation of gentianine from *Anthocleista procera* (Loganiaceae). *Chemistry and Industry*, 1963, 781–2.

Lechat, P., F. Bisseliches, F. Bournerias, H. Dechy, Y. Juillet, G. Lagier, C. Merignac, B. Rouveix, P. Sterin and S. Weber (1978) *Pharmacologie Médicale*, 3rd edn, 1 vol., 677 pp., Masson, Paris.

McQuillin, F. J. (1951) The structure of cyperone. *Journal of Organic Chemistry*, **1951**, 716–18.

Manavalan, R. and B. M. Mithal (1980) Constituents of the aerial parts of *Leptadenia pyrotechnica*. *Planta Medica*, **39**, 95.

Manesch, B., S. S. Gupta and N. K. Gupta (1967) Further observations on the anti-inflammatory effects of a few indigenous drugs in relation to histamine and H. T. depletion. *Indian Journal of Physiology and Pharmacology*, **1967**, 11–14.

Matsuda, S., K. Oda, M. Kawaguchi and H. Hayashi (1962) Histochemical studies on the effect of glycyrrhizin on corticoid metabolism. *Nisshin Igaku*, **49**, 464–8 (through *Chemical Abstracts*, **57**, 15762c).

Menssen, H. G. and G. Stapel (1973) Uber ein C_{28} steroid lacton aus der Wurzel von *Withania somnifera*. *Planta Medica*, **24**, 8–12.

Misra, A. N. and H. P. Tiwari (1971) Constituents of the roots of *Boerhavia diffusa*. *Phytochemistry*, **10**, 3318.

Mitra, C. (1957a) *Callophyllum inophyllum* L Part I. Chemical constituents of nut oil and the stem bark. *Journal of Scientific and Industrial Research*, **6B**, 120.

Mitra, C. (1957b) *Calophyllum inophyllum* L. Part II. Constitution of inophyllic acid. *Journal of Scientific and Industrial Research*, **16B**, 167.

Molnar, J. (1965) Die pharmakologischen Wirkungen des Capscicins. *Arzneimittel-forschung*, **15**, 718–26.

Mudgal, V. (1975) Studies on medicinal properties of *Convolvulus pluricaulis* and *Boerhavia diffusa*. *Planta Medica*, **28**, 62–8.

Mukerjee, S. K., T. Saroja and T. R. Seshadri (1971) Dalbergichromene, a new neoflavonoid from stembark and heartwood of *Dalbergia sissoo*. *Tetrahedron*, **27**, 799–803.

*Murti, V. V. S. and T. R. Seshadri (1964) Toxic aminoacids in plants. A review (87 references). *Current Science (India)*, **33**, (11) 323–9.

Naish, Ch. (1954) *Thyroid for lactation*. *Lancet*, 1077–8.

Nityanand, S., N. K. Kapoor and S. Dev (1973) Cholesterol lowering activity of the various fractions of *Commiphora mukul* (guggul). *Indian Journal of Pharmacology*, **5**, 259; also *Indian Journal of Experimental Biology* (1971) **9**, 376.

Noamesi, B. K. and S. O. A. Bamgbose (1980) The α-adrenoceptor blocking properties of cryptolepine on the rat isolated *vas deferens*. *Planta Medica*, **39**, 51–6.

Noamesi, B. K. and S. O. A. Bamgbose (1982) Preferential blockade of presynaptic alpha-adrenoceptors on the rat isolated *vas deferens* by cryptolepine. *Planta Medica*, **44**, 241–5.

Nucifora, T. L. and M. H. Malone (1971) Comparative psychopharmacologic investigation of Cryogenine, certain non-steroid anti-inflammatory compounds, lupine alkaloids and cycloheptadine. *Archives Internationales de Pharmacodynamie*, **191**, 345–56.

Odebiji, O. O. (1978) Preliminary phytochemical and antimicrobial examinations of leaves of *Securidaca longepedunculata*. *Nigerian Journal of Pharmacy*, **9**, 29.

Odutola, F. A. and D. E. U. Ekong (1968) The chemistry of some traditional antitussive drugs of Nigeria. Communication. *Interafrican Symposium of Traditional Pharmacopoeias and African Medicinal Plants*, Dakar 1968.

Okpaniyi, S. N. and G. C. Ezeukwu (1981) Anti-inflammatory and antipyretic activities of *Azadirachta indica*. *Planta Medica*, **41**, 34–9.

Oriowo, M. A. (1982) Anti-inflammatory activity of piperonyl-4-acrylic-isobutyl-amide, an extractive from *Zanthoxylum zanthoxyloides*. *Planta Medica*, **44**, 54–6.

Paris, R. and P. Delaveau (1977) Métabolisme et pharmacocinétique des flavonoides. *Plantes médicinales et Phytothérapie*, **11** (Suppl.), 198–204.

*Paris, R. R. and H. Moyse (1965) *Précis de Matière Médicale*, Vol. I, 416 pp., Masson & Cie, Paris.

*Paris, R. R. and H. Moyse (1967) *Précis de Matière Médicale*, Vol. II, 511 pp., Masson & Cie, Paris.

Paris, R. R. and H. Moyse (1971) *Précis de Matière Médicale*, Vol. III, 509 pp., Masson & Cie, Paris.

Parmar, N. S. and M. N. Gosh (1978) Anti-inflammatory activity of gossypin, a bioflavonoid from *Hibiscus vitifolius* L. *Indian Journal of Pharmacology*, **10**, 277–93.

Parmar, S. S., K. K. Tangri, P. K. Seth and K. P. Bhargave (1964) Biochemical basis for antiinflammatory effects of glycyrrhetic acid and its derivatives. *International Congress of Biochemistry*, **6**, 410.

Patil, V. D., U. R. Nayak and S. Dev (1972) Chemistry of Ayurvedic crude drugs I. Guggulu (resin from *Commiphora mukul*) I Steroidal constituents. *Tetrahedron*, **8**, 2341–52.

Paulus, H. E. and M. W. Whitehouse (1973) Non steroid anti-inflammatory agents. *American Review of Pharmacology*, **13**, 107–25.

Pelter, P. and P. I. Amenechi (1969) Isoflavonoid and pterocarpinoid extractives of *Lonchocarpus laxiflorus*. *Journal of the Chemical Society, C*, 887–96.

Pillay, N. R. and G. Santhakumari (1981) Anti-arthritic and anti-inflammatory actions of nimbidin. *Planta Medica*, **43**, 59–63.

Pillay, N. R., D. Seeganthan, C. Seshadri and G. Santhakumari (1978) Anti-gastric ulcer activity of nimbidin. *Indian Journal of Medical Research*, **68**, 169–75.

Polonsky, J. (1957) Structure chimique du calophyllolide de l'inophyllolide et de l'acide calophyllique, constituents des noix de *Calophyllum inophyllum*. *Bulletin de la Société chimique de France*, **1957**, 1079–87.

Powell, J. W. and W. B. Whalley (1969) Triterpenoid saponins from *Phytolacca dodecandra*. *Phytochemistry*, **8**, 2105–7.

Prabhakar, M. C., H. Bano, S. Kumar, M. H. Shamsi and M. S. Y. Khan (1981) Pharmacological investigations of vitexin. *Planta Medica*, **43**, 396–403.

Prema, P. (1968) Pharmacological studies of *Desmodium gangeticum* DC. M.D. Thesis, University of Kerali, Trivandrum, India.

Quartey, J. A. K. (1963) Chemical examination of the fruit of *Afraegle paniculata* (Schum. & Thonn.) Engl. III The coumarin component. *Indian Journal of Applied Chemistry*, **26**, 17–18.

Quevauviller, A. and O. Blanpin (1960) Propriétés pharmacodynamiques comparées de l'Holaphylline et de l'holaphyllamine, alcaloides steroïdiques de l'*Holarrhena floribunda*. *Thérapie, 1960*, **15**, 1212–21.

Ramjilal, L., R. S. Rathor, R. Chakravarty and P. K. Das (1972) Preliminary studies on the anti-inflammatory and anti-arthritic activity of *Crataeva nurvalla (C. religiosa)*. *Indian Journal of Pharmacology*, **4**, 122.

Ramprasad, C. and M. Sirsi (1956) Studies on Indian medicinal plants: *Curcuma longa* L. Effect of curcumin and of essential oil of *C. longa* on bile secretion. *Journal of Scientific and Industrial Research*, **15C**, 262.

Rao, G. S., J. E. Sinsheimer and H. M. McIlhenny (1972) Structure of gymnamine a trace alkaloid of *Gymnema sylvestre* leaves. *Chemistry and Industry (London)*, 537–8.

Reisch, J., I. Mester and E. A. Sofowora (1980) Seltene Cumarine aus *Citrus nobilis*. *Planta Medica (Suppl.)*, 56–9.

Rigaud, A., J. Brisou and R. Babin (1956) Thérapeutiques locales à base de papaîne. *Presse Médicale*, **64**, 722.

Robinson, M. (1947) Hormones and lactation. Dried thyroid gland. *Lancet*, 385–7.

Roshchin, Y. V. and G. I. Geraschenko (1973) Anti-inflammatory activity of some flavonoids. *Vopr. Farm. Dal'nem Vostola*, **1**, 134–5 (through *Chemical Abstracts*, **83**, 376837).

Sadritdinov, F. (1971) Antiinflammatory activity of bisbenzylisoquinoline alkaloids from *Thalictrum* plants in relation to their chemical structure. *Farmakol. Alkaloidov. Serdechnych Glykosidov*, **1971**, 122–5 (through *Chemical Abstracts*, **78**, 79555b).

Santhakumari, G., M. L. Gurjal and K. Sareen (1964) Further studies on the anti-arthritic and anti-inflammatory activities of gum gugul. *Indian Journal of Physiology and Pharmacology*, **8**, 36.

Satoda, I. and E. Yoshi (1962) Structure of photosantoninic acid. *Tetrahedron Letters*, **1962**, 331–7.

Satyavati, G. V., C. Dwarkanath and S. N. Tripathi (1969) Experimental studies on the hypocholesterinic effect of *Commiphora mukul* Engl. *Indian Journal of Medical Research*, **57**, 1950–62.

Saxena, R. C., R. Nath, G. Palit, S. K. Nigam and K. P. Bhargava (1982) Effect of calophyllide, a non steroidal anti-inflammatory agent, on capillary permeability. *Planta Medica*, **44**, 246–8.

Sayed, I. Z. and D. D. Kanga (1936) *Proceedings of the Royal Academy of Sciences*, **4A**, 255 (through Bhattacharya *et al.*, 1980).

Shankaranarayan, D., C. Gopalukrishnan and C. Kameswaran (1979) Pharmacology of mangiferin. *Indian Journal of Pharmaceutical Sciences*, **41**, 78–9 and *Archives Internationales de Pharmacologie et Thérapie*, **239**, 257–9.

Shellard, E. J. (1962) A pharmacognostical comparison of the rhizomes and roots of *Ipomea turpethum* R. Br. and *Ipomea biloba* Forsk. (*I. pes caprae*). *Proceedings of the XXI International Congress of Pharmaceutical Sciences*, Federation Internationale de Pharmacie, Pisa, 1961.

Shin, K. H., W. S. Woo and C. K. Lee (1979) Antiinflammatory action of phytolaccosides. *Soul Taehakkyo Saengyak Yonguso Opjukjip*, **18**, 90–4 (through *Chemical Abstracts*, **93** (1980) 215485t).

Shohat, B., I. Kirson and D. Lavie (1978) Immunosuppressive activity of two plant steroidal lactones, withaferine A and withanolide E. *Biomedicine*, **28**, 18–24.

Singh, R. H. and Chaturvedi, G. N. (1966) Further studies on the anti-arthritic effect of an indigenous drug *Dalbergia lanceolaria*. *Indian Journal of Medical Research*, **54**, 363–7.

Singh, R. H. and K. N. Udupa (1972a) Studies on the Indian indigenous drug Punarnava (*Boerhavia diffusa* L.) Part II. Preliminary phytochemical studies. *Journal of Research on Indian Medicine*, **7**, 13.

Singh, R. H. and K. N. Udupa (1972b) Studies on the Indian indigenous Punarnava (*B. diffusa* L.) Part III. Experimental and pharmacological studies. *Journal of Research on Indian Medicine*, **7**, 17.

Singh, R. H. and K. N. Udupa (1972c) Studies on the Indian indigenous drug Punarnava (*B. diffusa* L.) Part IV. Preliminary controlled clinical trial in nephrotic syndrome. *Journal of Research on Indian Medicine*, **7**, 28.

Sofia, R. D., S. D. Nalepa, H. B. Vassar and L. C. Knobloch (1974) Comparative antiphlogistic activity of Δ^9-tetrahydrocannabinol, hydrocortisone and aspirin in various rat paw edema models. *Life Science*, **15**, 251.

Srimal, R. C. and C. N. Dhawan (1973) Pharmacology of diferuloylmethane (curcumin), a non steroidal anti-inflammatory agent. *Journal of Pharmacy and Pharmacology*, **25**, 447–52.

Srivasta, D. N., R. H. Singh and K. N. Udupa (1972) Studies on the Indian indigenous drug Punarnava (*Boerhavia diffusa* L.) Part V. Isolation and identification of a steroid. *Journal of Research on Indian Medicine*, **7**, 34.

Subramanian, S. S. and S. Ramakrishnan (1965) Chemical differences between *Boerhavia diffusa* and *B. punarnava*. *Indian Journal of Pharmacy*, **27**, 41–2.

Tangri, K. K., P. K. Seth, S. S. Parmar and K. P. Bhargava (1965) Biochemical study of anti-inflammatory and anti-arthritic properties of glycyrrhetic acid. *Biochemistry and Pharmacology*, **14**, 1277–81.

Torto, F. G. (1961) The composition of *Afraegle paniculata* mucilage. *Journal of the Chemical Society*, 1961, 5234–6.

Tripathi, S. N. and P. Kishore (1967) Studies on the anti-inflammatory activity of a phytogenic principle of *Dalbergia lanceolaria* (a preliminary report). *Journal of Research on Indian Medicine*, **1**, 155.

Tripathi, R. M., S. S. Gupta and D. Chandra (1973) Anti-trypsin and anti-hyaluronidase activity of *Curcuma longa*. *Indian Journal of Pharmacology*, **5**, 260.

Wasuwat, S. (1970) Extract of *Ipomea pes-caprae* (Convolvulaceae) antagonistic to histamine and to jellyfish poison. *Nature (London)*, **225**, 758.

Wehmer, C. (1935) *Die Pflanzenstoffe*, Vols I and II, 1930, Fischer, Jena; Ergänzungsband, 1934; Supplement, 1935.

Whitehouse, M. W. (1965) Some biochemical and pharmacological properties of anti-inflammatory drugs. *Progress in Drug Research*, **8**, 301–429.

Whitehouse, M. W., P. D. G. Dean and T. G. Halshall (1967) Uncoupling of oxidative phosphorylation by glycyrrhetic acid, fusidic acid and some related triterpenoid acids. *Journal of Pharmacy and Pharmacology*, **19**, 533–44.

Winter, C. A., E. A. Risley and C. W. Nuss (1962) Carrageenan induced edema in hind paw of the rat as an assay for anti-inflammatory drugs. *Proceedings of the Society for Experimental Biology and Medicine*, **111**, 544–7.

Woo, W. S. and S. S. Kang (1978) Constituents of *Phytolacca* species III. Components of over-ground parts and callus tissues. *Chemical Abstracts*, **88**, 47504z.

Woo, W. S. and K. H. Shin (1976) Antiinflammatory action of *Phytolacca* saponins. *Yakhac Hoe Chi* (20) **3**, 149–55 (through *Chemical Abstracts*, **86**, 50644j).

Woo, W. S., K. H. Shin and S. S. Kang (1976) Constituents of *Phytolacca* species I–Anti-inflammatory saponins. *TaehakkeyoSaengyak Yonguso Opjukjip*, **15**, 103–6 (through *Chemical Abstracts*, **88**, (1978) 146072p.

Woodbury, D. M. and E. Fingl (1975) Analgesic-antipyretics, anti-inflammatory agents and drugs employed in the therapy of gout. In *The Biological Basis of Therapeutics*, 5th edn, ed. L. S. Goodman and A. Gilman, pp. 325–58, Macmillan, New York, 1975.

Xaio, P. (1983) Recent developments on medicinal plants in China. *Journal of Ethnopharmacology*, **7**, 95–109.

Yamahara, J., T. Sawada, M. Kozuka and H. Fujimara (1974) Pharmacological action of the alkaloids of Menispermaceous plants I. Pharmacological action of tetrandine. *Shoyakuguku Zàsshi*, **28**, 83–95.

Yarington, C. T. and M. Bestler (1964) A double blind evaluation of enzyme preparations in post-operative patients. *Clinical Medicine, USA*, **71**, 710–12.

Chapter 6: Sex hormones and thyroid hormones

Abraham, A., I. Kirson, D. Lavie and E. Glotter (1975) Constituents of *Withania somnifera* XIV. Withanolides of *W. somnifera*, chemotypes I and II. *Phytochemistry*, **14**, 189–94.

Abu-Mustafa, E. A., M. B. E. Fayez, A. M. Gad and F. Osman (1960) Isolation of β-cholesterol from Chufa (*Cyperus esculentus*) tubers. *Journal of Organic Chemistry*, **25**, 1269.

Ad Elbary, A. and S. A. Nour (1979) Correlation between the spermicidal activity and the hemolytic index of certain plant saponins. *Pharmazie*, **34**, 560–1.

Adams, N. R. (1977) Morphological changes in the organs of ewes grazing on oestrogenic subterranean clover. *Research in Veterinary Sciences*, **22**, 216.

Adrian, J. and R. Jacquot (1968) *Valeur Alimentaire de l'Arachide et de ses Dérives*, Maisonneuve et Larose, Paris, 1 vol., 274 pp.

Agrawal, S. K. and P. R. Rastogi (1971) Triterpenoids of *Hibiscus rosa-sinensis* L. *Indian Journal of Pharmacy*, **33**, 41–2.

Agrawal, S. S., N. Ghatak and R. B. Arora (1969) Antioestrogenic activity of alcoholic extract of the roots of *A. precatorius* L. *Indian Journal of Pharmacy*, **31**, 175.

Agrawal, S. S., N. Ghatak and R. B. Arora (1970) Antifertility activity of the roots of *Abrus precatorius* L. *Pharmacological Research Communication, London*, **2**, 159–63.

Ahmed, Z. T., C. J. Zufall and G. L. Jenkins (1949) Chemistry and toxicology of the root of *Phytolacca americana*. *Journal of the American Pharmaceutical Association*, **38**, 443–8.

Ainslie, J. R. (1937) *List of plants used in native medicine in Nigeria*, Oxford University Press, 109 pp.

Aizikov, M. I. and A. C. Kurmukov (1973) Toxicology of gossypol. *Lokl Akad. Nauk Uzb. SSR*, **30**, 28 (through *Chemical Abstracts*, **83**, 1923d).

Alamelu, I. and C. J. Bhuwan (1974) An iridoid glycoside from *Kigelia pinnata*, *Herba polonica*, **20**, 319.

Appa Rao, M. V. R., S. A. Rajagopalan, V. R. Srinavasan and R. Sarangan (1969) Study of Mandookaparni (*Centella asiatica*) for the anabolic effect on normal healthy adults. *Nagarjun*, **12**, 33.

Arora, R. B., N. Ghatak and S. P. Gupta (1971) Antifertility effect of *Embelia ribes*. *Journal of Research on Indian Medicine*, **6**, 107.

Astwood, E. B., M. A. Greer and M. G. Ettlinger (1949) Antithyroid factor of yellow turnip (rutabaga). *Science*, 109, 631.

Atal, C. K. (1980) Chemistry and pharmacology of vasicine A, a new oxytocic and abortifacient. *Regional Research Laboratory, Jammu-Tawi India*, 93–103.

Atal, C. K. and M. Sethi (1962) Proteolytic activity of some Indian plants II. Isolation, properties and kinetic studies of calotropain. *Planta Medica*, **10**, 77–90.

Ayoub, S. M. H. and A. B. Svendsen (1981) Medicinal and aromatic plants in the Sudan. *Fitoterapia*, **52**, 243–6.

Bacharach, A. L. (1940) Vitamin E and habitual abortion. *British Medical Journal*, 890.

Baer, V. and R. Haensel (1982) Immunomodulating properties of 5,20,α (R) dihydroxy 6α,7α-epoxy-1-oxo-(5α)witha-2,24-dienolide and solasodine. *Planta Medica*, **44**, 32–3.

Banerji, R., G. Masera and S. K. Nigam (1978a) Protobassic acid from *Mimusops littoralis* bark. *Fitoterapia*, **49**, 104–5.

Banerji, R., G. Masera and S. K. Nigam (1979a) Constituents of *Mimusops littorali* seeds. *Fitoterapia*, **50**, 53–4.

Banerji, R., G. Misra, S. K. Nigam, S. Singh and R. C. Saxena (1978b) Steroid and triterpenoid saponins: Possible spermicidal agents. *Journal of Steroid Biochemistry*, **9**, 864.

Banerji, R., A. K. Srivastava, G. Misra, S. K. Nigam, S. Singh, S. C. Nigam and R. Saxena (1979b) Steroid and triterpenoid saponins as spermicidal agents. *Indian Drugs*, **17**, 6–8.

Barnes, C. S., J. C. Price and R. L. Hughes (1975) An examination of some reputed antifertility plants. *Lloydia*, **38**, 135–40.

Batta, S. K. and G. Santhakumari (1970) The antifertility effect of *Ocimum sanctum* and *Hibiscus rosa-sinensis*. *Indian Journal of Medical Research*, **59**, 777–81.

Bennets, H. W. and E. J. Underwood (1951) Oestrogenic effects of subterranean clover. Uterine maintenance in the ovariectomized ewe on clover grazing. *Australian Journal of Experimental Biology and Medical Sciences*, **29**, 249–53.

Bezanger-Beauquesne, L., M. Pinkas, M. Torck and F. Trotin (1980) *Les Plantes Médicinales des régions tempérées*. Maloine S.A., Paris, 1 vol., 439 pp.

*Bhaduri, A., C. R. Ghose, A. N. Bose, B. K. Moza and U. P. Basu (1968) Antifertility activity of some medicinal plants. *Indian Journal of Experimental Biology*, **6**, 252–3.

*Bhakuni, D. S., M. L. Dhar, M. M. Dhar, B. N. Dhawan and B. N. Mehrotra (1969) Screening of Indian Plants for biological activities Part II. *Indian Journal of Experimental Biology*, **7**, 250–62.

Bhatia, B. B. and S. Lal (1933) The pharmacological action of *Plumbago zeylanica* and its active principle (plumbagin). *Indian Journal of Medical Research*, **20**, 777–88.

Bianchi, M. (1962) Interaction entre la reserpine et l'axe hypo-physo-medullo-surrénalien. *Comptes Rendus de la Société de Biologie*, **156**, 179.

Biggers, J. D. and D. H. Curnow (1954) Oestrogenic activity of subterranean clover. *Biochemical Journal*, **58**, 278–87.

Bingel, A. S. and N. R. Farnsworth (1980) Botanical sources of fertility regulating agents: Chemistry and pharmacology. In M. Briggs and A. Corbin, *Progress in Hormone Biochemistry and Pharmacology*, Eden Press, St Albans, 1980, pp. 149–225.

Bodhankar, S. L., S. K. Carg and V. S. Mathur (1974) Antifertility screening Part IX Effect of five indigenous plants on early pregnancy in female albino rats. *Indian Journal of Medical Research*, **62**, 831–7.

Boiteau, P., A. R. Ratsimanga and B. Pasich (1964) *Les Triterpenoides en Physiology Végétale et Animale*, Gauthier Villars, Paris, 1370 pp.

Booth, A. N., E. F. Bickoff and G. O. Kohler (1960) Estrogenic activity in vegetable oils and mill products. *Science*, **131**, 1807.

Bouquet, A. (1972) *Plantes Médicinales du Congo Brazzaville*. Travail et documents de l'ORSTOM, **13**, 1 vol., 112 pp.

Bouqet, A., M. M. Debray, J. C. Dauguet, A. Girre, J.-F. Leclair, M. Le Naour and R. Patay (1967) A propos de l'action pharmacologique de l'ecorce de *Combretodendron africanum* (Welw.) Exell. et particulièrement de son pouvoir abortif et perturbateur du cycle oestral. *Thérapie*, **22**, 325–36.

Bradbury, R. B. and D. E. White (1951) Chemistry of subterranean clover. Isolation of formononetin and genistein. *Journal of the Chemical Society*, 3447–9.

Briggs, M. (1976) Biochemical effects of oral contraceptives. *Advances in Steroid Biochemistry and Pharmacology*, **5**, 65–160; **7**, 347–67.

Briggs, M. H. and M. Briggs (1981) Metabolic effects of hormonal contraceptives. In C. C. Fen, D. Griffin and A. Woolman *Recent Advances in Fertility Regulation*, Proceedings of the Beijing Symposium September 1980, 1 vol., pp. 83–111, Atar, Geneva, 399 pp.

Brondegaard, V. J. (1973) Contraceptive plant drugs. *Planta Medica*, **23**, 167–72.

Burgen, A. S. V. and J. P. Mitchell (1972) *Gaddum's Pharmacology*, 7th edn, 1 vol, Oxford University Press, New York and Toronto, 251 pp.

Busson, F. (1965) *Plantes Alimentaires de l'Ouest Africain*, 1 vol., Leconte, Marseille, 568 pp.

Bustos-Obregon, E. and R. Feito (1974) The effect of vinblastine sulfate on rat spermatogenesis. *Archives Biologie (Bruxelles)*, **85**, 353.

Butenandt, A. and H. Jacobi (1933) Female sexual hormone preparation from a plant, tokokinin, and its identification with the α-follicular hormone. *Zeitschrift für Physiologische Chemie*, **218**, 104.

Buxton, J., M. Grundy, D. C. Wilson and D. G. Jamison (1954) *British Journal of Nutrition*, **8**, 170.

Casey, R. D. C. (1960) Alleged antifertility plants of India. *Indian Journal of Medical Sciences*, **14**, 590–600.

Chandra, D. and S. S. Gupta (1972) Anti-inflammatory and anti-arthritic activity of volatile oil of *Curcuma longa* (Haldi). *Indian Journal of Medical Research*, **60**, 138–42.

Chandrasekar, N. V., C. S. Vaydyanathan and M. Sirsi (1961) Some pharmacological properties of the blood anticoagulant obtained from the latex of *Carica papaya*. *Journal of Scientific and Industrial Research, India*, **20**, 213–15.

Chang, M. C., Z. P. Gu and S. K. Saksena (1980) Effects of gossypol on the fertility of male rats, hamsters and rabbits. *Contraception*, **21**, 461–9.

Chaudhury, R. R. (1966) *Plants with possible antifertility activity*. Indian Council of Medical Research, New Delhi 1966, Special Report, Ser. 55, B3.

Chesney, A. M. and B. Webster (1928) Endemic goiter in rabbits. III. Effect of administration of iodine. *Johns Hopkins Hospital Bulletin*, **43**, 291–308.

Chopra, R. N., S. L. Nayar and I. C. Chopra (1956) *Glossary of Indian Medicinal Plants*, Council of Scientific and Industrial Research, New Delhi, 330 pp.

Cooke, R. A., A. Nikles and H. P. Roeser (1978) A comparison of the antifertility effects of alkylating agents and Vinca alkaloids in male rats. *British Journal of Pharmacology*, **63**, 677.

Crout, H. G., C. H. Hassal and T. L. Jones (1964) Cardenolides of *Calotropis procera* IV Uscharidin, calotropin and calotoxin. *Journal of the Chemical Society*, 2187–94.

Dalziel, J. M. (1937) *The Useful Plants of West Tropical Africa*. Crown Agents, 1 vol., 612 pp.

Dar, R. N., L. C. Garg and R. D. Pathak (1965) Anthelmintic activity of *Carica papaya* seeds. *Indian Journal of Pharmacy*, **27**, 335–6.

Das, R. P. (1980) Effect of papaya seeds on the genital organs and fertility of male rats. *Indian Journal of Experimental Biology*, **18**, 408–9.

David Nes, W. and E. Heftman (1981) A comparison of triterpenoids with steroids as membrane components. *Journal of Natural Products*, **44**, 377–400.

Desai, R. V. and E. N. Rupawala (1966) Antifertility activity of the steroidal oil derived from the seed of *Abrus precatorius* L. on rats and mice. *Indian Journal of Pharmacy*, **28**, 344.

Desai, R. V. and E. N. Rupawala (1967) Antifertility of the steroidal oil of the seed of *Abrus precatorius* L. *Indian Journal of Pharmacy*, **29**, 235–7.

Desai, V. B. and M. Sirsi (1966) Chemical and pharmacological investigations on *Abrus*

precatorius L. (Leguminosae) XVIIIe Indian Pharmaceutical Congress 1966. *Indian Journal of Pharmacy*, **28**, 340.
Deshpande, V. Y., K. N. Mendulkar and N. L. Sadre (1980) Male antifertility activity of *Azadirachta indica* in male mice. *Journal of Postgraduate Medicine*, **26**, 167–70; Abstract of 4th Asian Symposium on Medicinal Plants and spices, Bangkok, September 1980, p. 64 (in Farnsworth and Waller, 1982).
Dhalla, N. S., K. C. Gupta, M. S. Sastry and C. L. Malhotra (1961) Chemical composition of the fruit of *Momordica charantia*. *Indian Journal of Pharmacy*, **23**, 128–9.
Dhar, M. L., M. M. Dhar, B. N. Dhawan, B. N. Mehrothra and C. Ray (1968) Screening of Indian plants for biological activity. Part I. *Indian Journal of Experimental Biology*, **6**, 232–47.
Dijkman, M. J., M. L. Boss, W. Lichter, M. M. Sigel, J. E. O'Connor and R. Search (1966) Cytotoxic substances from tropical plants. *Cancer Research*, **26**, 1121–30.
Dixit, N. P. and R. S. Gupta (1983) Antispermatogenic and antiandrogenic activity of *Sapindus trifoliatus*. *Planta Medica*, **46**, 242–6.
Dixit, V. P., P. Khanna and S. K. Bhargava (1978) Effects of *Momordica charantia* L. fruit extract on the testicular function of the dog. *Planta Medica*, **34**, 280–6.
Dlusi, S. O., O. L. Oke and A. Odusote (1979) Effect of cyanogenic agents on reproduction and neonatal development in rats. *Biology of the Neonate*, **36**, 233–43.
Dutta, T. and Basu, D. K. (1967) Terpenoids IV. Isolation and identification of asiatic acid from *Centella asiatica* L. *Indian Journal of Chemistry*, **5**, 586.
Ekong, D. E. U., C. O. Fakunle, A. K. Fasina and J. I. Okogun (1969) The Meliacins (limonoids). Nimbolin A and B, two meliacincinnamates from *Azadirachta indica* L. and *Melia azedarach* L. *Journal of the Chemical Society 20 D*, 1166–8.
El Ridi, M. S. (1960) Gonadotropic hormones in pollen grains of Date palm. *Zeitschrift für Naturforschung*, **15B**, 45.
El Sayyad, S. M. (1981) Flavonoids of the leaves and fruits of *Kigelia* pinnata. *Fitoterapia*, **52**, 189–91.
Ermans, A. M. (1979) Anti-thyroid effects of Cassava and thiocyanate. *Bulletin Mémoires Académie Royale de Médecine Belge*, **134**, 137–53.
Ermans, A. M., J. Kinthaut, M. van den Velden and P. Bourdoux (1980) Studies on the antithyroid effects of cassava and thiocyanate in rats. *International Development Research Centre*, **136**, 93–110, 167–82.
Eshiett, N. O., A. A. Ademosum and T. A. Omole (1980) Effect of feeding cassava root meal on reproduction and growth of rabbits *Planta Medica*, **39**, 226B.
Farnsworth, N. R. and D. P. Waller (1982) Current status of plant products reported to inhibit sperm. *Research Frontiers of Fertility Regulation*, **2**, 1–16.
Farnsworth, N. R., H. H. S. Fong and E. Diczfalusy (1983) New fertility regulating agents of plant origin. *International Symposium on Research on the Regulation of Human Fertility–Needs of Developing Countries and Priorities for the future*, Stockholm, February 1983.
Farnsworth, N. R., A. S. Bingel, G. A. Cordell, F. A. Crane and H. H. S. Fong (1975a) Potential value of plants as sources of new antifertility agents Part I. *Journal of Pharmaceutical Sciences*, **64**, 535–98.
Farnsworth, N. R., A. S. Bingel, G. A. Cordell, F. A. Crane and H. H. S. Fong (1975b) Potential value of plants as sources of new antifertility agents Part II. *Journal of Pharmaceutical Sciences*, **64**, 717–53.
Fen, C. C., D. Griffin and A. Woolman (1981) *Recent Advances in Fertility Regulation. Proceedings of The Beijing Symposium, 1980*, Atar, Geneva, 1 vol., 399 pp.
Fern, V. H. (1963) Congenital malformations in hamster embryos after treatment with vinblastine and vincristine. *Science*, **141**, 426.
Folman, Y. and G. S. Pope (1966) The interaction in the immature mouse of potent oestrogens with coumestrol, genistein, and other utero-vaginotropic compounds of low potency. *Journal of Endocrinology*, **34**, 215.

Furuya, M. and A. W. Galston (1965) Flavonoid complexes in *Pisum sativum*. I. Nature and distribution of the major compounds. *Phytochemistry*, **4**, 285–96.

Garg, S. K. (1974) Effect of *Curcuma longa* rhizomes on fertility in experimental animals. *Planta Medica*, **26**, 225–7.

Garg, A. (1979) Effect of *Calotropis procera* (Ait.) R.Br. flower extract on testicular function of the Indian desert male gerbil (*Merriones hurrianae* Jerdon). A biochemical and histological study. *Indian Journal of Experimental Biology*, **17**, 859–62.

Garg, S. K. and J. P. Garg (1971) Antifertility screening VII. Effect of five indigenous plants on early pregnancy in albino rats. *Indian Journal of Medical Research*, **59**, 302–6.

Garg, L. S. and G. C. Prasar (1965) Effect of *Withania somnifera* on reproduction in mice. *Planta Medica*, **13**, 46–7.

Garg, S. K., V. S. Mathur and R. R. Chaudhury (1978) Screening of Indian plants for antifertility activity. *Indian Journal of Experimental Biology*, **16**, 1077–9.

Garg, S. K., S. K. Saxena and R. R. Chaudhuri (1970) Antifertility screening of plants Part IV. Effect of 5 indigenous plants on early pregnancy in albino rats. *Indian Journal of Medical Research*, **58**, 1285.

Garg, H. S., B. S. Setty, V. P. Kamboj and N. M. Khanna (1976) Spermicidal saponins from plants. Indian Patent 141 240. *Contraception*, **14**, 571–8 (through *Chemical Abstracts*, **92** (1980), 99566y).

Ghosal, S., A. K. Srivastava, R. S. Srivastava, Y. Chattopadhyay and M. Mastra (1981) Justicisaponin, a new triterpenoid saponin from *Justicia simplex*. *Planta Medica*, **42**, 279–83.

Goodman and Gilman (1980) Hypothalamic control of the anterior pituitary. In *The Pharmacological Basis of Therapeutics*, 6th edn, Macmillan, New York, 1 vol., 1843 pp., pp. 1389–92.

Gopalachari, R. and M. L. Dhar (1958) Studies in the constitution of saponins from the seeds of *Achyranthes aspera*. Identification of the sapogenin. *Journal of Scientific and Industrial Research, India*, **17B**, 276–8.

Govindachari, T. R., S. J. Patankar and N. Viswanathan (1971) *Kigelia pinnata*. *Phytochemistry*, **10**, 603.

Gran, L. (1973a) On the effect of a polypeptide isolated from 'Kalata-kalata' (*Oldenlandia affinis* DC.) on the oestrogen dominated uterus. *Acta Pharmacologica et Toxicologica*, **33**, 400.

Gran, L. (1973b) The utero-active principles of Kalata kalata (*Oldenlandia affinis* DC.) *Edn Noregs Bolag*, Oslo, 1 vol.

Gran, L. (1973c) Oxytocic principles of *Oldenlandia affinis*. *Lloydia*, **36**, 174–8.

Gran, L. (1973d) Isolation of oxytocic peptides from *Oldenlandia affinis* by solvent extraction of tetraphenyl borate complexes and chromatography. *Lloydia*, **36**, 207–8.

Gran, L. (1973e) On the isolation of tetramethyl-putrescine from *Oldenlandia affinis*. *Lloydia*, **36**, 209–11.

Greer, M. A. and E. R. Astwood (1948) *Endocrinology*, **43**, 105.

Gupta, J. C., P. K. Roy, G. K. Ray and A. Dutta (1950) Pharmacological action of an active constituent isolated from *Daemia extensa* L. *Indian Journal of Medical Research*, **38**, 75–82.

Gupta, K. (1972) *Aloes* compound (a herbal drug) in functional sterility. XVI. Indian Obstetrics and Gynaecology Congress, 1972, New Delhi (in Indian Council of Medical Research, 1976).

Gupta, M. L., T. K. Gupta and K. P. Bhargava (1971) A study of antifertility effects of some indigenous drugs. *Indian Journal of Medical Research*, **6**, 112–17.

Gupta, N. C., B. Singh and D. S. Bhakuni (1968) Steroids and triterpenes from *Alangi lamarkii*, *Abrus precatorius* and *Holoptelea integrifolia*. *Phytochemistry*, **8**, 791–2.

Hadley, M. A., Y. C. Lin and M. Dym (1981) Effects of gossypol on the reproductive system of male rats. *Journal of Andrology*, **2**, 190.

Hahn, D. W., C. Rusticus, A. Probst, R. Homm and A. N. Johnson (1981) Antifertility and endocrine activities of gossypol in rodents. *Contraception*, **24**, 97.

*Harborne, J. B., T. J. Mabry and H. Mabry (1974) *The Flavonoids*. Chapman and Hall, London.

Hariharan, V. and Ranjaswani, S. (1970) Structure of saponins A and B from the seeds of *Achyranthes aspera* L. *Phytochemistry*, **9**, 409.

Hartwell, J. L. (1976) Types of anti-cancer agents isolated from plants. *Cancer Treatment Report*, **60**, 1031.

Hassan, A. and H. M. A. El Waffa (1947) Estrogenic substance in pollen-grains of date-palm tree. *Nature, (London)*, **159**, 409.

Hayashi, H. K. (1944) Several anthocyanins containing cyanidin as the aglycone. *Acta Phytochemica, Japan*, **14**, 55–74.

Heftman, E. (1967) Steroid hormones in plants. *American Perfumes and Cosmetics*, **82**, 47–9.

Heftman, E., S. T. Ko and R. D. Bennet (1966) Identification of oestrone in pomegranate seeds. *Phytochemistry*, **5**, 1337–9.

Indian Council of Medical Research (1976) *Medicinal Plants of India*, Vol. I, New Delhi.

Indira, M., P. S. N. Murthy and M. Sirsi (1956b) Estrogenic activity, antibacterial action and some pharmacodynamic properties of *Cyperus rotundus*. *Journal of the Mysore Medical Association*, **21**, 1.

Indira, M., M. Sirsi, S. Radomir and K. L. Suhdev (1956a) The occurrence of some estrogenic substances in plants Part I. Estrogenic activity of *Cyperus rotundus* L. *Journal of Scientific and Industrial Research*, **15C**, 202.

Jamwall, K. S. and N. K. Anand (1964) Preliminary screening of some reputed abortifacient indigenous plants. *Indian Journal of Medical Research*, **24**.

Joshi, M. S. and R. Y. Ambaye (1968) Effect of alkaloids from *Vinca rosea* L. on spermatogenesis in male rats. *Indian Journal of Experimental Biology*, **6**, 256–7.

Kagihara, A. (1980) Action and uses of antithyroid drugs. *Kojonsengaku*, 1980, 108–17 (through *Chemical Abstracts*, 96, 456772).

Kalla, N. R. and M. Vasudev (1980) Studies on male antifertility agent, gossypol acetic acid: *in vitro* studies on the effect of gossypol acetic acid on human spermatozoa. *IRCS Medical Science Library Comp.* **8**, 375 (in Farnsworth and Waller, 1982).

Kamboj, V. P. and B. N. Dhawan (1982) Research on plants for fertility regulation in India. *Journal of Ethnopharmacology*, **6**, 191–226.

Kapoor, M., S. K. Garg and V. S. Mathur (1974) Anti-ovulatory activity of five indigenous plants in rabbits. *Indian Journal of Medical Research*, **62**, 1225–7.

Kemper, F. and A. Loeser (1957) Untersuchungen zur Gewinnung anti-hormonal wirksamer Inhaltstoffe aus *Lithospermum officinale*. *Arzneimittel-forschung*, **7**, 81.

Kennedy, W. P., H. H. van der Ven, D. P. Waller, K. L. Polakosky and L. J. D. Zaneveld (1982) Gossypol inhibition of oocyte penetration and sperm acrosin. *Biological Reproduction*, **26**, Supplement 118.

Kerharo, J. and J. G. Adam (1974) *La Pharmacopée Sénégalaise Traditionelle*, Vigot, Paris, 1011 pp.

Khanna, P., S. C. Jain, A. Panagariya and V. P. Dixit (1981) Hypoglycaemic activity of polypeptide from a plant source. *Journal of Natural Products*, **44**, 648–65.

Kholkute, S. D. (1977) Effect of *Hibiscus rosa-sinensis* on spermatogenesis and accessory reproductive organs in rats. *Planta Medica*, **31**, 127–35.

Kholkute, S. D. and K. N. Udupa (1974) Antifertility properties of *Hibiscus rosa-sinensis*. *Journal of Research on Indian Medicine*, **9**, 99.

Kholkute, S. D. and K. N. Udupa (1976) Effects of *Hibiscus rosa-sinensis* on pregnancy of rats. *Planta Medica*, **29**, 320–9.

Kholkute, S. D. and K. N. Udupa (1978a) Biological profile of total benzene extract of *Hibiscus rosa-sinensis* flowers. *Planta Medica*, **29**, 151–5.

Kholkute, S. D. and K. N. Udupa (1978b) Biological profile of total benzene extract of *Hibiscus rosa-sinensis* flowers. *Journal of Research on Indian Medicine, Yoga, Homeopathy*, **13**, 107.

Kholkute, S. D., C. Chatterji and K. N. Udupa (1976a) Effect of *Hibiscus rosa-sinensis* on estrus cycle and reproductive organs in rats. *Indian Journal of Experimental Biology*, **14**, 703–4.

Kholkute, S. D., V. Mudgal and P. G. Deshpande (1976b) Screening of indigenouss medicinal plants for antifertility potentiality. *Planta Medica*, **29**, 151–3.

Kholkute, S. D., V. Mudgal and K. N. Udupa (1977) Studies on the antifertility potentiality of *Hibiscus rosa-sinensis*. *Planta Medica*, **31**, 35–9.

Kholkute, S. D., S. Chatterjee, D. N. Srivastava and K. N. Udupa (1972) Antifertility effect of the alcoholic extract of Japa (*Hibiscus rosa-sinensis*). *Journal of Research on Indian Medicine*, **7**, 72–4.

King, T. J. and L. B. de Silva (1968) Optically active gossypol from *Thespesia populnea*. *Tetrahedron Letters*, **1968**, 261.

Knobil, E. (1973) On the regulation of the primate corpus luteum. *Biology and Reproduction*, **8**, 246–58.

Ko, K. (1933) On the pharmacological action of plumbagin. *Japanese Journal of Medical Sciences*, **6**, 259.

Kokwaro, J. O. (1981) A review of research on plants for fertility regulation in Africa. *Korean Journal of Pharmacognosy*, **12**, 149–52.

Kopcewicz, J. (1971) Estrone, estriol and 17α-estradiol in *Phaseolus vulgaris*. *Phytochemistry*, **10**, 1423.

Kraus, S. D. and A. Kaminskis (1969) Antioestrogenic action of β-glycyrrhetinic acid. *Experimental Medicine and Surgery*, **27**, 411–20.

Kumar, M. A., S. S. Mandrekar and U. K. Sheth (1962) Aldosterone antagonist in Arachis oil. *Lancet*, 1053.

Langlois, J. (1941) Présence de tocophérol dans l'huile de ricin. Titre de l'huile en cette vitamine. *Comptes Rendus de l'Academie des Sciences*, **213**, 845–8.

Lechat, P., F. Bisseliches, F. Bournerias, H. Dechy, Y. Juillet, G. Lagier, C. Nerignac, B. Rouveix, P. Sterin and S. Weber (1978) *Pharmacologie Médicale*, 3rd edn, Masson, Paris, 1 vol., 677 pp.

Lee, C. Y. and H. V. Malling (1981) Selective inhibition of sperm-specific lactate dehydrogenase-X by an anti-fertility agent, gossypol. *Federation Proceedings*, **40**, 718.

Lightfoot, R. J., K. P. Croker and H. G. Neil (1967) Failure of sperm transport in relation to ewe infertility following prolonged grazing on oestrogenic pastures. *Australian Journal of Agricultural Research*, **18**, 755.

Lipton, A. (1967) Abortifacient and toxic actions of the glycoside 'albitocin' extracted from some *Albizia* species. *Journal of Pharmacy and Pharmacology*, **19**, 792.

Liu, Z. Q., G. Z. Liu, L. S. Hei, R. A. Zhang and C. Z. Yu (1981) Clinical trial of gossypol as a male antifertility agent. In C. C. Fen, D. Griffin and A. Weelman, *Recent Advances in Fertility Regulation*, pp. 160–3, Beijing, Sept. 1980, Atar, S.A., Geneva, 1 vol., 399 pp.

Livingstone, A. L. (1978) Forage plant oestrogens. *Journal of Toxicology and Environmental Health*, **4**, 301.

Luna, C. A. (1963) The effect of Kao Haole (*Leucaena glauca*) on the fertility of rats. MSc Thesis, University of Hawai, 33 pp. (in Farnsworth and Waller, 1982).

Luscombe, D. K. and S. A. Taha (1974) Pharmacological studies on the leaves of *Azadirachta indica*. *Journal of Pharmacy and Pharmacology*, **26**, 110–11.

McIlroy, R. J. (1951) *The Plant Glycosides*, 1 vol., Arnold, London, 138 pp.

Matin, M. A., G. P. Tewari and D. K. Kalani (1969) Pharmacological effects of 'paniculatin' a glycoside isolated from *Ipomea digitata* L. *Indian Journal of Pharmaceutical Sciences*, **58**, 757–8.

Mehrotra, P. K. and V. P. Kamboj (1978) Hormonal profile of coronaridine hydrochloride, an antifertility agent of plant origin. *Planta Medica*, **33**, 345–9.

Menssen, H. C. and Stapel, G. (1973) Uber ein C28 steroid lacton aus der Wurzel von *Withania somnifera*. *Planta Medica*, **24**, 8–12.

Mishra, S. S. and K. C. Datta (1962) A preliminary pharmacological study of *Ipomea digitata* L. *Indian Journal of Medical Research*, **50**, 43–5.

Mishra, A., J. U. V. Dogra, J. N. Singh and O. P. Jha (1979) Post-coital antifertility activity of *Annona squamosa* and *Ipomea fistulosa*. *Planta Medica*, **35**, 283–5.

Morris, J. M. and G. van Wagenen (1973) Interception: The use of post-ovulatory oestrogens to prevent implantation. *American Journal of Obstetrics and Gynecology*, **115**, 101.

Morton, J. F. (1967) The Balsam pear – an edible, medicinal and toxic plant. *Economic Botany*, **21**, 57–68.

Mudgal, N. R., E. Raghupaty and P. S. Sharma (1958) Studies on goitrogenic agents in food. Goitrogenic action of some glycosides isolated from edible nuts. *Journal of Nutrition*, **66**, 291–303.

Mudgal, N. R., V. Srinivasan and P. S. Sharma (1957) Studies on goitrogenic agents in food. Goitrogenic action of arachidoside. *Journal of Nutrition*, **61**, 97–101.

Munshi, S. R., T. A. Shetye and R. K. Nair (1977) Antifertility activity of three indigenous plant preparations. *Planta Medica*, **31**, 73–5.

Murti, V. V. S., T. R. Seshadri and T. A. Venkita Subramayan (1964) Effet goitrigene des choux et d'autres Crucifères. *Phytochemistry*, **3**, 73.

Nadkarni, A. K. and K. M. Nadkarni (1954) *Indian Materia Medica*, 3rd edn, Popular Book Depot, Bombay.

Naish, C. (1954) Thyroid for lactation. *Lancet*, 1077–8.

Naqvi, R. H. and J. C. Warren (1971) Interceptives: drugs interrupting pregnancy after implantation. *Steroids*, **18**, 731.

Olaniji, A. A. (1975) A neutral constituent of *Momordica foetida*. *Lloydia*, **38**, 361–2.

Oliver-Bever, B. (1971) Vegetable drugs for cancer therapy. *Quarterly Journal of Crude Drug Research*, **11**, 1665–83.

Osuntokun, B. O., G. L. Monekosso and J. Wilson (1969) Relationship of a degenerative tropical neuropathy to diet, report of a field survey. *British Medical Journal*, 1969, 547–50.

Osuntokun, B. O., S. P. Singh and F. D. Martinson (1970) Deafness in tropical nutritional ataxic neuropathy. *Tropical Geographical Medicine Pays Bas*, **22**, 281–8.

Pakrashi, S. C., B. Achari and P. C. Majumdar (1975a) Studies on Indian Medicinal plants Part XXXII Constituents of *Ananas comosus* (L.) Merr. leaves. *Indian Chemical Journal*, **13**, 754–6.

Pakrashi, A. and N. Bhattacharya (1977) *Achyranthes aspera* L. Uterotonic activity. *Indian Journal of Experimental Biology*, **15**, 856–8.

Pakrashi, A., B. Basak and N. Mookerji (1975b) Search for antifertility agents from indigenous medicinal plants. *Indian Journal of Medical Research*, **63**, 378–81.

Paris, R. R. and H. Moyse (1965–71) *Precis de Matière Médicale*, Vol. I, 1965; Vol. II, 1967; Vol. III, 1971, Masson et Cie, Paris, 3 vols., 1436 pp.

Paris, R. and J. P. Theallet (1961) Nouvelles recherches sur la composition chimique et l'activité ocytocique de divers *Grewias* (Tiliacées) d'origine africaine. *Annales de Pharmacie françaises*, **19**, 20–3.

Parkhurst, R. M. and S. J. Stolzenberg (1975) Saponin containing spermatocidal composition. U.S. Patent 3886 272 (through *Chemical Abstracts*, **82**, 116078).

Parvinen, L. M., K. O. Söderstrom and M. Parvinen (1978) Early effects of vinblastine and vincristine on rat spermatogenesis: analysis by a new transillumination-phase contrast microscopic method. *Experimental Pathology*, **15**, 85.

Peyster, A. de and Y. Y. Wang (1979) Gossypol, proposed contraceptive for men, passes the Ames test. *New England Journal of Medicine*, **301**, 275.

Pillay, N. R., M. Alam and K. K. Pursotham (1977) Studies on the antifertility activity of oleanolic acid-3β-glucoside. *Journal of Research on Indian Medicine*, **12**, 26–9.

Planchon, L. and P. Breton (1946) *Précis de Matière Médicale*, 2 vols., edn Maloine, Paris.

Prakash, A. O. and R. Mathur (1979) Studies on the oestrus cycle of albino rats. Response to *Embelia ribes* extracts. *Planta Medica*, **36**, 134–41.

Prasad, M. R. N. and Y. E. Diczfalusy (1981) Gossypol. Proceedings of the Second International Congress of Andrology, Tel Aviv, June 28–30, 1981, *International Journal of Andrology*, Supplement 5 (1981) 53–70 (in Farnsworth and Waller, 1982).

Premkumari, P., K. Rathinam and G. Santhakumari (1977) Antifertility activity of plumbagin. *Indian Journal of Medical Research*, **65**, 828–38.

Qian, S. Z. (1980) Effect of gossypol on potassium and prostaglandin metabolism and mechanism of action of gossypol. In C. C. Fen, D. Griffin and A. Woolman, *Recent Advances in Fertility Regulation (Proceedings of the Beijing Symposium)* Atar, Geneva, 1 vol., 399 pp., pp. 152–9.

Qian, S. Z., J. H. Hu, L. X. Ho, M. X. Sun, Y. Z. Huang and J. H. Fang (1981) The first clinical trial of gossypol on male antifertility. In P. Turner, *Clinical Pathology and Therapeutics*, Macmillan, New York, 489–92.

Rao, P. S. and T. R. Seshadri (1969) Variations in the chemical composition of Indian samples of *Centella asiatica*. *Current Science*, **38**, 77–9.

Ridley, A. J. and L. Blasco (1981a) Testosterone and gossypol effects on human sperm motility. *Fertility and Sterility*, Suppl., **35**, 244.

Ridley, A. J. and L. Blasco (1981b) Testosterone and gossypol effects on human sperm motility. *Fertility and Sterility*, Suppl., **36**, 638.

Robinson, M. (1947) Hormones and lactation. Dried thyroid gland. *Lancet*, 385–7.

Ross, S. A., S.-M. El-Sayyad, A. A. Ali and N. E. El-Keltawy (1982) Phytochemical studies on *Jasminum sambac*. *Fitoterapia*, **53**, 91–5.

Sakkawala, O. P., O. P. Shukla and C. R. Krishnamurti (1962) Bacteriolytic activity of plant latexes. *Journal of Scientific and Industrial Research*, **21**, 279–89.

Santhakumari, G. and D. Sujantham (1980) Antigonadotropic action of plumbagone from root of *Plumbago rosea*. *Planta Medica*, **39**, 244.

Sanyal, S. N. (1956) *Pisum sativum m*-xylohydroquinone as an oral contraceptive. A critical evaluation. *Acta Endocrinologica* Supplement, 72–82.

Sanyal, S. N. (1958) Oral contraceptive *m*-xylohydroquinone. Biological studies on males. *Journal of Internal Medicine, Abstract*, **22**, 19.

Sanyal, S. N. and M. Rana (1959) Oral contraceptive. Clinical trial with human males with *m*-xylohydroxyquinone. A preliminary note. *Journal of Internal Medicine Abstract*, **23**, 33.

Schaub, F., H. Kaufman, W. Stöcklin and T. Reichstein (1968) Die Pregnanglykoside der oberirdischen Teile von *Sarcostemma viminale* (L.) R.Br. *Helvetica Chimica Acta*, **51**, 738–67 and 767–72.

Setty, B. S., V. P. Kamboj, H. S. Garg and N. M. Khanna (1976) Spermicidal potential of saponins isolated from Indian medicinal plants. *Contraception*, **14**, 571–9.

Setty, B. S., V. P. Kamboj and N. M. Khanna (1977) Screening of Indian plants for biological activity. VII Spermicidal activity of Indian plants. *Indian Journal of Experimental Biology*, **15**, 231–2.

Shaaban, A. H. and Z. F. Ahmed (1959) A new spermicidal principle from *Phytolacca americana* L. *Gazette Egyptian. Society of Gynecology and Obstetrics*, **9**, 27.

Shah, C. C. and K. D. Mody (1967) Estimation of barbaloin in Indian Aloes. *Indian Journal of Pharmacy*, **29**, 10.

Sharma, S. C., N. Chadha and M. N. Bunrjoree (1972) The effect of *Aloes indica* on the fertility of female rabbits XVI. Indian Obstetrics and Gynecology Congress, New Delhi, 1972 (in Indian Council of Medical Research, 1976).

Sharma, V. N. and K. P. Saksena (1959) Spermicidal action of sodium nimbinate. *Indian Journal of Medical Research*, **47**, 322.

Shukla, B. and C. R. Krishnamurti (1961) Bacteriolytic activity of plant latexes. *Journal of Scientific and Industrial Research*, **20C**, 109–12.

Shukla, O. P. and C. R. Krishnamurti (1961) Properties from the latex of *Calotropis procera*. *Journal of Scientific and Industrial Research, India*, **20C**, 225–6.

Shutt, D. A. (1976) The effect of plant oestrogens on animal reproduction. *Endeavor*, **35**, 110.

Singh, M., J. N. Sharma, R. B. Arora and B. R. Kocher (1973) Beneficial effect of *Aloe vera* in the healing of thermal burns and radiation injury in albino rats. *Indian Journal of Pharmacy*, 5, 258.

Singh, M. P., R. H. Singh and K. N. Udupa (1982) Anti-fertility activity of a benzene extract of *Hibiscus rosa-sinensis* flowers on female albino rats. *Planta Medica*, **44**, 171–4.

Singh, N., S. P. Singh, J. N. Sinha and R. P. Kohli (1978) An analysis of hypotensive response to *Sapindus trifoliatus*. *Quarterly Journal of Crude Drug Research*, **66**, 96–102.

Skarzynski, B. (1933) Estrogenic substance from plants. *Nature (London)*, **131**, 766.

Smith, E. R. J. (1963) Reserpine. Les rapports avec l'adrenaline et la noradrenaline. *Journal de Pharmacologie Experimentale et Thérapie*, **139**, 321.

Soejarto, D. D., A. S. Bingel, M. Slaytor and N. R. Farnsworth (1978) Fertility regulating agents from plants. *Bulletin of the World Health Organization*, **56**, 343–52.

Steinegger, E. and R. Hänsel (1968) *Lehrbuch des allgemeinen Pharmacognosie; Pflanzen und Pflanzenstoffen mit hormonähnlicher Wirkung*, pp. 492–511, 2nd edn, Springer, Berlin.

Stepka, W., K. E. Wilson and G. E. Madge (1974) Antifertility investigation on *Momordica*. *Lloydia*, **37**, Proceedings, 645.

Steyn, D. G. (1937) Recent investigations into the toxicity of known and unknown poisonous plants in South Africa. *Onderstepoort Journal of Veterinary Sciences*, **9**, 573–82.

Stöckel, K., K. Hürzeler and T. Reichstein (1969) Viminolon, Struktur beweis. *Helvetica Chimica Acta*, **52**, 1089–91.

Stolzenberg, S. J. and R. M. Parkhurst (1974) Spermicidal actions of extracts and compounds from *Phytolacca dodecandra*. *Contraception*, **10**, 135–43.

Stolzenberg, S. J., R. M. Parkhurst and E. J. Riest (1975) Blasticidal and contraceptive actions of saponins from *Phytolacca dodecandra* L. *Federation Proceedings*, **34**, 339.

Sucrow, W. (1966) Ueber Steringlukoside und ein neues Stigmastadienol aus *Momordica charantia*. *Tetrahedron Letters*, **26**, 2217–21.

Tannous, R. I. and S. N. Nayfeh (1969) Effect of feeding lupin seeds on spermatogenesis in the rat. *Australian Journal of Biological Sciences*, **22**, 1071.

Tewari, P. V. (1974) Preliminary clinical trial of *Hibiscus rosa-sinensis* as an oral contraceptive agent. *Journal of Indian Medicine, Yoga. Homeopathy*, **9**, 96–8.

Tewari, P. V., S. K. Sharma and K. Basu (1976) Clinical trial of an indigenous drug as oral contraceptive. *Journal of the National Integrated Medical Association*, **18**, 115–18.

Thompson, I. B. and C. C. Anderson (1978) Cardiovascular effects of *Azadirachta indica* extract. *Journal of Pharmaceutical Sciences*, **67**, 1476–8.

Train *et al.* (1941) *Medicinal Uses of Plants by Indian Tribes of Nevada*. Bureau of Plant Industry, Washington (through Steinegger and Hänsel, 1963).

Tripathi, W. J. and B. Dasgupta (1974) Neutral constituents of *Albizia lebbek*. *Current Science*, **43**, 46.

Varshney, I. P. and M. Khanna (1978) Partial structure of a new saponin, samanin D, from the flowers of *Pithecellobium saman* Benth. *Indian Journal of Pharmaceutical Sciences*, **40**, 60.

Varshney, I. P., P. Vyas, H. C. Srivastava and P. P. Singh (1979) Study of *Albizzia lebbeck* Benth. wood saponin lebbekanin E. *National Academy of Science Letters (India)*, **2**, 135 (through *Chemical Abstracts*, **91**, 120369c).

Ven, H. H. van der, W. P. Kennedy, A. K. Bhattacharya, D. P. Waller, K. L. Polakoski and I. D. J. Zaneveld (1982) Gossypol inhibition of human sperm acrosin and oocyte penetration, submitted for publication in 1982 (through Farnsworth and Waller, 1982).

Vilar, O. (1974) Effect of cytostatic drugs on human testicular function. In *Male Fertility and Sterility*, Proceedings of the Serono Symposium, vol. 5, ed. R. E. Mancini and L. Martini, Academic Press, London, p. 423.

Virtanen, A. I. (1961) Uber die Chemie des Brassica Faktoren. Ihre Wirkung auf die Funktion der Schilddrüse und ihr Ubergehen in die Milch. *Experientia*, **17**, 241–51.

Vohora, S. B., S. K. Garg and R. R. Chaudury (1969) Antifertility screening of plants Part III. *Indian Journal of Medical Research*, **57**, 893–9.

Waller, D. P., H. H. S. Fong and L. J. D. Zaneveld (1981) Spermicidal composition. US Patent 4 297 431.

Waller, D. P., L. J. D. Zaneveld and H. H. S. Fong (1980) In vitro spermicidal activity of gossypol. *Contraception*, **22**, 183–7.

Watt, J. M. and M. G. Breyer-Brandwijk (1962) *The Medicinal and Poisonous Plants of Southern and Eastern Africa*, Livingstone, Edinburgh and London, 2nd edn, 1 vol., 1057 pp.

Weniger, B., M. Haag-Berturier and R. Anton (1980) Plantes d'Haiti et anti-fecondité. *Planta Medica*, **39**, 260.

Weniger, R., M. Haag-Berrurier and R. Anton (1982) Plants of Haiti used as antifertility agents. *Journal of Ethnopharmacology*, **6**, 67–84.

Willett, E. L., L. A. Henke and C. Maruyama (1945) Roughage for brood sows. Hawaii Agricultural Experimental Station Biennal Report 1942–44, 95 (through Farnsworth and Waller, 1982).

Wu, X. R. (1972) Study of antifertility action of cottonseed and the effective component, gossypol. National Conference of Recent Advances of Family Planning Research, Beijing, pp. 5–20 (in Farnsworth and Waller, 1982).

Xue, S. P. (1981) Studies on the antifertility effect of gossypol, a new contraceptive for males. In C. C. Fen, D. Griffin and A. Woolman (1981), *Recent Advances in Fertility Regulation*, pp. 122–46, Beijing Symposium, Atar S.A., Geneva.

Xiao, P. (1983) Recent developments on medicinal plants in China. *Journal of Ethnopharmacology*, **7**, 95–109.

Yamada, Y. K., K. Hagiwara, Y. Iguchi and Hsu H. Takahasi (1978) Cucurbitacins from *Anagallis arvensis*. Triterpenoids, cucurbitacins B, D, E, I, L and R, bitter principles. *Phytochemistry*, **17**, 1798.

Zatuchni, G. I. and C. K. Osborne (1981) Gossypol a possible male antifertility agent. Report of a workshop. *Research Frontiers of Fertility Regulation*, **I**, (4) 1 (through Farnsworth and Waller, 1982).

Zhou, L. F. and H. P. Lei (1981) Recovery of fertility in rats after gossypol treatment. In C. C. Fen, D. Griffin and A. Woolman, *Recent Advances in Fertility Regulation*, Atar S.A., Geneva, pp. 147–51.

Chapter 7: Oral hypoglycaemic action

Adesina, S. K. and J. B. Harborne (1978) The occurrence and identification of flavonoids in *Thaumatococcus danielli* Benth. *Planta Medica*, **34**, 323–7.

Ahmed, Z. F., F. M. Hammouda, A. M. Rizk and S. I. Ismail (1971) Phytochemical studies of certain *Centaurea* species. *Planta Medica*, **19**, 264–9; **18**, 227–31.

Allen, F. M. (1927) Blueberry leaf extracts. Physiological and clinical properties in relation to carbohydrate metabolism. *Journal of the American Medical Association*, **89**, 1577.

Ambike, S. H. and M. R. Rajarama Rao (1967) Studies on a phytosterolin from the bark of *Picus religiosa*. *Indian Journal of Pharmacy*, **29**, 91–2.

Ashurst, P. R. (1971) Toxic substances of Akee. Review. *Journal of the Scientific Research Council, Jamaica*, **2**, 4–16.

Athar, M. A. (1979) Effect of *Momordica charantia* L. on blood sugar level of normal and alloxan diabetic rabbits. MSc Thesis, University of Agriculture, Faisalabad, India.

Athar, M. S., M. A. Athar and M. Yaqub (1981) Effect of *Momordica charantia* on the blood glucose level of normal and alloxan diabetic rabbits. *Planta Medica*, **42**, 205–12.

Augusti, K. T. (1975) Studies on the effect of allicin (diallyl disulphide oxide) on alloxan diabetes. *Experientia*, **31**, 1263.

Augusti, K. T. (1976a) Gas chromatographic analysis of onion principles and a study of their hypoglycaemic action. *Indian Journal of Experimental Biology*, **14**, 110–12.

Augusti, K. T. (1976b) Chromatographic identification of certain sulphoxides of cysteine present in onion (*Allium cepa* L.) *Current Sciences*, **45**, 863–4.

Augusti, K. T. and M. E. Benaim (1974) Effect of essential oil of onion (allyl-propyldisulphide) on blood glucose, free fatty acid and insulin levels of normal subjects. *Clinical Chimica Acta*, **60**, 121–3.

Augusti, K. T., V. C. M. Roy and M. Semple (1975) Effect of allyl propyl disulphide isolated from onion (*Allium cepa* L.) on glucose tolerance of alloxan diabetic rabbits. *Experientia*, **30**, 119.

Bapat, S. K., K. N. Ansari, A. C. Jauhari and V. Chandra (1970) Hypoglycaemic effect of two indigenous plants. *Indian Journal of Pharmacology*, **14**, 28–34.

Basu, D. K. and S. Rakhit (1957) Chemical investigation of *Hygrophila spinosa*. *Indian Journal of Pharmacy*, **19**, 205, 282, 285.

Ben-David, M., E. Menczel and F. G. Sulman (1963) The hypoglycemic effect of azacyclonol and its mechanism. *Archives Internationales de Pharmacodynamie et Thérapie*, **145**, 309–20.

Bhandahari, P. R. and B. Mukerje (1959) Garlic (*Allium sativum*) and its medicinal values. *Nagarjun, India*, 121.

Brahmachari, H. D. and K. T. Augusti (1961a) Hypoglycaemic agents from Indian indigenous plants. *Journal of Scientific and Industrial Research*, **13**, 381.

Brahmachari, M. D. and K. T. Augusti (1961b) Orally effective hypoglycaemic agents from plants. Orally effective principle from *Allium cepa*. *Journal of Pharmacy and Pharmacology*, **13**, 128.

Brahmachari, M. D. and K. T. Augusti (1962a) Orally effective hypoglycaemic principles from *Allium sativum* and *Picus religiosa*. *Journal of Pharmacy and Pharmacology*, **14**, 254.

Brahmachari, M. D. and K. T. Augusti (1962b) Effects of orally effective agent from plants on alloxan diabetes. *Journal of Pharmacy and Pharmacology*, **14**, 617.

Brahmachari, M. D. and K. T. Augusti (1963) Orally effective hypoglycaemic principles from *Coccinia indica*. *Journal of Pharmacy and Pharmacology*, **15**, 411.

Brouver, J. N., H. van der Wel, A. Francke and J. Hennig (1968) Miraculin, the sweetness producing protein from miracle fruit. *Nature, London*, **220**, 373.

Burkhard, M., H. G. Hormonsky and E. Boehli (1968) Mechanism of prostaglandin E induced hyperglycaemia. *Zeitschrift der Gesellschaft für experimentelle Medizin*, **148**, 99–107.

Busson, F. (1965) *Plantes Alimentaires de l'Ouest Africain*, 1 vol. Lecomte, Marseille, 588 pp.

Chatterjee, K. P. (1963) On the presence of an antidiabetic principle in *Momordica charantia*. *Indian Journal of Physiology and Pharmacology*, **7**, 240.

Chatterjee, M. S. and D. Roy (1965) Pharmacological studies of the seeds of *Securinega virosa*. Effect on normal blood sugar of cat and rabbit. *Bulletin of the Calcutta School of Tropical Medicine*, **3**, 12–14.

Chaudhury, R. R. and S. B. Vohora (1966) *Advances in Research in Indian Medicine. Plants with Possible Hypoglycaemic activity*. Indian Council of Medical Research, New Delhi, pp. 57–75.

Cochran, K. and H. F. Maasab (1970) Inhibition of a cold variant of influenza virus by selected chemicals. *Archives for Environmental Health*, **21**, 312–15.

Collip, J. B. (1923) Glukokinin, a new hormone in plant tissue. *Journal of Biological Chemistry, Baltimore*, **56**, 513–31.

Currie, A. L. and T. E. Timell (1959) Constitution of a methylglucuron-oxylan from kapok (*Ceiba pentandra*). *Canadian Journal of Chemistry*, **37**, 922–59.

De, U. V. and B. Mukherjee (1963) Effect of *Coccinia indica* Wright and Arn on alloxan diabetes in rabbits. *Indian Journal of Medical Sciences*, **7**, 665–72.

Deshpande, V. H. (1968) Four analogues of artocarpin and cycloartocarpin from *Morus alba*. *Tetrahedron. Letters*, 1968, 1715.

Dhalla, N. S., K. C. Gupta, M. S. Sastry and C. L. Malhotra (1961) Chemical composition of the fruit of *Momordica charantia*. *Indian Journal of Pharmacy*, **23**, 128–9.

Dhar, M. L., M. M. Dhar, B. N. Bhawan, B. N. Mehrothra and C. Ray (1968) Screening of Indian plants for biological activity Part I. *Indian Journal of Experimental Biology*, **6**, 232.

Donard, E. and H. Labbé (1933) Sur les propriétés hypoglycémiantes du maltose et de la mannite contenus dans les extraits de radicelles d'orge avant et après le fermentation. *Comptes Rendus de la Société de Biologie*, **112**, 1675.

Dupaigne, P. (1974) Quelques édulcorants naturels à fort pouvoir sucrant. *Plantes Médicinales et Phytothérapie*, **8**, 104–8.

Frèrejacque, M. and M. Durgat (1954) Poisons digitaliques de graines de jute. *Comptes Rendus de l'Académie des Sciences*, **238**, 507.

Garcia, F. (1941) Distribution and deterioration of the insulin-like principle in *Lagerstroemia speciosa*. *Acta Medica Philippensis*, **3**, 99 (through *Chemical Abstracts*, 1942, 560).

Garcia, F. and J. Colin (1926) Study on the antidiabetic properties of *Tecoma mollis*. *Preliminary Report*. *Journal of the American Pharmaceutical Association*, **15**, 556–60.

Ghosal, S., S. Singh and S. K. Bhattacharya (1971) Alkaloids of *Mucuna pruriens*. Chemistry and Pharmacology. *Planta Medica*, **24**, 434–40.

Githens, T. S. (1949) *Drug Plants of Africa*, 1 vol., ed. University of Pennsylvania Press, 125 pp.

Goldner, M. G. (1958) Oral hypoglycaemic agents past and present. *Archives of International Medicine*, **102**, 830–40.

Govidachari, T. R. *et al.* (1957) *Asteracantha longifolia* constituents. *Journal of Scientific and Industrial Research, India*, **16B**, 72.

Guerra, F. (1946) Farmacologia de plantas antidiabeticas mexicanas. Accion de la tronadora (*Tecoma mollis*) en la glucemia normal y la hiperglucemia diabetica. *Rev. Inst. salubr. y enfermed. Trop. Mexico*, **7**, 213–20.

Gueye, M. S. (1973) Contribution à l'étude pharmacodynamique d'une plante antidiabétique (*Sclerocarya birrea*). Thèse Pharmacie, Dakar.

Gupta, S. S. (1961) Inhibitory effect of *Gymnema sylvestre* (Gurmar) on adrenaline induced hyperglycaemia in rats. *Indian Journal of Medical Sciences*, **15**, 883–7.

Gupta, S. S. (1963a) Effect of indigenous antidiabetic drugs against acute hyperglycaemic response of anterior pituitary in glucose fed albino rats. *Indian Journal of Medical Research*, **51**, 716–24.

Gupta, S. S. (1963b) Effect of *Gymnema sylvestre* and *Pterocarpus marsupium* on glucose tolerance in albino rats. *Indian Journal of Medical Sciences*, **17**, 501–5.

Gupta, S. S. and M. C. Vatiyar (1964) Experimental studies on pituitary diabetes. III. Effect of *Gymnema Sylvestre* and *Coccinia indica* against hyperglycaemic response of some somatrophin and corticotrophin hormones. *Indian Journal of Medical Research*, **52**, 200–7.

Gupta, S. S., I. S. Jonathan and A. Ahmad (1966) Experimental studies on pituitary diabetes. Effects of *Ficus bengalensis* and pituitary extract on glucose tolerance in rats. *Indian Journal of Medical Research*, **54**, 354–62.

Hammouda, Y. and M. S. Amer (1966) Antidiabetic effect of tecomine and tectostanine. *Journal of Pharmaceutical Sciences*, **55**, 1452–4.

Hammouda, Y. and N. Khallafallah (1971) Stability of tecomine, the major antidiabetic factor of *Tecoma stans*. Juss. *Journal of Pharmaceutical Sciences*, **60**, 1142–5.

Hammouda, Y. and Motawi (1959) Alkaloids and triterpenes of *Tecoma stans*. *Proceedings of the Pharmaceutical Society of Egypt*, **16**, 73–179.

Hammouda, Y., M. M. Plat and J. Le Men (1963) Un nouvel alcaloide du *Tecoma stans:* la tecostanine. *Annales de Pharmacie françaises*, **21**, 699–702.

Hammouda, Y., A. Rashid and M. S. Amer (1964) Hypoglycaemic properties of tecomine and tecostanine. *Journal of Pharmacy and Pharmacology*, **16**, 833–4.

*Harborne, J. B., T. J. Mabry and H. Mabry (1974) *The Flavonoids*, Chapman and Hall, London.

Hardman, R. and F. R. Y. Fazli (1972) Steroid sapogenins from *Trigonella foenum graecum*. *Planta Medica*, **21**, 131, 188, 322.

Hartleb, C. (1932) Experimentelle und klinische Untersuchungen zur Frage der peroralen Diabetes Behandlung mit Insulinähnlichen pflanzlichen Stoffen (Phaseolan). *Münchener Medizinische Wochenschrift*, 1932, 1795–9.

Hericz *et al.* (1964) Alkaloids of *Securinega* species. *Chemical Pharmaceutical Bulletin Tokyo*, **122**, 1118.
Hermann, K. (1956, 1958) *Allium cepa* L. Flavonoiden. *Naturwissenschaften*, **43**, 158; *Archiv der Pharmazie*, **291**, 238.
Holt, C. von, L. von Holt and H. Bühm (1966) Metabolic effects of hypoglycin and methylenecyclopropane-acetic acid. *Biochemica & Biophysica Acta*, **125**, 11.
Holt, C. von, W. Leppla, B. Kroener and L. von Holt (1956) Zur chemischen Kennzeichnung der Hypoglycine. *Naturwissenschaften*, **43**, 279.
Hood, A. M. and E. J. L. Lowburry (1954) Anthocyanins in bananas. *Nature, London*, **173**, 402–3.
Iketobosin, G. D. and D. W. Mathieson (1963) The isolation of hordenine and norsecurinine from *Securinega virosa* Baill. *Journal of Pharmacy and Pharmacology*, **15**, 810.
Inglett, G. E. and J. F. May (1968) Tropical plants with unusual taste properties. *Economic Botany*, **22**, 326.
Inglett, G. E. and J. F. May (1969) Serendipity berries, source of a new intense sweetener. *Journal of Food Sciences*, **34**, 408
Iwu, M. M. (1980) Antidiabetic properties of *Bridelia ferruginea* leaves. *Planta Medica*, **39**, 247.
Iwu, M. M. (1983) The hypoglycaemic properties of *Bridelia ferruginea*. *Fitoterapia*, **54**, 243–8.
Jain, S. R. (1968) *Musa sapientum*. *Planta Medica*, **16**, 43–7.
Jain, S. R. (1969) *Musa sapientum* triterpenes. *Planta Medica*, **17**, 99–100.
Jain, S. R. and M. R. Jain (1972) Therapeutic utility of *Ocimum basilicum* var. *album*. *Planta Medica*, **22**, 136–9.
Jain, S. R. and S. N. Sharma (1967) Hypoglycaemic drugs of Indian indigenous origin. *Planta Medica*, **15**, 439–42.
Jain, R. C. and C. R. Vyas (1974) Antidiabetic-like activity of onion extracts. *British Medical Journal*, ii, 730.
Johnson, A. E., H. E. Nursten and A. A. Williams (1971) Organic disulphides in plants. *Chemistry and Industry*, 1971, 556.
Jones, G., F. H. M. Fales and W. C. Wildman (1963) The structure of tecomanine. *Tetrahedron Letters*, 397–401.
Kerharo, J. and J. G. Adam (1974) *La Pharmacopée Sénégalaise Traditionelle*, Vigot, Paris, 1011 pp.
Khanna, P., S. C. Jain, A. Panagariya and V. P. Dixit (1981) Hypoglycaemic activity of a polypeptide from a plant source. *Journal of Natural Products*, **44**, 648–65.
Kjaer, A. and P. Friis (1962) Isothiocyanates from *Putranjeva roxburghii* Wall including S-2-methylbutyl isothiocyanate, a new mustard oil of natural derivation. *Acta Chemica Scandinavica*, **16**, 936–46.
Krishnamurti, G. V. and T. R. Seshadri (1946) Bitter principle of *Phyllanthus niruri* L. *Proceedings of the Indian Academy of Sciences*, 1946, p. 357 (through *Chemical Abstracts* (1947), 2712).
Kulkarni, R. D. and B. B. Gaitonde (1962) Potentiation of tolbutamide action by jasad bhasma and karela (*Momordica*). *Indian Journal of Medical Research*, **50**, 717.
Kurihari, Y. (1969) Antisweet activity of gymnemic acid A_1 and its derivatives. *Life Sciences*, 1969, 537–43.
Kurihari, Y. and L. M. Beidler (1968) Taste modifying protein from miracle fruit. (*Synsepalum dulcificum*). *Science*, **161**, 1241.
Kurihari, Y. and L. M. Beidler (1969) Mechanism of the action of a taste-modifying protein. *Nature, London*, **222**, 1176.
Labbé, M. (1936) The vegetable insulinoides and their therapeutic indications. *Journal of the Canadian Medical Association*, **34**, 141–4.
Labo, B., J. V. Ma and A. C. Puig (1953) Hypoglycaemic effect of *Centaurea aspera*. *Farmacognosia*, **13**, 223.
Laroche Navaron, Patent 1968. *Chemical Abstracts* (1972) 158336t.

Laurens, A. and R. Paris (1976) Sur les polyphenols d'Anacardiacées africaineset malgaches; *Poupartia* spp. et *Anacardium occidentale*. *Plantes Médicinales et Phytothérapie*, **11**, 16–24.

Leclerc, H. (1934) Action hypoglycémiante de la feuille du murier noir (*Morus nigra* L.) *Presse Médicale*, **42**, 1522.

Lewis, J. J. (1950) Cabbage extracts and insulin-like activity. *British Journal of Pharmacology*, **5**, 21–4.

Lotlikar, M. M. and M. R. Rajarama (1960) Note on a hypoglycaemic principle isolated from the fruits of *Momordica charantia* L. *Journal of the University of Bombay*, **29**, 223–4.

Lotlikar, M. M. and M. R. Rajarama Rao (1966) Pharmacology of a hyperglycaemic principle isolated from the fruits of *Momordica charantia* L. *Indian Journal of Pharmacy*, **28**, 129–33.

Luscombe, D. K. and S. A. Taha (1974) Pharmacological studies on the leaves of *Azadirachta indica*. *Journal of Pharmacy and Pharmacology*, **26**, 110–11.

Lyass, M. A. and V. L. Vovski (1932) Kidney bean extract with properties similar to insulin for therapy of diabetes. *Sov. Klin.* **17**, 240.

MacDonald, A. D. and M. Wislicki (1938) Effect of cabbage extracts on carbohydrate metabolism. *Journal of Physiology, London*, **94**, 249.

McMillan (1954) *Tropical Planting and Gardening Sect. IV. Medicinal Plants*, Macmillan, London.

Majumdar, D. N. and C. D. Zalani (1953) *Mucuna pruriens* DC. Alkaloid constituents III. Isolation of water soluble alkaloids and a study of their chemical and physiological characterisations. *Indian Journal of Pharmacy*, **15**, 62–6.

Marquis, V. O., T. A. Adanlawo and A. A. Olaniyi (1977) The effect of foetidin from *Momordica foetida* on blood glucose level of albino rats. *Planta Medica*, **31**, 367–74.

Martindale (1958) *The Extra Pharmacopoeia*, 1 vol., 24th edn, The Pharmaceutical Press, London.

Masso, J. L., M. N. Bertran and T. Adzet (1979) Contribution à l'étude chimique et pharmacologique de quelques espèces de *Centaurea* (Composées). *Plantes Médicinales et Phytothérapie*, **13**, 41–5.

Matthew, P. T. and K. T. Augusti (1975) Hypoglycaemic effects of onion on diabetes mellitus. *Indian Journal of Physiology and Pharmacology*, **19**, 213–17.

Menczel, E., J. Mishinsky and F. G. Sulman (1965) Trigonelline in hyperglycaemia. *Proceedings of the Israeli Physiological and Pharmacological Society*, **1**, 47.

Mercier, F. and B. Bonnafous (1940) Action hypoglycémiante d'*Eugenia jambolana*. *Comptes Rendus des Séances de la Sociéte de Biologie, Paris*, **133**, 150.

Milhet, Y., F. Ferron and C. Costis (1978) Quelques résultats sur la physiologie d'*Abrus precatorius* L. *Plantes Médicinales et Phytothérapie*, **12**, 151–6.

Mishinsky, J., B. Joseph, F. G. Sulman and A. Goldschmied (1967) Hypoglycaemic effect of trigonelline. *Lancet*, 1311–12.

Mitra, P., P. Chakraborty and T. Ganguely (1975) Hypoglycaemic effect of indigenous drugs. *Bulletin of the Calcutta School of Tropical Medicine*, **23** (1–4), 6–7.

Modak, A. T. and M. R. Rajarama Rao (1966) Phytosterolin from *Ficus religiosa*. *Indian Journal of Pharmacy*, **28**, 105.

Monya, M. and G. Racz (1974) Recherches concernant le contenu en flavonosides de certaines espèces du genre *Centaurea* L. Recherches chromatographiques. *Plantes Médicinales et Phytothérapie*, **10**, 78–84.

Morris, J. A. and N. Juscy (1976) Taste modifying agents. *Lloydia*, **39**, 25–38.

Mukerjee, K. and N. C. Ghosh (1972) *Coccinia indica* as potential hypoglycaemic agent. *Indian Journal of Experimental Biology*, **10**, 347–9.

Mukherjee, S. K., U. N. De and B. Mukherjee (1963) Contribution in the field of diabetes research in the last decade. *Indian Medical Gazette*, **3**, 97.

Nair, A. G. R. and S. Subramanian (1962) Eugenia triterpene A and Eugenia triterpene B in *Eugenia jambolana*. *Journal of Scientific and Industrial Research, India*, **21B**, 437.

Naito, K. (1968) Moracetin from the leaves of the mulberry tree. *Agriculture, Biology and Chemistry, Tokyo*, **32**, 33a.

Nara, F. K., J. Gleye, E. de C. Lavergne and E. Stanislas (1977) Flavonoides de *Phyllanthus niruri* L. et *Phyllanthus* spp. *Plantes Médicinales et Phytothérapie*, **11**, 82–6.

Nath, M. C. (1943) Investigations on a new antidiabetic principle (amellin) occurring in nature; studies on some of its biochemical properties. *Annals of Biochemistry and Experimental Medicine*, **3**, 55–62.

Nath, M. C. and S. R. Bannerjee (1943) Amellin; its effect on glycosuria and hyperglycaemia in cases of human diabetes. *Annals of Biochemistry and Experimental Medicine*, **3**, 63–84.

Nath, M. C. and N. K. Chowdurry (1943) Amellin; its effect on utilization of inorganic phosphate in blood of diabetics. *Annals of Biochemistry and Experimental Medicine*, **3**, 121–30. Amellin; its influence in causing relief in hypercholesterolemia. *Ibid*, **3**, 147–56.

Nath, M. C. and N. K. Chowdurry (1945) Amellin; its role in prevention of excessive protein catabolism in diabetes. *Annals of Biochemistry and Experimental Medicine*, **5**, 11–16.

Nath, M. C., M. K. Chakravorty and S. R. Bannerjee (1943) Amellin; its influence in increasing haematopoietic activity in diabetes. *Annals of Biochemistry and Experimental Medicine*, **3**, 107–20.

Nath, M. C., M. K. Chakravorty and M. D. Brahmachari (1945) Amellin; its role in reduction of acetone bodies and increase of alkali reserve of blood nitrogen of diabetics. *Annals of Biochemistry and Experimental Medicine*, **5**, 101–4.

Noble, R. L., G. T. Beer and J. H. Cutts (1958) Role of chance observation in chemotherapy, *Vinca rosea*. *Annals of the New York Academy of Sciences*, **76**, 882.

Nomora, T. and T. Fukai (1981) Constituents of the cultivated Mulberry tree VII. Isolation of three new isoprenoid flavanones from root bark of *Morus alba*. *Planta Medica*, **42**, 79–88.

Olaniji, A. A. (1975) A neutral constituent of *Momordica foetida*. *Lloydia*, **38**, 361–2.

Olaniji, A. A. and V. O. Marquis (1975) Phytochemical and preliminary pharmacological investigation of the alkaloid contained in *Momordica foetida*. *Journal of Pharmacy (Nigeria)*, **6**, 117–19.

Oliver-Bever, B. and G. R. Zahnd (1979) Plants with oral hypoglycaemic action. *Quarterly Journal of Crude Drug Research*, **17**, 139–96.

Osuntokun, B. O. (1975) Diabetes mellitus as the cause of atherosclerotic vascular disease in Nigeria. *West African Medical Journal*, **24**, 133–6.

Pant, M. C., I. Uddin, U. R. Bhardway and R. D. Tewari (1968) Blood sugar and total cholesterol lowering effect of *Glycine soja* (Sieb. & Succ.), *Mucuna pruriens* DC. and *Dolichos biflorus* L. seeds in normal fasting albino rats. *Indian Journal of Medical Research*, **561**, 1808–12.

Persaud, T. V. (1972) Effect of intraamniotic administration of hypoglycin B on foetal development in the rat. *Experimental Pathology*, **6**, 55–8.

Peters, G. (1957) Insulinersatzmittel pflanzlichen Ursprungs. *Deutsche Medizinische Wochenschrift*, **82**, 320–2.

Plouvier, V. (1948) Sur la recherche des itols (dulcitol) et du saccharose chez quelques Sapindales. *Comptes Rendus de l'Académie des Sciences*, **227**, 85–7.

Pourrat, H. (1977) Drogues à anthocyanes et maladies vasculaires. *Plantes Médicinales et Phytothérapie*, XL, n-spéc. 143–51.

Pourrat, H., P. Tronche and A. Pourrat (1977) Nouveau procédé d'extraction des glycosides d'anthocyanes. *Bulletin de la Société chimique de France*, 1966, 1918–20; and *Bulletin de Chimie et Thérapie*, 1967, 33.

Pourrat, H., J. P. Guichard, A. Pourrat and J. L. Malmaison (1978) Anthocyanes et flavones des feuilles de *Vaccinium corymbosum* L. *Plantes Médicinales et Phytothérapie*, **12**, 212–16.

Randle, P. J., P. B. Garland, C. N. Hales and A. E. Newsholme (1963) The glucose fatty acid cycle: its role in insulin sensitivity and the metabolic disturbances of diabetes mellitus. *Lancet*, i, 785–9.

Rivera, C. (1941, 1942) Preliminary chemical and pharmacological studies on 'cundeamor', *Momordica charantia*. *American Journal of Pharmacy*, **113**, 281; **114**, 72–8.
Rouffiac, R. and J. Perello (1969) Etude chimique des alcaloides de *Phyllanthus niruri* L. Présence de l'antipode optique de la norsecurinine. *Plantes Médicinales et Phytothérapie*, **3**, 220.
Sachser, J. A. (1961) An IAA oxidase inhibitor system in bean pods. *American Journal of Botany*, **48**, 820.
Saito, S., T. Tanaka *et al.* (1964a) Determination of securinine and its stereo-isomers in plants of the *Securinega* species and isolation of viroallosecurinine and virosine. *Journal of the Pharmaceutical Society, Japan*, **84**, 1126–33.
Saito, S. *et al.* (1964b) Structure and stereochemistry of norsecurinine and dihydrosecurinine. *Chemical and Pharmaceutical Bulletin, Tokyo*, **12**, 1520.
*Samilova, R. D. and T. A. Lagodich (1977) The glycoside olitoriside from *Corchorus olitorius*. *Vrach Delo*, 1977, 27–31 (through *Chemical Abstracts* (1978), **86**, 133178m).
Satoda, I. (1962) Studies on securinine and allosecurinine. *Tetrahedron Letters*, 1962, 1199–1202.
Sengupta, P. and B. P. Das (1965) Terpenoids and related compounds V. Tri-terpenoids from the flowers of *Eugenia jambolana* Lam. *Journal of the Indian Chemical Society*, **42**, 539.
Sepaha, G. C. and S. N. Bose (1956) Clinical observations on the antidiabetic properties of *Pterocarpus marsupium* and *Eugenia jambolana*. *Journal of the Indian Medical Association*, **27**, 388.
Shani, J. A., A. Goldschmied, B. Joseph, Z. Ahronson and F. G. Sulman (1974) Hypoglycaemic effect of *Trigonella foenum graecum and Lupinus terminis* seeds and their major alkaloids in alloxan diabetic and normal rats. Arch Int. Pharmacodyn. Ther. **210**, 27–37.
Shanmugasundaram, K. R., C. Panneerselvam, P. Samudram and E. R. B. Shanmugasumdarum, (1983) Enzyme changes of glucose utilisation in diabetic rabbits: The effect of *Gymnema sylvestre* R.Br. *Journal of Ethnopharmacology*, **7**, 205–34.
Sharaf, A. and M. Y. Mansour (1964) Pharmacological studies on the leaves of *Morus alba*, with special reference to its hypoglycemic activity. *Planta Medica*, **12**, 71–2.
Sherrat, H. S. A., P. C. Holland, J. Marley and A. E. Senior (1970) Mode of action of hypoglycin and related compounds. In: *Symposium on the Mechanism of Toxicity 1970*, ed. W. N. Aldrige, pp. 205–18. St. Martins, New York.
Shrothri, D. S., M. Kehar, V. K. Deshmukh and R. Aiman (1963) Investigations of the hypoglycaemic properties of *Vinca rosea*, *Cassia auriculata* and *Eugenia jambolana*. *Indian Journal of Medical Research*, **51**, 464.
Sicognau-Jagodzinski, M., P. Bibal-Prot, M. Chanez, P. Boiteau and R. Ratsimamanga (1966) *Eugenia jambolana*. *Comptes Rendus de l'Académie des Sciences*, **264**, D 1119–23.
Simmonds, N. W. (1954) Anthocyanins in bananas. *Nature, London*, **173**, 402.
Sinha *et al.* (1962) 5-Hydroxytryptamine in bananas. *Biological Abstracts*, **30**, 16587.
Sinsheimer, J. E., R. G. Subba and B. McIlhenny (1970) Constituents of *Gymnema sylvestre* leaves. V. Isolation and preliminary characterization of the gymnemic acids. *Journal of Pharmaceutical Sciences*, **59**, 622–8 and 629–32. Isolation and antiviral activity of gymnemic acids. *Experientia*, **24**, 302–3.
Stanislas, E., R. Rouffiac and J. J. Foyard (1967) Constituants de *Phyllanthus niruri* L. *Plantes Médicinales et Phytothérapie*, **1**, 13–141.
Stoecklin, W. (1968, 1969) Gymnestrogin, ein neues pentahydroxyterpen aus den Blättern von *Gymnema sylvestre*. *Helvetica Chimica Acta*, **51**, 1235–45; Structure. *Helvetica Chimica Acta*, **52**, 365–70.
Stoecklin, W., E. Weiss and T. Reichstein (1967) *Gymnema sylvestre*. *Helvetica Chimica Acta*, **50**, 474.
Sucrow, W. (1965) Uber Steringlucoside und ein neues Stigmastadienol aus *Momordica charantia*. *Tetrahedron Letters*, 26, 2217.

Sucrow, W. (1966) Inhaltstoffe von *Momordica charantia* L. I: $\Delta^{5.25}$ stigmastadeniol (3β-) und sein βD-glucosid. *Chemische Berichte*, **99**, 2765–78.

Sulman, F. G. and E. Menczel (1962) Antidiabetic plant products: Extracts of *Eragrostis bipinnata, Opuntia Ficus indica, O. vulgaris, Teucrium polium, Trigonella foenum graecum* and *Zea styles*. *Harokeach Haivre*, **9**, 6–26 (through *Chemical Abstracts*, **57**, 11308c).

Svoboda, G. H. (1969) Alkaloids of *Catharanthus roseus* in cancer chemotherapy. *Current Topics in Plant Science*, 303–5.

Svoboda, G. H., M. Gorman and M. A. Root (1964) Alkaloids of *Vinca rosea*. A preliminary report on hypoglycaemic activity. *Lloydia*, **27**, 361–3.

Talyshinski, G. M. (1967) Formation of rutin in leaves of mulberry trees. *Dokl. Akad. Nauk. Azerb.* **23**, 63–6 (through *Chemical Abstracts*, **67**, 71091z).

Tanaka, K., K. J. Isselbacher and U. Shils (1972) Isovaleric and methylbutyric acidemias induced by hypoglycin A. Mechanism of vomiting sickness. *Science*, **175**, 69–71.

Tella, A. F. and O. O. Ojihomon (1980) An extraction method for evaluating the seed proteins of cowpea (*Vigna unguiculata* (L.) Walp.) *Journal of Science, Food and Agriculture*, **31**, 1268–74 (through *Chemical Abstracts*, **94**, 188159b).

Trivedi, C. P. (1963) Observations on the effect of some indigenous drugs on the blood sugar level. *Physiology and Pharmacology*, **7**, 11.

Ucciani, E., J. P. Defretin, M. Boutoux and F. Busson (1964) The proteins and lipids of *Blighia sapida*. *Oléagineux*, **19**, 18–19, 563–9.

Uzan, M. and A. Dziri (1952) Influence sur le metabolisme glucidique d'un extrait des feuilles de *Corchorus olitorius*. *Semaines des Hôpitaux de Paris*, **28**, 2532.

Vad, B. G. (1960) Place of *Momordica charantia* in the treatment of diabetes. *Maharasthra Medical Journal*, **6**, 733.

Varsney, I. P. and S. C. Sharma (1968) Saponins and sapogenins XXXII. *Trigonella foenum graecum* seeds. *Journal of the Indian Chemical Society*, **43**, 564–7.

Venkateswarlu, G. J. (1952) Cyanidine rhamnoglucosides in *Eugenia jambolana*. *Indian Chemical Society*, **29**, 435.

Viguera, J. M. and A. Casabuena (1965) Hypoglycaemic activity in the genus *Centaurea*. VI. Fractionation of the active peptides by ammonium sulfate. *Farmacognosia, Madrid*, **25**, 89–102.

Vila, C. (1940) Sobre la pretendida accion antidiabetica del sarandi blanco (*Phyllanthus sellowianus* Müll.) *Revista del Medicina Rosario*, **30**, 921–30.

Vohora, S. B. (1970) Antidiabetic studies on *Ficus bengalensis* L. *Indian Journal of Pharmacy*, **32**, 68–70.

Vohora, S. B., M. Rizwan and J. A. Khan (1973) Medicinal uses of common Indian vegetables. *Planta Medica*, **23**, 381–93.

Wel, H. van der (1972) Isolation and characterisation of the sweet principle from *Dioscoreophyllum cumminsii* (Stapf) Diels. *FEBS Letters*, **21**, 88.

Wel, H. van der and K. Loeve (1972) Isolation and characterisation of Thaumatin I and II, the sweet tasting proteins of *Thaumatococcus daniellii*. *European Journal of Biochemistry*, **31**, 221.

Weniger, R., M. Haag-Berrurier and R. Anton (1982) Plants of Haiti used as antifertility agents. *Journal of Ethnopharmacology*, **6**, 67–84.

Whittacker, H. (1948) Amellin for diabetes. *British Medical Journal*, i, 546–7.

BOTANICAL AND GENERAL INDEX

Main entries are italicized, synonyms are not italicized.

Species	Common name	Family	Page
Butyrospermum paradoxum subsp. *parkii* (Don) Hepper	Shea butter tree	Sapotaceae	144
Caesalpinia bonduc (L.) Roxb. (C. crista L., C. bonducella (L.) Flem.)	Bonduc or physic (nicker) nut	Caesalpiniaceae	155, 158
Caesalpinia pulcherrima (L.) Sw. c.	Pride of Barbados	Caesalpiniaceae	158
Cajanus cajan (L.) Millsp.	Pigeon pea	Fabaceae	221
Callichilia barteri (Hook.) Stapf (see Hedranthera barteri)			
Callichilia monopodialis (Schum.) Stapf		Apocynaceae	12
Caloncoba echinata (Oliv.) Gilg	Gorli	Flacourtiaceae	130
Caloncoba glauca (P. Beauv.) Gilg		Flacourtiaceae	130
Caloncoba welwitschii (Oliv.) Gilg		Flacourtiaceae	130
Calophyllum inophyllum L. c.		Guttiferae	207
Calotropis procera (Ait.) Ait.	Mudar, sodom apple, swallow wort	Asclepiadaceae	18, 24, 144, 226
Calyptranthes guineensis Willd. (see Syzgium guineense var. guineense)			
Camellia thea Link	Tea plant	Ternstroemaceae	51
Canavallia ensiformis (L.) DC.	Sword or horse bean	Fabaceae	53, 158
Cannabis sativa L. var. *indica* Lam.	Indian hemp	Cannabinaceae	78, 144, 212
Canscora decussata (Roxb.) Roem. & Schult.		Gentianaceae	134, 212
Capparis decidua (Forsk.) Edgew. (C. aphylla Hayne ex Roth.)		Capparidaceae	136, 212
Capsicum annuum L.	Capsicum	Solanaceae	144, 204
Capsicum frutescens L.	African pepper	Solanaceae	204
Carapa procera DC. (C. guineensis Sweet ex Juss., C. touloucouna Guill. & Perr., C. gummiflua DC., C. velutina DC., C. microcarpa Chev.)	Crabwood, monkey cola	Meliaceae	164
Cardiospermum halicacabum L. (C. micro-carpum Kunth)	Balloon vine	Sapindaceae	122, 14
Carica papaya L.	Pawpaw, papaya	Caricaceae	42, 138, 145, 150, 167, 174, 212, 224, 227, 24
Carissa edulis Vahl (edulis forma pubescens (DC.) Pichon, C. pubescens DC.)		Apocynaceae	11, 17
Carpodinus umbellatus Schum. (see Hunteria umbellata)			
Cassia absus L.	Four-leaved senna	Fabaceae	93, 14
Cassia alata L. c. and nat.	Ringworm bush	Fabaceae	14
Cassia fistula L.	Indian laburnum	Fabaceae	15
Cassia occidentalis L.	Coffee senna, stinkweed	Fabaceae	14

Species	Common name	Family	Pag
Coffea arabica L.	Coffee	Rubiaceae	6
Cola acuminata (Beauv.) Schott & Endl. (C. pseudoacuminata Engl.)	Kola nut tree	Sterculiaceae	6
Cola nitida (Vent.) Schott & Endl. (Cola acuminata Engl.)		Sterculiaceae	6
Combretum floribundum Engl. & Diels, C. altum Perr.)			
Combretum micranthum Don (C. altum Perr.)	Kinkeliba	Combretaceae	13
Combretum racemosum Beauv.		Combretaceae	13
Combretum rhodanthum Engl. & Diels		Combretaceae	13
Commiphora africana (Rich.) Engl.	African Myrrh	Burseraceae	19
Conocarpus leiocarpus DC. (see Anogeissus leiocarpus)			
Conopharyngia durissima (Stapf) Stapf, C. crassa Benth. Stapf (see Tabernaemontana crassa)			
Conopharyngia pachysiphon (Stapf) Stapf (see Tabernaemontana pachysiphon)			
Conyza cinerea L. (see Vernonia cinerea (L.) Less.)			
Conyza pauciflora Willd. (see Vernonia pauciflora)			
Corchorus olitorius L.	Jute	Tiliaceae	25, 26
Corynanthe johimbe Schum. (see Pausinystalia johimbe)			
Corynanthe pachyceras Schum.		Rubiaceae	40, 62, 9
Costus afer Ker-Gawl., *C. obliterans* Schum.	Ginger lily	Zingiberaceae	19
Costus anomocalyx Schum., C. insularis Chev., C. lucanusciamis Chev.			
Cotyledon pinnata Lam. (see Bryophyllum pinnatum)			
Cracca purpurea L. (see Tephrosia purpurea)			
Crateva religiosa Forst. (C. adansonii Oliv.)		Capparidaceae	121, 145, 21
Crossopteryx febrifuga (Afzel. ex G. Don) Benth. (C. kotschyana Fenzl)		Rubiaceae	62, 16
Croton macrostachyus Hochst. ex Del. (Croton guerzesiensis Beille)		Euphorbiaceae	18
Croton oppositifolius Geisel (see Mallotus oppositifolius)			
Cryptolepis sanguinolenta (Lindl.) Schltr. (C. triangularis N.E.Br.)		Asclepiadaceae	18, 41, 131, 20
Cryptostegia grandiflora (Roxb.) R.Br. ex Lindley		Asclepiadaceae	18, 21, 2
Cucurbita maxima Duchesne	Squash gourd	Cucurbitaceae	17
Cucurbita pepo L.	Pumpkin	Cucurbitaceae	17

Species	Common name	Family	Pa
Elaeocarpus sphaericus Schum. (E. ganitrus Roxb.)		Tiliaceae	
Elaeis guineensis Jacq.	Oil palm	Palmae	2
Elcaja roka Forsk. (see Trichilia roka)			
Embelia schimperi Vatke (E. abyssinica Bak.)		Myrsinaceae	1
Enantia chlorantha Oliv.	African Yellow-wood	Annonaceae	1
Enantia polycarpa (DC.) Engl. & Diels		Annonaceae	1
Entada africana Guill. & Perr. (E. sudanica Schweinf., E. ubanguiensis de Wild., Entadopsis sudanica (Schweinf.) Gilb. & Boutique)		Mimosaceae	1
Epibaterium pendulum J. & G. Forst., Epibaterium leaeba (Del.) DC. (see Cocculus pendulus)			
Epinetrum cordifolium Mangenot & Miège		Menispermaceae	59, 1
Eriodendron anfractuosum DC. (see Ceiba pentandra)			
Erythrina excelsa Bak. (E. sereti de Wild.)		Fabaceae	1
Erythrina mildbraedii Harms (E. altissima Chev.)		Fabaceae	1
Erythrinas senegalensis DC.	Coral flower, parrot tree	Fabaceae	1
Erythrina sigmoidea Hua (E. dybowski Hua, E. eryotricha Harms)			1
Erythrina vogelii Hook. f.		Fabaceae	1
Erythrophleum guineense G. Don	Sassy bark, ordeal tree	Caesalpiniaceae	28,
Erythrophleum ivorense Chev. (E. micranthum Harms.)		Caesalpiniaceae	[illegible]
Erythrophleum suaveolens (Guill. & Perr.) Brenam		Caesalpiniaceae	[illegible]
Erythroxylum coca Lam.		Erythroxylaceae	[illegible]
Eucalyptus globulus Labill. c.	Blue gum tree	Myrtaceae	14
Eugenia jambolana Lam. (see Syzygium cumini)			
Eupatorium coloratum Willd. (see Vernonia colorata)			
Euphorbia hirta L. (E. pilulifera Chev.)	Asthma herb	Euphorbiaceae	113, 167, 1
Euphorbia prostrata Ait.		Euphorbiaceae	14
Euphorbia thymifolia L. (E. burmannia Gay partly; E. aegyptiaca Soiss.; E. scordifolia Jacq. of F.T.A. partly)		Euphorbiaceae	14
Euphorbia tirucalli L.		Euphorbiaceae	16
Fagara leprieuri (Guill. & Perr.) Engl.		Rutaceae	[illegible]
Fagara macrophylla Engl. (see Zanthoxylum gilletii (de Wild.) Watson)			

Species	Common name	Family	Page
Fagara rubescens (Planch. ex Hook. f. Engl.) (see Zanthoxylum rubescens Planch. ex Hook. f.) Watson			
Fagara zanthoxyloides Lam. (F. senegalensis (DC.) Chev. (see Zanthoxylum zanthoxyloides (Lam.) Watson)			
eretia apodanthera Del. (F. canthioides Hiern)		Rubiaceae	62
Fluggea klaineana Pierre ex Chev. (see Phyllanthus discoideus)			
Fluggea microcarpa Blume (see Securinega virosa)			
Fluggea obovata var. luxurians Beille (see Phyllanthus discoideus)			
Fluggea virosa (Roxb. ex Willd.) Baill. (see Securinega virosa)			
untumia africana (Benth.) Stapf, F. latifolia (Stapf) Schlechter	False rubber tree	Apocynaceae	13, 193, 218
untumia elastica (Preuss) Stapf	West African rubber tree	Apocynaceae	193
Gabunia glandulosa Stapf (see Tabernaemontana glandulosa)			
Glossopholis dinklagei (Engl.) Stapf (see Tiliacora dinklagei)			
Gnidia kraussiana Meisn. (see Lasiosiphon kraussianus)			
ngronema latifolium Benth.		Asclepiadaceae	23
Gossampinus buonopozensis (Beauv.) Bak. (see Bombax buonopozense)			
ssypium herbaceum L., *G. hirsutum* L., *G. barbadense* L.	Cotton plant	Malvaceae	156, 227, 232
ewia bicolor Juss. (G. salvifolia Heyne ex Roth)		Tiliaceae	116
ewia carpinifolia Juss.		Tiliaceae	116
ewia lasiodiscus Schum. (G. kerstingii Burrett)		Tiliaceae	116
Groutia celtidifolia Guill. & Perr. (see Opilia celtidifolia)			
iera senegalensis Gmell.		Combretaceae	89, 146
Guillandina bonduc L. (see Caesalpinia bonduc)			
nandropsis gynandra (L.) Briq. (G. pentaphylla DC.)	Cat's whiskers	Capparidaceae	175, 213
nnema sylvestre (Retz.) Schultes		Asclepiadaceae	23, 213, 262
Gymnosporia senegalensis (Lam.) Loes. (see Maytenus senegalensis)			
ematoxylum campechianum L. c.	Logwood	Caesalpiniaceae	151, 213

Species	Common name	Family	I
Harrisonia abyssinica Oliv. (H. occidentalis Engl.)		Simaroubaceae	
Hedranthera barteri (Hook. f.) Pichon		Apocynaceae	
Hedysarum gangeticum L. (see Desmodium gangeticum)			
Helmia dumetorum Kunth. (see Dioscorea dumetorum)			
Heptacyclum Engl. (see Penianthus zenkeri)			
Heudelotia africana Rich. (see Commiphora africana)			
Hibiscus rosa-sinensis L.		Malvaceae	224, 227,
Hibiscus vitifolius L.		Malvaceae	
Holarrhena floribunda (Don) Dur. & Schinz. var. floribunda (H. africana DC., H. wulfsbergii Stapf)	Conessi or kurchi bark	Apocynaceae	13, 38, 72, 84, 193,
Holalafia multiflora Stapf (see Alafia multiflora)			
Hordeum vulgare L.	Barley	Gramineae	
Hunteria eburnea Pichon		Apocynaceae	11,
Hunteria elliotii (Stapf) Pichon		Apocynaceae	
Hunteria umbellata (Schum.) Hallier	Erin tree	Apocynaceae	39,
Hydrocotyle asiatica L. (see Centella asiatica (L.) Urb.)			
Hygrophyla auriculata (Schum.) Heine (H. spinosa Anders.)		Acanthaceae	151,
Hylacium owariense Beauv. (see Rauvolfia vomitoria)			
Hymenocallis littoralis Salisb.	Spider lily	Amaryllidaceae	151,
Hymenocardia acida Tul.		Euphorbiaceae	
Hyptis suavolens (L.) Poit.	Bush tea-bush	Labiatae	
Ipomoea mauritiana Jacq. (I. digitata of F.T.A.)		Convolvulaceae	2
Ipomoea pes-caprae (L.) Sweet (I. pes-caprae Roth)	Goats foot Convolvulus	Convolvulaceae	2
Ipomoea purpurea (L.) Roth		Convolvulaceae	2
Isolona campanulata Engl. & Diels (I. leonensis Sprague & Hutch., I. soubreana Chev. ex Hutch. & Dalz.)		Annonaceae	1
Jasminum sambac (L.) Ait		Oleaceae	2
Jateorhiza macrantha (Hook. f.) Exell. & Mendonça (J. strigosa Miers)		Menispermaceae	60,
Jatropha curcas L.	Physic nut, Barbados nut	Euphorbiaceae	54, 1
Jatropha gossypifolia L.	Wild cassada	Euphorbiaceae	1
Jatropha podagrica Hook.		Euphorbiaceae	

Species	Common name	Family	Page
Justicia insularis Anders. (J. galeopsis Anders.)		Acanthaceae	76
Kalanchoe pinnata (Lam.) Pers. (see Bryophyllum pinnatum)			
Karlea berchemoides Pierre (see Maesopsis eminii Engl.)			
Khaya senegalensis (Desv.) Juss.	Dry zone mahogany cail cedrat	Meliaceae	85, 146, 168
Kigelia africana (Lam.) Benth. (K. pinnata (Jacq.) DC., K. aethiopica Decne., K. eliottii Sprague (16 synonyms))	Sausage tree	Bignoniaceae	240
Kolobopetalum auriculatum Engl. (K. veitchianum Diels)		Menispermaceae	60, 84
Kickxia africana, K. latifolia, K. zenkeri and K. elastica (see Funtumia)			
agerstroemia speciosa (L.) Pers. (L. flos regina Retz.)		Lythraceae	263
antana camara L. (L. antidotalis Thonn.)	Wild sage, Bahama tea	Verbenaceae	118
asiosiphon kraussianus (Meisn.) Burtt-Davy (L. kraussi Meisn., L. kerstingii Pearson, L. guineensis Chev.)		Thymelaceae	146
awsonia inermis L. (L. alba Lam.)	Henna, Egyptian privet	Lythraceae	53, 146
epidium sativum L. c. and nat.	Common cress	Cruciferae	151
eptactina densiflora Hook. f.		Rubiaceae	62
eptactina senegambica Hook. f.		Rubiaceae	62
eptadenia hastata (Pers.) Decne. (L. lancifolia (Schum. & Thonn.) Decne)		Asclepiadaceae	23
eptadenia pyrotechnica (Forsk.) Decne. (L. spartium Wight & Arn.)		Asclepiadaceae	213
ucena glauca (L.) Benth.	Wild or horse tamarind	Mimosaceae	227
Limonia Warnecki Engl. (see Afraegle paniculata)			
Lithospermum hispidissimum Sieber ex Lehm (see Arnebia hispidissima)			
Lochnera rosea Reichb. (see Catharanthus roseus)			
nchocarpus cyanescens (Schum. & Thonn.) Benth.	Indigo vine	Fabaceae	202
nchocarpus laxiflorus Guill. & Perr.		Fabaceae	202

Species	Common name	Family	Pa
Lonchocarpus sericeus (Poir.) H.B. & K.		Fabaceae	1
Lupinus tassilicus Maire (*L. termis* Forsk.)		Fabaceae	216, 227, 2
Maesopsis eminii Engl. (M. berchemoides (Pierre) Chev.)		Rhamnaceae	1
Mallotus oppositifolius (Geisel.) Müll. Arg. (M. beillei Chev.)	Kamala	Euphorbiaceae	1
Mallotus philippinensis c.		Euphorbiaceae	1
Malouetia heudelotii DC.		Apocynaceae	
Mangifera indica L.	Mango tree	Anacardiaceae	50, 1
Manihot esculenta Crantz (M. utilissima Pohl.)	Cassava, manioc	Euphorbiaceae	2
Mansonia altissima (Chev.) Chev. var. *altissima*		Sterculiaceae	
Marsdenia latifolia (Benth.) Schum. (see Gongronema latifolium)			
Massularia acuminata (Don) Bullock ex Holye		Rubiaceae	1
Medicago sativa L. c.		Fabaceae	2
Melia azadirachta L., Melia indica (Juss.) Brandis (see Azadirachta indica)			
Melia azedarach L.	Persian lilac, bead tree	Meliaceae	1
Mesua ferrea L. c.		Guttifereae	2
Milletia aboensis (Hook.) Bak. (M. macrophylla var. aboensis Hook. f.)		Fabaceae	1
Milletia barteri (Benth.) Dunn		Fabaceae	1
Mimosa farnesiana L. (see Acacia farnesiana)			
Mimosa glauca L. (see Leucena glauca)			
Mimosa glomerata Forsk., M. nutans Pers. (see Dichrostachys glomerata)			
Mimosa lebbeck L. (see Albizia lebbeck)			
Mimosa nilotica L. (see Acacia nilotica var. nilotica)			
Mimusops elengi L.		Sapotaceae	2
Mitragyna ciliata Aubrev. & Pellegr.		Rubiaceae	4
Mitragyna inermis (Willd.) Ktze. (M. africana (Willd.) Korth.)		Rubiaceae	40, 62, 16
Mitragyna stipulosa (DC.) Ktze. (M. macrophylla Hiern)	African linden	Rubiaceae	40, 62, 95, 16
Modecca cissampeloides Planch. ex Benth. (see Adenia cissampeloides)			
Momordica charantia L.	African cucumber, balsam pear	Cucurbitaceae	146, 184, 228, 23 247, 24

Species	Common name	Family	Page
Montanoa tomentosa (not in West Africa)		Compositae	
Morinda longiflora G. Don.		Rubiaceae	63
Morinda lucida Benth. (M. citrifolia Chev.)	Brimstone tree	Rubiaceae	44, 63
Morinda geminata DC.		Rubiaceae	62, 175
Moringa oleifera Lam. (M. pterygosperma Gaertn.)	Horseradish tree	Moringaceae	46, 136, 146, 214
Morus alba L., *Morus nigra* L.	Mulberry	Moraceae	257, 259
Morus excelsa Welw. (see Chlorophora excelsa)			
Mostuea hirsuta (Anders. ex Benth.) Baill. ex Bak.		Loganiaceae	44, 72
Mucuna pruriens (L.) DC. var. pruriens	Cowhage, Cowitch	Fabaceae	102, 263
Mundulea sericea (Willd.) Chev. (M. suberosa (DC.) Benth.)		Fabaceae	178
Musa sapientum L. (Musa paradisiaca var. sapientum Ktze.)	Banana	Musaceae	46, 257, 259
Myristica fragrans Houtt.	Nutmeg	Myristicaceae	79
Nauclea africana Willd. (see Mitragyna inermis)			
Nauclea diderrichii (de Wild. & Dur.) Merrill		Rubiaceae	63, 76
Nauclea latifolia Sm. (N. esculenta Afz. ex. Sab. Merrill)	African or Guinea	Rubiaceae	63, 75
Nauclea pobéguinii (Pob. ex Pellegr.) Petit		Rubiaceae	63
Nauclea stipulosa DC. (see Mitragyna stipulosa)			
Nerium obesum Forsk (see Adenium obesum)			
Nerium oleander L.	Oleander	Apocynaceae	14, 17
Newbouldia laevis (Beauv.) Seem. ex Bureau.		Bignoniaceae	117, 168
Nicotiana tabacum L.	Tobacco plant	Solanaceae	185
Nymphaea lotus and spp. (N. liberiensis Chev.)	Water lily	Nymphaceae	87, 168
Ocimum basilicum L. (O. americanum L., O. lanceolatum Schum. & Thonn., O. dichotomum Hochst ex Benth., O. menthaefolium Chev.)	Basil	Labiatae	66, 146, 185
Ocimum canum Sims (O. hispidum Schum. & Thonn.)		Labiatae	66, 146, 185
Ocimum gratissimum L. (O. viride Willd., O. guineense Schum. & Thonn.)		Labiatae	66

Species	Common name	Family	Pa
Olax latifolia Engl.		Olacaceae	1
Oldenlandia affinis DC.		Rubiaceae	2
Omphalogonus nigritans N.E.Br. (see Parquetina nigrescens)			
Oncoba echinata Oliv. (see Caloncoba echinata)			
Oncoba glauca (P. Beauv.) Hook. f. (see Caloncoba glauca)			
Ophiocaulon cissampeloides (Planch. ex Benth.) Mast. (see Adenia cissampeloides)			
Opilia celtidifolia (Guill. & Perr.) Endl. ex Walp. (O. amentaceae Chev.)		Opiliaceae	1
Ormosia laxiflora Benth. ex Bak. (see Afrormosia laxiflora)			
Oryza sativa L.		Gramineae	2
Pachycarpus lineolata (Decne.) Bullock (Asclepias lineolata (Decne) Schlechler partly)		Asclepiadaceae	[illegible]
Pachypodanthium staudtii Engl. & Diels		Annonaceae	1
Parquetina nigrescens (Afzel.) Bullock	Silk vine	Apocynaceae	18, 2
Passiflora edulis Sims.	Passion fruit	Passifloraceae	8
Passiflora foetida L.	Stinking passion flower	Passifloraceae	8
Paullinia pinnata L.		Sapindaceae	27, 165, 16
Pauridiantha viridiflora (Schweinf. ex Hiern) Hepper		Rubiaceae	63, 74, 16
Pausinystalia johimbe (K. Schum.) Pierre ex Beille (P. macroceras Kenn.)	Yohimbe tree	Rubiaceae	39, 62, 9
Pausinystalia pachyceras Schum. de Wild. (see Corynanthe pachyceras)			
Penianthus zenkeri (Engl.) Diels		Menispermaceae	6
Pentaclethra macrophylla Benth.	Oil bean tree	Mimosaceae	116, 18
Pergularia daemia (Forsk.) Chiov. (P. extensa N.E.Br.)		Asclepiadaceae	19, 24, 23
Pergularia sanguinolenta Lindl. (see Cryptolepis sanguinolenta)			
Periploca nigrescens Afzel, P. calophylla (Baill.) Roberty (see Parquetina nigrescens)			
Periploca sylvestre Retz. (see Gymnema sylvestre)			
Phaseolus vulgaris L. c.	Vegetable bean	Fabaceae	217, 21
Philenoptera cyanescens (Schum. & Thonn.) Roberty (see Lonchocarpus cyanescens)			
Phoenix dactylifera L.	Date palm	Palmae	21

Species	Common name	Family	Pa
Pseudocedrela kotschyi (Schweinf.) Harms (P. chevalieri DC.)	Dry zone cedar	Meliaceae	164, 18
Pseudocinchona africana Chev. ex Perr. (see Corynanthe pachyceras)			
Psidium guajava L. c.	Guava	Myrtaceae	13
Pueraria thunbergiana (Sieb. & Zucc.) Benth.		Fabaceae	21
Punica granatum L. c.	Pomegranate	Rubiaceae	17
Quassia africana (Baill.) Baill. c.		Simaroubaceae	18
Quisqualis indica L.	Rangoon creeper	Combretaceae	133, 17
Randia cladantha Schum. (see Porterandia cladantha)			
Rauvolfia caffra Sond. (R. welwitschii Stapf)		Apocynaceae	[illegible]
Rauvolfia macrophylla Stapf		Apocynaceae	[illegible]
Rauvolfia mannii Stapf (R. preussii Schum.)		Apocynaceae	3
Rauvolfia vomitoria Afzel. (R. senegambiae DC.)	Rauvolfia, Swizzle stick	Apocynaceae	13, 34, 36, 8
Rhigiocarya racemifera Miers (R. nervosa (Miers) Chev.)		Menispermaceae	60, 8
Rhizophora racemosa Mey.	Mangrove	Rhizophoraceae	25
Rhoeo spathaceae (Sw.) Stearn (Rhoeo discolor (l'Herit) Hance)		Commelinaceae	22
Ricinus communis L.	Castor oil plant	Euphorbiaceae	15
Ritchia longipedicilata Gilg		Capparidaceae	17
Robinia cyanescens Schum. & Thonn. (see Lonchocarpus)			
Robinia sericea Poir. (see Lonchocarpus sericeus)			
Rondeletia febrifuga Afzel. ex Don, R. africana Winterb. (see Crossopteryx febrifuga)			
Rondeletia floribunda Don (see Holarrhena floribunda)			
Salvadora persica L.	Salt bush, toothbrush tree	Salvadoraceae	147, 20
Samanea saman Jacq. & Merr.		Mimosaceae	22
Santalum album L. c.	Sandalwood	Santalaceae	18
Sapindus trifoliatus L.		Sapindaceae	44, 225, 228, 23
Sarcocephalus diderrichii de Wild. & Dur. (see Nauclea diderrichii)			
Sarcocephalus esculentus Afzel. ex Sabine (see Nauclea latifolia)			
Sarcocephalus pobeguini. Pobég. ex Pell. (see Nauclea pobéguinii)			

Species	Common name	Family	Pa
Stephania dinklagei (Engl.) Diels		Menispermaceae	59, 1
Sterculia acuminata Beauv. (see Cola acuminata).			
Sterculia nitida Vent. (see Cola nitida)			
Strophanthus gracilis Schum. & Pax		Apocynaceae	14,
Strophanthus gratus (Hook.) Franch.	Strophanthus	Apocynaceae	10,
Strophanthus hispidus DC.		Apocynaceae	14,
Strophanthus sarmentosus DC.		Apocynaceae	14,
Strychnos afzelii Gilg (S. zizyphoides Bak., S. erythrocarpa Gilg)		Loganiaceae	1
Strychnos camptoneura Gilg & Busse		Loganiaceae	
Strychnos decussata		Loganiaceae	
Strychnos dinklagei Gilg		Loganiaceae	
Strychnos dolichothyrsa Gilg ex Onochi & Hepper		Loganiaceae	
Strychnos spinosa Lam. (S. lokua Rich., S. laxa Solered., S. djalonensis Chev., S. buettneri Gilg, S. spinosa var. pubescens Bak.)		Loganiaceae	
Strychnos usambarensis Gilg (S. micans Moore, S. cooperi Hutch. & Moss)		Loganiaceae	
Swartzia madagascariensis Desv.		Caesalpiniaceae	55, 1
Swietenia senegalensis Desr. (see Khaya senegalensis)			
Synsepalum dulcificum (Schum. & Thonn.) Daniell		Sapotaceae	2
Syzygium cumini (L.) Skeels c.	Jambolan	Myrtaceae	249, 2
Syzygium guineense (Willd.) DC. var. guineense		Myrtaceae	1
Tabernaemontana crassa Benth. (T. durissima Stapf)		Apocynaceae	11, 7
Tabernaemontana glandulosa (Stapf) Pichon		Apocynaceae	14
Tabernaemontana nitida Stapf (see Picralima nitida)			
Tabernaemontana pachysiphon Stapf var. pachysiphon		Apocynaceae	12, 7
Teclea sudanica Chev.		Rutaceae	5
Tecoma stans (L.) H.B.K. c. (ornam.)		Bignoniaceae	25
Tephrosia purpurea (L.) Pers. (T. leptostachya DC.)	Fish poison bean	Fabaceae	5
Tephrosia vogelii Hook. f.	Fish poison bean	Fabaceae	17
Terminalia avicennoides Guill. & Perr. (T. lecardii Engl. & Diels, T. dictyoneura Diels)		Combretaceae	52, 14
Terminalia glaucescens Planch. ex Benth. (T. togoensis Engl. & Diels, T. baumannii Engl. & Diels, T. passargei Engl.)		Combretaceae	14

Species	Common name	Family	Page
Ximenia aegyptiaca (see Balanites aegyptiaca)			
Xylopia aethiopica (Dunal) Rich (X. emini Chev.)	Ethiopian pepper	Annonaceae	140, 214
Xysmalobium heudelotianum Decne.		Asclepiadaceae	19, 25
Zanthoxylum gilletii (de Wild.) Watson		Rutaceae	33
Zanthoxylum guineense Stapf (see Harrisonia abyssinica)			
Zanthoxylum rubescens (Planch. ex Hook. f.) Watson		Rutaceae	33
Zanthoxylum zanthoxyloides (Lam.) Watson (Zanthoxylum senegalense (DC.) Chev., Z. polyganum Schum.)	Prickly ash, candlewood, tooth-ache bark	Rutaceae	32, 131, 205